McGraw-Hill

My Math

Welcome to *My Math* — your very own math book! You can write in it – in fact, you are encouraged to write, draw, circle, explain, and color as you explore the exciting world of mathematics. Let's get started. Grab a pencil and finish each sentence.

My name is ______________________.

My favorite color is ______________________.

My favorite hobby or sport is ______________________.

My favorite TV program or video game is

______________________.

My favorite class is ______________________.

Math, of course!

Bothell, WA • Chicago, IL • Columbus, OH • New York, NY

connectED.mcgraw-hill.com

STEM McGraw-Hill is committed to providing instructional materials in Science, Technology, Engineering, and Mathematics (STEM) that give all students a solid foundation, one that prepares them for college and careers in the 21st century.

Send all inquiries to:
McGraw-Hill Education
STEM Learning Solutions Center
8787 Orion Place
Columbus, OH 43240

ISBN: 978-0-02-115022-9 *(Volume 1)*
MHID: 0-02-115022-2

Printed in the United States of America.

9 10 11 12 13 14 DOW 19 18 17 16 15 14

Our mission is to provide educational resources that enable students to become the problem solvers of the 21st century and inspire them to explore careers within Science, Technology, Engineering, and Mathematics (STEM) related fields.

Meet The Artists!

Abby Crutchley

The Market Math Winning this contest made me feel like I had butterflies in my stomach! I didn't think I would win out of the 72 finalists. *Volume 1*

Matt Gardner

Using Math to Build Math means everything to me. It is my favorite subject in school. I got the idea because I like to build and I thought a K'Nex vehicle would make a cool math book cover. *Volume 2*

Other Finalists

Jacob Alvarez
Beach Math-1

Emily Jiang
Math Angel

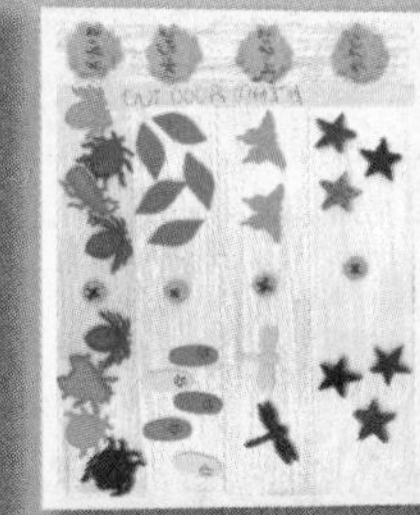

India Johnson
Math in Nature

Nathan Baal
Math in My Neighborhood

Sergio Reyes
Math With My Fingers

Kelsey Thompson
Doggy Shapes

Maddie Mathews
The Book of Math-ews

Kaya Ross
Math is in My Neighborhood

Yonaton Barkel
We Use Math Skills Everyday

Ellie Hull
Patterns
Qwirkle™ tiles reproduced with the permission of MindWare®.

Find out more about the winners and other finalists at www.MHEonline.com.

We wish to congratulate all of the entries in the 2011 *McGraw-Hill My Math* "What Math Means To Me" cover art contest. With over 2,400 entries and more than 20,000 community votes cast, the names mentioned above represent the two winners and ten finalists for this grade.

GO digital

it's all at
connectED.mcgraw-hill.com

Go to the Student Center for your eBook, Resources, Homework, and Messages.

Write your Username

Password

Get your resources online to help you in class and at home.

Vocab

Find activities for building vocabulary.

Watch

Watch animations of key concepts.

Tools

Explore concepts with virtual manipulatives.

Check

Self-assess your progress.

eHelp

Get targeted homework help.

Games

Reinforce with games and apps.

Tutor

See a teacher illustrate examples and problems.

Scan this QR code with your smart phone* or visit mheonline.com/stem_apps.

*May require quick response code reader app.

Contents in Brief

Organized by Domain

Chapter 1 Place Value

ESSENTIAL QUESTION
How can numbers be expressed, ordered, and compared?

Look for this! Watch
Click online and you can watch videos that will help you learn the lessons.

connectED.mcgraw-hill.com

Chapter 2 Addition

ESSENTIAL QUESTION
How can place value help me add larger numbers?

Getting Started

Lessons and Homework

Wrap Up

connectED.mcgraw-hill.com

Chapter 3 Subtraction

ESSENTIAL QUESTION
How are the operations of subtraction and addition related?

Getting Started

Lessons and Homework

Wrap Up

Look for this! eHelp
Click online and you can get more help while doing your homework.

connectED.mcgraw-hill.com

Chapter 4 Understand Multiplication

ESSENTIAL QUESTION
What does multiplication mean?

Getting Started

Lessons and Homework

Wrap Up

Yummy!

connectED.mcgraw-hill.com

Chapter 5 Understand Division

ESSENTIAL QUESTION
What does division mean?

Getting Started

Lessons and Homework

Wrap Up

connectED.mcgraw-hill.com

Look for this!
Tools
Click online and you can find tools that will help you explore concepts.

Chapter 6 Multiplication and Division Patterns

ESSENTIAL QUESTION
What is the importance of patterns in learning multiplication and division?

BLAST OFF!

Getting Started

Lessons and Homework

Wrap Up

connectED.mcgraw-hill.com

Chapter 7 Multiplication and Division

ESSENTIAL QUESTION
What strategies can be used to learn multiplication and division facts?

Getting Started

Lessons and Homework

Wrap Up

Look for this!
Click online and you can watch a teacher solving problems.
Tutor

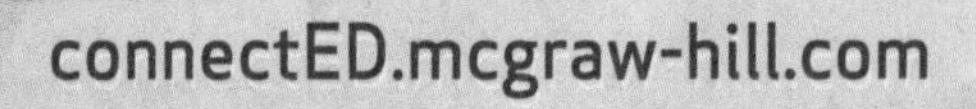

Chapter 8 Apply Multiplication and Division

ESSENTIAL QUESTION
How can multiplication and division facts with smaller numbers be applied to larger numbers?

Getting Started

Lessons and Homework

Wrap Up

connectED.mcgraw-hill.com

Chapter 9 Properties and Equations

ESSENTIAL QUESTION
How are properties and equations used to group numbers?

Getting Started

Lessons and Homework

Wrap Up

connectED.mcgraw-hill.com

Look for this!
Click online and you can find activities to help build your vocabulary.

Vocab

Chapter 10 Fractions

ESSENTIAL QUESTION
How can fractions be used to represent numbers and their parts?

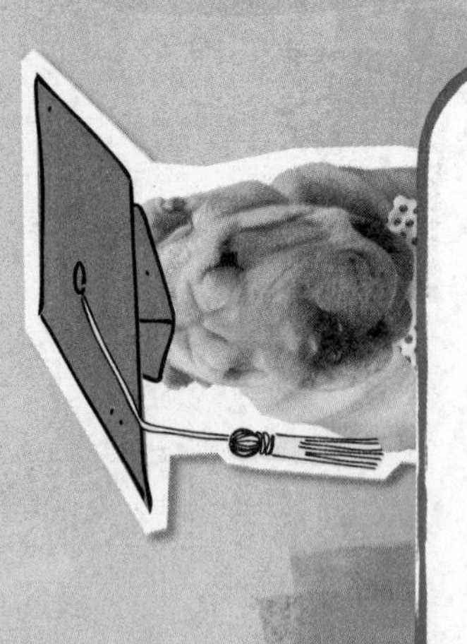

Getting Started

Lessons and Homework

Wrap Up

I've got the answer!

connectED.mcgraw-hill.com

Chapter 11 Measurement

ESSENTIAL QUESTION
Why do we measure?

Getting Started

Lessons and Homework

Wrap Up

Look for this! Check
Click online and you can check your progress.

connectED.mcgraw-hill.com

Chapter 12 Represent and Interpret Data

ESSENTIAL QUESTION
How do we obtain useful information from a set of data?

Getting Started

Lessons and Homework

Wrap Up

connectED.mcgraw-hill.com

Chapter 13 Perimeter and Area

ESSENTIAL QUESTION
How are perimeter and area related and how are they different?

I dig this area!

Getting Started

Lessons and Homework

Wrap Up

connectED.mcgraw-hill.com

Chapter 14 Geometry

ESSENTIAL QUESTION
How can geometric shapes help me solve real-world problems?

Getting Started

Lessons and Homework

Wrap Up

connectED.mcgraw-hill.com

Chapter 1 Place Value

ESSENTIAL QUESTION

How can numbers be expressed, ordered, and compared?

Watch a video!

Watch

MY Common Core State Standards

CCSS

Number and Operations in Base Ten

CCSS

3.NBT.1 Use place value understanding to round whole numbers to the nearest 10 or 100.

3.NBT.2 Fluently add and subtract within 1000 using strategies and algorithms based on place value, properties of operations, and/or the relationship between addition and subtraction.

3.NBT.3 Multiply one-digit whole numbers by multiples of 10 in the range 10-90 (e.g., 9×80, 5×60) using strategies based on place value and properties of operations.

Cool! This is what I am going to be doing!

Standards for Mathematical PRACTICE

1. Make sense of problems and persevere in solving them.
2. Reason abstractly and quantitatively.
3. Construct viable arguments and critique the reasoning of others.
4. Model with mathematics.
5. Use appropriate tools strategically.
6. Attend to precision.
7. Look for and make use of structure.
8. Look for and express regularity in repeated reasoning.

= focused on in this chapter

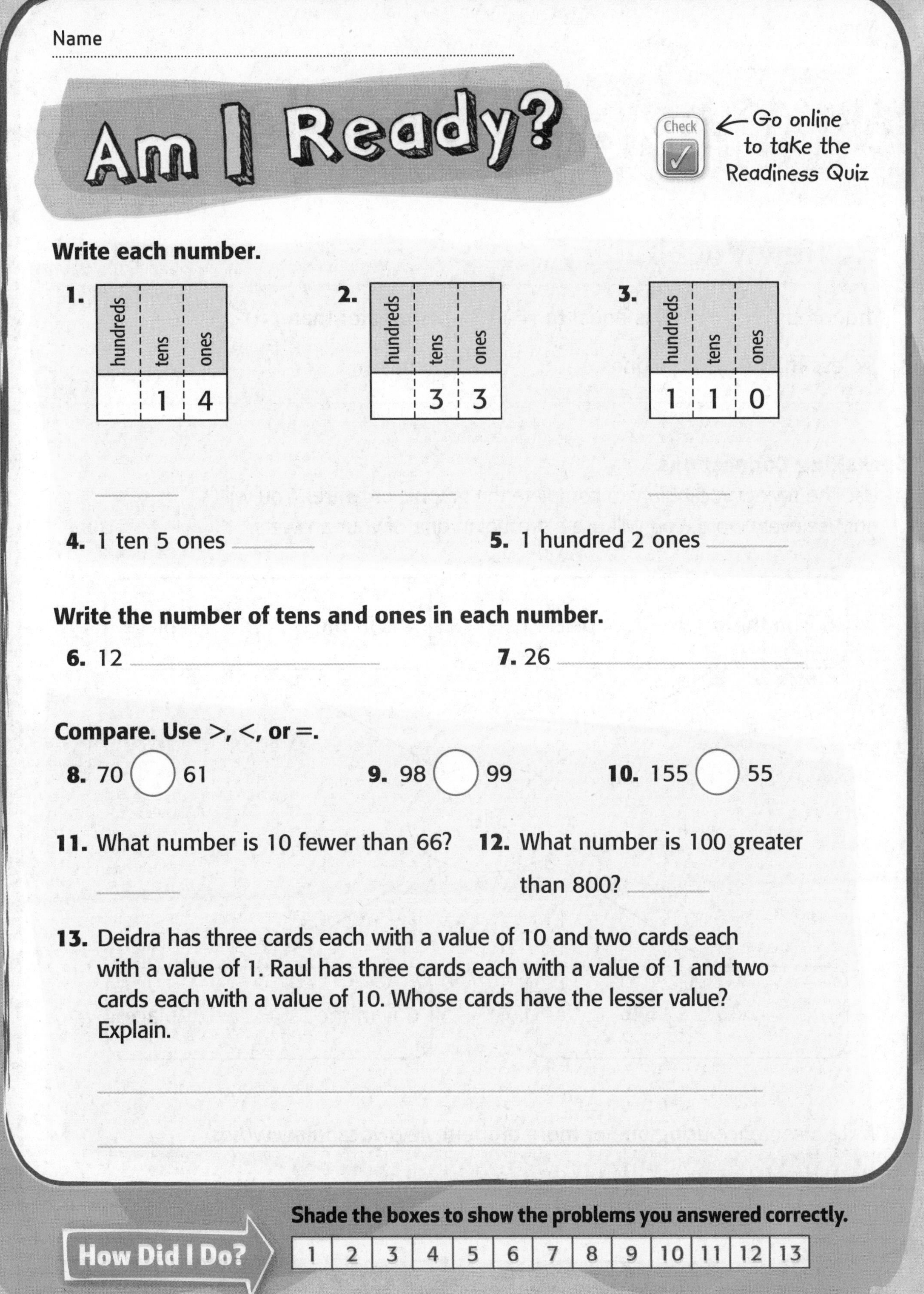

Name ....

Am I Ready?

Check ← Go online to take the Readiness Quiz

Write each number.

1.

hundreds	tens	ones
	1	4

______

2.

hundreds	tens	ones
	3	3

______

3.

hundreds	tens	ones
1	1	0

______

4. 1 ten 5 ones ______

5. 1 hundred 2 ones ______

Write the number of tens and ones in each number.

6. 12 __________________

7. 26 __________________

Compare. Use >, <, or =.

8. 70 ◯ 61

9. 98 ◯ 99

10. 155 ◯ 55

11. What number is 10 fewer than 66?

______

12. What number is 100 greater than 800? ______

13. Deidra has three cards each with a value of 10 and two cards each with a value of 1. Raul has three cards each with a value of 1 and two cards each with a value of 10. Whose cards have the lesser value? Explain.

____

____

How Did I Do?

Shade the boxes to show the problems you answered correctly.

1	2	3	4	5	6	7	8	9	10	11	12	13

Name

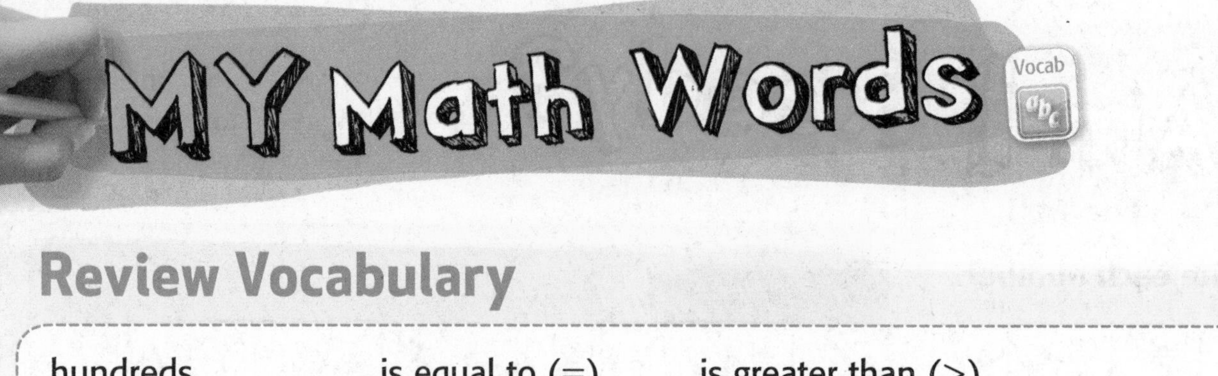

Review Vocabulary

hundreds	is equal to (=)	is greater than (>)
is less than (<)	ones	tens

Making Connections

Use the review vocabulary to complete the graphic organizer. You will not use every word. You will use a symbol in one of your answers.

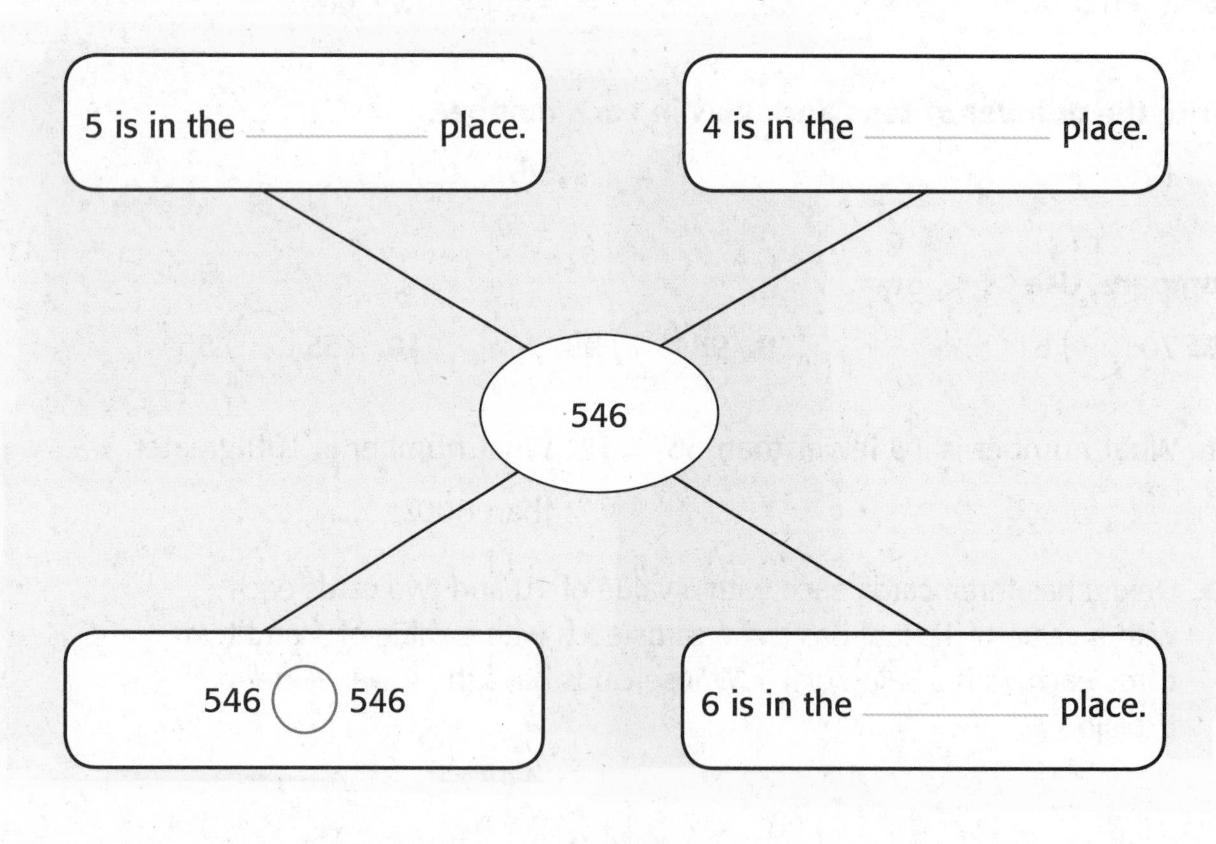

Write a sentence using one or more of the review vocabulary words.

____

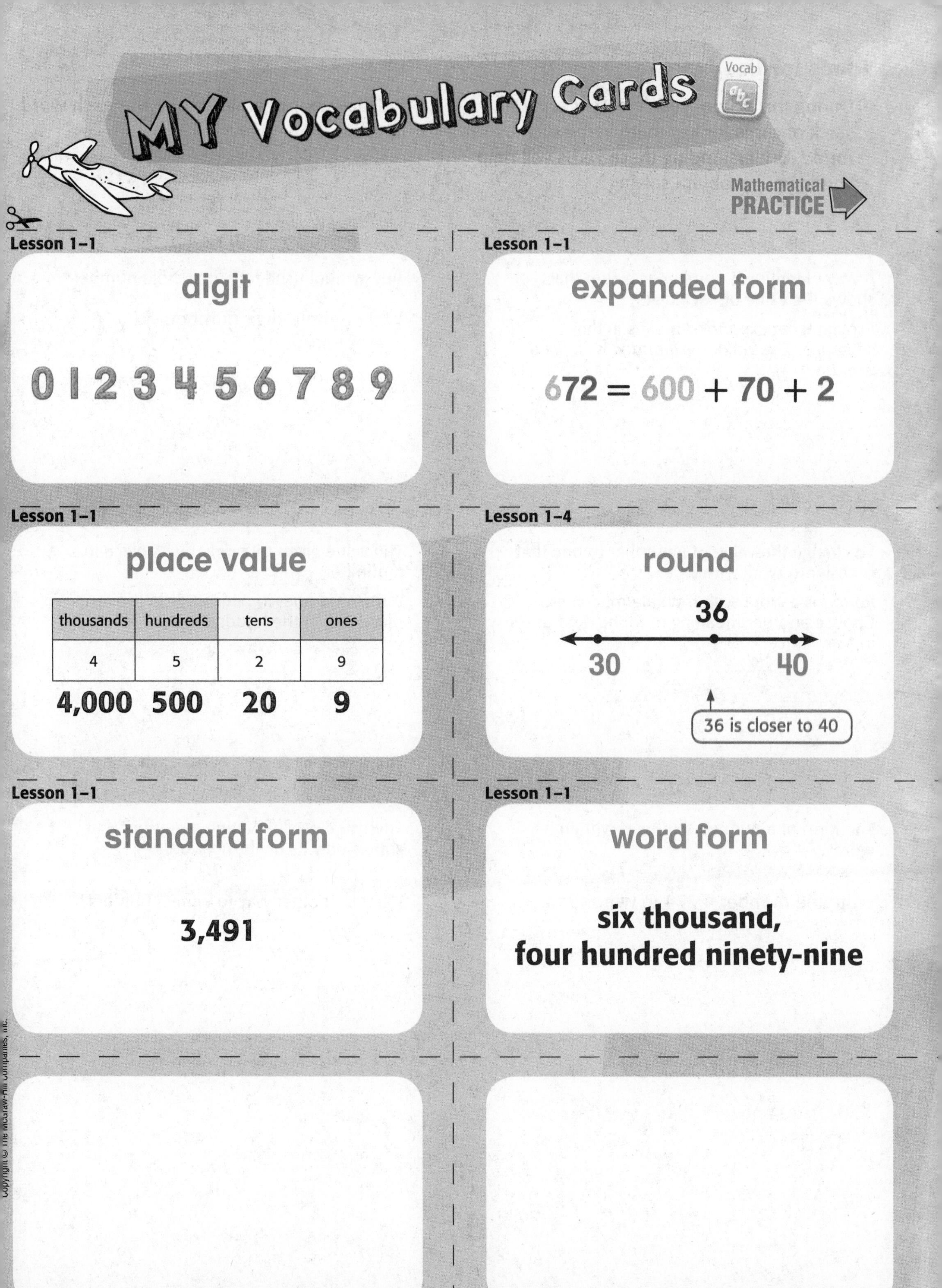

MY Vocabulary Cards

Vocab

abc

Mathematical PRACTICE

Lesson 1–1

digit

0 1 2 3 4 5 6 7 8 9

Lesson 1–1

expanded form

672 = 600 + 70 + 2

Lesson 1–1

place value

thousands	hundreds	tens	ones
4	5	2	9
4,000	500	20	9

Lesson 1–4

round

Lesson 1–1

standard form

3,491

Lesson 1–1

word form

six thousand,
four hundred ninety-nine

Ideas for Use

- During this school year, create a separate stack of cards for key math verbs such as *round*. Understanding these verbs will help you in your problem solving.
- Practice your penmanship! Write each word in cursive.

A way of writing a number as a sum that shows the value of each digit.

Explain what *expanded* means in this sentence: *The balloon expanded as it filled with air.*

Any symbol used to write whole numbers.

Write a three-digit number.

To change the value of a number to one that is easier to work with.

Round is a word with multiple meanings. Choose another meaning of *round,* and use it in a sentence.

The value given to a *digit* by its place in a number.

Write a number in which 6 is in the tens place and in the hundreds place.

The form of a number that uses written words.

Write the number 4,274 in word form.

The usual way of writing a number that shows only its digits, no words.

What is another way to write a number?

MY Foldable

Follow the steps on the back to make your Foldable.

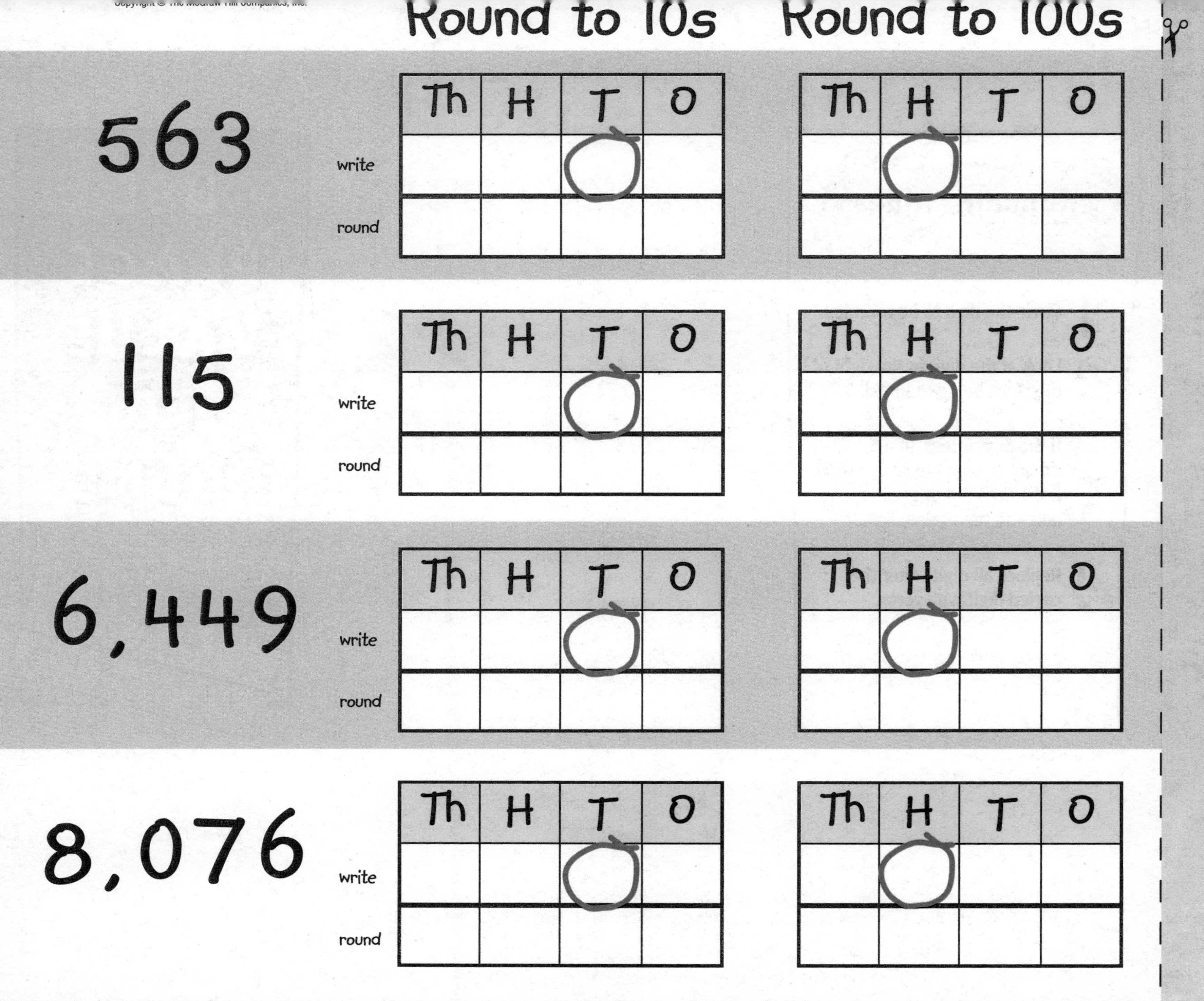

FOLDABLES® Study Organizer

Rounding Rules

1. Circle the digit to be rounded.
2. Look at the digit to the right of the place being rounded.
3. If the digit is less than 5, do not change the circled digit. If the digit is 5 or greater, add 1 to the circled digit.
4. Replace all digits after the circled digit with zeros.

USE PLACE VALUE TO ROUND

Th	H	T	O
	7	3	7
	7	4	0

Name ....

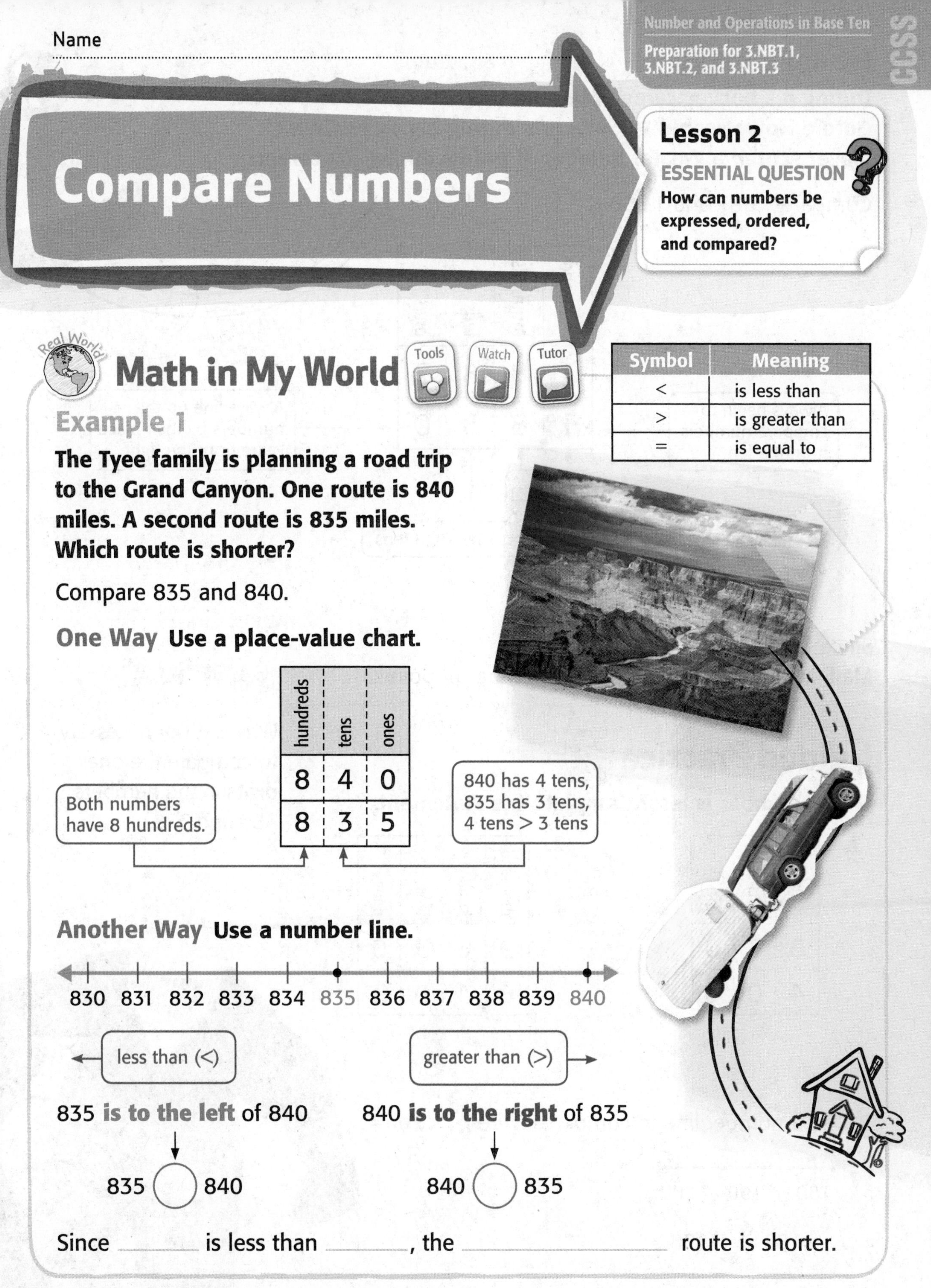

Compare Numbers

Lesson 2

ESSENTIAL QUESTION
How can numbers be expressed, ordered, and compared?

Math in My World

Tools Watch Tutor

Symbol	Meaning
<	is less than
>	is greater than
=	is equal to

Example 1

The Tyee family is planning a road trip to the Grand Canyon. One route is 840 miles. A second route is 835 miles. Which route is shorter?

Compare 835 and 840.

One Way Use a place-value chart.

hundreds	tens	ones
8	4	0
8	3	5

Both numbers have 8 hundreds.

840 has 4 tens, 835 has 3 tens, 4 tens > 3 tens

Another Way Use a number line.

830 831 832 833 834 835 836 837 838 839 840

less than (<) greater than (>)

835 **is to the left** of 840 840 **is to the right** of 835

835 ◯ 840 840 ◯ 835

Since ______ is less than ______, the ______________ route is shorter.

Example 2 Tutor

During his hockey career, Mark Messier scored 1,887 points. Gordie Howe scored 1,850 points during his career. Which player scored a greater number of points during his career?

Compare 1,887 and 1,850.

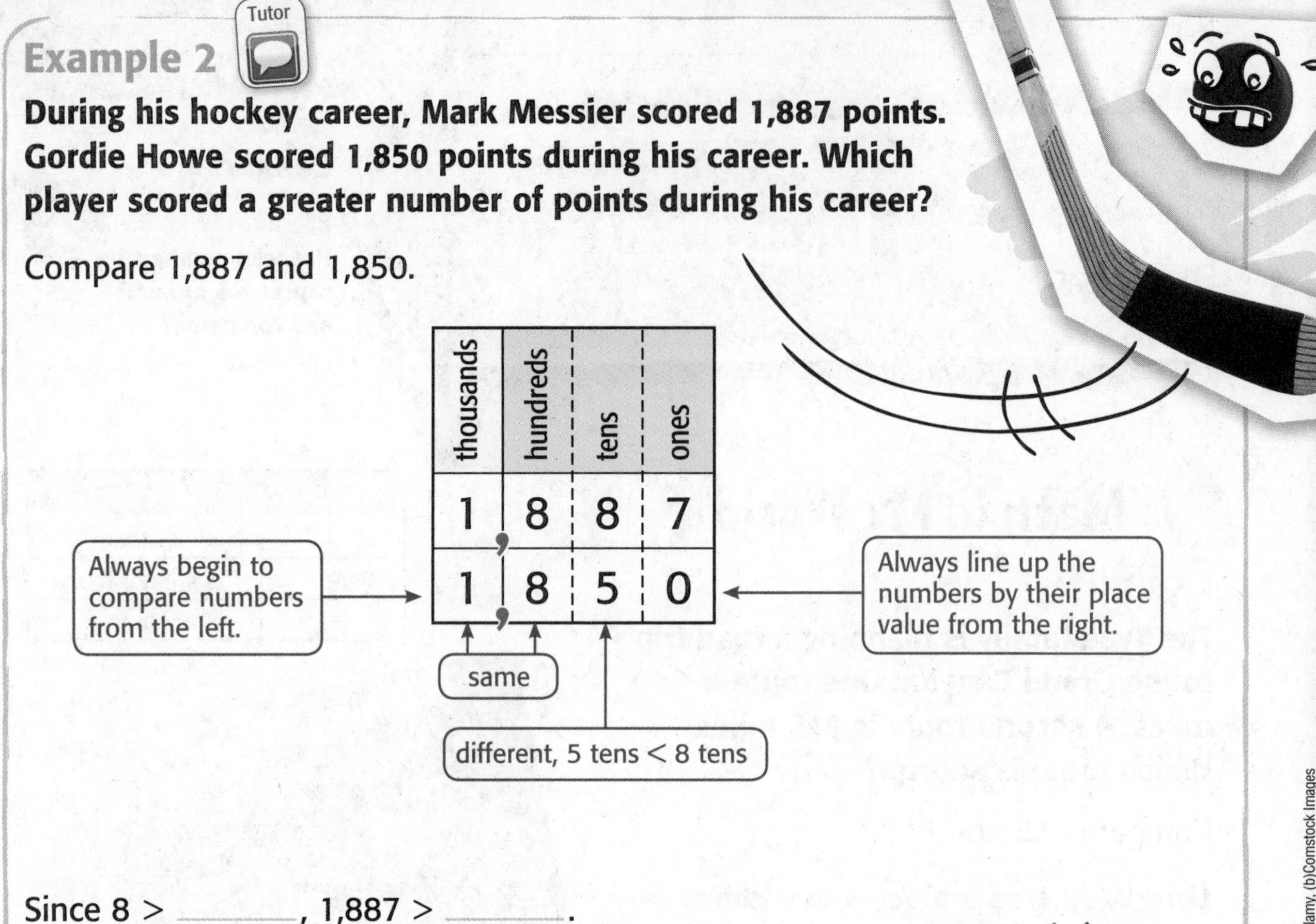

thousands	hundreds	tens	ones
1,	8	8	7
1,	8	5	0

Since 8 > ______, 1,887 > ______.
Mark Messier scored the greater number of points.

Talk MATH

Why is it not necessary to compare the ones digits in the numbers 365 and 378?

Guided Practice Check

Which number is less? Complete the statement.

1.

hundreds	tens	ones
8	7	0
4	0	0

______ < ______

2.

thousands	hundreds	tens	ones
9,	6	3	0
6,	4	0	3

______ < ______

3. Use a number line to compare. Write >, <, or =.

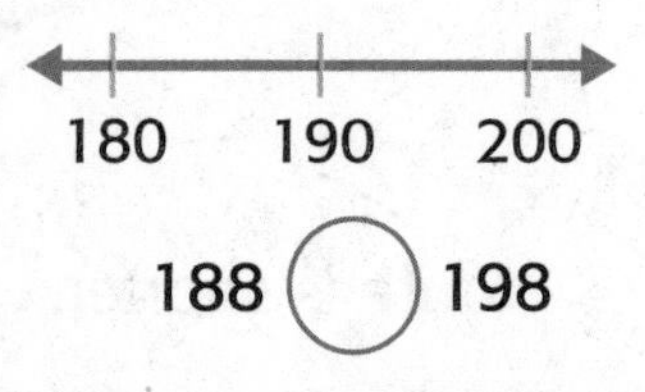

188 ◯ 198

Name ....

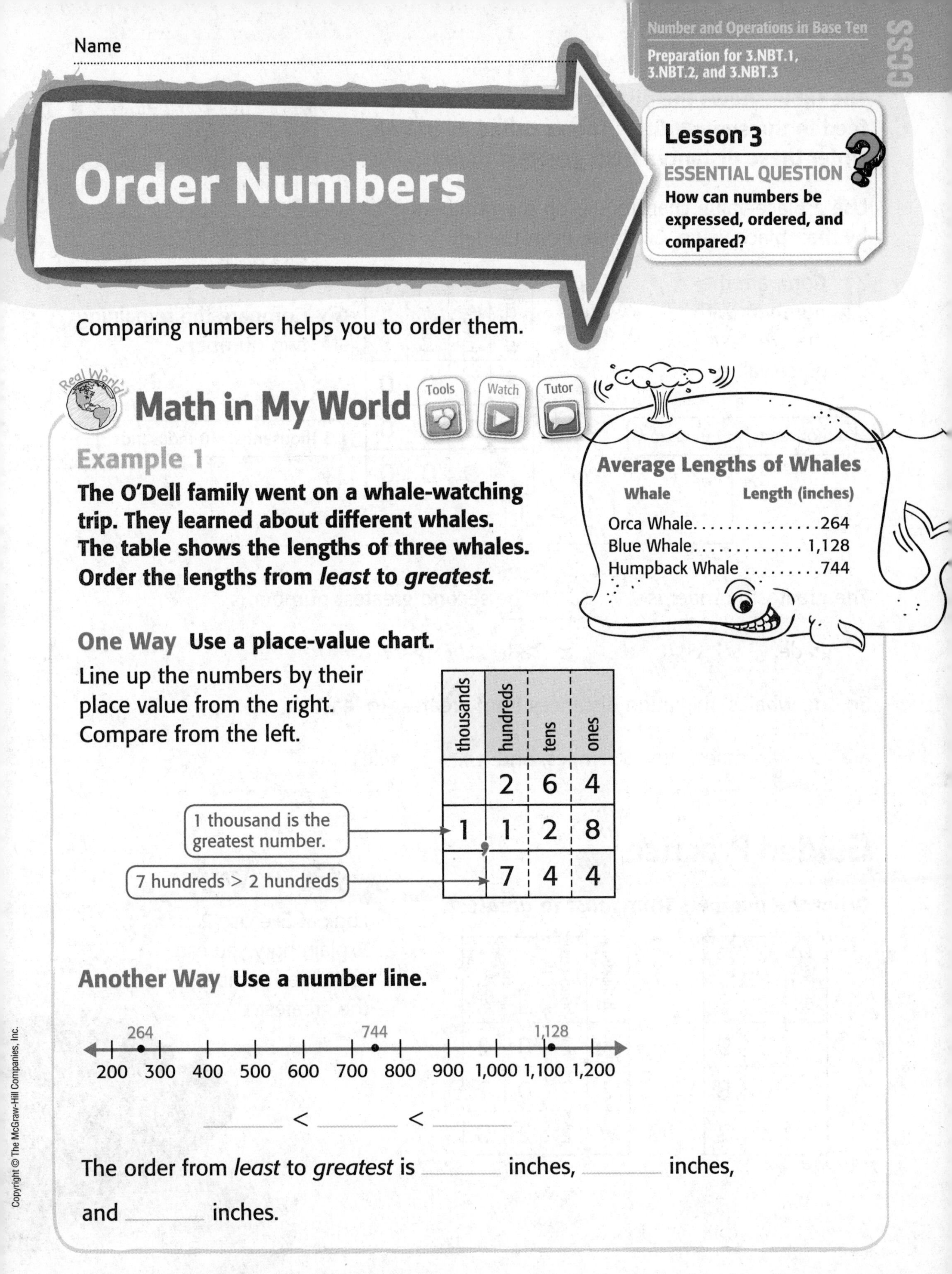

Order Numbers

Lesson 3

ESSENTIAL QUESTION

How can numbers be expressed, ordered, and compared?

Comparing numbers helps you to order them.

Math in My World

Tools Watch Tutor

Example 1

The O'Dell family went on a whale-watching trip. They learned about different whales. The table shows the lengths of three whales. Order the lengths from *least* to *greatest*.

Average Lengths of Whales

Whale	Length (inches)
Orca Whale	264
Blue Whale	1,128
Humpback Whale	744

One Way **Use a place-value chart.**

Line up the numbers by their place value from the right. Compare from the left.

thousands	hundreds	tens	ones
	2	6	4
1,	1	2	8
	7	4	4

1 thousand is the greatest number.

7 hundreds > 2 hundreds

Another Way **Use a number line.**

_____ < _____ < _____

The order from *least* to *greatest* is _____ inches, _____ inches, and _____ inches.

Example 2

Tutor

The table shows the distances whales travel to feed in the summertime. This is called migration. Order these distances from *greatest* to *least*.

Whale Migration	
Whale	Distance (miles)
Humpback Whale	3,500
Gray Whale	6,200
Orca Whale	900

Use a place-value chart to line up the numbers by their place value. Compare from the left.

1 Compare the numbers with the greatest place value.

thousands	hundreds	tens	ones
3,	5	0	0
6,	2	0	0
	9	0	0

6 thousands > 3 thousands

2 Compare the remaining two numbers.

3 thousands > 0 thousands

The greatest number is ______. The second greatest number is ______.

______ > ______ > ______

So, the whales' migration distances from greatest to least

are ______ miles, ______ miles, and ______ miles.

Guided Practice

Check

Order the numbers from *least* to *greatest*.

1.

hundreds	tens	ones
	3	9
	6	8
	3	2

______; ______; ______

2.

thousands	hundreds	tens	ones
	2	0	2
2,	2	0	2
	2	2	0

______; ______; ______

Talk MATH

Look at Exercise 2. Explain how you can tell which number is the greatest.

MY Homework

Lesson 3
Order Numbers

Homework Helper

eHelp

Need help? connectED.mcgraw-hill.com

The table shows the number of each type of car sold. Order the cars sold from *least* to *greatest*.

Type of Car	Cars Sold
Sports Utility	1,309
Sedan	1,803
Compact	1,117

thousands	hundreds	tens	ones
1,	1	1	7
1,	3	0	9
1,	8	0	3

Use a place-value chart.

Line up the numbers from the right.

Begin to compare from the left.

Number lines are another way to order numbers.

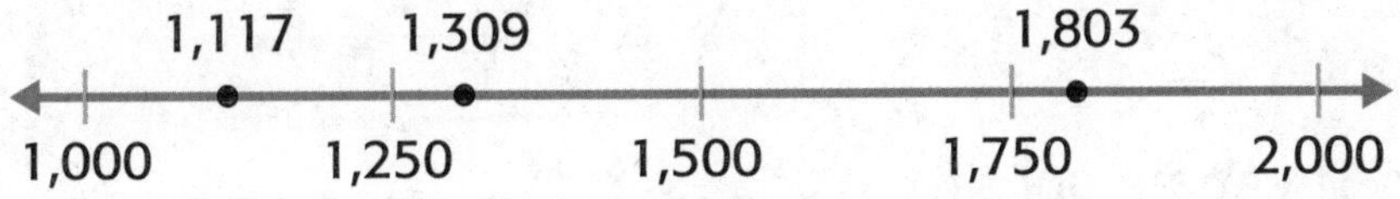
1,117
1,309
1,803
1,000
1,250
1,500
1,750
2,000

1,117 < 1,309 < 1,803

The order from *least* to *greatest* is compact, sports utility, and sedan.

Practice

Order the numbers from *least* to *greatest*.

1. 210; 182; 153

____

2. 1,692; 1,687; 1,685

____

3. 9,544; 9,455; 9,564

____

4. 653; 535, 335

____

Order the numbers from *greatest* to *least*.

5. Electronics for Sale

6. Miles Traveled

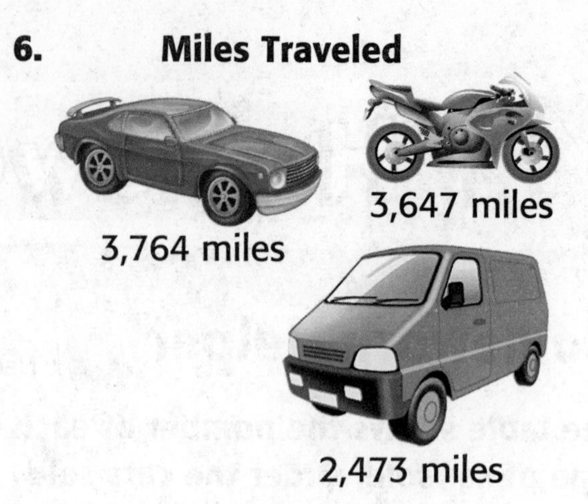

_______________ _______________

Problem Solving

7. Mathematical PRACTICE 1 **Make Sense of Problems**
The Jacksons, Chens, and Simms each went on vacation. The Jacksons drove 835 miles. The Simms drove 947 miles and the Chens drove 100 miles further than the Jacksons. Which family drove the furthest on vacation?

____

8. The addresses on Plum Street are out of order. Place them in order from least to the greatest.

7867, 8112, 7831

____

My Work!

Test Practice

9. Which set of numbers is correctly ordered from *least* to *greatest*?

Ⓐ 7,659; 7,668; 8,985; 9,887

Ⓑ 9,887; 8,985; 7,668; 7,659

Ⓒ 8,985; 9,887; 7,668; 7,659

Ⓓ 9,887; 8,985; 7,659; 7,668

Name ....

Round to the Nearest Ten

Lesson 4

ESSENTIAL QUESTION
How can numbers be expressed, ordered, and compared?

When you **round**, you change the value of a number to one that is easier to work with.

Math in My World

Tools Tutor

Example 1

There are 32 people in line ahead of Cassandra to buy popcorn. About how many people is that? Round to the nearest ten.

Use a place-value chart.

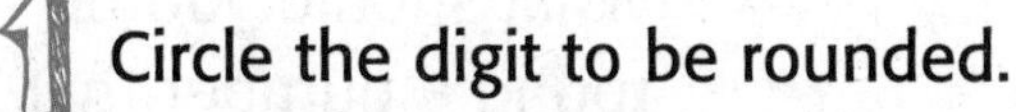

1. Circle the digit to be rounded.

hundreds	tens	ones
	(3) →	2

2. Look at the digit to its right.

3. If the digit is less than 5, do not change the circled digit.
   2 < 5

hundreds	tens	ones

4. Replace all the digits after the circled digit with zeros.

Use a number line.

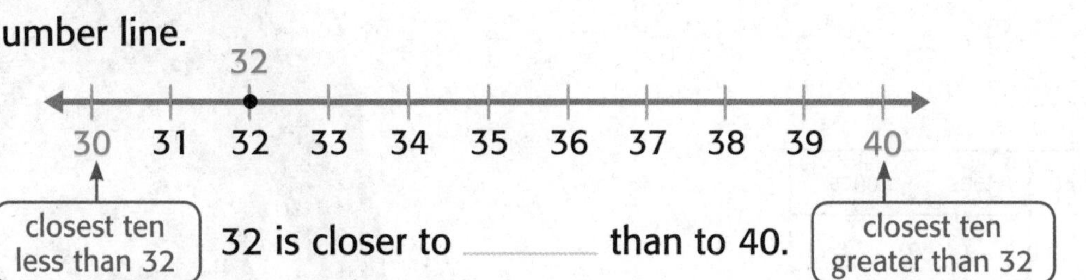

32 is closer to ______ than to 40.

Either method you use, the answer is the same. Cassandra has about ______ people in front of her.

Example 2

Sonja sent 165 text messages on her family's cell phone. About how many messages did Sonja send?

Use a place-value chart to round 165 to the nearest ten.

1. Circle the digit to be rounded.

hundreds	tens	ones
1	(6) →	5

2. Look at the digit to its right.

3. If the digit is 5 or greater, add 1 to the circled digit.
   5 = 5

hundreds	tens	ones
1		

4. Replace all the digits after the circled digit with zeros.

So, Sonja sent about 170 text messages.

Talk MATH

What should you do to round a number that ends in 5, which is exactly halfway between two numbers?

Guided Practice 

Round to the nearest ten.

1.

hundreds	tens	ones
	5	8

____

2.

hundreds	tens	ones
	8	5

____

3.

hundreds	tens	ones
	7	2

____

Name ________________________________

Round to the Nearest Hundred

Lesson 5

ESSENTIAL QUESTION
How can numbers be expressed, ordered, and compared?

There can be more than one reasonable rounded number.

ZOOM!

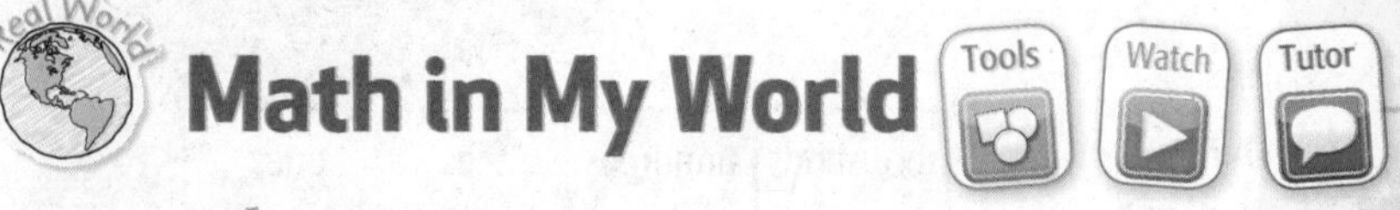

Example 1

A hydroplane is an extremely fast motor boat. In California, a record speed was set of 213 miles per hour. What is the record speed rounded to the nearest ten and rounded to the nearest hundred?

Round to the nearest ten.

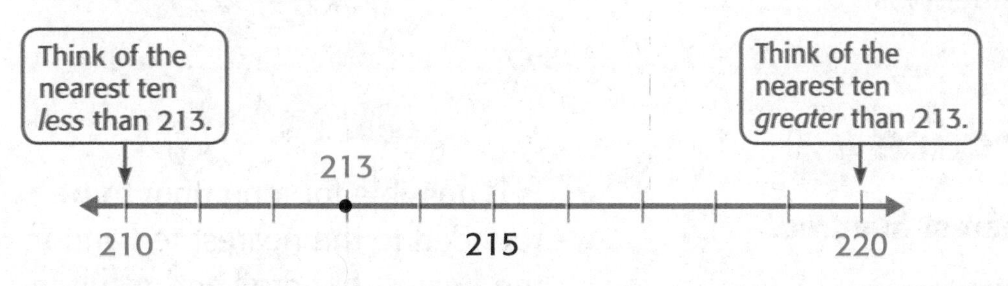

Rounded to the nearest ten, the hydroplane's speed was ______ miles per hour.

Round to the nearest hundred.

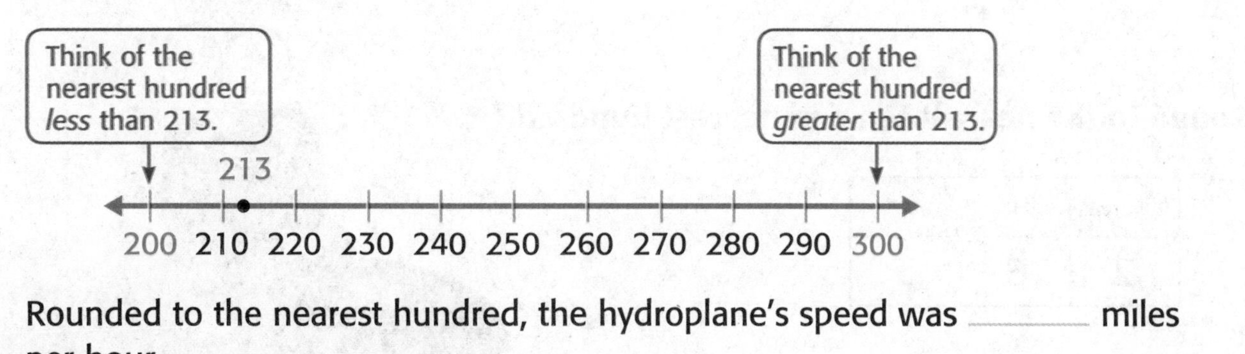

Rounded to the nearest hundred, the hydroplane's speed was ______ miles per hour.

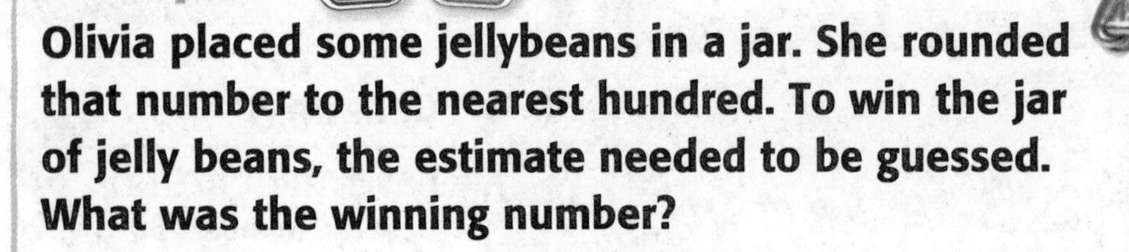

Example 2

Olivia placed some jellybeans in a jar. She rounded that number to the nearest hundred. To win the jar of jelly beans, the estimate needed to be guessed. What was the winning number?

Use a place-value chart to round.

1. Circle the digit to be rounded.

thousands	hundreds	tens	ones
1,	(4) →	8	3

2. Look at the digit to the right of the place being rounded.

3. If the digit is less than 5, do not change the circled digit. If the digit is 5 or greater, add 1 to the circled digit.

thousands	hundreds	tens	ones
1,			

4. Replace all digits after the circled digit with zeros.

So, 1,483 rounded to the nearest hundred is 1,500.

The winning number was ______.

Guided Practice

Round to the nearest hundred.

1.

hundreds	tens	ones
6	2	2

______

Round to the nearest ten and nearest hundred.

2.

hundreds	tens	ones
2	6	5

ten ______ hundred ______

Is it possible for a number to be rounded to the nearest ten and to the nearest hundred and result in the same rounded number?

Name Rich Nguyen

CCSS

Independent Practice

Round to the nearest hundred.

3. 750 800

4. 1,368 1,400

5. 618 600

6. 372 400

7. 509 500

8. 1,216 1,200

Round to the nearest ten and nearest hundred.
Cross out the rounded number that does not belong.

9. 453

450 ~~460~~ 500

10. 6,333

~~7,000~~ 6,330 6,300

11. 5,037

~~5,000~~ 5,040 5,100

12. 4,776

~~4,700~~ 4,800 4,780

Round each number to the nearest hundred. Circle the row or column in which three rounded numbers are the same.

13.

113	279	367
404	321	223
189	291	363

14.

1,925	4,782	2,295
850	3,815	3,795
4,723	4,689	4,717

Circle whether each number is rounded to the nearest ten or hundred.

15. 557 rounds to 560 ten hundred

16. 415 rounds to 400 ten hundred

17. 89 rounds to 100 ten hundred

18. 75 rounds to 80 ten hundred

Problem Solving

My Work!

19. A passenger train traveled 687 miles. To the nearest hundred, how many miles did the train travel?

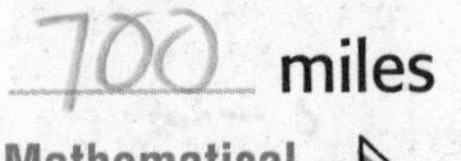

miles

20. Mathematical PRACTICE 2 **Reason** Myron has 179 postcards. He says he has about 200 cards. Did he round the number of cards to the nearest ten or nearest hundred? Explain.

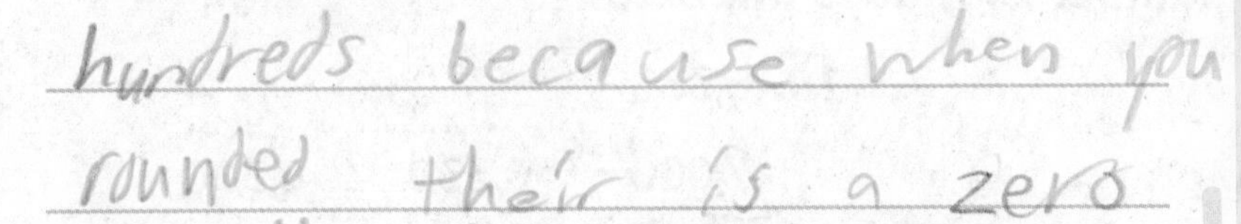

21. To the nearest hundred, what will the cost be for the third grade to take a trip to the zoo?

1,400

HOT Problems

22. Mathematical PRACTICE 4 **Model Math** Mrs. Jones is thinking of a number that when rounded to the nearest hundred is 400. What is the number? Explain.

23. **Building on the Essential Question** Why might you want to round to the nearest hundred rather than the nearest ten?

Name ....

Real World

Problem-Solving Investigation

STRATEGY: Use the Four-Step Plan

Lesson 6

ESSENTIAL QUESTION
How can numbers be expressed, ordered, and compared?

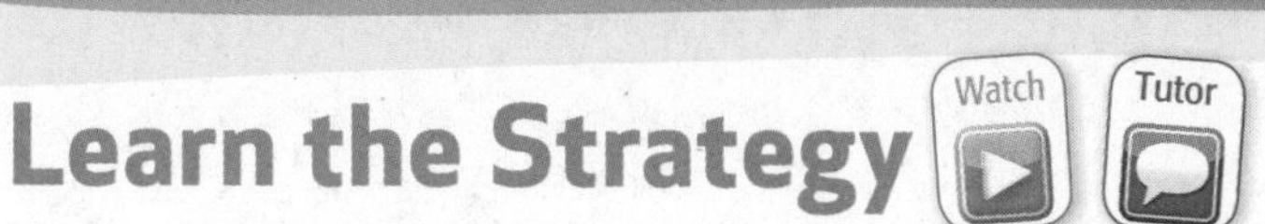

Learn the Strategy

Watch Tutor

Dina's family went to a zoo on their vacation. They learned that a roadrunner is 1 foot tall. An African elephant is 12 feet tall. How much taller is an African elephant than a roadrunner?

I may be small but I'm fast!

1 Understand

What facts do you know?

The roadrunner is ______ foot tall.

The African elephant is ______ feet tall.

What do you need to find?

how much taller an African elephant is than a ____________

2 Plan

______ the roadrunner's height from the African elephant's height.

3 Solve

$$\begin{array}{r} 12 \\ -\ 1 \\ \hline \square \end{array}$$

12 ← height of elephant
1 ← height of roadrunner

So, the elephant is ______ feet taller than the roadrunner.

4 Check

Does your answer make sense? Explain.

____

Practice the Strategy

Alex wants to buy a train ticket to his grandfather's house. A one-way ticket costs $43. A round-trip ticket costs $79. Is it less expensive to buy 2 one-way tickets or one round-trip ticket?

1 Understand

What facts do you know?

____

What do you need to find?

____

2 Plan

____

____

3 Solve

4 Check

Does your answer make sense? Explain.

____

Name

Apply the Strategy

Solve each problem by using the four-step plan.

1. The table shows the number of play tickets four friends sold on Saturday.

Louise 10 + 7	**Malcolm** fourteen
Bobby 20 − 5	**Shelly** 19 + 2

Write the number of tickets sold by each friend in standard form. Then order the numbers from *greatest* to *least.*

My Work!

2. Cameron and Mara walked 2 blocks. Then they turned a corner and walked 4 blocks. If they turned around now and returned home the way they came, how many blocks will they have walked altogether?

3. Mathematical **PRACTICE** 5 **Use Math Tools** Follow the directions to find the correct height of the CN Tower in Toronto, Canada. Start with 781 feet. Add 100. Add 1,000. Subtract 7 tens, and add 4 ones.

4. In 1,000 years from now, what year will it be? What year will it be 100 years from now? 10 years from now?

Review the Strategies

Use any strategy to solve each problem.

- Use the four-step plan.
- Guess, check, and revise.
- Act it out.

5. The blue number cubes represent hundreds. The red number cubes represent tens. Write the standard form for each person's set of number cubes.

Robin's number cubes

Percy's number cubes

Byron's number cubes

Order these numbers from *greatest to least*.

My Work!

6. Mathematical PRACTICE 1 **Plan Your Solution** Two students traveled during their summer vacation. Tina traveled 395 miles. Carl traveled 29 more miles than Tina. How many miles did Carl travel?

7. Mathematical PRACTICE 6 **Be Precise** A group of hikers traveled to the Adirondack Mountains. To the nearest ten, about how many feet did they hike if they climbed Big Slide Mountain and Algonquin Peak?

Adirondack Mountains	
Mt. Marcy	5,344 ft
Algonquin Peak	5,114 ft
Whiteface Mt.	4,867 ft
Big Slide Mt.	4,240 ft

Is there any difference in the distance if you round to the nearest hundred instead? Use <, >, or = to explain.

Name ....

MY Homework

Lesson 6
Problem Solving: Use the Four-Step Plan

Homework Helper

eHelp Need help? connectED.mcgraw-hill.com

There were 418 tickets sold for Conrad's piano recital. About how many tickets were sold? Round to the nearest hundred.

1 Understand **What facts do you know?**
There were 418 tickets sold.
What do you need to find?
the number of tickets sold to the nearest hundred.

2 Plan **Use a number line.**

3 Solve

418

400 410 420 430 440 450 460 470 480 490 500

418 is closer to 400 than to 500. About 400 tickets were sold.

4 Check Check using a place-value chart.
The answer is correct.

hundreds	tens	ones
(4) →	1	8
4	0	0

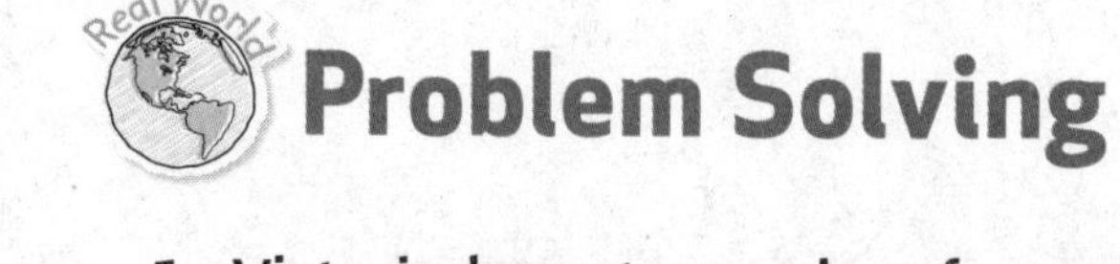

Problem Solving

1. Victoria buys two pairs of sunglasses for $8 each. About how much did she pay for both pairs of glasses? Round to the nearest ten. Solve the problem using the four-step plan.

Solve each problem using the four-step plan.

My Work!

2. Ridgeway Elementary School collected *eight thousand, five hundred thirty-one* cans of food. Park Elementary School collected *eight thousand, six hundred forty-two* cans. Which school collected more cans of food? Write the greater number of cans in expanded form.

____

____

3. Mathematical PRACTICE 1 **Make a Plan** There are 398 students at Rebecca's school and 462 students at Haley's school. There are 10 more students at Ronnie's school than there are at Rebecca's school. Order the number of students at each school from least to greatest.

____

4. Mathematical PRACTICE 3 **Justify Conclusions** The table shows the points each player scored in a video game. Who scored more points? Circle the player's name.

Player	Points
Josh	2,365
Jake	2,475

In which place are the digits that helped you solve this problem?

____

5. The distance from Los Angeles, California to New York, New York is 2,782 miles. What is the distance between the two cities rounded to the nearest hundred miles?

____

Review

Chapter 1

Place Value

Vocabulary Check

Vocab

Read each clue. Fill in the matching section of the crossword puzzle to answer each clue. Use the words in the word bank.

digit	**expanded form**	**place value**
round	**standard form**	**word form**

Across

1. A form which shows the sum of the value of the digits.

2. The usual way of writing numbers that shows only its digits.

Down

3. The value given to a digit by its place in a number.

4. The form of a number that uses written words.

5. Any symbol used to write whole numbers.

6. To find the nearest value of a number based on a given place value.

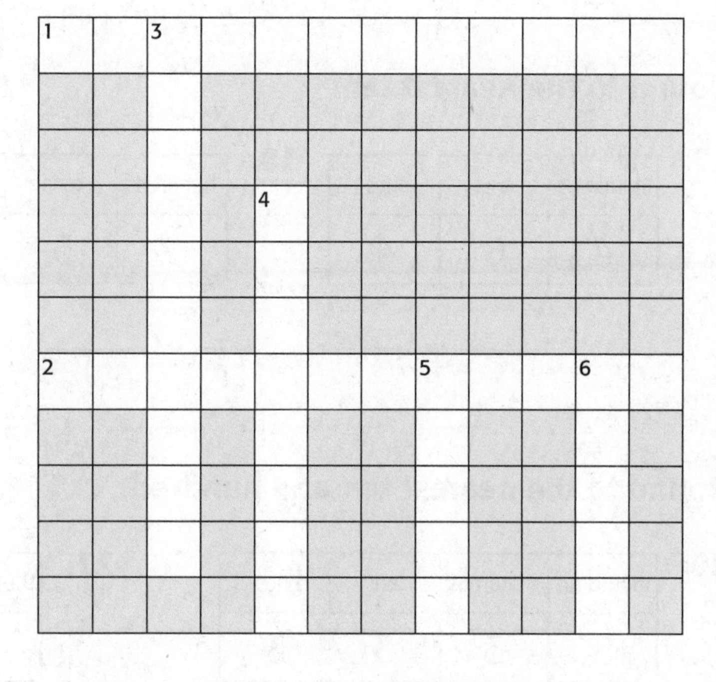

Concept Check

Write the highlighted digit's place and value.

7. 945 ______ ______

8. 4,731 ______ ______

9. 5,409 ______ ______

Write each number in standard form.

10. 300 + 40 + 7 ______

11. *two thousand, six hundred twenty-two* ______

Write 3,651 in expanded form and word form.

12. Expanded Form: ______ + ______ + ______ + ______

13. Word Form: ______

Compare. Use >, <, or =.

14. 268 ◯ 298

15. 3,499 ◯ 3,499

16. 2,675 ◯ 2,567

Round to the nearest ten.

17.

hundreds	tens	ones
4	8	4

______

18.

hundreds	tens	ones
2	5	9

______

19.

hundreds	tens	ones
7	1	2

______

Round to the nearest ten and hundred.

20.

thousands	hundreds	tens	ones
5,	3	3	3

ten ______

hundred ______

21.

thousands	hundreds	tens	ones
2,	7	8	7

ten ______

hundred ______

Name

Problem Solving

Real World

22. Lindsey uses the digits 3, 8, 0, and 1. She uses each digit only once. Find the greatest whole number she can make.

23. Alicia wrote 5,004 in word form. Find and correct her mistake.

Five hundred four

24. Keith's family bought a computer for $1,200. Margareta's family bought a computer for $1,002. Which computer cost less? Explain.

Test Practice

25. The Knights track team scored 117 points at last week's meet. This week, they scored 10 more points than last week. About how many points did The Knights team score this week? Round to the nearest ten.

Ⓐ 120 Ⓒ 130

Ⓑ 127 Ⓓ 137

My Work!

Reflect

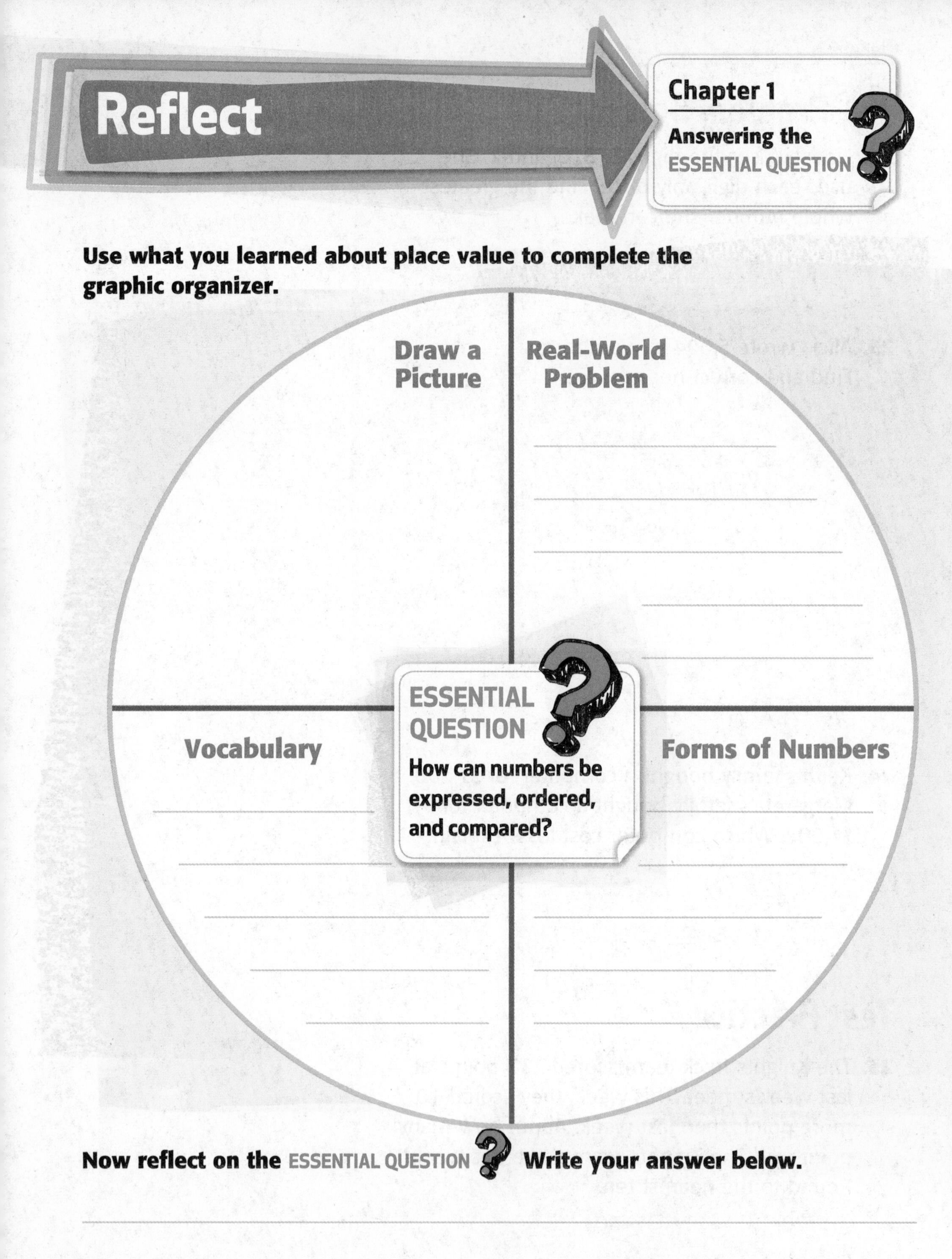

Use what you learned about place value to complete the graphic organizer.

Now reflect on the ESSENTIAL QUESTION **Write your answer below.**

Chapter

2 Addition

ESSENTIAL QUESTION

How can place value help me add larger numbers?

My Transportation

Watch a video!

Watch

MY Common Core State Standards

CCSS

Number and Operations in Base Ten

CCSS

3.NBT.2 Fluently add and subtract within 1000 using strategies and algorithms based on place value, properties of operations, and/or the relationship between addition and subtraction.

Operations and Algebraic Thinking *This chapter also addresses this standard:*

3.OA.9 Identify arithmetic patterns (including patterns in the addition table or multiplication table), and explain them using properties of operations.

I'll be able to get this—no problem!

Standards for Mathematical PRACTICE

1. Make sense of problems and persevere in solving them.
2. Reason abstractly and quantitatively.
3. Construct viable arguments and critique the reasoning of others.
4. Model with mathematics.
5. Use appropriate tools strategically.
6. Attend to precision.
7. Look for and make use of structure.
8. Look for and express regularity in repeated reasoning.

= focused on in this chapter

Name

Am I Ready?

Check — Go online to take the Readiness Quiz

Add.

1. $5 + 4$

2. $6 + 7$

3. $9 + 6$

4. $4 + 8$

5. 9 + 2 = ______ **6.** 4 + 6 = ______ **7.** 9 + 8 = ______ **8.** 7 + 7 = ______

9. 24 + 11 = ______

10. 65 + 12 = ______

11. What number is 10 more than 66? ______

12. What number is 100 more than 800? ______

Round to the nearest ten.

13. 72 ______ **14.** 17 ______ **15.** 63 ______ **16.** 88 ______

Round to the nearest hundred.

17. 470 ______ **18.** 771 ______ **19.** 301 ______ **20.** 149 ______

How Did I Do?

Shade the boxes to show the problems you answered correctly.

1	2	3	4	5	6	7	8	9	10	11	12	13	14	15	16	17	18	19	20

Name ....

Review Vocabulary

addend   addition sentence   sum

Making Connections

Use the review vocabulary to complete the graphic organizer. You will use numbers in some of your answers.

Bella and Tyler looked for birds at the park. Bella saw 10 tundra swans. Tyler saw 30 cardinals. How many birds did they see altogether?

Complete the ____________ ____________ to solve the word problem.

10 + ______ = ______

The ____________ are 10 and 30.

The ________ of 10 and 30 is 40.

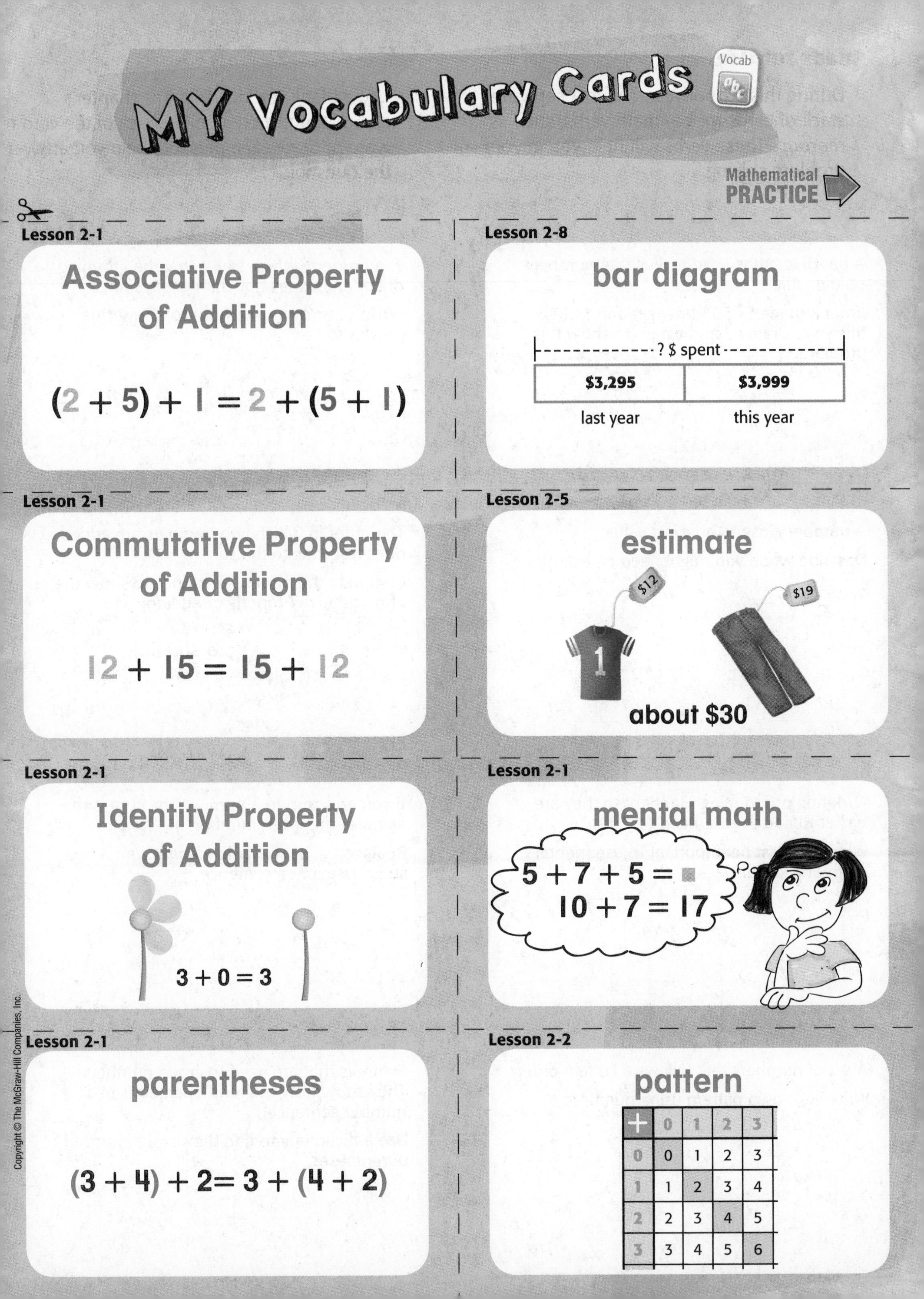

MY Vocabulary Cards

Vocab

Mathematical PRACTICE

Lesson 2-1

Associative Property of Addition

$(2 + 5) + 1 = 2 + (5 + 1)$

Lesson 2-8

bar diagram

Lesson 2-1

Commutative Property of Addition

$12 + 15 = 15 + 12$

Lesson 2-5

estimate

about $30

Lesson 2-1

Identity Property of Addition

$3 + 0 = 3$

Lesson 2-1

mental math

Lesson 2-1

parentheses

$(3 + 4) + 2 = 3 + (4 + 2)$

Lesson 2-2

pattern

+	0	1	2	3
0	0	1	2	3
1	1	2	3	4
2	2	3	4	5
3	3	4	5	6

Ideas for Use

- During this school year, create a separate stack of cards for key math verbs, such as *regroup*. These verbs will help you in your problem solving.
- Use a blank card to write this chapter's essential question. Use the back of the card to write or draw examples that help you answer the question.

A bar diagram is used to illustrate number relationships.

Jimena made $1,595 last year and $1,876 this year. Draw a bar diagram to show this problem.

The property which states that the grouping of addends does not change the sum.

Write your own example to show this property.

A number close to an exact value.

Describe when you might need an estimate.

The order in which two numbers are added does not change the sum.

Complete the number sentence to show the Commutative Property of Addition.

$11 + 7 =$ ______

Ordering or grouping numbers so they are easier to add in your head.

When might it be important to use mental math at a store?

If you add zero to a number, the sum is the same as the given number.

Replace the suffix *-ity* in *identity* with a new suffix. Use it in a sentence.

A set of numbers that follows a certain order.

Write your own pattern using numbers.

Symbols that are used to group numbers. They show how to group operations in a number sentence.

Use a dictionary to find the singular form of *parentheses*.

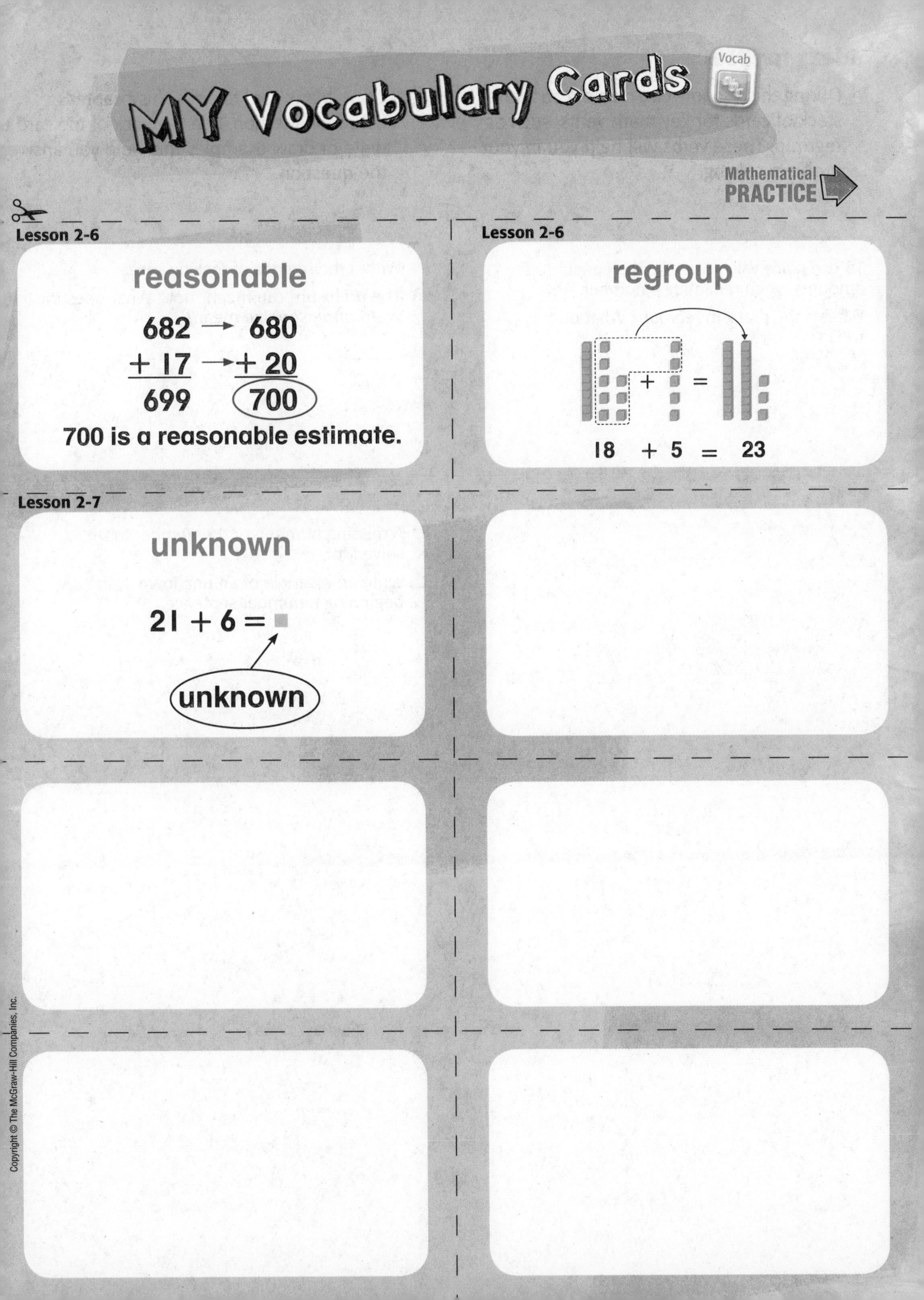

MY Vocabulary Cards

Vocab

Mathematical PRACTICE

Lesson 2-6

reasonable

682 → 680
+ 17 → + 20
699 → 700

700 is a reasonable estimate.

Lesson 2-6

regroup

18 + 5 = 23

Lesson 2-7

unknown

21 + 6 = ■

unknown

Ideas for Use

- During this school year, create a separate stack of cards for key math verbs, such as regroup. These verbs will help you in your problem solving.
- Use a blank card to write this chapter's essential question. Use the back of the card to write or draw examples that help you answer the question.

To use place value to exchange equal amounts when renaming a number.

What is the prefix in *regroup*? What does it mean?

Within the bounds of making sense.

The prefix un- can mean "not." What does the word *unreasonable* mean?

A missing number, or the number to be solved for.

Write an example of an unknown at the beginning a number sentence.

MY Foldable

FOLDABLES® Follow the steps on the back to make your Foldable.

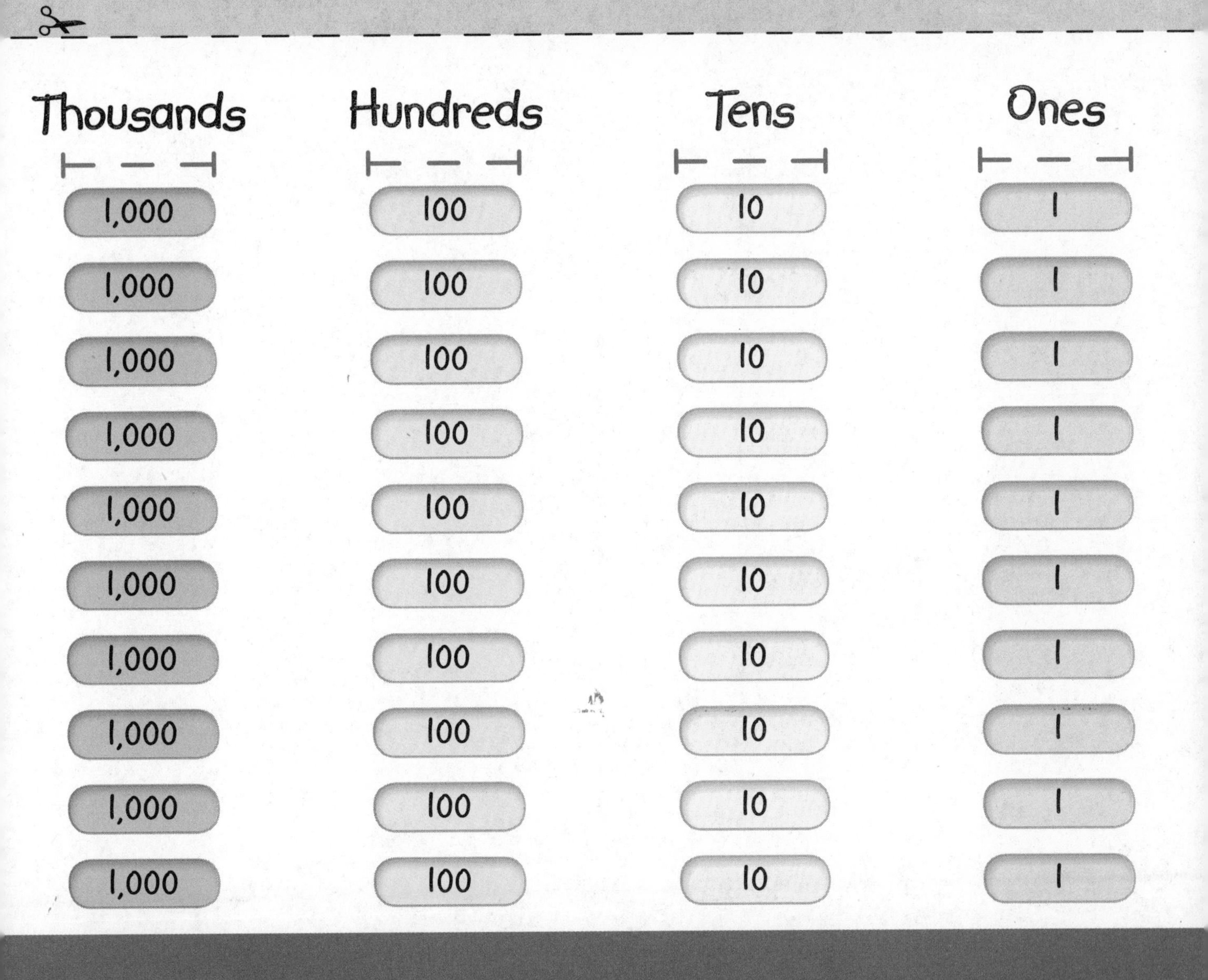

FOLDABLES
Study Organizer
1
2
3

Name

CCSS
Number and Operations in Base Ten
3.OA.9

Patterns in the Addition Table

Lesson 2

ESSENTIAL QUESTION
How can place value help me add larger numbers?

Study the addition table for number **patterns.**
Look for sets of numbers that follow a certain order.

Make a pattern!

Example 1

Danny colored a pattern of squares from left to right on the downward diagonal in yellow. Describe the pattern.

columns of addends

rows of addends

+	0	1	2	3	4	5	6	7	8	9	10
0	0	1	2	3	4	5	6	7	8	9	10
1	1	2	3	4	5	6	7	8	9	10	11
2	2	3	4	5	6	7	8	9	10	11	12
3	3	4	5	6	7	8	9	10	11	12	13
4	4	5	6	7	8	9	10	11	12	13	14
5	5	6	7	8	9	10	11	12	13	14	15
6	6	7	8	9	10	11	12	13	14	15	16
7	7	8	9	10	11	12	13	14	15	16	17
8	8	9	10	11	12	13	14	15	16	17	18
9	9	10	11	12	13	14	15	16	17	18	19
10	10	11	12	13	14	15	16	17	18	19	20

Even Numbers

Finish Danny's pattern of even numbers. Color the squares.

0, 2, 4, 6, 8, 10, 12, 14, 16, 18, 20

There is a pattern of add ______.

When you add ______ to an even number, the sum is an ______ number.

Odd Numbers

Start with the green square. Color the pattern of odd numbers on the downward diagonal in green. Write the numbers.

1, ____, ____, ____, ____, ____, ____, ____, ____, ____

There is a pattern of add ____. When you add ____ to an odd number, the sum is an ______ number.

Example 2

1 What pattern of numbers do you see in the diagonal of yellow boxes?

____

2 Look at the circled sum. Follow left and above to the circled addends.

3 Draw a triangle around the sum in the addition table that has the same addends. Follow left and above to its addends. Complete the number sentences.

+	0	1	2	3	4	5	6	7	8	9	10
0	0	1	2	3	4	5	6	7	8	9	10
1	1	2	3	4	5	6	7	8	9	10	11
2	2	3	4	5	6	7	8	9	10	11	12
3	3	4	5	6	7	8	9	10	11	12	13
4	4	5	6	7	8	9	10	11	12	13	14
5	5	6	7	8	9	10	11	12	13	14	15
6	6	7	8	9	10	11	12	13	14	15	16
7	7	8	9	10	11	12	13	14	15	16	17
8	8	9	10	11	12	13	14	15	16	17	18
9	9	10	11	12	13	14	15	16	17	18	19
10	10	11	12	13	14	15	16	17	18	19	20

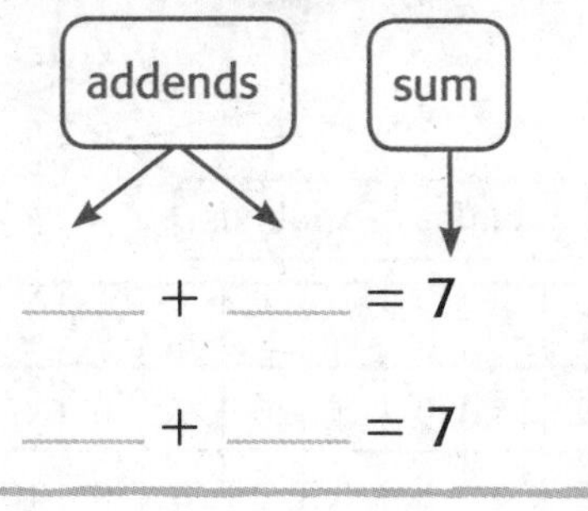

____ + ____ = 7

____ + ____ = 7

The two number sentences are an example of the

_______________ Property.

Guided Practice

Describe Danny's new pattern in the addition table below.

1. When ______ is added to a number, the sum is that number.

2. This is an example of the ______ Property of Addition.

+	0	1	2	3	4	5	6
0	0	1	2	3	4	5	6
1	1	2	3	4	5	6	7
2	2	3	4	5	6	7	8
3	3	4	5	6	7	8	9
4	4	5	6	7	8	9	10
5	5	6	7	8	9	10	11
6	6	7	8	9	10	11	12

Talk MATH

How do you find patterns in numbers?

Name ....

CCSS

Independent Practice

Use the addition table.

3. Shade a diagonal of numbers **blue** that show the sums equal to 8.

4. Shade a diagonal of numbers **green** that show the sums equal to 5.

5. Shade a row of numbers **yellow** that represent sums with one addend of 4.

6. Shade a column of numbers **pink** that represent sums with one addend of 6.

+	0	1	2	3	4	5	6	7	8	9	10
0	0	1	2	3	4	5	6	7	8	9	10
1	1	2	3	4	5	6	7	8	9	10	11
2	2	3	4	5	6	7	8	9	10	11	12
3	3	4	5	6	7	8	9	10	11	12	13
4	4	5	6	7	8	9	10	11	12	13	14
5	5	6	7	8	9	10	11	12	13	14	15
6	6	7	8	9	10	11	12	13	14	15	16
7	7	8	9	10	11	12	13	14	15	16	17
8	8	9	10	11	12	13	14	15	16	17	18
9	9	10	11	12	13	14	15	16	17	18	19
10	10	11	12	13	14	15	16	17	18	19	20

7. Shade two squares **purple** that each represent the sum of 3 and 9. What property does this show?

____

8. Circle two squares that each represent a sum of 0 and 10. What two properties does this show?

____

____

9. Shade two addends **red** that have a sum of 8. Complete the number sentence. Write the greater addend first.

_______ + _______ = 8

Use the Commutative Property of Addition and shade the other two addends **red**. Complete the other addition sentence.

_______ + _______ = 8

Problem Solving

Use the addition table.

+	0	1	2	3	4	5	6	7	8	9	10
0	0	1	2	3	4	5	6	7	8	9	10
1	1	2	3	4	5	6	7	8	9	10	11
2	2	3	4	5	6	7	8	9	10	11	12
3	3	4	5	6	7	8	9	10	11	12	13
4	4	5	6	7	8	9	10	11	12	13	14
5	5	6	7	8	9	10	11	12	13	14	15
6	6	7	8	9	10	11	12	13	14	15	16
7	7	8	9	10	11	12	13	14	15	16	17
8	8	9	10	11	12	13	14	15	16	17	18
9	9	10	11	12	13	14	15	16	17	18	19
10	10	11	12	13	14	15	16	17	18	19	20

10. Mathematical PRACTICE 5 **Use Math Tools** Marlo stacked 8 boxes. She had no more boxes to stack. Find the total number of boxes stacked. Shade two squares that each represent the sum. Write two number sentences.

____

What two properties does this show?

____

____

11. Pedro ran 3 miles on Sunday and 2 miles on Monday. Find the total number of miles he ran. Shade two squares that each represent the sum. Write two number sentences.

____

What property does this show?

____

HOT Problems

12. Mathematical PRACTICE 4 **Model Math** Write a real-world problem for which you can use the addition table and the Commutative Property of Addition to solve. Then solve.

____

____

13. **Building on the Essential Question** How can addition patterns help me add mentally?

____

____

MY Homework

Lesson 2
Patterns in the Addition Table

Homework Helper

eHelp

Need help? connectED.mcgraw-hill.com

Eddie colored the top row of an addition table blue. What is the pattern?

The sum of any number and zero is the number. This shows the Identity Property of Addition.

Using green, Eddie started at 2 and colored a downward diagonal pattern. What is the pattern?

Adding 2 to an even number shows a pattern of even numbers.

Using purple, Eddie started at 5 and colored a downward diagonal. What is the pattern?

Adding 2 to an odd number shows a pattern of odd numbers.

+	0	1	2	3	4	5	6	7	8	9	10
0	0	1	2	3	4	5	6	7	8	9	10
1	1	2	3	4	5	6	7	8	9	10	11
2	2	3	4	5	6	7	8	9	10	11	12
3	3	4	5	6	7	8	9	10	11	12	13
4	4	5	6	7	8	9	10	11	12	13	14
5	5	6	7	8	9	10	11	12	13	14	15
6	6	7	8	9	10	11	12	13	14	15	16
7	7	8	9	10	11	12	13	14	15	16	17
8	8	9	10	11	12	13	14	15	16	17	18
9	9	10	11	12	13	14	15	16	17	18	19
10	10	11	12	13	14	15	16	17	18	19	20

Practice

Use the addition table above.

1. Shade a diagonal of odd numbers red.

2. Shade a diagonal of even numbers yellow.

Use the addition table.

3. Circle two squares that each represent a sum of 3 and 4. This shows the Commutative Property of Addition.

   3 + 4 = ____ and 4 + 3 = ____

4. Circle the two addends which make a sum of the shaded 12. Write the number sentence.

   ____

5. Shade a diagonal of numbers green that show the sums equal to 9.

6. Shade a row of numbers yellow that represents sums with one addend of 10.

+	0	1	2	3	4	5	6	7	8	9	10
0	0	1	2	3	4	5	6	7	8	9	10
1	1	2	3	4	5	6	7	8	9	10	11
2	2	3	4	5	6	7	8	9	10	11	12
3	3	4	5	6	7	8	9	10	11	12	13
4	4	5	6	7	8	9	10	11	12	13	14
5	5	6	7	8	9	10	11	12	13	14	15
6	6	7	8	9	10	11	12	13	14	15	16
7	7	8	9	10	11	12	13	14	15	16	17
8	8	9	10	11	12	13	14	15	16	17	18
9	9	10	11	12	13	14	15	16	17	18	19
10	10	11	12	13	14	15	16	17	18	19	20

Real World

Problem Solving

7. Mathematical PRACTICE 3 **Justify Conclusions** Jasmine had 11 friends over at her house. Every time the doorbell rang, 2 more friends arrived. The doorbell rang 3 times. How many friends did Jasmine have over altogether?

   ____

8. Steve colors the following sums on an addition table. If he continues the pattern, will the numbers continue to be even? Explain.

   12, 14, 16, 18

   ____

   ____

Test Practice

9. Danielle is saving for a bicycle. Her last 4 bank deposits are shown in the table. If the pattern continues, how much will her next bank deposit be?

Bank Deposits
$9
$11
$13
$15
?

Ⓐ $2 Ⓑ $16 Ⓒ $17 Ⓓ $19

Name ....

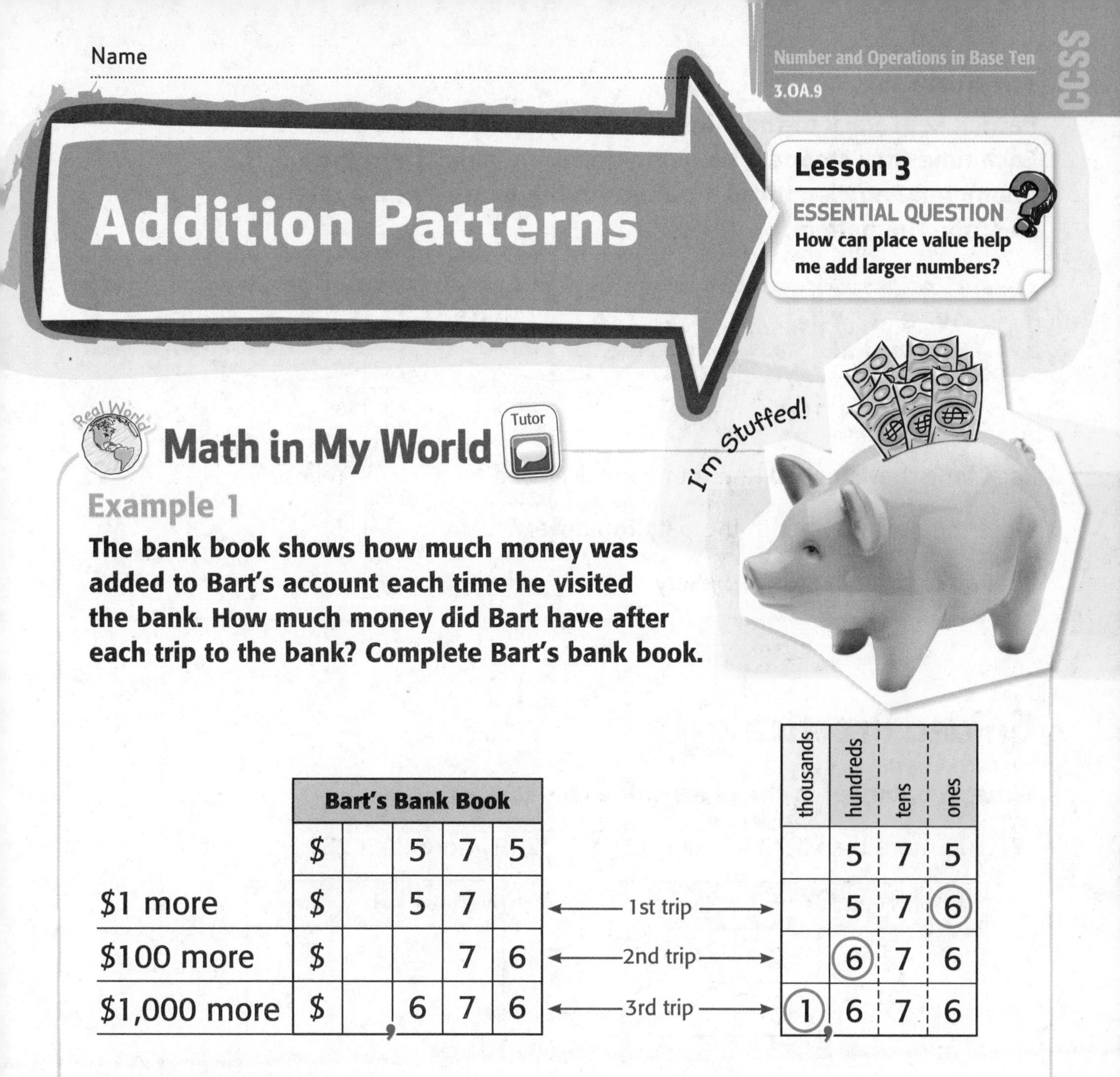

Addition Patterns

Lesson 3

ESSENTIAL QUESTION
How can place value help me add larger numbers?

Math in My World

Tutor

Example 1

The bank book shows how much money was added to Bart's account each time he visited the bank. How much money did Bart have after each trip to the bank? Complete Bart's bank book.

	Bart's Bank Book				
	$		5	7	5
$1 more	$		5	7	
$100 more	$			7	6
$1,000 more	$	,	6	7	6

1st trip, 2nd trip, 3rd trip

	thousands	hundreds	tens	ones
		5	7	5
1st trip		5	7	(6)
2nd trip		(6)	7	6
3rd trip	(1),	6	7	6

The circled digits show which place's value changed each time. So, Bart had $ ________ after the first trip, $ ________ after the second trip, and $ ________ after the third trip.

Bart added some money to the $1,676 he already had in his account. He now has $1,686. Complete the number sentence to show how much he added.

$1,676 + ________ = $1,686

Example 2

Patrick kept track of the miles his family traveled on a trip. Each time they stopped, he wrote down the miles from the odometer. Patrick noticed a pattern in the numbers he wrote. Describe the pattern.

		9	6	7	4

		9	7	7	4

		9	8	7	4

+ 100 ______ ______

Each time they stopped, the numbers increased by ______ miles.

Write the next number in the pattern above.

So, Patrick's family stopped every ______ miles.

Guided Practice

Write the number in the place-value chart.

1. 100 more than 3,728

thousands	hundreds	tens	ones
3,	7	2	8

2. 1 more than 281

hundreds	tens	ones
2	8	1

3. Complete the number sentence.

thousands	hundreds	tens	ones
6,	3	2	5
6,	4	2	5

6,325 + ______ = 6,425

Talk MATH Tell what happens to the digits in the number 1,057 if 100 is added to that number.

Name

CCSS

Independent Practice

Write the number.

4. 1 more than 972 ______

5. 1,000 more than 374 ______

6. 10 more than 310 ______

7. 1,000 more than 8,993 ______

8. 10 more than 1,437 ______

9. 100 more than 2,819 ______

10. 100 more than 173 ______

11. 10 more than 6,910 ______

Complete the number sentence.

12. 974 + ______ = 975

13. 1,234 + ______ = 2,234

14. 8,264 + ______ = 9,264

15. 1,038 + ______ = 1,138

16. 6,123 + ______ = 6,223

17. 8,877 + ______ = 8,887

Identify and complete the number pattern.

18. 6,282; 7,282; ______; 9,282

The number pattern is ______.

19. 9,379; ______; 9,381; 9,382

The number pattern is ______.

20. 7,874; 7,884; ______; 7,904; ______; ______

The number pattern is ______.

21. Add to move up the stairs.

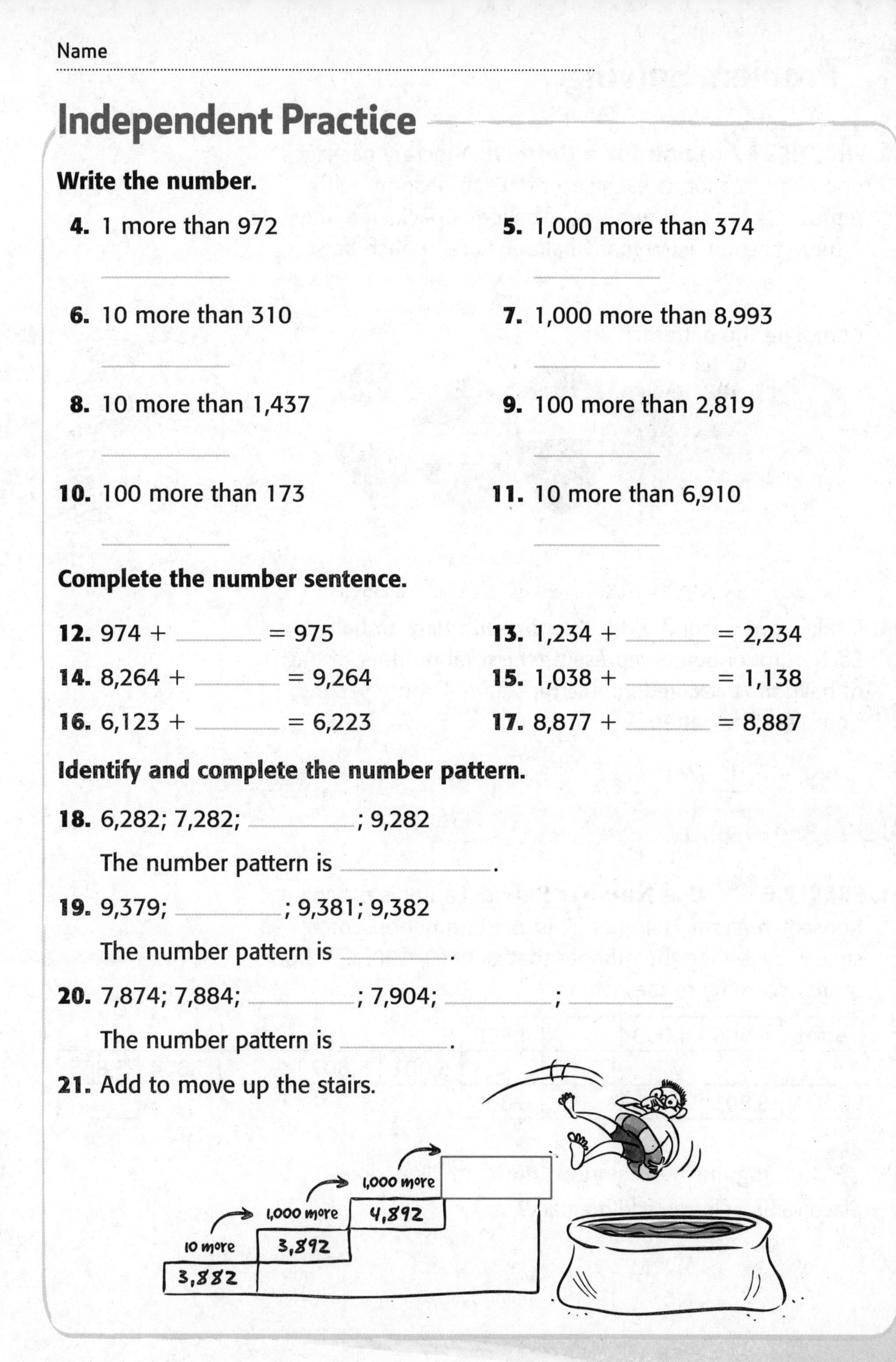

Problem Solving

22. Mathematical PRACTICE 8 **Look for a Pattern** A factory packages one bag of balloons each second. Each balloon below represents the total number of balloons packaged after 1 more second. How many balloons are in each bag?

____

Complete the pattern.

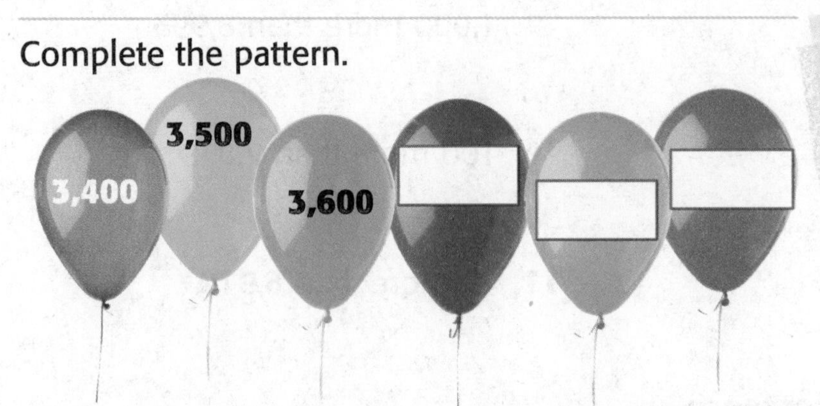

23. It takes one second to fill a carton with bags of balloons. Each number below represents the total number of bags of balloons packaged into cartons after 1 more second. Complete the pattern.

4,720; 4,730; 4,740; ______; ______; ______

HOT Problems

24. Mathematical PRACTICE 2 **Use Number Sense** Laurel is playing a hopscotch game. Help her fill in the empty hopscotch spaces by writing the number that is 1, 10, 100, or 1,000 more according to the pattern.

Start	4,500	4,600		4,800					
					5,801	5,802		5,804	5,805
6,804	6,803			6,800					

25. **Building on the Essential Question** How does place value help me add mentally?

____

____

MY Homework

Lesson 3
Addition Patterns

Homework Helper

eHelp

Need help? connectED.mcgraw-hill.com

Over several years, Mrs. Bowers' students made a total of 2,367 paper cranes. This year, she challenged her students to make the new total 2,467 paper cranes. How many paper cranes do they need to make this year to meet the challenge?

Use a place-value chart to find which place changed in value.

thousands	hundreds	tens	ones
2,	(3)	6	7
2,	(4)	6	7

The new number is 100 more.

2,367 + 100 = 2,467

So, this year's class needs to make 100 paper cranes.

Practice

Write the number in the place-value chart.

1. 10 more than 567

hundreds	tens	ones
5	6	7

2. 1 more than 358

hundreds	tens	ones
3	5	8

3. 1,000 more than 1,529

thousands	hundreds	tens	ones
1,	5	2	9

4. 100 more than 5,834

thousands	hundreds	tens	ones
5,	8	3	4

Complete the number sentence.

5.

thousands	hundreds	tens	ones
1,	2	7	1
2,	2	7	1

1,271 + ______ = 2,271

6.

thousands	hundreds	tens	ones
4,	2	4	4
4,	3	4	4

4,244 + ______ = 4,344

Write the number.

7. 10 more than 1,465

______

8. 100 more than 8,699

______

9. 1,000 more than 3,007

______

Identify and complete the number pattern.

10. 2,378; 2,478; 2,578; ______; 2,778; 2,878

The number pattern is ______.

11. 5,903; 5,913; 5,923; ______; 5,943; 5,953

The number pattern is ______.

Real World

Problem Solving

12. Mathematical PRACTICE 4 **Model Math** A baby horse weighed 104 pounds at birth. In one month, it gained 100 pounds. How much does the horse weigh now? Complete the number sentence to show the change.

104 + ______ = ______

Test Practice

13. Which pattern shows 100 more?

Ⓐ 1,456; 1,556; 1,656; 1,756

Ⓑ 4,987; 4,887; 4,787; 4,687

Ⓒ 5,832; 5,833; 5,834; 5,835

Ⓓ 6,001; 7,001; 8,001; 9,001

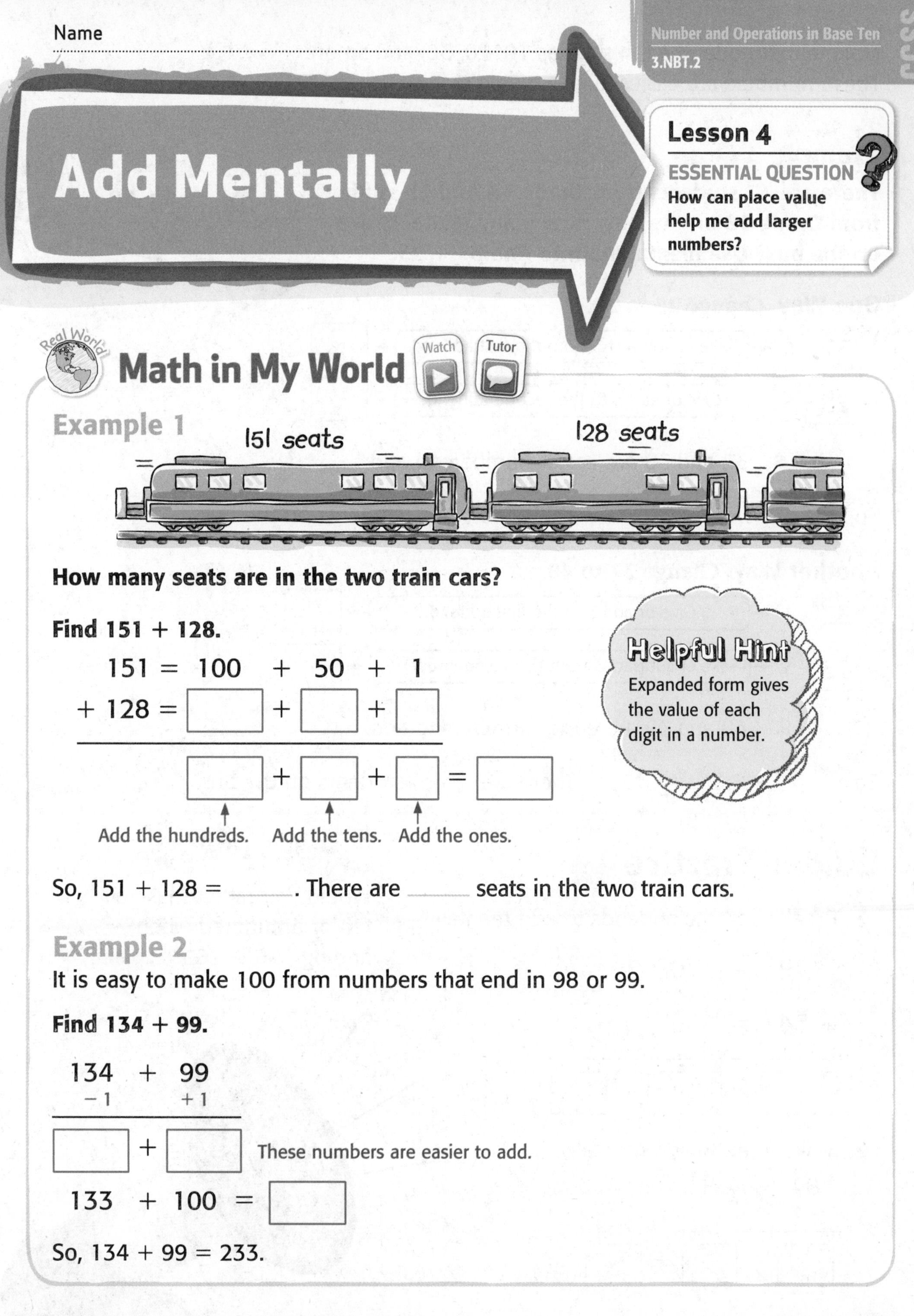

Name ....

Add Mentally

Lesson 4

ESSENTIAL QUESTION
How can place value help me add larger numbers?

Math in My World

Watch Tutor

Example 1

151 seats 128 seats

How many seats are in the two train cars?

Find 151 + 128.

$151 = 100 + 50 + 1$

$+\ 128 = \square + \square + \square$

$\square + \square + \square = \square$

Add the hundreds. Add the tens. Add the ones.

Helpful Hint
Expanded form gives the value of each digit in a number.

So, 151 + 128 = ______. There are ______ seats in the two train cars.

Example 2

It is easy to make 100 from numbers that end in 98 or 99.

Find 134 + 99.

$134 + 99$
$-1 \quad +1$

$\square + \square$ These numbers are easier to add.

$133 + 100 = \square$

So, 134 + 99 = 233.

Make either addend a ten such as 10, 20, 30, and so on. These numbers are easier to add mentally.

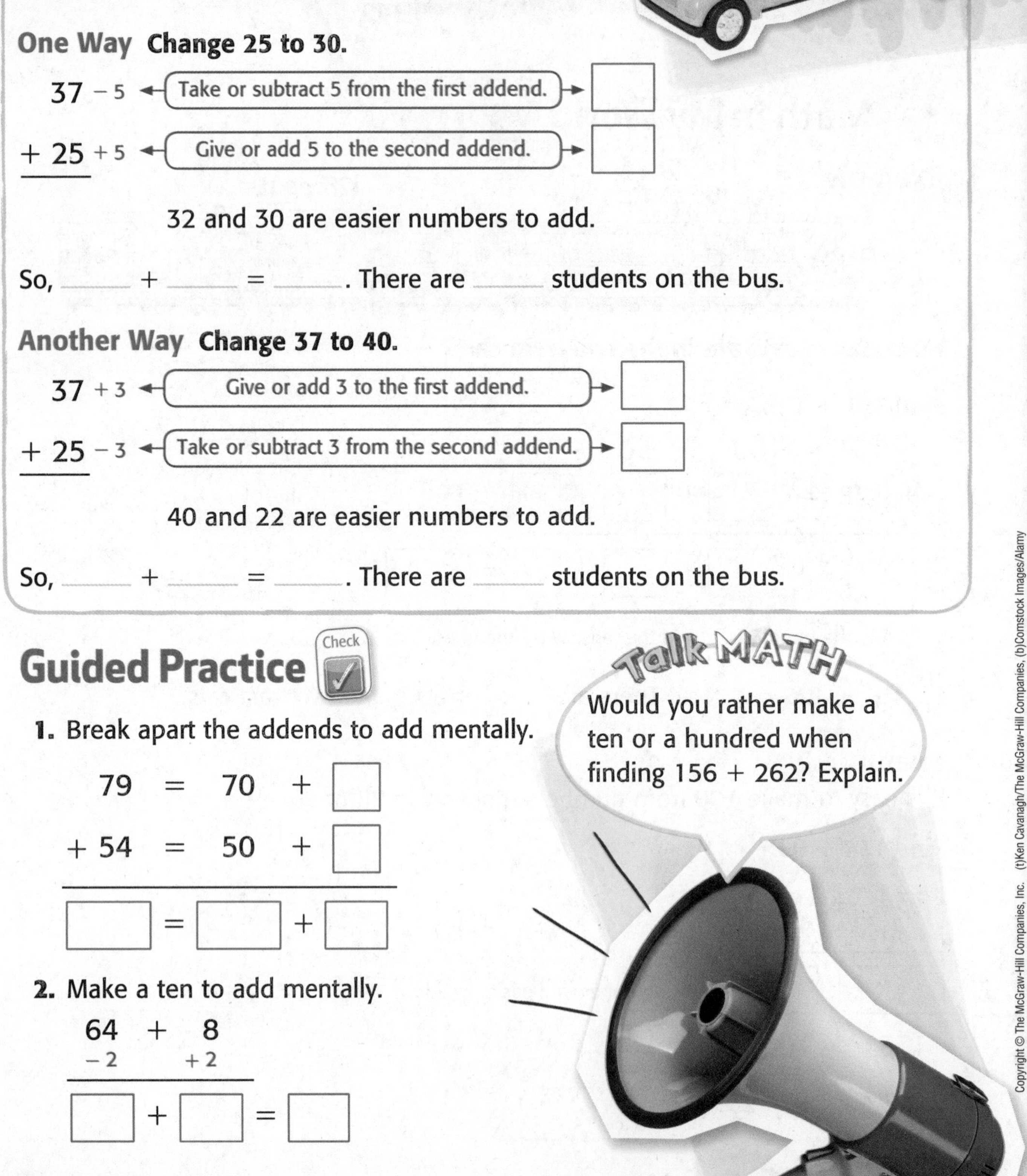

Example 3

There are 37 students from Grade 3A and 25 students from Grade 3B on the bus. How many students are on the bus? Use mental math to find 37 + 25.

One Way **Change 25 to 30.**

37 − 5 ← Take or subtract 5 from the first addend. → ☐

+ 25 + 5 ← Give or add 5 to the second addend. → ☐

32 and 30 are easier numbers to add.

So, ____ + ____ = ____. There are ____ students on the bus.

Another Way **Change 37 to 40.**

37 + 3 ← Give or add 3 to the first addend. → ☐

+ 25 − 3 ← Take or subtract 3 from the second addend. → ☐

40 and 22 are easier numbers to add.

So, ____ + ____ = ____. There are ____ students on the bus.

Guided Practice

Check

Talk MATH

Would you rather make a ten or a hundred when finding 156 + 262? Explain.

1. Break apart the addends to add mentally.

79 = 70 + ☐

+ 54 = 50 + ☐

☐ = ☐ + ☐

2. Make a ten to add mentally.

64 + 8

−2 +2

☐ + ☐ = ☐

Name

Independent Practice

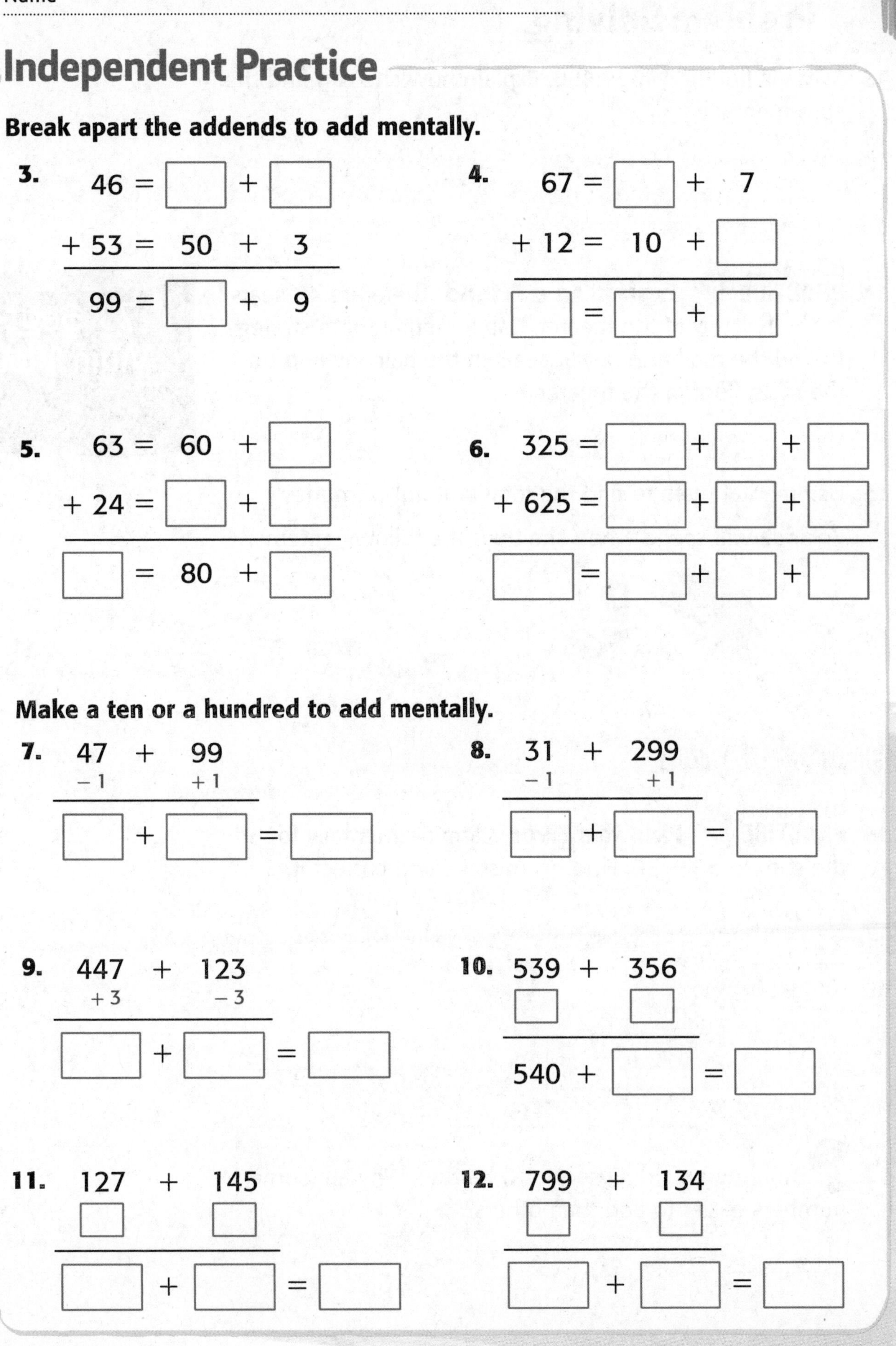

Break apart the addends to add mentally.

3. 46 = ☐ + ☐
+ 53 = 50 + 3
99 = ☐ + 9

4. 67 = ☐ + 7
+ 12 = 10 + ☐
☐ = ☐ + ☐

5. 63 = 60 + ☐
+ 24 = ☐ + ☐
☐ = 80 + ☐

6. 325 = ☐ + ☐ + ☐
+ 625 = ☐ + ☐ + ☐
☐ = ☐ + ☐ + ☐

Make a ten or a hundred to add mentally.

7. 47 + 99
−1 +1
☐ + ☐ = ☐

8. 31 + 299
−1 +1
☐ + ☐ = ☐

9. 447 + 123
+3 −3
☐ + ☐ = ☐

10. 539 + 356
☐ ☐
540 + ☐ = ☐

11. 127 + 145
☐ ☐
☐ + ☐ = ☐

12. 799 + 134
☐ ☐
☐ + ☐ = ☐

Problem Solving

13. Sylvia is finding 135 + 456. Explain how she can find the sum mentally.

14. Mathematical PRACTICE 6 **Explain to a Friend** There are 49 seats in the balcony of the theater. Use a mental math strategy to find the total number of seats in the balcony and on the main floor of the theater.

15. Use mental math to find the total amount of money Yolanda will spend when she buys the following items. ______

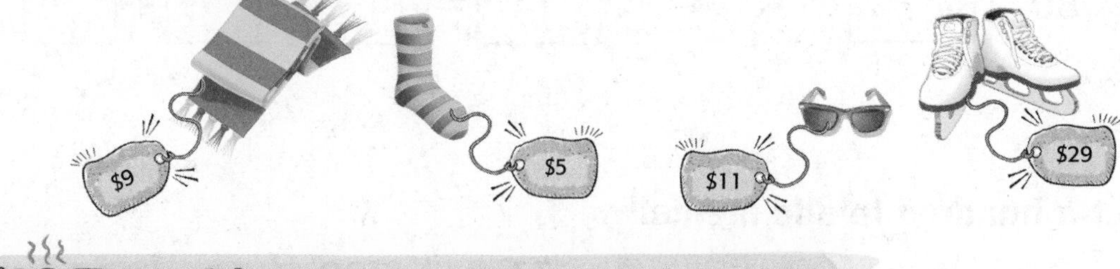

HOT Problems

16. Mathematical PRACTICE 3 **Find the Error** Carmine mentally found the sum to 56 + 36. Find his mistake and correct it.

$$\begin{array}{l} 56 + 36 \\ \underline{+\,4\qquad} \\ 60 + 36 = 96 \end{array}$$

17. **Building on the Essential Question** Why are some numbers easier to add than others?

Lesson 4
Add Mentally

Homework Helper

eHelp Need help? connectED.mcgraw-hill.com

There are 58 passengers on a subway train. At the next stop, 33 more passengers board the train. How many passengers are there altogether?

You can add mentally.

Make one addend a ten such as 10, 20, 30, and so on.

58 + 2 ← Give, or add, 2 to the first addend. → 60

+ 33 − 2 ← Take, or subtract, 2 from the second addend. → + 31

91

So, 60 + 31 = 91. There are 91 passengers.

Practice

Break apart the addends to add mentally.

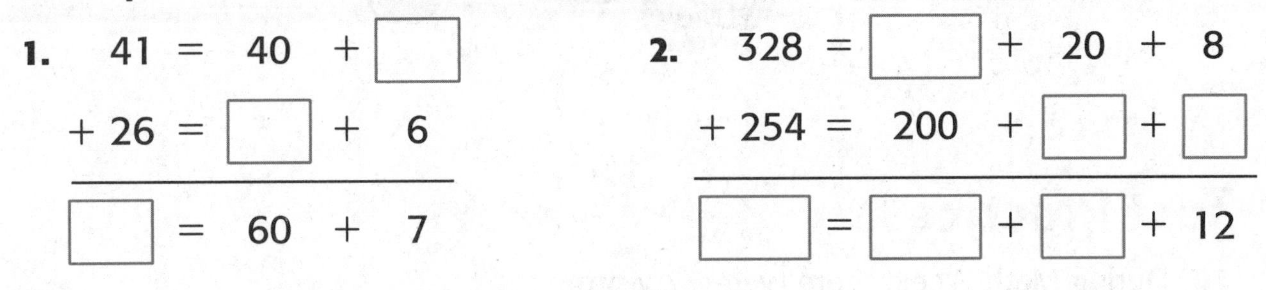

1. 41 = 40 + ☐
+ 26 = ☐ + 6
☐ = 60 + 7

2. 328 = ☐ + 20 + 8
+ 254 = 200 + ☐ + ☐
☐ = ☐ + ☐ + 12

Make a ten or a hundred to add mentally.

3. 76 + 7
−3 +3
☐ + ☐ = ☐

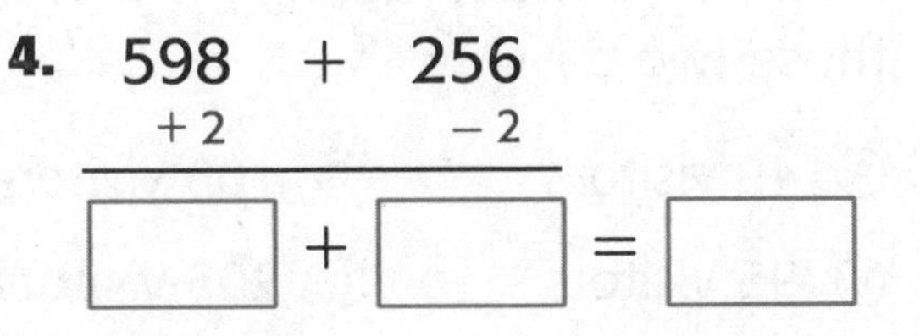

4. 598 + 256
+2 −2
☐ + ☐ = ☐

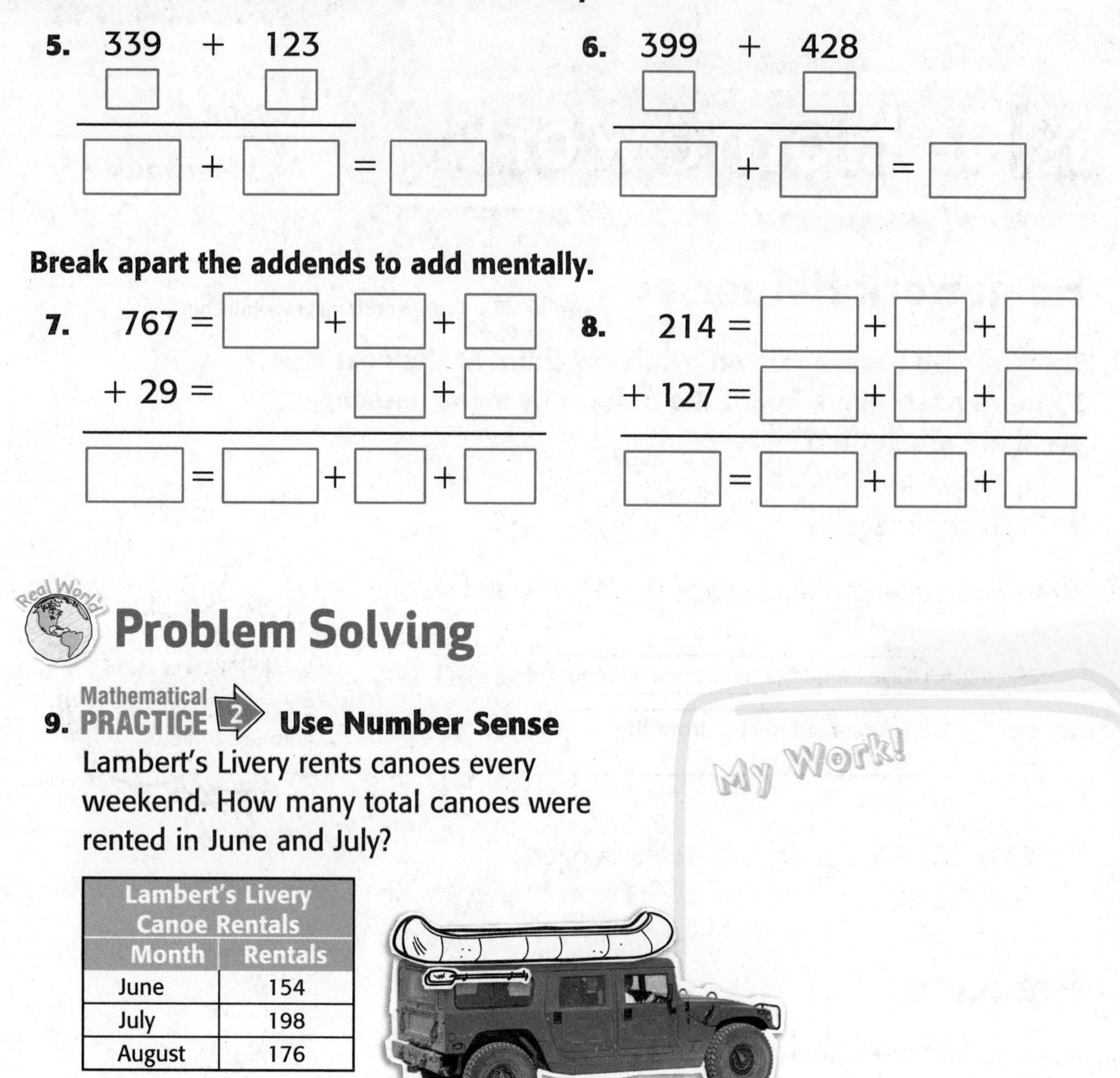

Make a ten or a hundred to add mentally.

5. 339 + 123

☐ ☐

☐ + ☐ = ☐

6. 399 + 428

☐ ☐

☐ + ☐ = ☐

Break apart the addends to add mentally.

7. 767 = ☐ + ☐ + ☐

+ 29 = ☐ + ☐

☐ = ☐ + ☐ + ☐

8. 214 = ☐ + ☐ + ☐

+ 127 = ☐ + ☐ + ☐

☐ = ☐ + ☐ + ☐

Real World

Problem Solving

9. Mathematical PRACTICE 2 **Use Number Sense**

Lambert's Livery rents canoes every weekend. How many total canoes were rented in June and July?

Lambert's Livery Canoe Rentals	
Month	**Rentals**
June	154
July	198
August	176

My Work!

Test Practice

10. During Math Week, there were 77 visitors on Monday and 28 visitors on Tuesday. How many visited during Math Week those two days?

Ⓐ 49 visitors

Ⓑ 95 visitors

Ⓒ 105 visitors

Ⓓ 205 visitors

Check My Progress

Vocabulary Check

Choose the correct word(s) to complete the sentence.

Associative	Commutative	Identity
mental math	parentheses	pattern

1. The ________________ Property of Addition states that the order in which addends are added does not change the sum.

2. A set of numbers that follow a certain order is a ________________.

3. You can use ________________ to add numbers in your head.

4. The ________________ Property of Addition states that the way addends are grouped does not change the sum.

5. ________________ show which numbers to add first.

6. The ________________ Property of Addition states that the sum of any number and zero is that number.

Concept Check

Make a ten or a hundred to add mentally.

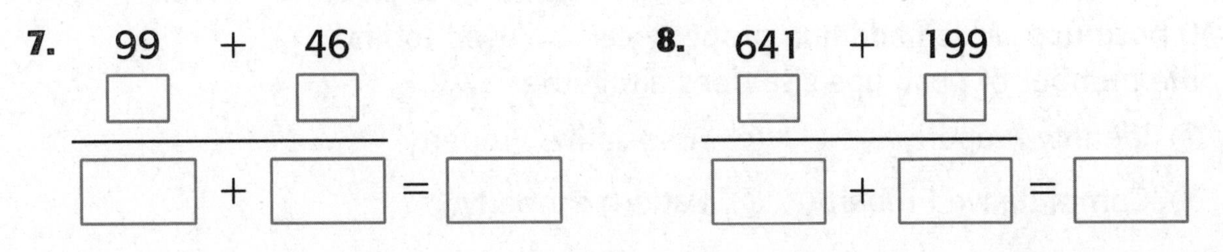

Concept Check

Break apart the addends to add mentally.

9.

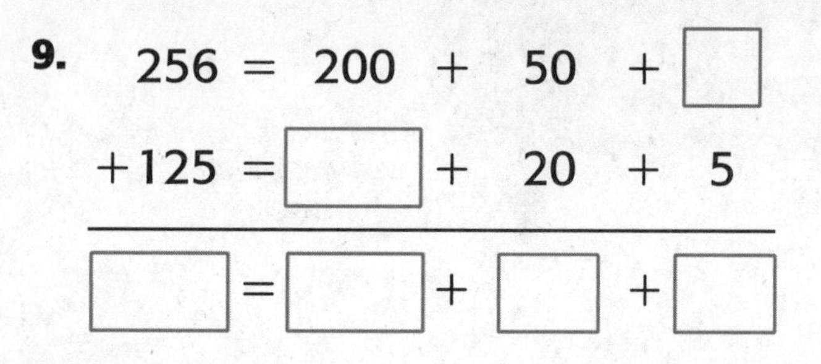

Identify and complete the number pattern.

10. 573; 673; __________; 873

The number pattern is __________.

11. 2,930; __________; 2,950; 2,960

The number pattern is __________.

Write the number in the place-value chart.

12. 1,000 more than 2,491

thousands	hundreds	tens	ones
2,	4	9	1

13. 100 more than 8,762

thousands	hundreds	tens	ones
8,	7	6	2

Real World! Problem Solving

14. At a book sale, Emily buys 4 books. Paul buys 8 books. Genie buys 6 books. Use an addition property to find how many books Emily, Paul, and Genie buy altogether. Write the addition property.

____

Test Practice

15. On Monday, Lisa does 9 push ups. On Tuesday, she does 0 push ups. Which addition property can be used to find the number of push ups Lisa does altogether?

Ⓐ Identity Property

Ⓑ Commutative Property

Ⓒ Associative Property

Ⓓ Pattern Property

Hands On
Use Models to Add

Lesson 6

ESSENTIAL QUESTION
How can place value help me add larger numbers?

Use a place-value chart and base-ten blocks to model three-digit addition with regrouping. To **regroup** means to rename a number using place value.

Build It

Tools

While on a trip, Rosa counted 148 red cars and 153 green cars. How many total cars did Rosa count?

Find 148 + 153.

1 Estimate the sum →

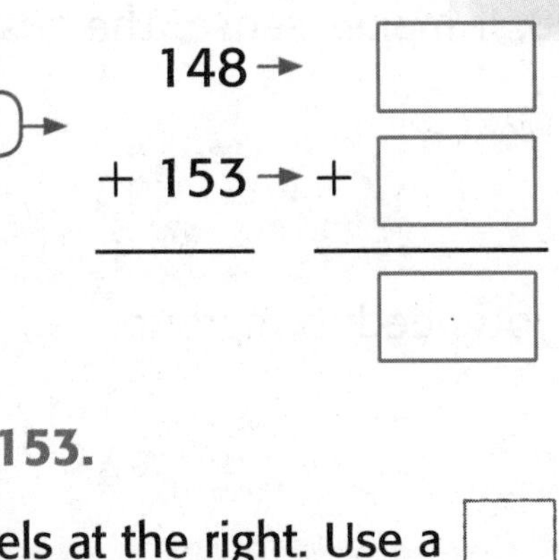

$$\begin{array}{r} 148 \rightarrow \square \\ +\ 153 \rightarrow +\ \square \\ \hline \square \end{array}$$

hundreds	tens	ones

2 **Model 148 + 153.**

Draw your models at the right. Use a □ for hundreds, ▯ for tens, and ▫ for ones.

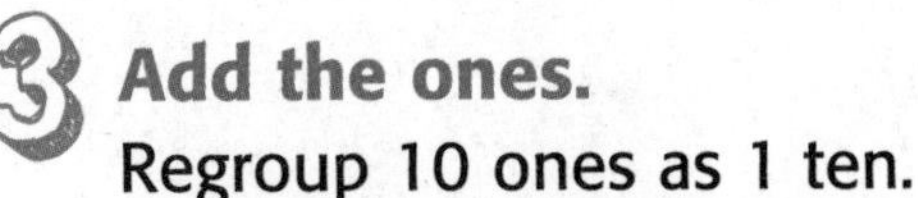

3 **Add the ones.**
Regroup 10 ones as 1 ten.

hundreds	tens	ones

Draw your models.

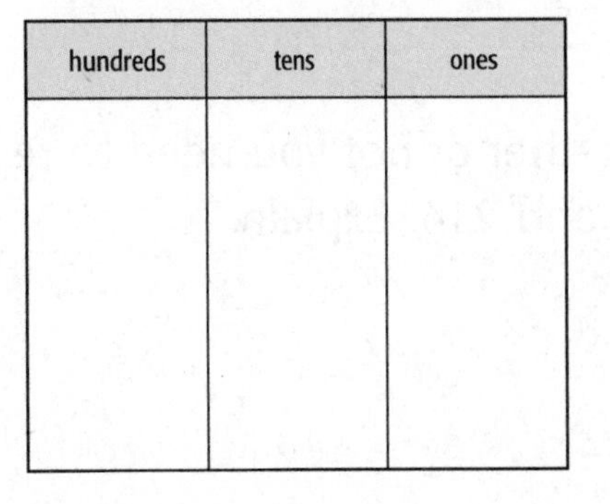

hundreds	tens	ones

4 Add the tens and the hundreds.

Regroup 10 tens as ______ hundred. Draw your models below.

hundreds	tens	ones

There are ______ hundreds, ______ tens, and ______ one.

So, 148 + 153 = ______.

Check for Reasonableness
Ask yourself if the answer makes sense. Is your answer **reasonable**?

301 is close to the estimate of 300. It makes sense. The answer is reasonable.

Talk About It

1. Explain how you know when you need to regroup.

______

______

2. Mathematical PRACTICE 6 **Be Precise** Why were the ones and tens regrouped?

______

______

3. Tell whether or not you need to regroup when finding the sum of 147 and 214. Explain.

______

______

Name ....

Add Three-Digit Numbers

Lesson 7

ESSENTIAL QUESTION
How can place value help me add larger numbers?

When you add, you may need to regroup. To **regroup** means to rename a number using place value.

Math in My World

Watch Tutor

Looking for me?

Example 1

During a weekend trip to Vermont, the Wildlife Club saw 127 wrens and 58 eagles. How many wrens and eagles did the Wildlife Club see?

Find 127 + 58.

	hundreds	tens	ones
	1	2	7
+		5	8

Estimate 127 + 58 ⟶ ______ + ______ = ______

1 Add the ones.
7 ones + 8 ones = 15 ones
Regroup 15 ones as 1 ten and 5 ones.

2 Add the tens and hundreds.
1 ten + 2 tens + 5 tens = 8 tens
1 hundred + 0 hundreds = 1 hundred

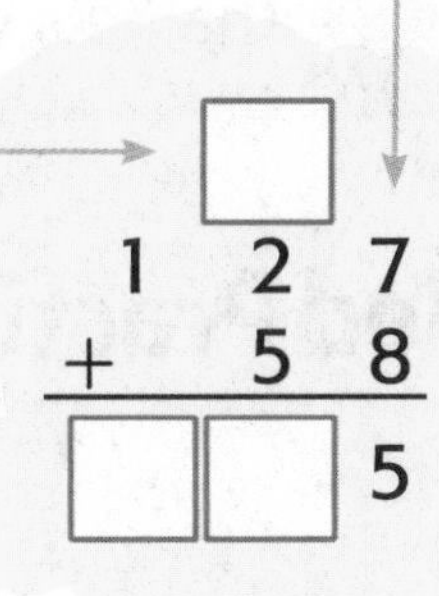

Check for Reasonableness

185 is close to the estimate of ______. It makes sense.
The answer is **reasonable.**

127 + 58 = ______

So, the Wildlife Club saw ______ wrens and eagles.

You can write a number sentence to find the **unknown,** or missing number.

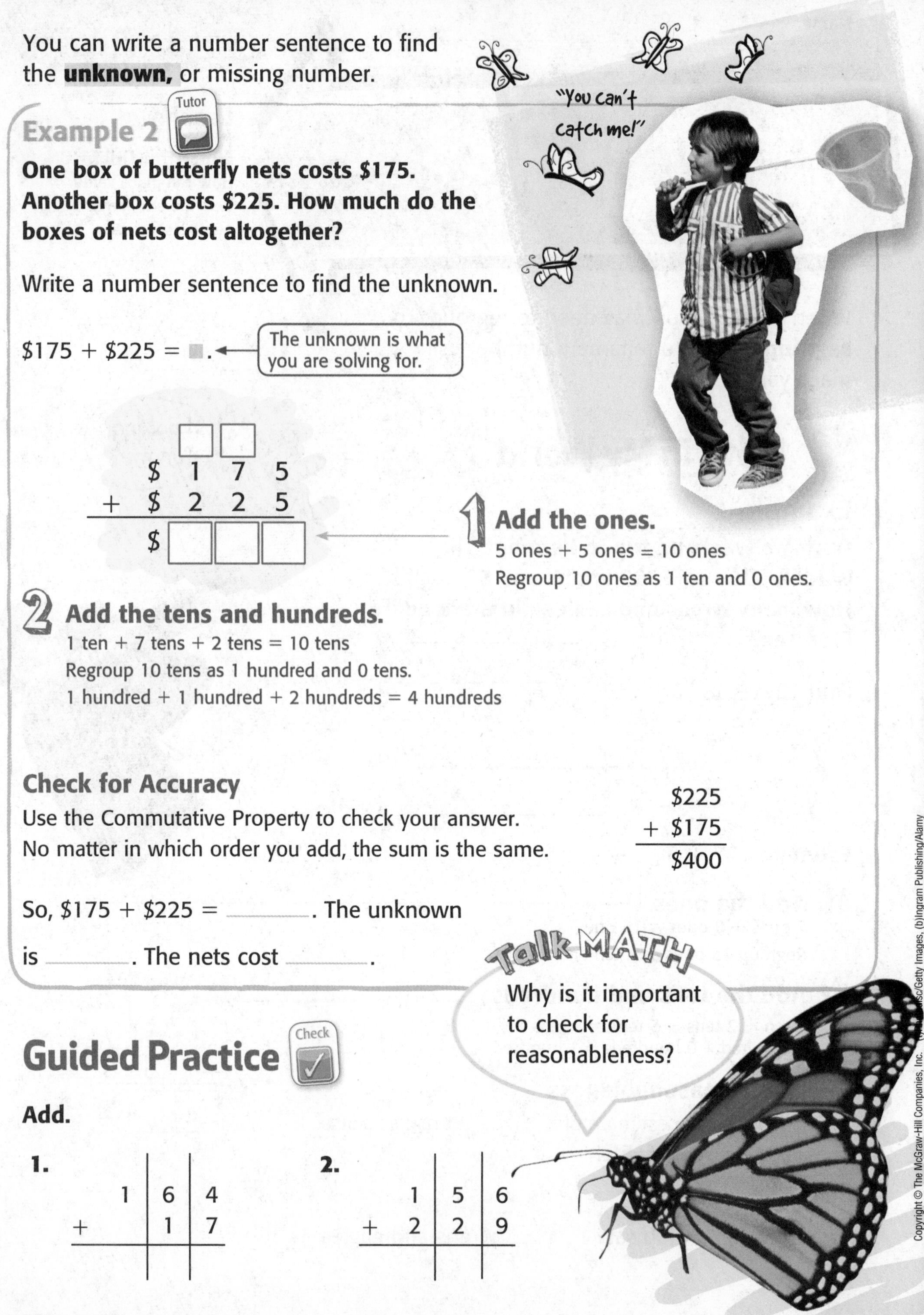

Example 2

Tutor

One box of butterfly nets costs $175. Another box costs $225. How much do the boxes of nets cost altogether?

Write a number sentence to find the unknown.

$175 + $225 = ■. ← The unknown is what you are solving for.

	□	□	
	$ 1	7	5
+	$ 2	2	5
	$ □	□	□

1 Add the ones.
5 ones + 5 ones = 10 ones
Regroup 10 ones as 1 ten and 0 ones.

2 Add the tens and hundreds.
1 ten + 7 tens + 2 tens = 10 tens
Regroup 10 tens as 1 hundred and 0 tens.
1 hundred + 1 hundred + 2 hundreds = 4 hundreds

Check for Accuracy

Use the Commutative Property to check your answer. No matter in which order you add, the sum is the same.

	$225
+	$175
	$400

So, $175 + $225 = ______. The unknown is ______. The nets cost ______.

Talk MATH

Why is it important to check for reasonableness?

Guided Practice

Check

Add.

1.

	1	6	4
+		1	7

2.

	1	5	6
+	2	2	9

Name ....

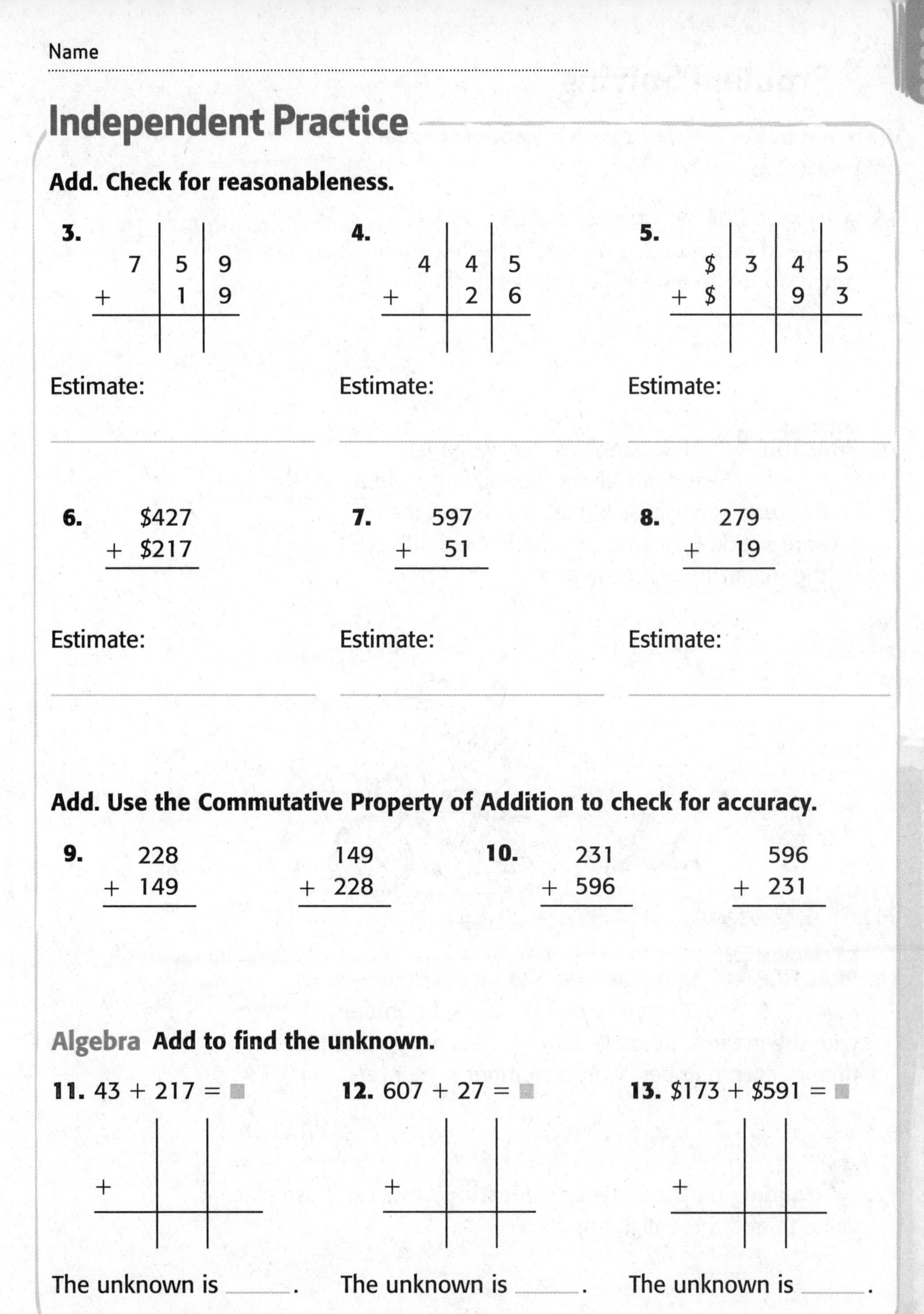

Independent Practice

Add. Check for reasonableness.

3.

7	5	9
+	1	9

Estimate: ________

4.

4	4	5
+	2	6

Estimate: ________

5.

$	3	4	5
+ $		9	3

Estimate: ________

6. $427 + $217

Estimate: ________

7. 597 + 51

Estimate: ________

8. 279 + 19

Estimate: ________

Add. Use the Commutative Property of Addition to check for accuracy.

9. 228 + 149   149 + 228

10. 231 + 596   596 + 231

Algebra **Add to find the unknown.**

11. 43 + 217 = ■

The unknown is ____.

12. 607 + 27 = ■

The unknown is ____.

13. $173 + $591 = ■

The unknown is ____.

Problem Solving

Write a number sentence with a symbol for the unknown. Then solve.

My Work!

14. A 10-speed bike is on sale for $199, and a 12-speed racing bike is on sale for $458. How much do the two bikes cost altogether?

____

15. 

Use Algebra A newspaper surveyed 475 students about their favorite sport. A magazine surveyed 189 students about their favorite snack. How many students were surveyed by the magazine and newspaper?

____

HOT Problems

16. Mathematical PRACTICE 2 **Use Number Sense** Use the digits 3, 5, and 7 to make two three-digit numbers with the greatest possible sum. Use each digit one time in each number. Write a number sentence.

____

17. **Building on the Essential Question** How can I use place value to add three-digit numbers?

____

____

Name

MY Homework

Lesson 7
Add Three-Digit Numbers

Homework Helper

eHelp Need help? connectED.mcgraw-hill.com

A toy store sold 223 robots last year. This year, they sold 198 robots. How many robots did they sell over the two years?

Find 223 + 198.

Estimate 223 + 198 ⟶ 200 + 200 = 400

1 Add the ones.
3 ones + 8 ones = 11 ones
Regroup 11 ones as 1 ten and 1 one.

2 Add the tens and hundreds.
1 ten + 2 tens + 9 tens = 12 tens
Regroup 12 tens as 1 hundred and 2 tens.
1 hundred + 2 hundreds + 1 hundred = 4 hundreds

$$\begin{array}{r} {}^{1}{}^{1} \\ 223 \\ +\ 198 \\ \hline 421 \end{array}$$

Check for reasonableness
421 is close to the estimate of 400. The answer is reasonable.

So, 223 + 198 = 421.

The store sold 421 toy robots over the two years.

Practice

Add. Check for reasonableness.

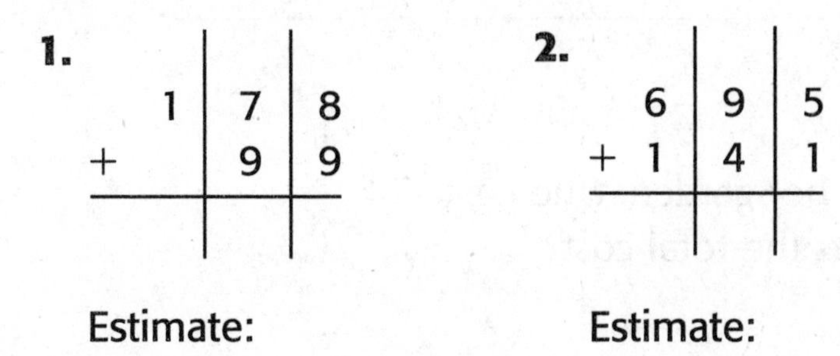

1.

	1	7	8
+		9	9

Estimate: ____________

2.

	6	9	5
+	1	4	1

Estimate: ____________

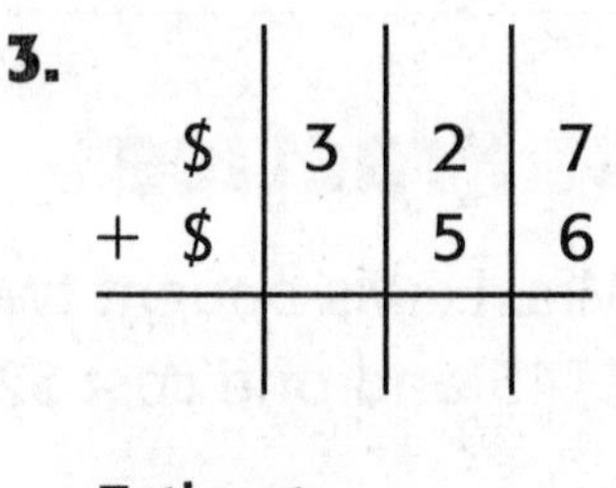

3.

$	3	2	7
+ $		5	6

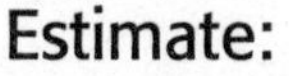

Estimate: ____________

Add. Use the Commutative Property to check for accuracy.

4.

$	3	5	0
+ $	4	6	5
$			

$	4	6	5
+ $	3	5	0
$			

5.

1	9	6
+ 2	8	6

2	8	6
+ 1	9	6

Algebra **Add to find the unknown.**

6. 661 + 99 = ■

7. $258 + $337 = ■

8. $739 + $81 = ■

The unknown is ☐.

The unknown is ☐.

The unknown is ☐.

Problem Solving

9. Mathematical PRACTICE 1 **Make a Plan** The principal ordered 215 muffins and 155 bagels. How many muffins and bagels were ordered in all?

____

Vocabulary Check

Choose the correct word to complete each sentence.

reasonable regroup unknown

10. To ______ means to rename a number using place value.

11. A missing number is the ______.

12. If an answer makes sense, it is ______.

Test Practice

13. Mrs. Lewis bought two statues for her garden. One cost $145 and one cost $262. What was the total cost?

Ⓐ $117 Ⓒ $407

Ⓑ $317 Ⓓ $410

Add Four-Digit Numbers

Lesson 8

ESSENTIAL QUESTION
How can place value help me add larger numbers?

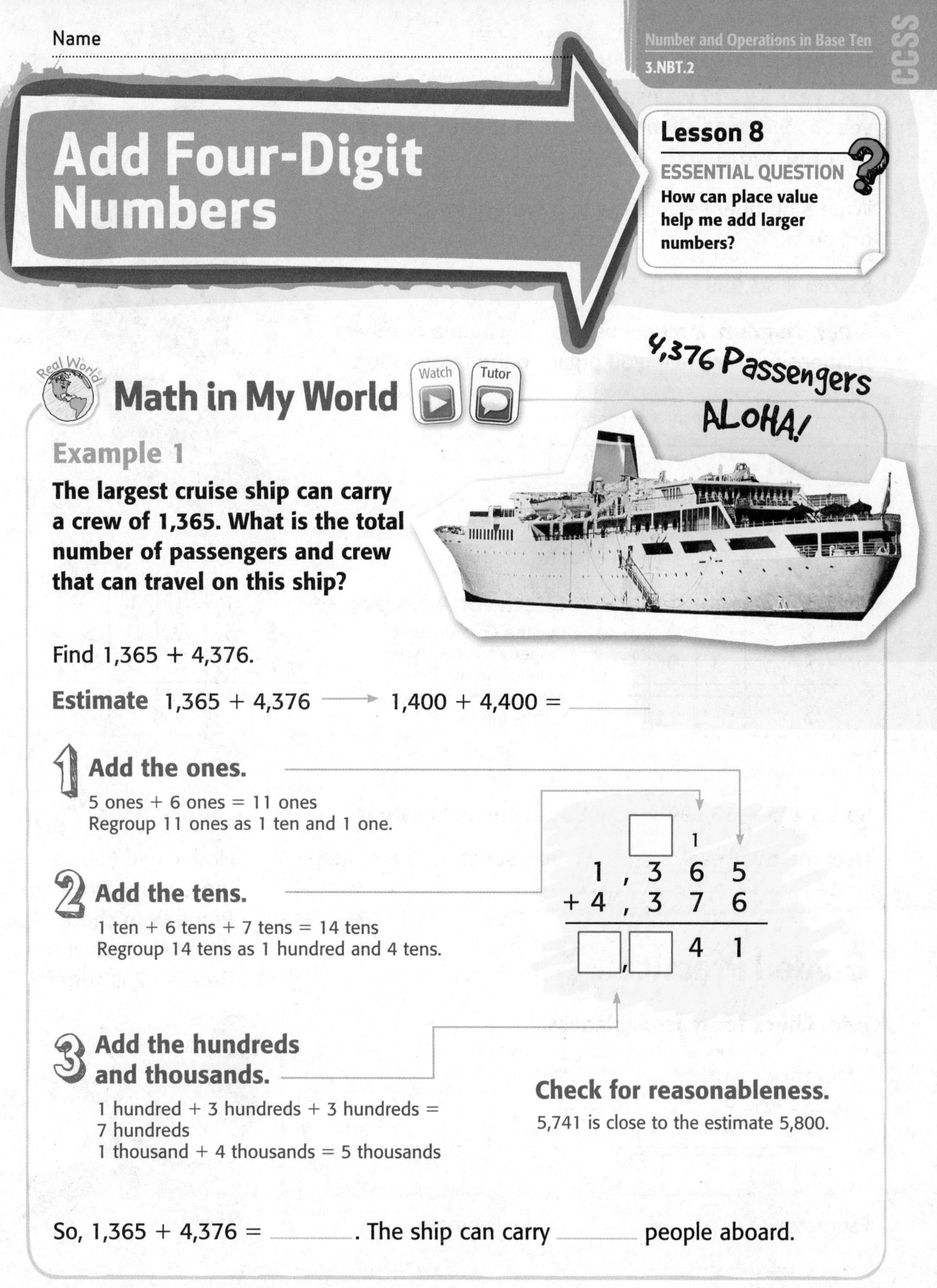

Math in My World

Watch Tutor

Example 1

The largest cruise ship can carry a crew of 1,365. What is the total number of passengers and crew that can travel on this ship?

Find 1,365 + 4,376.

Estimate 1,365 + 4,376 ⟶ 1,400 + 4,400 = ______

1 Add the ones.

5 ones + 6 ones = 11 ones
Regroup 11 ones as 1 ten and 1 one.

2 Add the tens.

1 ten + 6 tens + 7 tens = 14 tens
Regroup 14 tens as 1 hundred and 4 tens.

```
    [ ] 1
  1 , 3 6 5
+ 4 , 3 7 6
-----------
 [ ],[ ] 4 1
```

3 Add the hundreds and thousands.

1 hundred + 3 hundreds + 3 hundreds = 7 hundreds
1 thousand + 4 thousands = 5 thousands

Check for reasonableness.
5,741 is close to the estimate 5,800.

So, 1,365 + 4,376 = ______. The ship can carry ______ people aboard.

Example 2

Last year \$3,295 was spent on a skate park. This year \$3,999 was spent. How much money was spent over the two years?

Write a number sentence with a symbol for the unknown.

\$3,295 + \$3,999 = ■

A **bar diagram**, a model used to illustrate a number relationship, may help you organize the information.

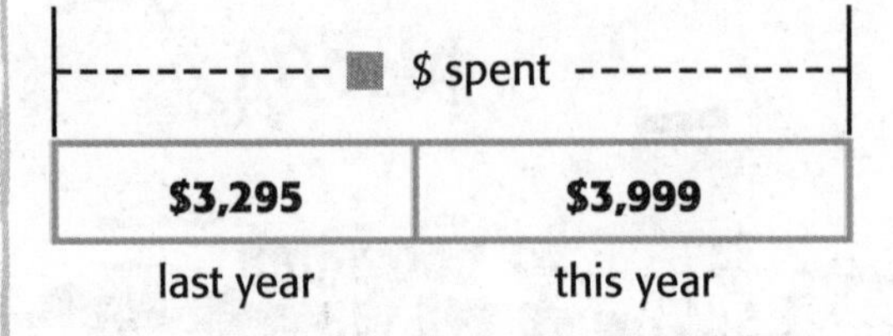

Add.

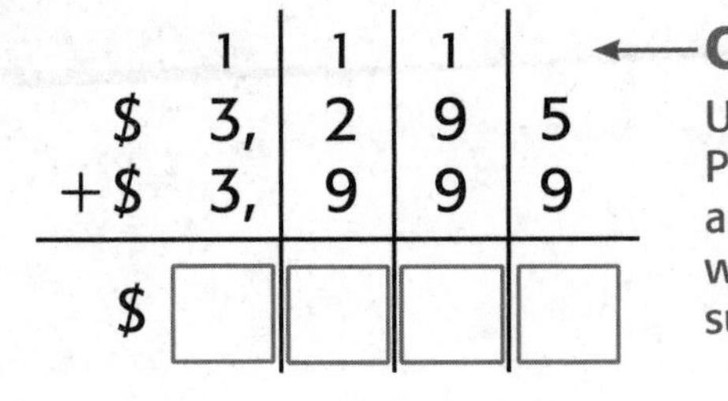

Check for Accuracy
Use the Commutative Property to check your answer. No matter in which order you add, the sum is the same.

		1	1	1
\$	3,	9	9	9
+\$	3,	2	9	5
\$	☐	☐	☐	☐

So, \$3,295 + \$3,999 = ________. The unknown is ________.

Over the two years, ________ was spent on the skate park.

Talk MATH

How could you use the Commutative Property to check that your answer to Exercise 2 is correct?

Guided Practice

Add. Check for reasonableness.

1.

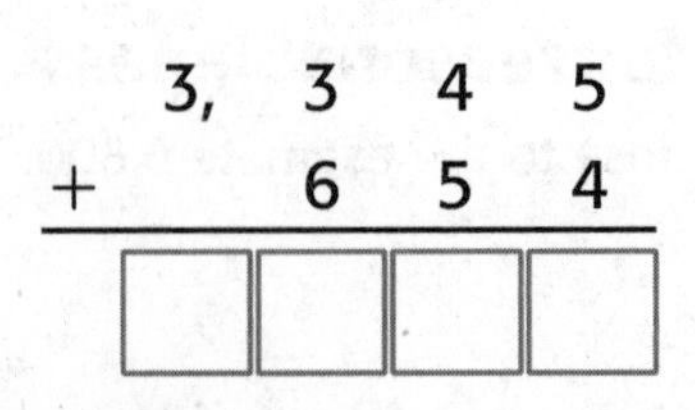

Estimate:

_____ + _____ = _____

2.

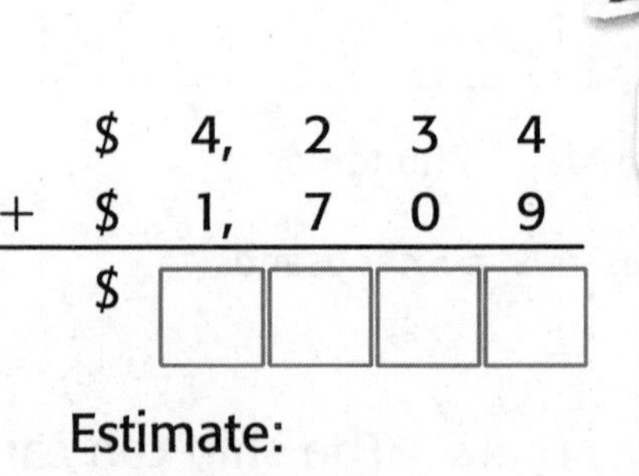

Estimate:

_____ + _____ = _____

Name ....

CCSS

Independent Practice

Add. Check for reasonableness.

3.
```
  6,499
+   543
```

4.
```
  1,998
+   300
```

5.
```
  $2,503
+ $2,899
```

6.
```
  $8,285
+ $1,456
```

7.
```
  2,390
+ 3,490
```

8.
```
  $5,555
+ $3,555
```

Add. Use the Commutative Property of Addition to check for accuracy.

9.
```
  1,734        2,882
+ 2,882      + 1,734
```

10.
```
  2,333        5,977
+ 5,977      + 2,333
```

Algebra **Add to find the unknown.**

11. 2,865 + 5,522 = ■ **12.** 3,075 + 5,640 = ■ **13.** 1,603 + 3,509 = ■

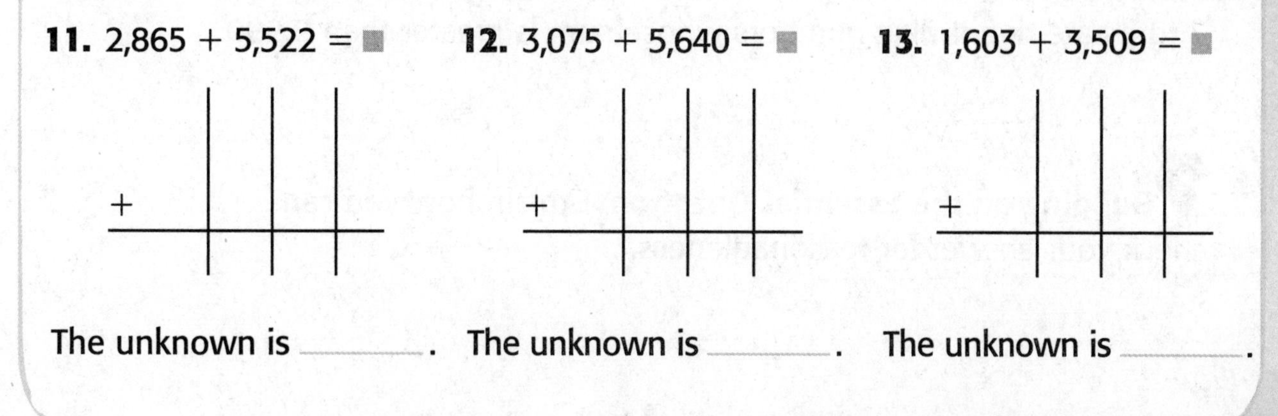

The unknown is ______. The unknown is ______. The unknown is ______.

Problem Solving

14. In one year, Lou's dad used 1,688 gallons of gas in his car. His mom's car used 1,297 gallons. Use a bar diagram to find the total gallons of gas used. Write a number sentence with a symbol for the unknown. Then solve.

■ total gallons

My Work!

15. 

Draw a Conclusion About how many people were surveyed about their favorite summer place? Is the estimate greater than or less than the exact answer? Explain.

Beach	2,311
Amusement Park	2,862

HOT Problems

16. Mathematical PRACTICE 2 **Use Number Sense** Use the digits 0 through 7 to create two 4-digit numbers whose sum is greater than 9,999.

17. **Building on the Essential Question** Explain how you can check your answer for reasonableness.

MY Homework

Lesson 8
Add Four-Digit Numbers

Homework Helper

eHelp Need help? connectED.mcgraw-hill.com

The circus had an attendance of 7,245 people in the bleachers and 1,877 people in the box seats for the Friday night show. How many people attended the circus altogether?

Find 7,245 + 1,877.

Estimate 7,200 + 1,900 = 9,100

1 Add the ones.
5 ones + 7 ones = 12 ones
Regroup 12 ones as 1 ten and 2 ones.

2 Add the tens.
1 ten + 4 tens + 7 tens = 12 tens
Regroup 12 tens as 1 hundred and 2 tens.

3 Add the hundreds and thousands.
1 hundred + 2 hundreds + 8 hundreds = 11 hundreds
Regroup 11 hundreds as 1 thousand and 1 hundred.
1 thousand + 7 thousands + 1 thousand = 9 thousands

$$\begin{array}{r} {}^{1}\ \ {}^{1}{}^{1}\ \ \\ 7,245 \\ +\ 1,877 \\ \hline 9,122 \end{array}$$

Check for reasonableness.
9,122 is close to the estimate of 9,100. The answer is reasonable.

7,245 + 1,877 = 9,122.
On Friday night, 9,122 people attended the circus.

Practice

Add. Check for reasonableness.

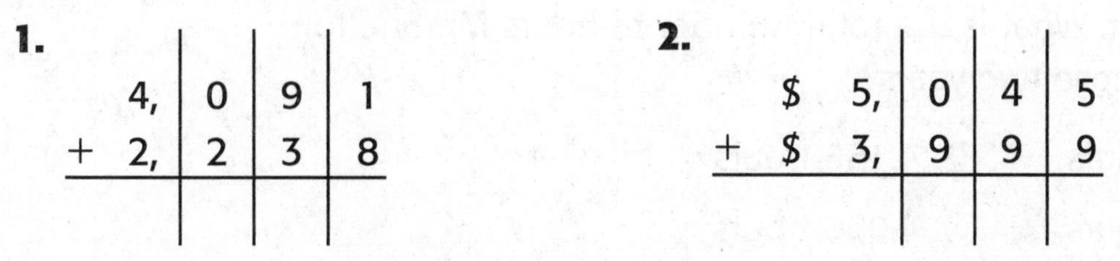

1.

4,	0	9	1
+ 2,	2	3	8

2.

$ 5,	0	4	5
+ $ 3,	9	9	9

Add. Check for reasonableness.

3.

2,	0	8	8
+ 6,	3	4	6

4.

4,	4	6	3
+ 4,	8	1	9

5.

3,	8	6	6
+ 4,	7	2	7

6. Use the Commutative Property to check your answer to Exercise 3.

$$\begin{array}{r} 6,346 \\ +\,2,088 \\ \hline \end{array}$$

Algebra Add to find each unknown.

7. 7,028 + 2,578 = ■ **8.** 5,724 + 2,197 = ■ **9.** 4,999 + 4,265 = ■

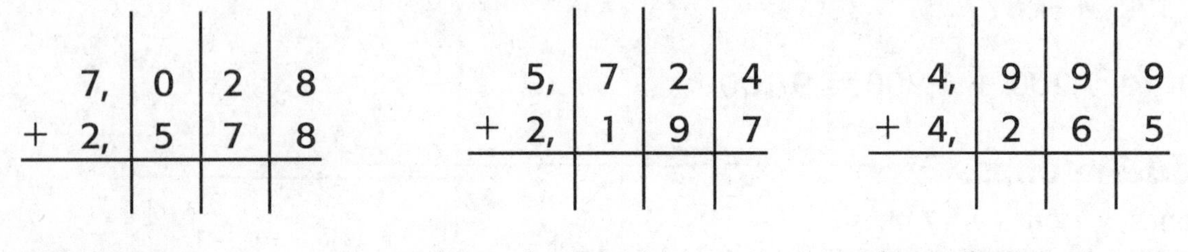

The unknown is ________. The unknown is ________. The unknown is ________.

Problem Solving

10. Mathematical PRACTICE 2 **Use Algebra** Rachael ran 3,012 meters on Monday and 5,150 meters on Wednesday. How many meters did she run altogether? Write a number sentence with a symbol for the unknown. Then solve.

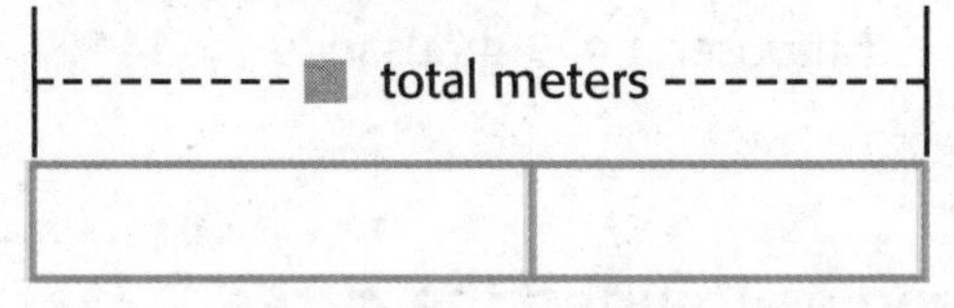

Test Practice

11. Mr. Shelton worked 1,976 hours one year and 2,080 hours the next year. What is the total number of hours Mr. Shelton worked in those two years?

Ⓐ 3,056 hours Ⓒ 4,156 hours

Ⓑ 3,956 hours Ⓓ 4,056 hours

Name ....

Real World

Problem-Solving Investigation

STRATEGY: Reasonable Answers

Lesson 9

ESSENTIAL QUESTION
How can place value help me add larger numbers?

Learn the Strategy

Tutor

Is it reasonable to say that Wallace's family has traveled about 2,200 miles this year?

1 Understand

This Year's Travel	
Dad	398 miles
Wallace	737 miles
Brother	1,106 miles

What facts do you know?

I know Wallace's family has traveled

______ miles, ______ miles, and ______ miles.

What do you need to find?

I need to find if the family traveled about ______ miles.

2 Plan

I will estimate the sum. Then I will compare the ________.

3 Solve

Estimate
$$\begin{array}{rcr} 1{,}106 & \longrightarrow & 1{,}100 \\ 737 & \longrightarrow & 700 \\ +\ 398 & \longrightarrow & +\ 400 \\ \hline & & \boxed{} \end{array}$$

The estimated sum of ______ miles is the same as the estimate in the problem. So, the answer is reasonable.

4 Check

Does your answer make sense? Explain.

Yes. 398 + 737 + 1,106 = 2,241, which is close to the estimated sum of ______.

Practice the Strategy

Anson swam 28 laps last week and 24 laps this week. He says he needs to swim the same number of laps each week for two more weeks to swim a total of about 100 laps. Is this a reasonable estimate? Explain.

1 Understand

What facts do you know?

____

____

What do you need to find?

____

____

2 Plan

____

____

3 Solve

4 Check

Does your answer make sense? Explain.

____

CCSS

Apply the Strategy

Determine a reasonable answer for each problem.

1. A bank account has $3,701 in it on Monday. On Tuesday, $4,294 is added to the account. Is it reasonable to say that now there is about $7,000 in the account? Explain.

2. Last year, 337 books were sold at the book fair. This year, 217 more books were sold than last year. Is it reasonable to say that more than 500 books were sold this year? Explain.

3. Juliana estimated that she needs to make 100 favors for the family reunion. Is this a reasonable estimate if 67 family members come on Friday and 42 family members come on Saturday? Explain.

4. Mathematical PRACTICE 3 **Justify Conclusions** A goal of 9,000 people in attendance was set for the State fair Friday and Saturday.

State Fair Attendance	
Day	**Attendance**
Friday	5,653
Saturday	4,059

Is it reasonable to say that the goal was met? Explain.

Review the Strategies

Use any strategy to solve each problem.

- Use the four-step plan.
- Determine reasonable answers.

5. How much will all of the flowers cost?

Mr. White's Garden Shop		
Flower	Quantity	Cost Each
Daisy	7	$5
Rose	3	$10
Lily	4	$6
Petunia	9	$4
Marigold	9	$3

____

6. Mathematical PRACTICE 2 **Use Symbols** The cards show the number of points Sylvia and Meli have in a game.

Sylvia's cards

Meli's cards

How many points do Sylvia and Meli each have? Who has the greater number of points? Use < or >.

____

____

7. Val spent $378 at the mall. Her sister spent $291. About how much did the sisters spend together?

____

My Work!

Review

Chapter 2

Addition

Vocabulary Check

Vocab

Choose the correct word(s) below to complete each sentence.

Associative Property	**bar diagram**	**Commutative Property**
estimate	**Identity Property**	**mental math**
parentheses	**pattern**	**reasonable**
regroup	**unknown**	

1. The ______________________ of Addition states that you can change the order of the addends and still get the same sum.

2. ______________________ tell you which operations to group in a number sentence.

3. ______________________ is a way of doing math in my head.

4. A number close to its actual value is a(n) ______________________.

5. To use place value to exchange equal amounts when renaming a number is to ______________________.

6. A ______________________ is a set of numbers that follow a certain order.

7. The ______________________ of Addition states when you add zero to any number, the sum is that number.

8. When an answer makes sense, it is ______________________.

9. The missing number or what you are solving for in a problem is the ______________________.

Concept Check

Find each sum. Identify the property of addition. Write *Commutative, Associative,* or *Identity.*

10. 8 + 0 = ______

______

11. 2 + (4 + 3) = ______

(2 + 4) + 3 = ______

______

12. 5 + 7 = ______

7 + 5 = ______

______

Write the number.

13. 1 more than 375

______

14. 1,000 more than 2,184

______

Make a ten or a hundred to add mentally.

15. 198 + 132

☐ ☐

______ + ______ = ______

16. 1,274 + 3,599

☐ ☐

______ + ______ = ______

Estimate. Round each addend to the indicated place value.

17. 725 + 229; tens

______ + ______ = ______

18. 8,291 + 1,101; hundreds

______ + ______ = ______

Add. Check for reasonableness.

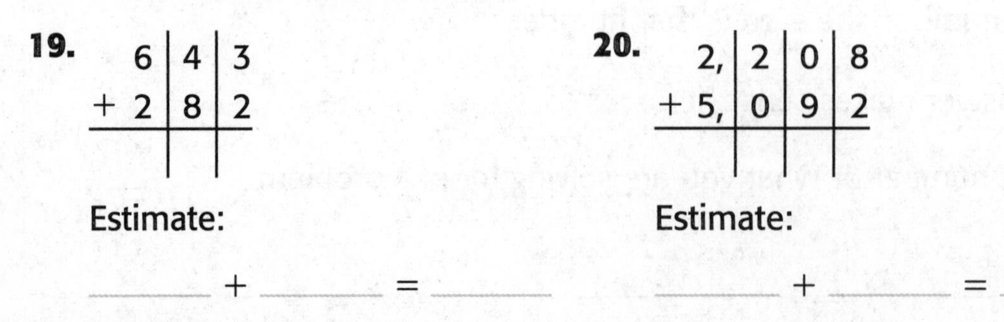

19.

	6	4	3
+	2	8	2

Estimate:

______ + ______ = ______

20.

	2,	2	0	8
+	5,	0	9	2

Estimate:

______ + ______ = ______

Name

Problem Solving

Write a number sentence with a symbol for the unknown. Then solve.

21. A horse weighs 1,723 pounds. A smaller horse weighs 902 pounds. How much do both horses weigh together?

____

22. A golf ball has 336 dimples. How many dimples would two golf balls have altogether?

____

23. Samuel walked one mile before lunch and one mile after lunch. The length of a mile is 1,760 yards. In yards, what is the total distance Samuel walked?

____

24. There are two different mountains in the state of Washington that get record snowfall. One winter season, Mount Ranier received 1,224 inches of snow. Mount Baker received 1,140 inches of snow. What is the total snowfall of these two mountains?

____

Test Practice

25. Dwight drove 792 miles from Pensacola, Florida, to Key West, Florida, and then back again. What is the distance he traveled to the nearest hundred miles?

Ⓐ 800 miles
Ⓑ 1,580 miles
Ⓒ 1,584 miles
Ⓓ 1,600 miles

My Work!

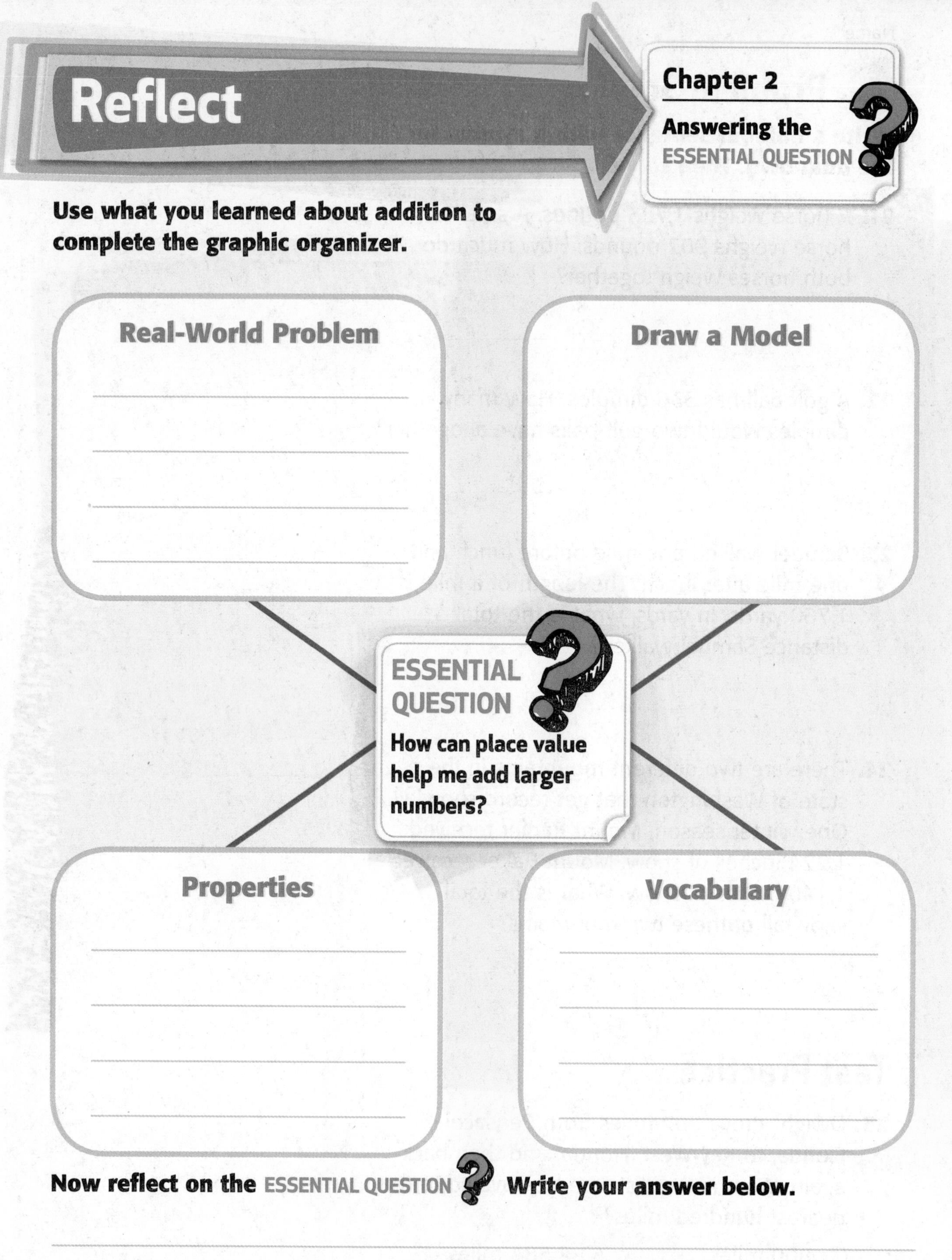

Now reflect on the ESSENTIAL QUESTION Write your answer below.

Chapter

3 Subtraction

ESSENTIAL QUESTION

How are the operations of subtraction and addition related?

Activities I Do for Fun

Watch a video!

Watch

MY Common Core State Standards

CCSS

Number and Operations in Base Ten

CCSS

3.NBT.2 Fluently add and subtract within 1000 using strategies and algorithms based on place value, properties of operations, and/or the relationship between addition and subtraction.

Operations and Algebraic Thinking *This chapter also addresses this standard:*

3.OA.8 Solve two-step word problems using the four operations. Represent these problems using equations with a letter standing for the unknown quantity. Assess the reasonableness of answers using mental computation and estimation strategies including rounding.

Cool! This is what I am going to be doing!

Standards for Mathematical PRACTICE

1. Make sense of problems and persevere in solving them.
2. Reason abstractly and quantitatively.
3. Construct viable arguments and critique the reasoning of others.
4. Model with mathematics.
5. Use appropriate tools strategically.
6. Attend to precision.
7. Look for and make use of structure.
8. Look for and express regularity in repeated reasoning.

= focused on in this chapter

Name ....

Am I Ready?

Check ✓ ← Go online to take the Readiness Quiz

Subtract.

1. $\begin{array}{r} 9 \\ -\ 4 \\ \hline \end{array}$

2. $\begin{array}{r} 12 \\ -\ 7 \\ \hline \end{array}$

3. $\begin{array}{r} 15 \\ -\ 8 \\ \hline \end{array}$

4. $\begin{array}{r} 11 \\ -\ 6 \\ \hline \end{array}$

5. $13 - 7 =$ ____

6. $10 - 6 =$ ____

7. $9 - 6 =$ ____

8. $16 - 8 =$ ____

Use the base-ten blocks to find each difference.

9.

$24 - 11 =$ ______

10.

$65 - 24 =$ ______

Round to the nearest ten.

11. 454 ______

12. 689 ______

13. 712 ______

Round to the nearest hundred.

14. 377 ______

15. 409 ______

16. 1,335 ______

How Did I Do?

Shade the boxes to show the problems you answered correctly.

1	2	3	4	5	6	7	8	9	10	11	12	13	14	15	16

Name

Review Vocabulary

add	addend	difference
equal	equals sign (=)	estimate
minus sign (−)	plus sign (+)	subtract
subtraction sentence	sum	

Making Connections

Use the review vocabulary to complete the Venn diagram.

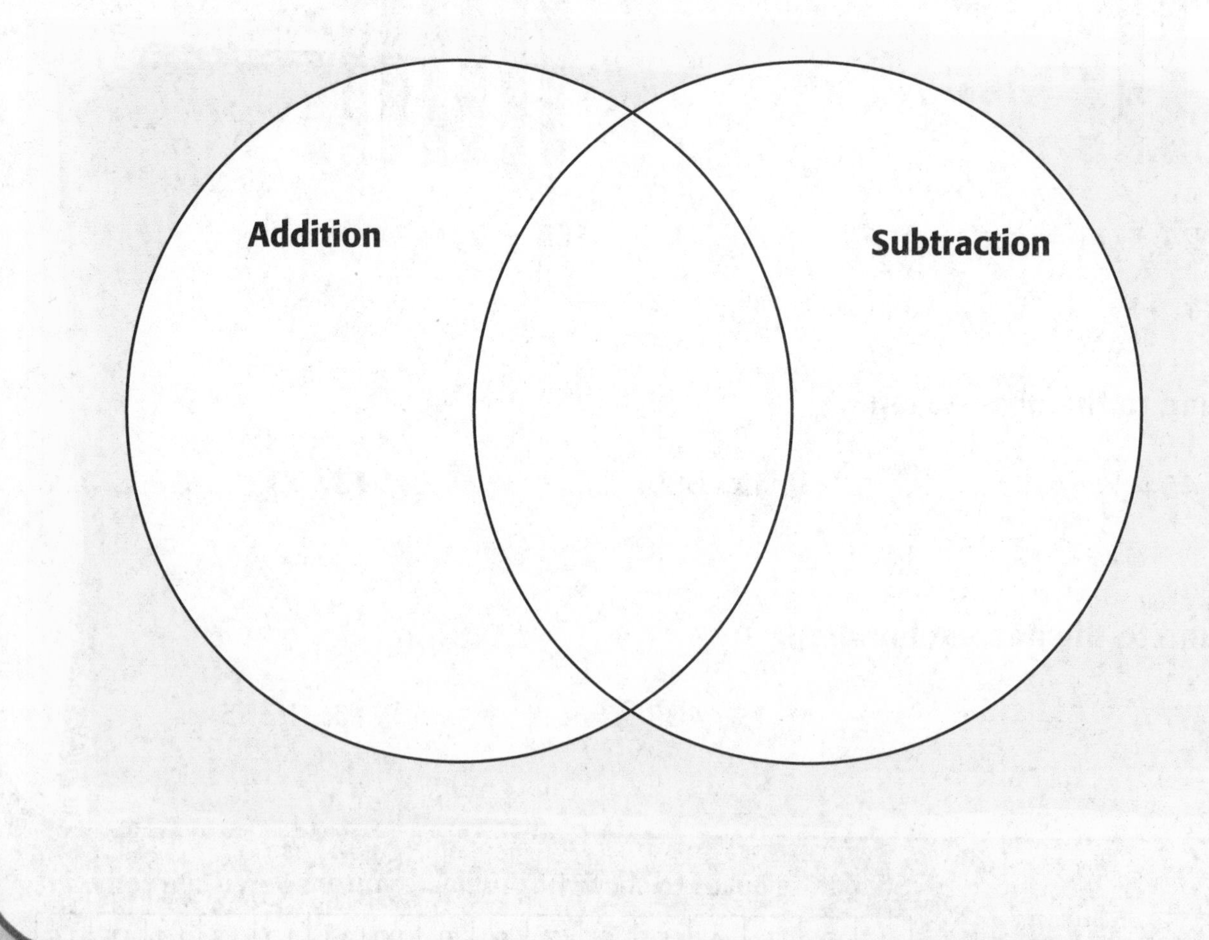

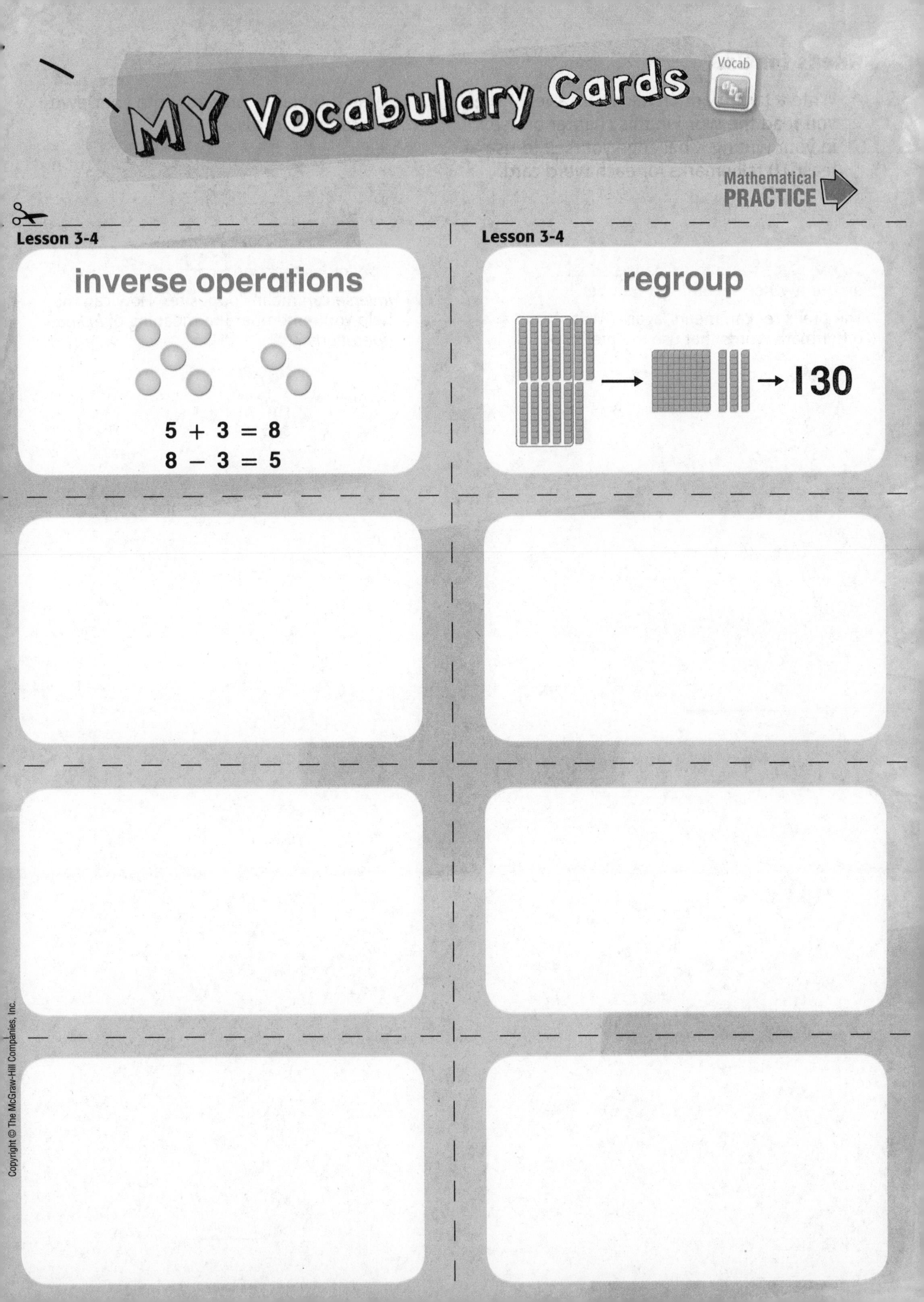
MY Vocabulary Cards
Vocab
Mathematical PRACTICE
Lesson 3-4
inverse operations
5 + 3 = 8
8 − 3 = 5
Lesson 3-4
regroup
130

Ideas for Use

- Write a tally mark on each card every time you read the word in this chapter or use it in your writing. Challenge yourself to use at least 10 tally marks for each word card.
- Use the blank cards to write your own vocabulary cards.

To use place value to exchange equal amounts when renaming a number.

The prefix *re-* can mean "again." Write two other math words that use the prefix *re-*.

Operations that undo each other.

Inverse can mean "opposite." How can this help you remember the meaning of *inverse operations ?*

MY Foldable

FOLDABLES® Follow the steps on the back to make your Foldable.

Subtracting
Across Zeros

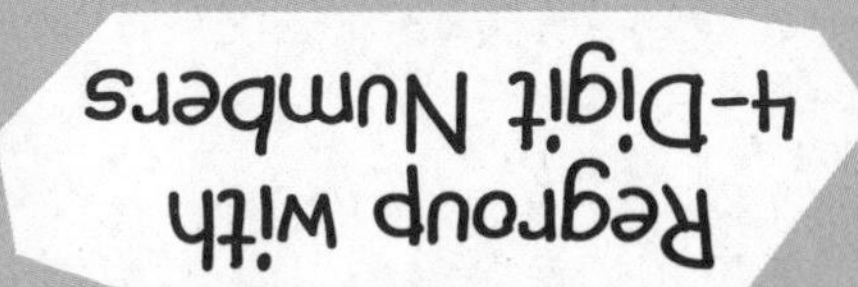

Regroup with
3-Digit Numbers

Estimating
Difference

There are 5,395 men and women in a race.

There are 2,697 men.

How many runners are women?

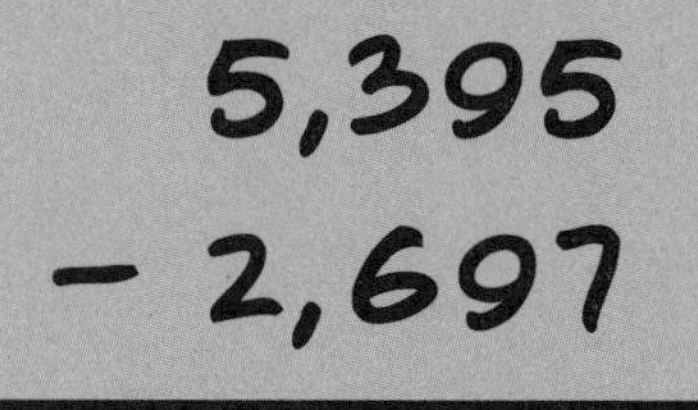

On Saturday, there were 1,000 balloons at a hot air balloon festival. On Sunday, there were 752 balloons.

How many more balloons were there on Saturday than on Sunday?

$$\begin{array}{r} 1{,}000 \\ -\ 752 \\ \hline \end{array}$$

The school's lacrosse game was attended by 244 students and 117 parents.

To the nearest ten, about how many more students attended the game?

244 → 240

117 → −120

There were 381 students who voted for a field trip to the aquarium and 125 students who voted for a field trip to the museum.

How many more students voted for the aquarium?

$$\begin{array}{r} 381 \\ -\ 125 \\ \hline \end{array}$$

Name ....

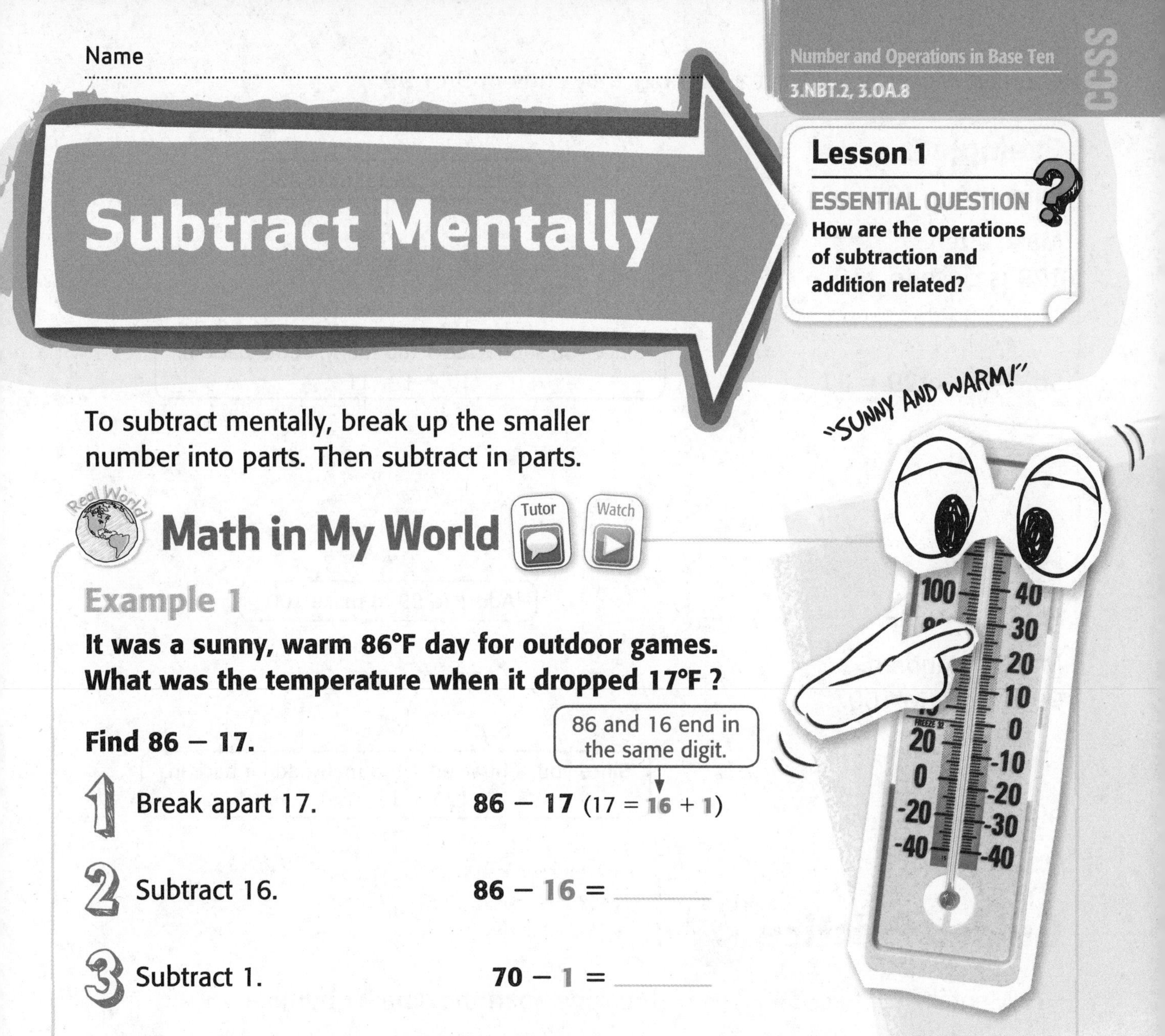

Subtract Mentally

Lesson 1

ESSENTIAL QUESTION
How are the operations of subtraction and addition related?

To subtract mentally, break up the smaller number into parts. Then subtract in parts.

Math in My World

Tutor Watch

Example 1

It was a sunny, warm 86°F day for outdoor games. What was the temperature when it dropped 17°F ?

Find 86 − 17.

86 and 16 end in the same digit.

1. Break apart 17. **86 − 17** (17 = 16 + 1)
2. Subtract 16. **86 − 16 =** ______
3. Subtract 1. **70 − 1 =** ______

So, 86 − 17 = 69. The temperature was 69°F by the end of the day.

You can also use subtraction rules to subtract mentally.

Subtraction Rules	
Subtracting a number from itself equals zero.	367 − 367 = 0
Subtracting zero from a number equals itself.	545 − 0 = 545

Example 2

Find 417 − 417.

Subtracting a number from itself equals ______.

So, 417 − 417 = ______.

You can mentally subtract a number that ends in 9 or 99.

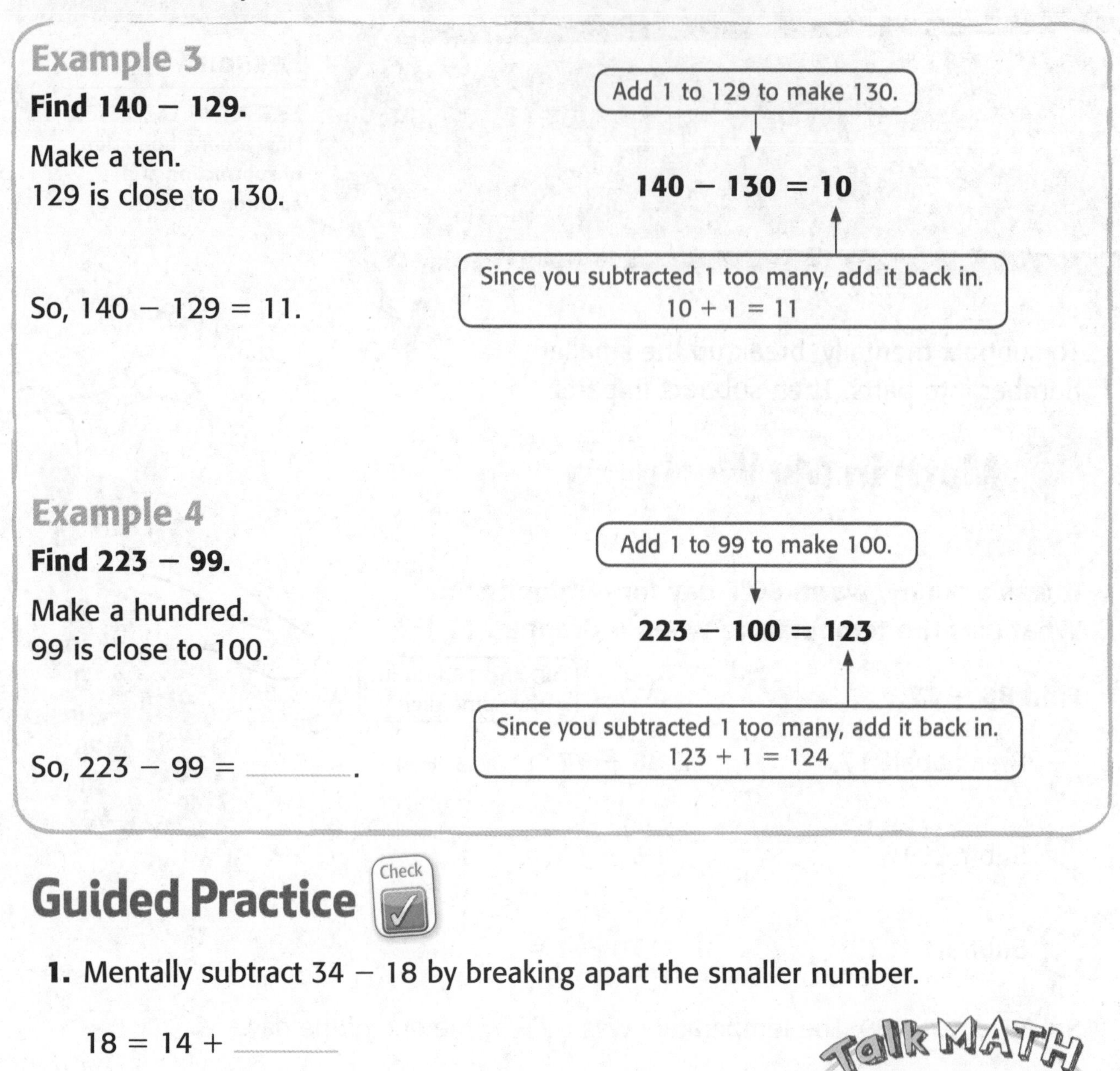

Example 3

Find 140 − 129.

Make a ten.
129 is close to 130.

Add 1 to 129 to make 130.

140 − 130 = 10

Since you subtracted 1 too many, add it back in.
10 + 1 = 11

So, 140 − 129 = 11.

Example 4

Find 223 − 99.

Make a hundred.
99 is close to 100.

Add 1 to 99 to make 100.

223 − 100 = 123

Since you subtracted 1 too many, add it back in.
123 + 1 = 124

So, 223 − 99 = ______.

Guided Practice

Check

1. Mentally subtract 34 − 18 by breaking apart the smaller number.

18 = 14 + ______

34 − 14 = ______

______ − 4 = ______

So, 34 − 18 = ______.

2. Mentally subtract 94 − 59 by making a ten.

94 − 60 = ______

______ + 1 = ______

So, 94 − 59 = ______.

Talk MATH

What mental subtraction strategy could you use to find 234 − 29?

Name ____________________

Independent Practice

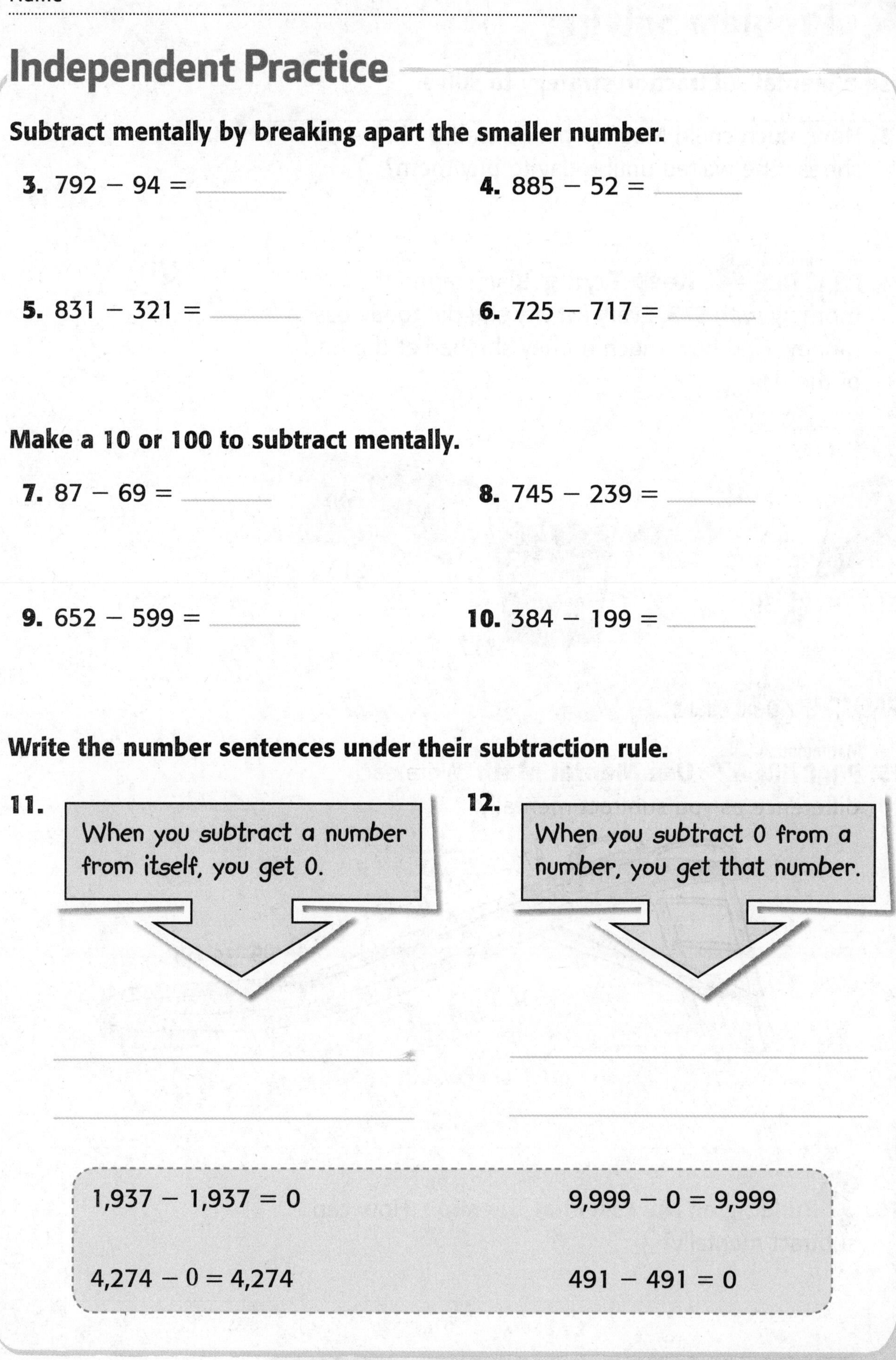

Subtract mentally by breaking apart the smaller number.

3. 792 − 94 = ______

4. 885 − 52 = ______

5. 831 − 321 = ______

6. 725 − 717 = ______

Make a 10 or 100 to subtract mentally.

7. 87 − 69 = ______

8. 745 − 239 = ______

9. 652 − 599 = ______

10. 384 − 199 = ______

Write the number sentences under their subtraction rule.

11. When you subtract a number from itself, you get 0.

____

____

12. When you subtract 0 from a number, you get that number.

____

____

1,937 − 1,937 = 0

9,999 − 0 = 9,999

4,274 − 0 = 4,274

491 − 491 = 0

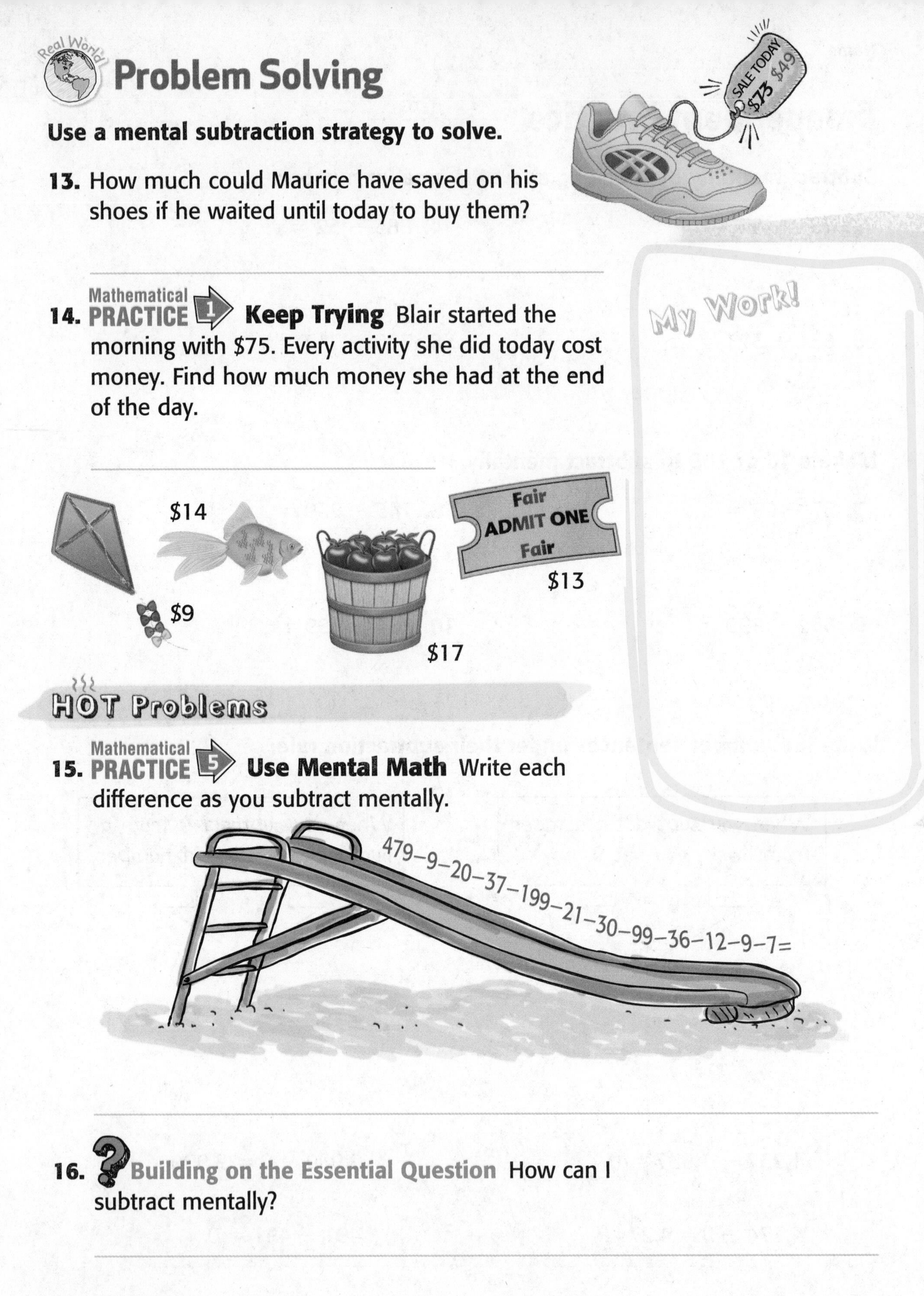

Problem Solving

Use a mental subtraction strategy to solve.

13. How much could Maurice have saved on his shoes if he waited until today to buy them?

SALE TODAY $73 $49

14. Mathematical PRACTICE 1 **Keep Trying** Blair started the morning with $75. Every activity she did today cost money. Find how much money she had at the end of the day.

$14

$9

$17

Fair ADMIT ONE Fair

$13

My Work!

HOT Problems

15. Mathematical PRACTICE 5 **Use Mental Math** Write each difference as you subtract mentally.

479–9–20–37–199–21–30–99–36–12–9–7=

16. **Building on the Essential Question** How can I subtract mentally?

Name ....

Lesson 1

Subtract Mentally

Homework Helper

eHelp Need help? connectED.mcgraw-hill.com

Mia wants a guitar that costs \$96. So far, she has saved up \$48. How much more money does she need to save in order to buy the guitar?

Find \$96 − \$48. Subtract in parts.

1. Break apart 48. $\$96 - \$48 \; (\$48 = \$46 + \$2)$

2. Subtract one part. $\$96 - \$46 = \$50$

3. Subtract the other part. $\$50 - \$2 = \$48$

So, Mia needs \$48 more.

Practice

Subtract mentally by breaking apart the smaller number.

1. 82 − 47 ______

2. 165 − 26 = ______

3. 387 − 308 = ______

4. 674 − 426 = ______

Make a 10 or 100 to subtract mentally.

5. 76 − 59 = ______

6. 120 − 39 = ______

7. 554 − 199 = ______

8. 453 − 19 = ______

Problem Solving

Use any mental math strategy to solve.

9. Mathematical PRACTICE 5 **Use Mental Math** The red team has 522 fans in the bleachers. The blue team has 425 fans in the bleachers. How many fewer fans does the blue team have?

____

10. Mathematical PRACTICE 2 **Use Number Sense** There are 172 houses in Kyle's neighborhood. He delivers papers to 99 houses. To how many houses does Kyle *not* deliver papers?

____

Vocabulary Check

Draw a line to match the word to its definition or meaning.

11. subtract	• The answer to a subtraction problem.
12. difference	• An operation that tells the difference between two numbers.

Test Practice

13. Avery has a digital camera with a memory card that can hold 284 pictures. If Avery has taken 159 pictures, how many pictures can the memory card still hold?

Ⓐ 124 pictures

Ⓑ 125 pictures

Ⓒ 135 pictures

Ⓓ 443 pictures

Name ....

Real World

Problem-Solving Investigation

STRATEGY: Estimate or Exact Answer

Lesson 3

ESSENTIAL QUESTION
How are the operations of subtraction and addition related?

Learn the Strategy

Tutor

To celebrate Arbor Day, Brita's school district planted trees. The high school students planted 1,536 trees. The elementary students planted 1,380 trees. About how many more trees did the high school students plant?

I dig Arbor Day!

1 Understand

What facts do you know?

The high school students planted ______ trees.

The elementary students planted ______ trees.

What do you need to find?

about how many more trees the ____________ students planted

2 Plan

An exact answer is not needed. I will estimate.

3 Solve

Round each number.

1,536 → 1,500
1,380 → 1,400

Round each to the nearest hundred.

Then, subtract.

$$\begin{array}{r} 1{,}500 \\ -\ 1{,}400 \\ \hline 100 \end{array}$$

About 100 more trees were planted by the high school students.

4 Check

Does your answer make sense? Explain.

____

Practice the Strategy

Students in grades 2 and 3 wrote 61 stories to celebrate author day. The second graders wrote 26 stories. How many stories did the third graders write?

1 Understand

What facts do you know?

____

____

What do you need to find?

____

2 Plan

____

____

3 Solve

4 Check

Does your answer make sense? Explain.

____

Name

Apply the Strategy

Determine whether each problem requires an estimate or an exact answer. Then solve.

My Work!

1. Mathematical PRACTICE 4 **Model Math** Michelann cut two lengths of rope. One was 32 inches long. The other was 49 inches long. Will she have enough rope for a project that needs 76 inches of rope? Explain.

2. Mathematical PRACTICE 2 **Use Number Sense** The number 7 septillion has 24 zeros after it. The number 7 octillion has 27 zeros after it. How many zeros is that altogether? What is the difference between the number of zeros in 7 septillion and 7 thousand?

3. Mathematical PRACTICE 3 **Justify Conclusions** Three buses can carry a total of 180 students. Is there enough room on the 3 buses for 95 boys and 92 girls? Explain.

4. Mrs. Carpenter received a bill for car repairs. About how much was the oil change if the other repairs were $102?

Car Repair: $134.00

Review the Strategies

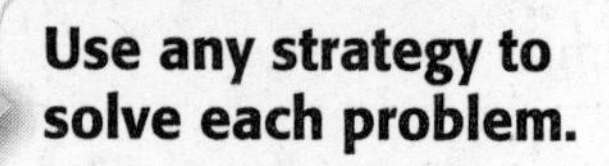

Use any strategy to solve each problem.

- Determine reasonable answers.
- Determine an estimate or exact answer.
- Use the four-step plan.

5. Some children took part in a penny hunt. About how many more pennies did Pat find than either of his two friends?

Penny Hunt	
Cynthia	133
Pat	182
Garcia	125

My Work!

6. Mathematical PRACTICE 1 **Check for Reasonableness** A bank account has $320 in it on Monday. On Tuesday, $629 is added to the account. On Wednesday, $630 is taken out of the account. Is it reasonable to say that there is about $100 in the account after Wednesday? Explain.

7. When the school first opened, its library had 213 books. Today it has more than 650 books. About how many books has the school bought since it first opened? Explain.

8. The Bonilla family spent $1,679 on their vacation. The Turner family spent $983. About how much less did the Turner family spend? Explain.

Name ....

MY Homework

Lesson 3

Problem Solving: Estimate or Exact Answer

Homework Helper

eHelp

Need help? connectED.mcgraw-hill.com

What is the difference in the total number of ribbons the girls earned?

Swim Team Ribbons		
Swimmer	Last Year	This Year
Mackenzie	26	31
Zoe	19	33

1 Understand

What facts do you know?

The number of ribbons each girl earned each year.

What do you need to find?

The difference in the total number of ribbons between the girls.

2 Plan

An exact answer is needed. I will add to find the total.
Then I will subtract to find the difference.

3 Solve

Mackenzie	Zoe
26	19
+ 31	+ 33
57	52

$57 - 52 = 5$

The difference between the total number of ribbons for each girl is 5.

4 Check

Does your answer make sense? Explain.

Yes; Use the inverse operation of addition.
$5 + 52 = 57$ shows the answer is correct.

Problem Solving

Determine whether each problem requires an estimate or an exact answer. Then solve.

My Work!

1. Gavin saved $53 in August and $15 in September. Is it reasonable to say he needs $50 more to pay for a karate class that costs $100? Explain.

______

______

______

2. There were 4,569 fans at a soccer game. The visiting team had 1,604 fans cheering. About how many fans cheered for the home team? Explain.

______

______

3. Mathematical PRACTICE 1 **Make Sense of Problems** Luke buys 2 small pumpkins and 1 large pumpkin. How much does Luke spend?

Pumpkin Sale

Small Pumpkins	$4
Large Pumpkins	$7

______

4. Peyton learned that the brown bear at a local zoo weighs 1,578 pounds. About how much more does the brown bear weigh than the polar bear? Explain.

______

______

______

Check My Progress

Vocabulary Check

1. Match each word with its definition. Finish drawing the puzzle piece of each word so that it fits the puzzle piece of its definition.

subtract	a number close to the exact value
difference	the answer to a subtraction problem
estimate	to take some or all away

Concept Check

2. Mentally find the difference between the numbers that are connected. Write the difference in the box below the arrow. The last number is given so that you can check your work.

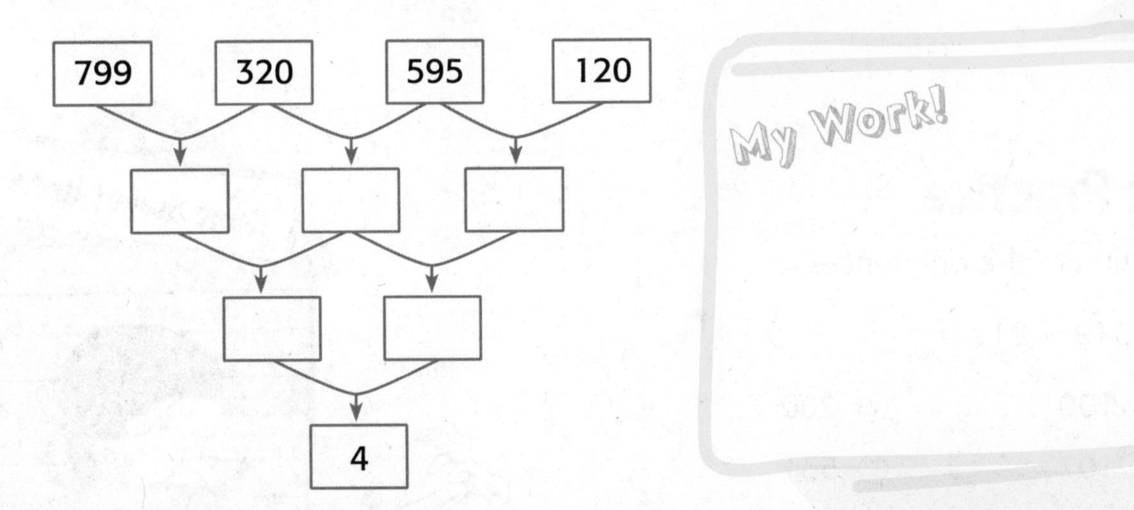

Problem Solving

3. A tent costs $499. It is on sale for $75 off of the original price. Mentally subtract to find the sale price.

My Work!

4. The table shows how much money was made at a movie theater last week. About how much more was earned on Friday than on Sunday?

Movie Sales	
Day	**Amount**
Friday	$432
Saturday	$721
Sunday	$184

Determine if an estimate or an exact answer is needed. Then solve.

5. This year, the third grade raised $379 for a dog shelter. Last year, they raised $232. Mentally subtract to find how much more money was raised this year than last year.

Test Practice

6. Estimate the difference.

319 − 212 = ■

Ⓐ 100　　Ⓒ 200

Ⓑ 107　　Ⓓ 531

Name

Hands On
Subtract with Regrouping

Lesson 4

ESSENTIAL QUESTION
How are the operations of subtraction and addition related?

Sometimes you need to **regroup** when subtracting. You will use place value to exchange equal amounts to rename a number.

Build It

Tools

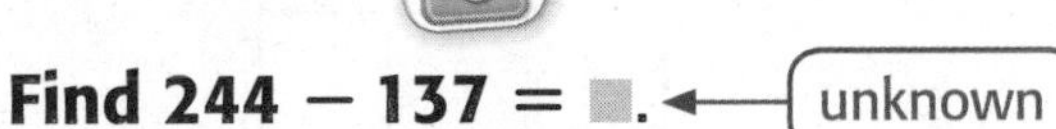

Find 244 − 137 = ■. ← unknown

Estimate 244 − 137 ⟶ 200 − 100 = 100

1 Model 244.
Use base-ten blocks. Draw your model.

My Drawing!

2 Subtract ones.
You cannot subtract 7 ones from 4 ones.
Regroup 1 ten as 10 ones.
There are now 14 ones.
Subtract 7 ones.
14 ones − 7 ones = ☐ ones.
Draw the result.

3 Subtract 3 tens.

3 tens − 3 tens = ______ tens

My Drawing!

4 Subtract hundreds.

2 hundreds − 1 hundred =

______ hundred

Draw the result.

So, 244 − 137 = ______.

The unknown is ______.

Check

Addition and subtraction are **inverse operations** because they undo each other.

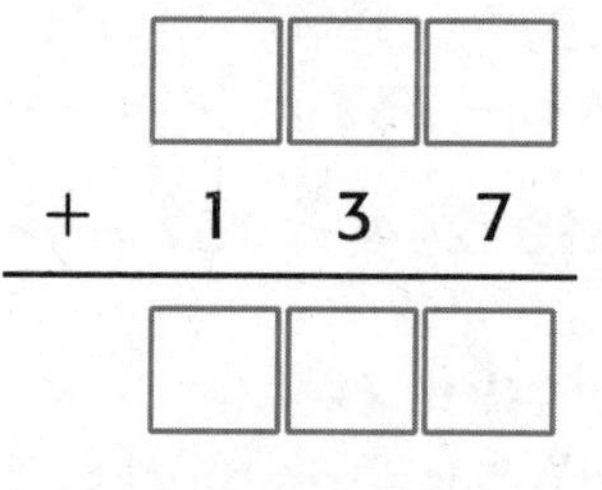

Talk About It

1. In Step 2, why did you regroup 1 ten as 10 ones?

______

2. What did you notice about the tens in Step 3 when you subtracted them?

______

______

3. Mathematical PRACTICE 2 **Stop and Reflect** Suppose, after subtracting the ones, there were not enough tens from which to subtract. What do you think you may have to do?

______

4. Why can you use addition to check your answer to a subtraction problem?

______

Name

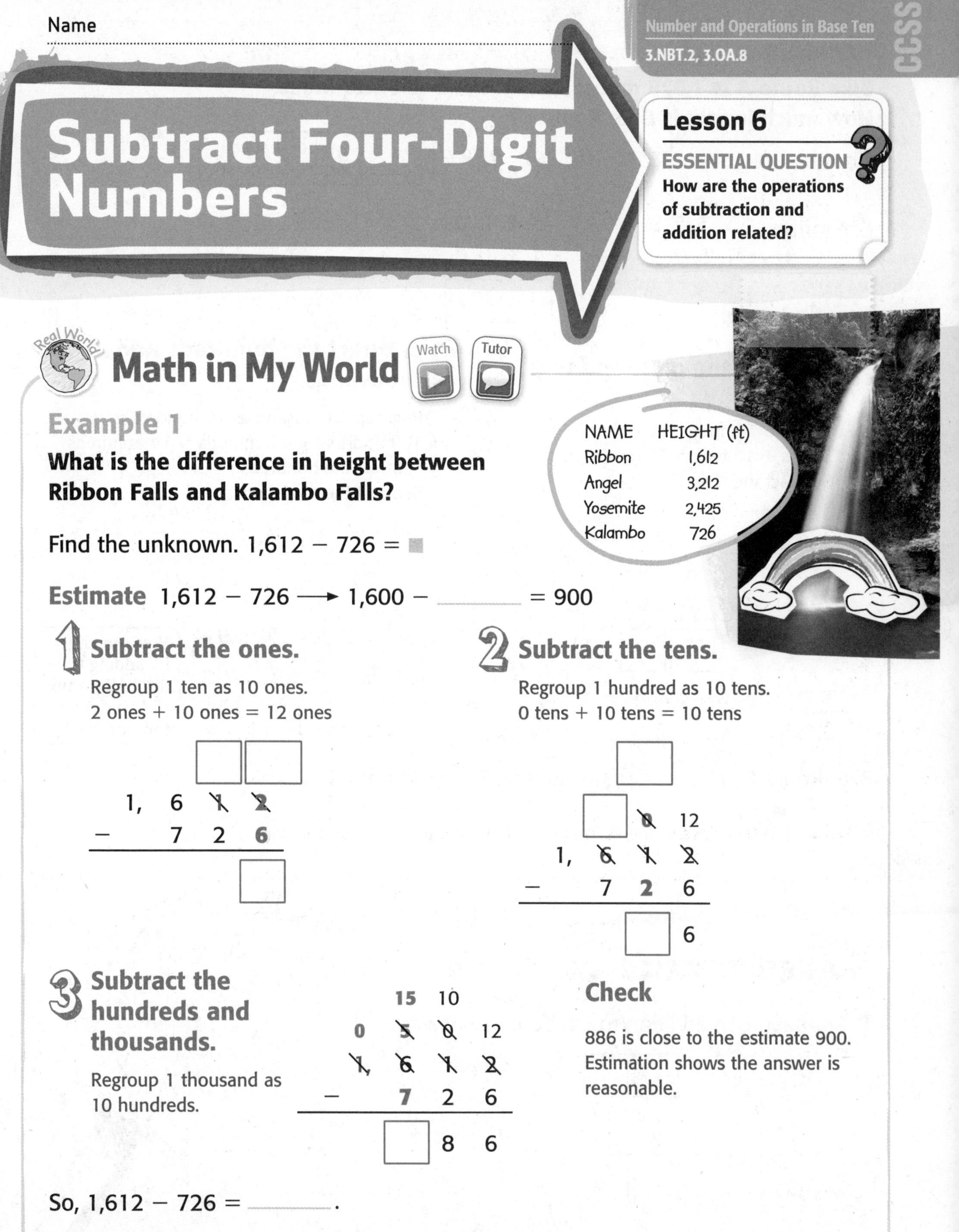

Subtract Four-Digit Numbers

Lesson 6

ESSENTIAL QUESTION
How are the operations of subtraction and addition related?

Math in My World

Watch Tutor

Example 1

What is the difference in height between Ribbon Falls and Kalambo Falls?

NAME	HEIGHT (ft)
Ribbon	1,612
Angel	3,212
Yosemite	2,425
Kalambo	726

Find the unknown. 1,612 − 726 = ■

Estimate 1,612 − 726 ⟶ 1,600 − ______ = 900

1 Subtract the ones.

Regroup 1 ten as 10 ones.
2 ones + 10 ones = 12 ones

$$\begin{array}{r} 1,\ 6\ 1\ 2 \\ -\ \ \ 7\ 2\ 6 \\ \hline \square \end{array}$$

2 Subtract the tens.

Regroup 1 hundred as 10 tens.
0 tens + 10 tens = 10 tens

$$\begin{array}{r} 0\ 12 \\ 1,\ 6\ 1\ 2 \\ -\ \ \ 7\ 2\ 6 \\ \hline \square\ 6 \end{array}$$

3 Subtract the hundreds and thousands.

Regroup 1 thousand as 10 hundreds.

$$\begin{array}{r} 15\ 10\ \ \\ 0\ 5\ 0\ 12 \\ 1,\ 6\ 1\ 2 \\ -\ \ \ 7\ 2\ 6 \\ \hline \square\ 8\ 6 \end{array}$$

Check

886 is close to the estimate 900. Estimation shows the answer is reasonable.

So, 1,612 − 726 = ______.

Ribbon Falls is ______ feet taller than Kalambo Falls.

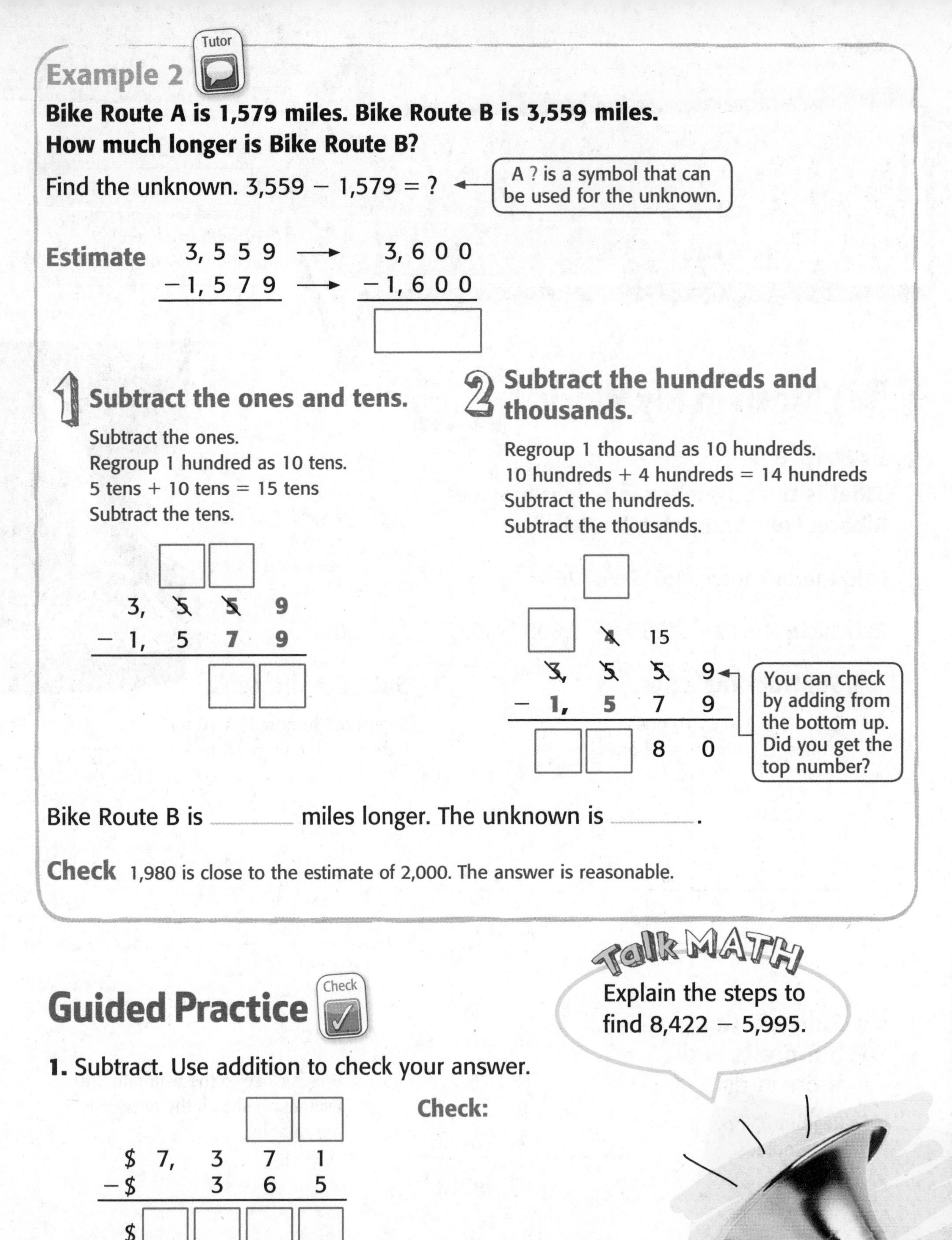

Example 2

Bike Route A is 1,579 miles. Bike Route B is 3,559 miles. How much longer is Bike Route B?

Find the unknown. 3,559 − 1,579 = ? ← A ? is a symbol that can be used for the unknown.

Estimate

$$\begin{array}{r} 3,559 \\ -1,579 \\ \hline \end{array} \longrightarrow \begin{array}{r} 3,600 \\ -1,600 \\ \hline \square \end{array}$$

1 Subtract the ones and tens.

Subtract the ones.
Regroup 1 hundred as 10 tens.
5 tens + 10 tens = 15 tens
Subtract the tens.

$$\begin{array}{rrrr} & \square & \square & \\ 3, & \not{5} & \not{5} & 9 \\ -\ 1, & 5 & 7 & 9 \\ \hline & & \square & \square \end{array}$$

2 Subtract the hundreds and thousands.

Regroup 1 thousand as 10 hundreds.
10 hundreds + 4 hundreds = 14 hundreds
Subtract the hundreds.
Subtract the thousands.

$$\begin{array}{rrrr} & \square & & \\ \square & \not{4} & 15 & \\ \not{3}, & \not{5} & \not{5} & 9 \\ -\ 1, & 5 & 7 & 9 \\ \hline \square & \square & 8 & 0 \end{array}$$

You can check by adding from the bottom up. Did you get the top number?

Bike Route B is ______ miles longer. The unknown is ______.

Check 1,980 is close to the estimate of 2,000. The answer is reasonable.

Guided Practice

1. Subtract. Use addition to check your answer.

$$\begin{array}{rrrr} & & \square & \square \\ \$\ 7, & 3 & 7 & 1 \\ -\$ & 3 & 6 & 5 \\ \hline \$\ \square & \square & \square & \square \end{array}$$

Check:

Talk MATH

Explain the steps to find 8,422 − 5,995.

Name

CCSS

Independent Practice

Subtract. Use addition to check your answer.

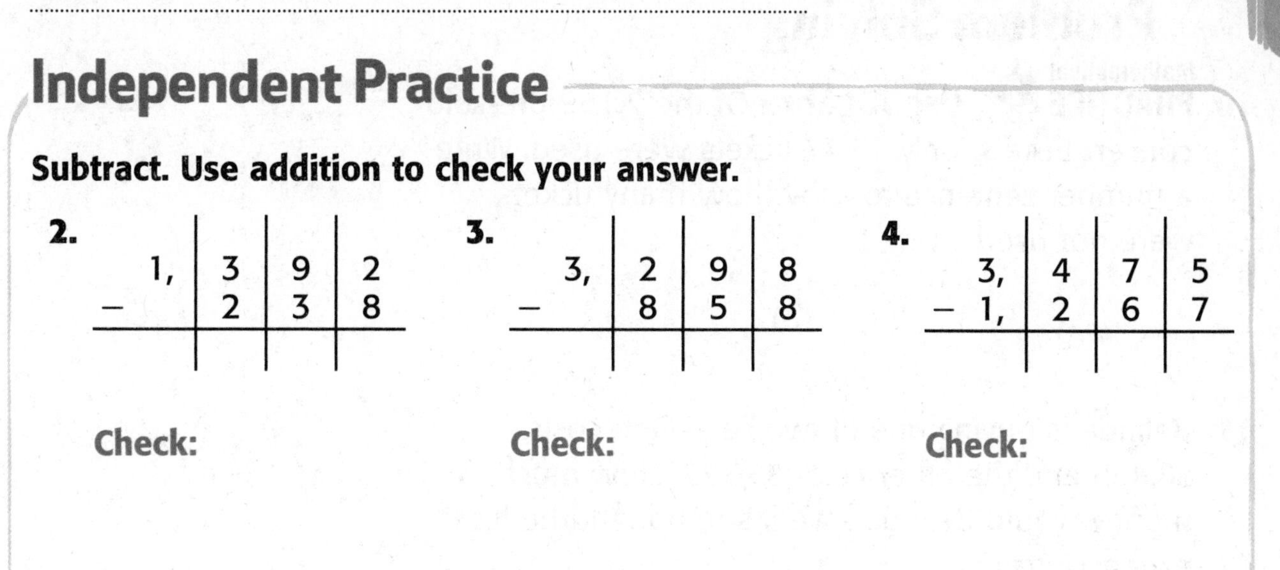

2.
$$\begin{array}{r} 1,392 \\ -\ \ \ 238 \\ \hline \end{array}$$

Check:

3.
$$\begin{array}{r} 3,298 \\ -\ \ \ 858 \\ \hline \end{array}$$

Check:

4.
$$\begin{array}{r} 3,475 \\ -\ 1,267 \\ \hline \end{array}$$

Check:

Algebra Subtract to find the unknown.

5. $4,875 − $3,168 = ?

The unknown is ______.

6. $6,182 − $581 = ?

The unknown is ______.

7. 6,340 − 3,451 = ?

The unknown is ______.

Algebra Compare. Use >, <, or =.

8. 1,543 − 984 ◯ 5,193 − 4,893

9. 2,116 − 781 ◯ 5,334 − 3,999

Real World

Problem Solving

10. Mathematical PRACTICE 2 **Use Algebra** Of the 2,159 pre-sold concert tickets, only 1,947 tickets were used. Write a number sentence to show how many tickets were not used.

____

11. Belinda is buying one of two cars. One costs $8,463 and the other costs $5,322. How much money would Belinda save if she bought the less expensive car?

$______ − $______ = $______

My Work!

HOT Problems

12. Mathematical PRACTICE 2 **Reason** A group of students used 6,423 cans to create a sculpture. Another group made a sculpture using 2,112 cans. What is the difference in the number of cans used for the sculptures? How do you know your answer is correct?

____

____

____

13. **Building on the Essential Question** Explain how subtracting four-digit numbers is like subtracting three-digit numbers.

____

____

____

Name ....

MY Homework

Lesson 6
Subtract Four-Digit Numbers

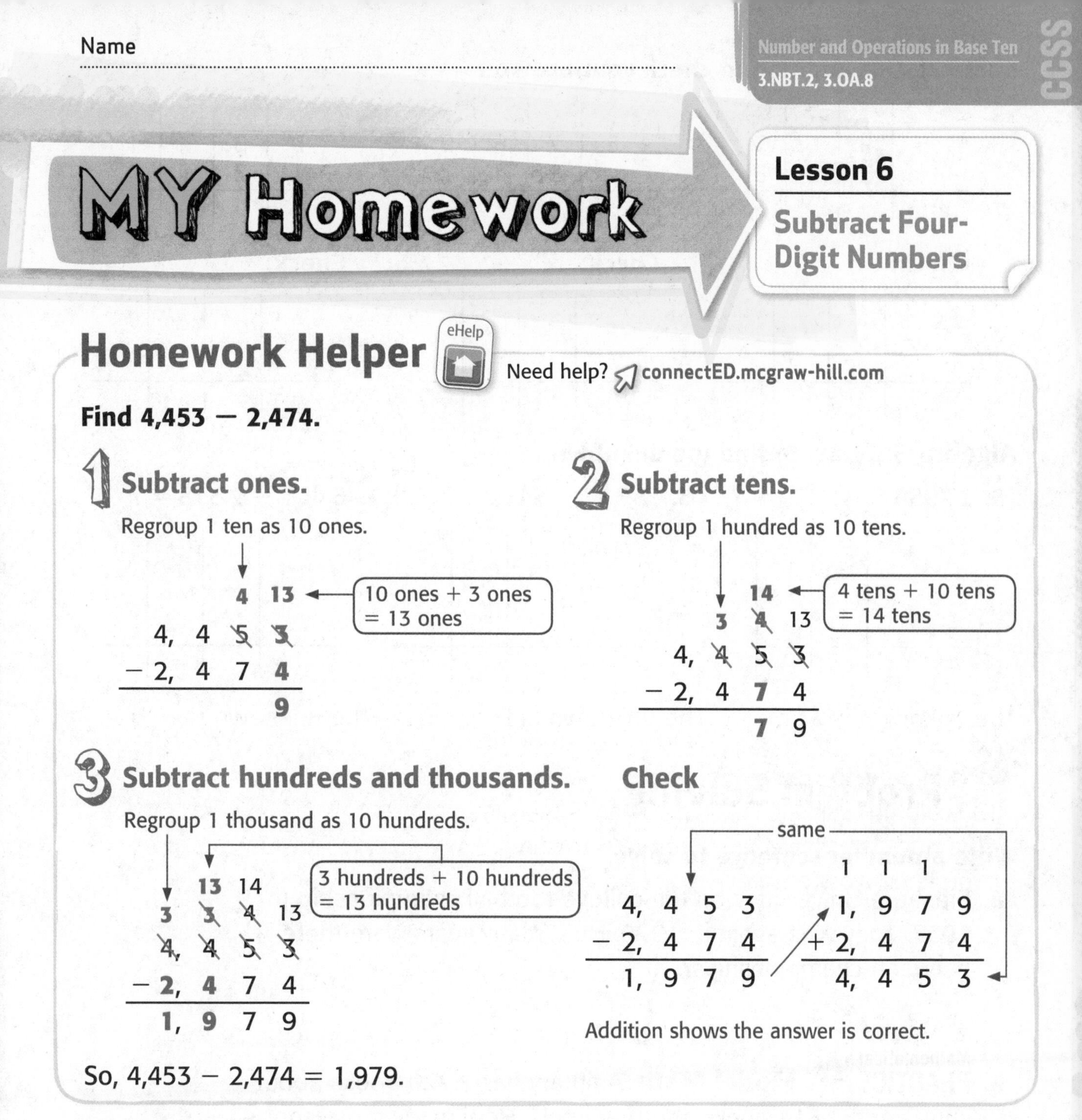

Practice

1. Subtract. Use addition to check your answer.

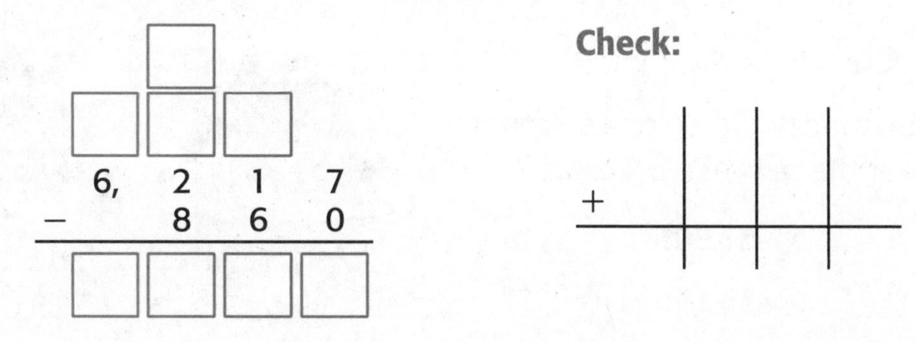

Subtract. Use addition to check your answer.

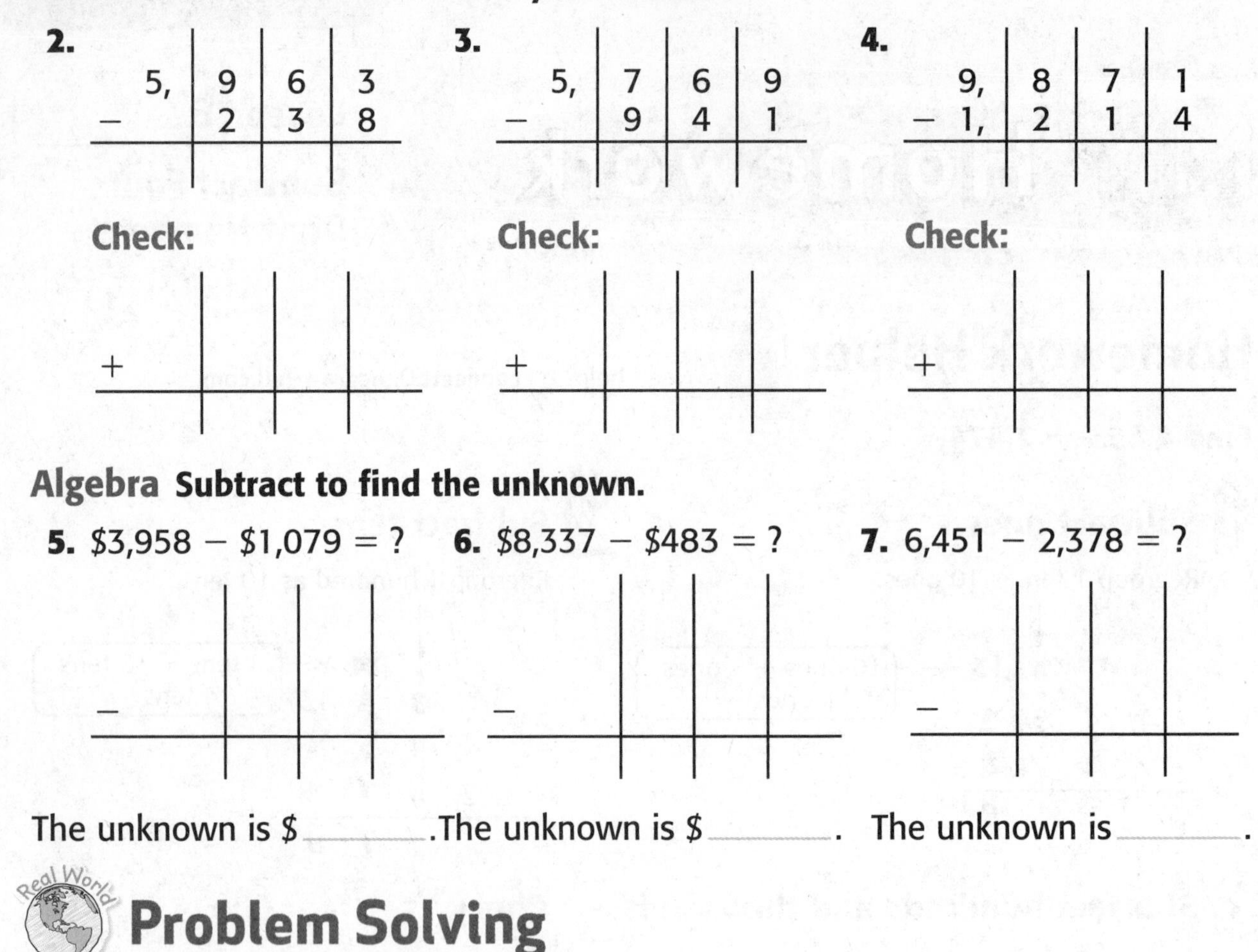

2.

5,	9	6	3
−	2	3	8

Check:

+

3.

5,	7	6	9
−	9	4	1

Check:

+

4.

9,	8	7	1
− 1,	2	1	4

Check:

+

Algebra **Subtract to find the unknown.**

5. \$3,958 − \$1,079 = ?

−

The unknown is \$______.

6. \$8,337 − \$483 = ?

−

The unknown is \$______.

7. 6,451 − 2,378 = ?

−

The unknown is ______.

Real World Problem Solving

Write a number sentence to solve.

8. Pittsburg University won the college football championship in 1937. They won again in 1976. How many years were there between championships?

____

9. Mathematical PRACTICE 4 **Model Math** A library has 2,220 books about sports and 1,814 books about animals. How many more sports books are there than animal books?

____

Test Practice

10. How much less money did Selena's school raise this year at the pancake breakfast?

Ⓐ \$900

Ⓑ \$905

Ⓒ \$1,905

Ⓓ \$8,145

LAST YEAR \$4,525

THIS YEAR \$3,620

Name

Subtract Across Zeros

Lesson 7

ESSENTIAL QUESTION
How are the operations of subtraction and addition related?

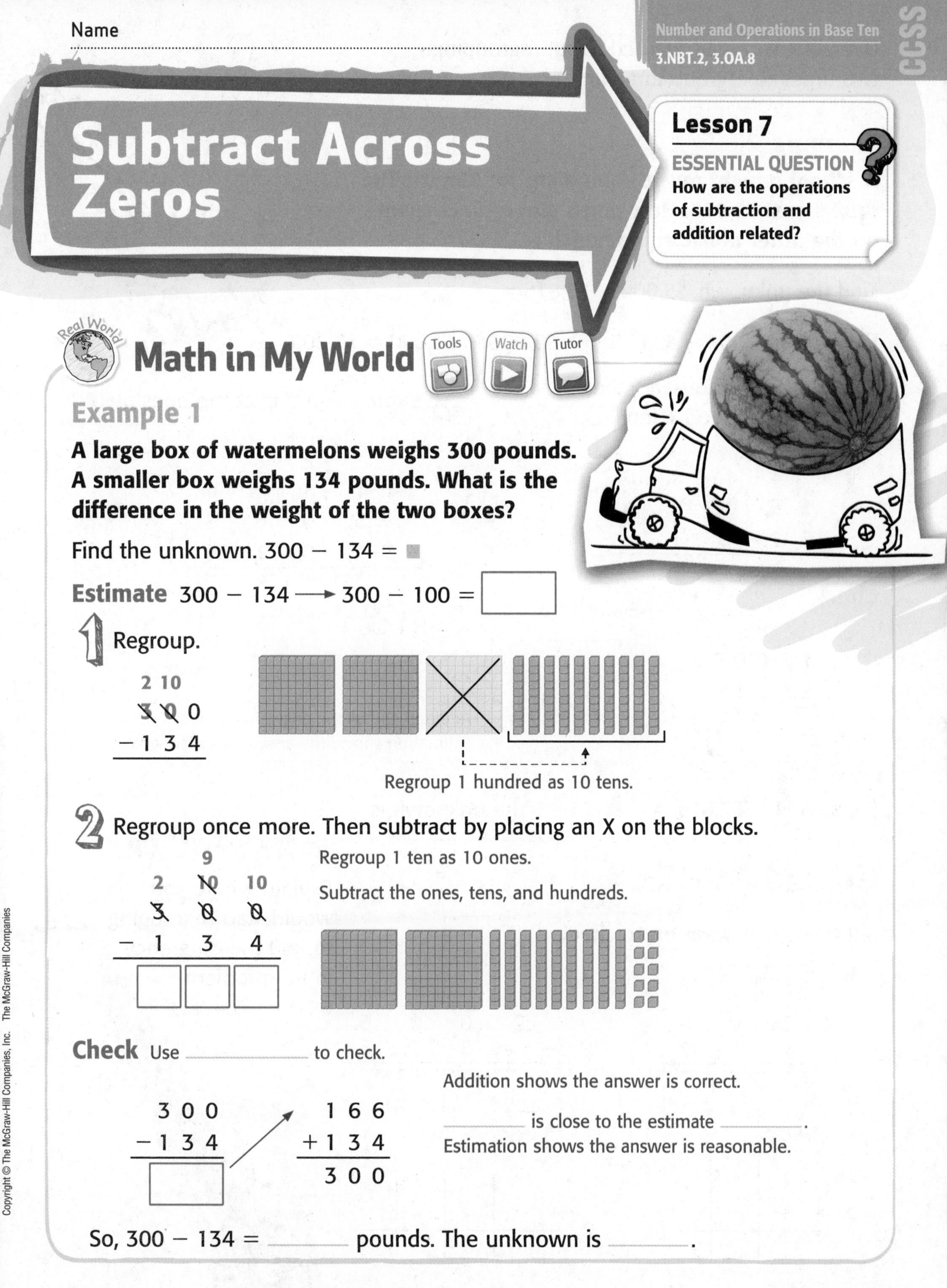

Math in My World

Tools Watch Tutor

Example 1

A large box of watermelons weighs 300 pounds. A smaller box weighs 134 pounds. What is the difference in the weight of the two boxes?

Find the unknown. 300 − 134 = ■

Estimate 300 − 134 ⟶ 300 − 100 = ☐

1 Regroup.

```
  2 10
  3̸ 0̸ 0
− 1 3 4
```

Regroup 1 hundred as 10 tens.

2 Regroup once more. Then subtract by placing an X on the blocks.

```
     9
  2 1̸0̸ 10
  3̸  0̸  0̸
− 1  3  4
  ☐  ☐  ☐
```

Regroup 1 ten as 10 ones.

Subtract the ones, tens, and hundreds.

Check Use ________ to check.

```
  300        166
− 134      + 134
  ☐          300
```

Addition shows the answer is correct.

________ is close to the estimate ________.
Estimation shows the answer is reasonable.

So, 300 − 134 = ________ pounds. The unknown is ________.

Use what you have learned about regrouping two times to regroup three times.

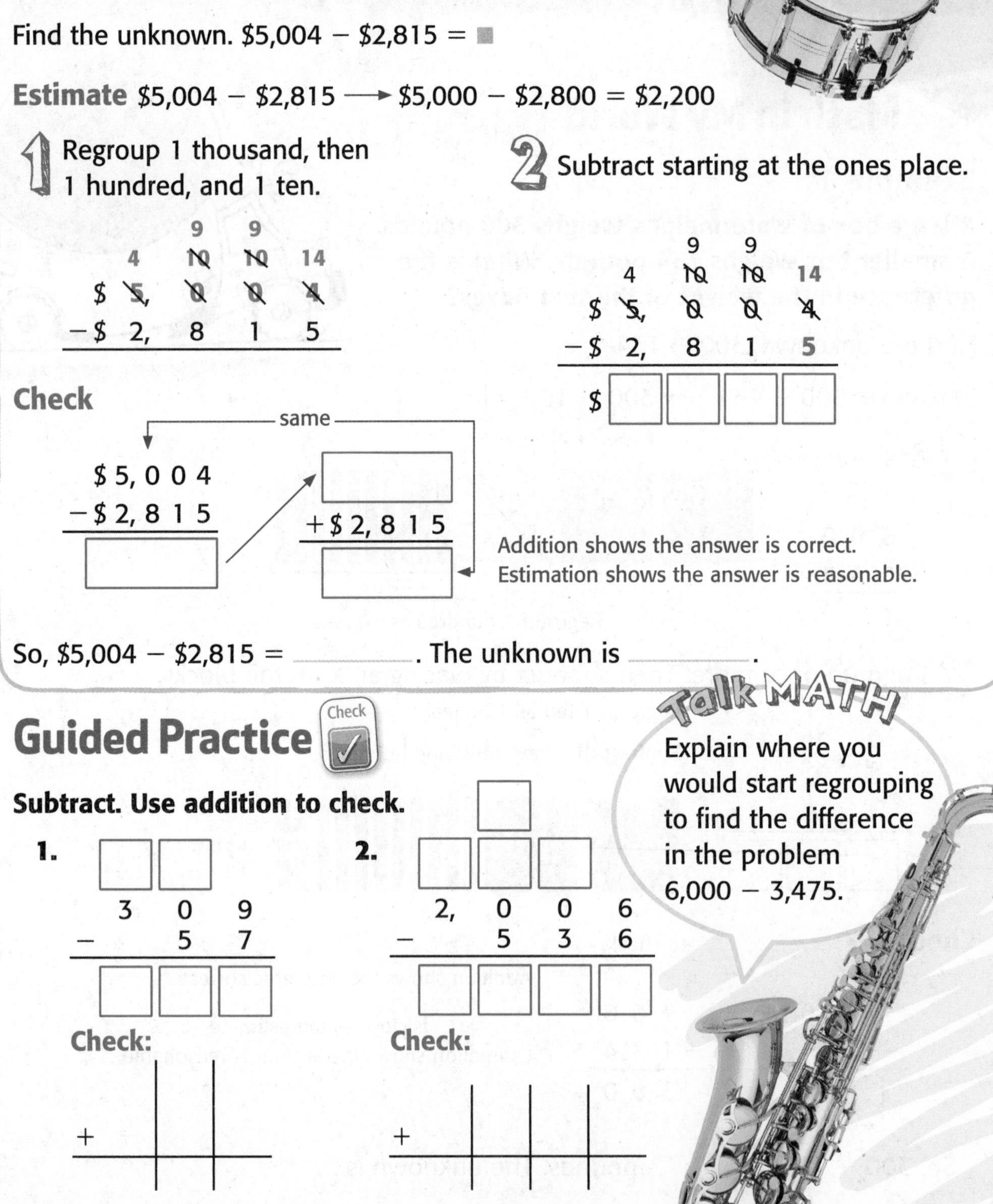

Example 2

A school bought music equipment for $5,004. The drums cost $2,815. How much money was spent on the other music equipment?

Find the unknown. $5,004 − $2,815 = ■

Estimate $5,004 − $2,815 ⟶ $5,000 − $2,800 = $2,200

1 Regroup 1 thousand, then 1 hundred, and 1 ten.

```
          9    9
    4    10   10   14
  $ 5,    0    0    4
− $ 2,    8    1    5
```

2 Subtract starting at the ones place.

```
          9    9
    4    10   10   14
  $ 5,    0    0    4
− $ 2,    8    1    5
  $ [ ]  [ ]  [ ]  [ ]
```

Check

```
  $5,004            [     ]
− $2,815          + $2,815
  [     ]           [     ]
```

same

Addition shows the answer is correct. Estimation shows the answer is reasonable.

So, $5,004 − $2,815 = ________. The unknown is ________.

Guided Practice

Check

Subtract. Use addition to check.

1.

```
  [ ][ ]
   3  0  9
−     5  7
  [ ][ ][ ]
```

Check:

+

2.

```
      [ ]
  [ ][ ][ ]
  2, 0  0  6
−    5  3  6
  [ ][ ][ ][ ]
```

Check:

+

Talk MATH

Explain where you would start regrouping to find the difference in the problem 6,000 − 3,475.

Name

Independent Practice

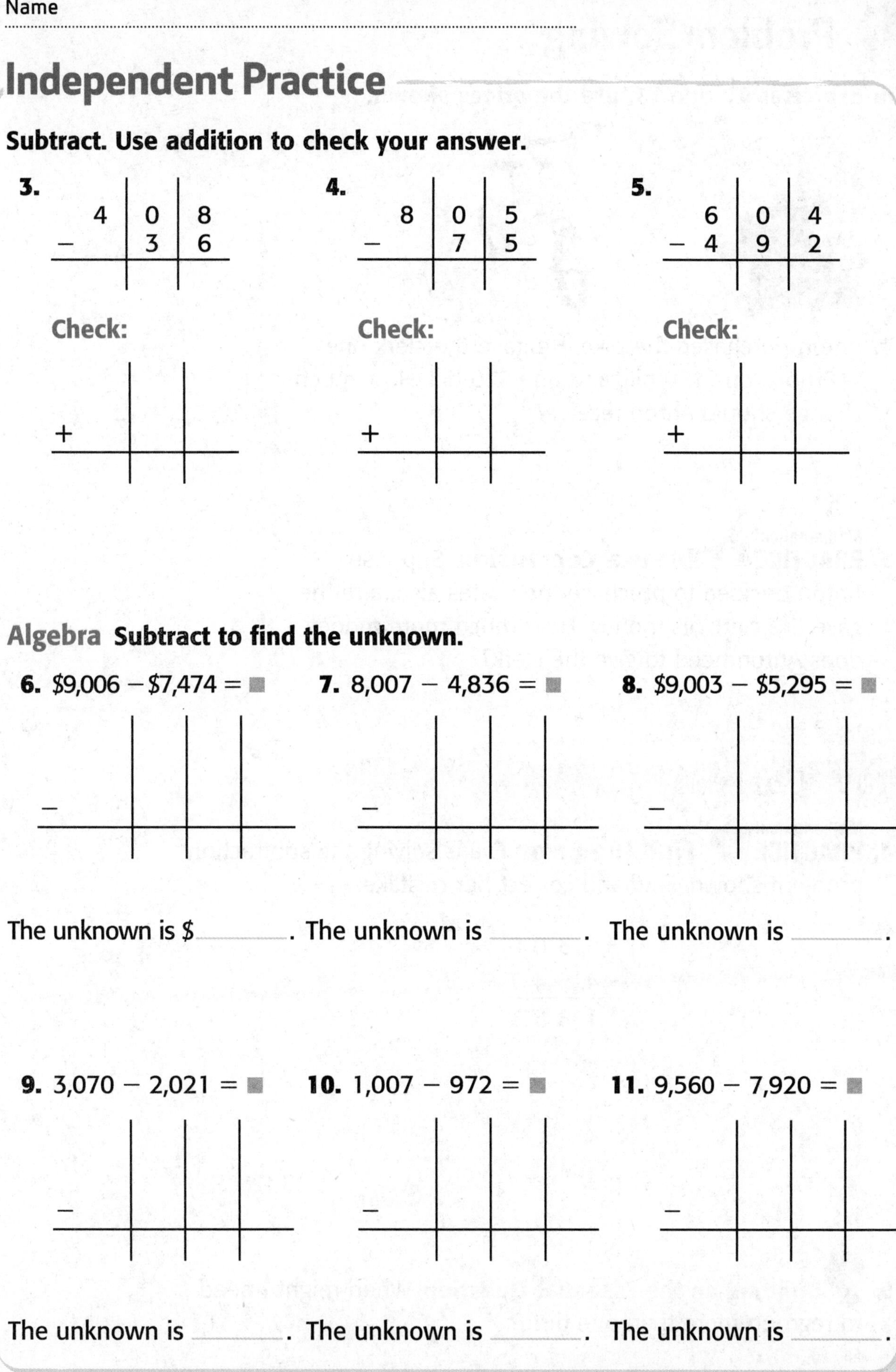

Subtract. Use addition to check your answer.

3.

4	0	8
−	3	6

Check:

+

4.

8	0	5
−	7	5

Check:

+

5.

6	0	4
− 4	9	2

Check:

+

Algebra Subtract to find the unknown.

6. \$9,006 − \$7,474 = ■

−

The unknown is \$ ______.

7. 8,007 − 4,836 = ■

−

The unknown is ______.

8. \$9,003 − \$5,295 = ■

−

The unknown is ______.

9. 3,070 − 2,021 = ■

−

The unknown is ______.

10. 1,007 − 972 = ■

−

The unknown is ______.

11. 9,560 − 7,920 = ■

−

The unknown is ______.

Problem Solving

For Exercises 12 and 13, use the prices shown.

My Work!

12. Anton purchased the bike. He gave the clerk one \$100-bill, one \$50-bill, and one \$20-bill. How much change should Anton receive?

____

13. Mathematical PRACTICE 3 **Draw a Conclusion** Suppose Anton decided to purchase the skates also, after he gave the clerk his money. How much more money does Anton need to give the clerk?

____

HOT Problems

14. Mathematical PRACTICE 3 **Find the Error** Eva is solving the subtraction problem shown. Find and correct her mistake.

$$\begin{array}{r} 5,300 \\ -4,547 \\ \hline 1,853 \end{array}$$

____

____

____

15. **Building on the Essential Question** When might I need to regroup more than one time?

____

Name

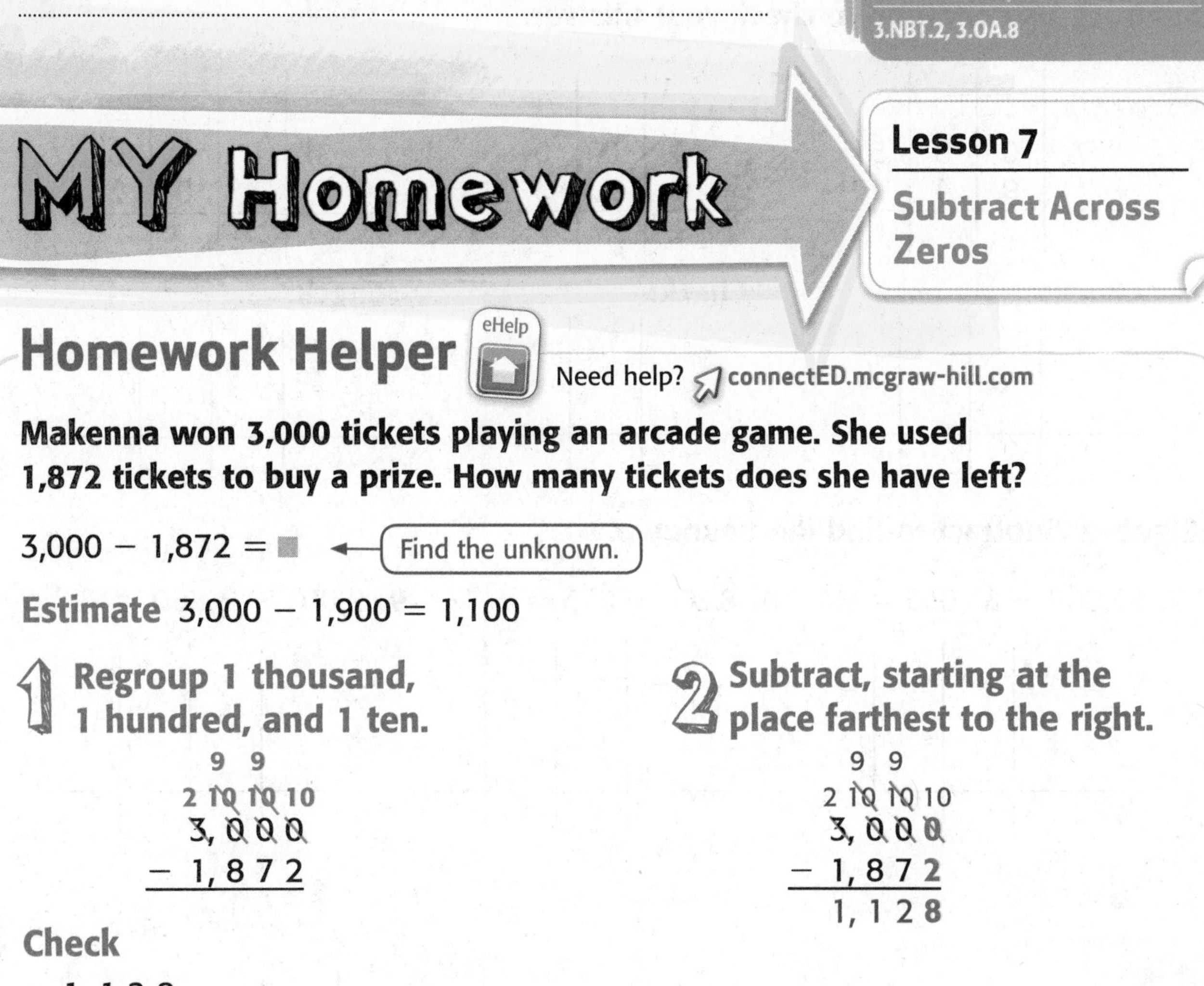

MY Homework

Lesson 7
Subtract Across Zeros

Homework Helper

Need help? connectED.mcgraw-hill.com

Makenna won 3,000 tickets playing an arcade game. She used 1,872 tickets to buy a prize. How many tickets does she have left?

3,000 − 1,872 = ■ ← Find the unknown.

Estimate 3,000 − 1,900 = 1,100

1 Regroup 1 thousand, 1 hundred, and 1 ten.

```
   9  9
 2 10 10 10
   3, 0 0 0
 − 1, 8 7 2
```

2 Subtract, starting at the place farthest to the right.

```
   9  9
 2 10 10 10
   3, 0 0 0
 − 1, 8 7 2
   1, 1 2 8
```

Check

```
   1, 1 2 8
 + 1, 8 7 2
   3, 0 0 0
```

Addition shows the answer is correct.

1,128 is close to the estimate of 1,100. Estimation shows the answer is reasonable.

Practice

Subtract. Use addition to check.

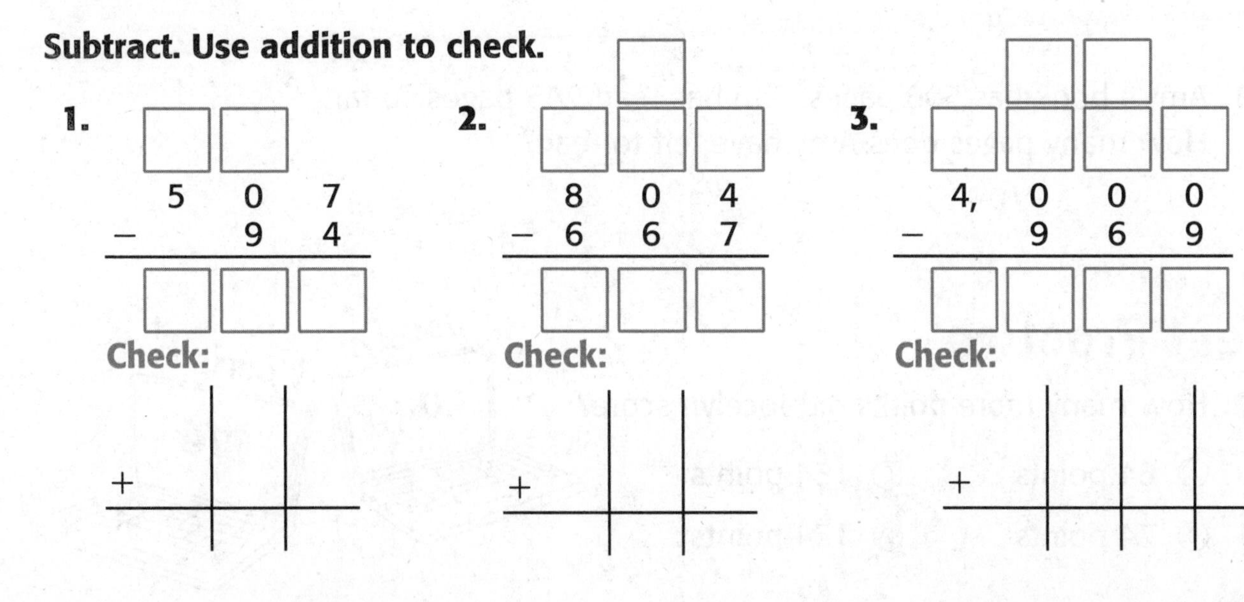

1. 5 0 7 − 9 4

Check: +

2. 8 0 4 − 6 6 7

Check: +

3. 4, 0 0 0 − 9 6 9

Check: +

Subtract. Use addition to check your answer.

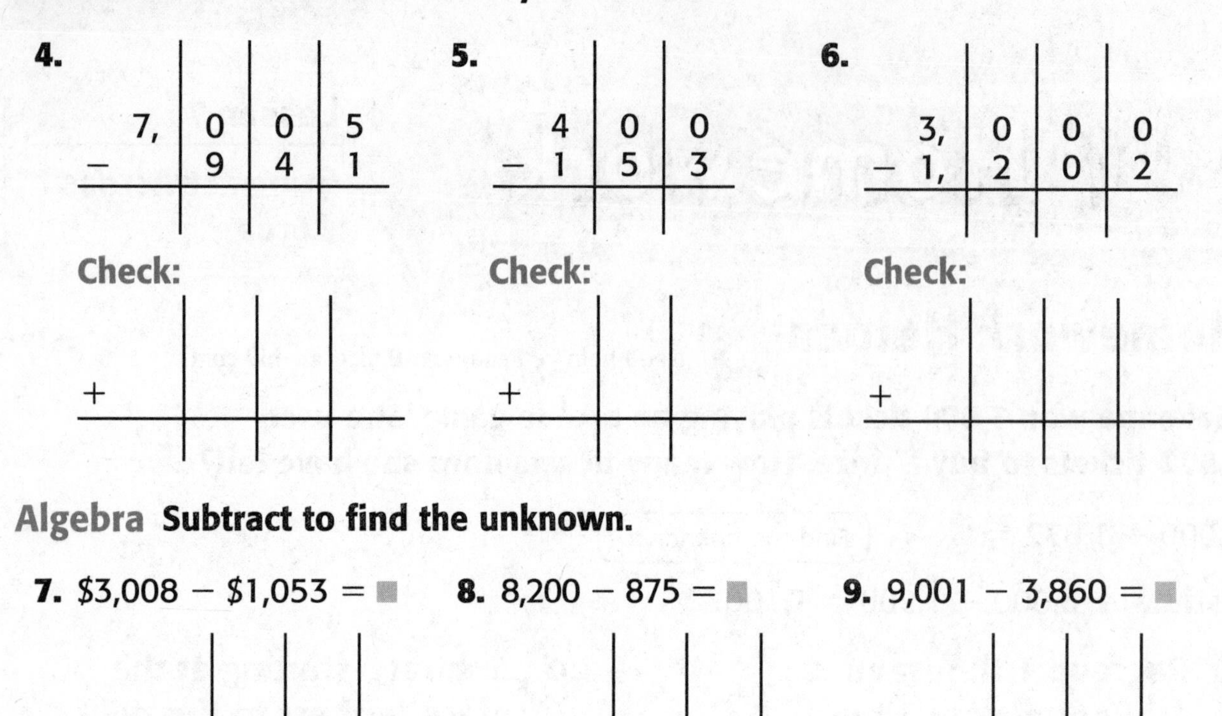

Algebra **Subtract to find the unknown.**

7. $3,008 − $1,053 = ■

−

■ = ________

8. 8,200 − 875 = ■

−

■ = ________

9. 9,001 − 3,860 = ■

−

■ = ________

Problem Solving

Mathematical PRACTICE 2 **Understand Symbols** **Write a number sentence.**

10. A bag holds 5,300 seeds. Brandon plants 790 of the seeds. How many seeds are left?

____

11. Amy's book has 500 pages. She has read 245 pages so far. How many pages does Amy have left to read?

____

Test Practice

12. How many more points did Jocelyn score?

Ⓐ 64 points
Ⓑ 74 points
Ⓒ 164 points
Ⓓ 174 points

Review

Chapter 3

Subtraction

Vocabulary Check

Vocab

Choose the correct word(s) to complete each sentence.

difference	**estimate**	**inverse operations**
regroup	**subtract**	**subtraction sentence**

1. The answer to a subtraction problem is called the ______________.

2. When I ______________, I take away the lesser number from the greater number.

3. I can ____________ to trade equal amounts when renaming a number.

4. When I do not need an exact answer I can ______________ to find one that is close.

5. A number sentence in which one quantity is taken away from another quantity is a ______________________.

6. Two operations that can undo each other, as in addition and subtraction, are called ______________________.

Concept Check

Subtract mentally by breaking apart the smaller number.

7. 884 − 51 = ________ **8.** 283 − 171 = ________ **9.** 724 − 616 = ________

Estimate. Round each number to the indicated place value.

10. 765 − 121; tens

_____ − _____ = _____

11. 2,219 − 1,109; hundreds

_____ − _____ = _____

Subtract. Use addition to check your answer.

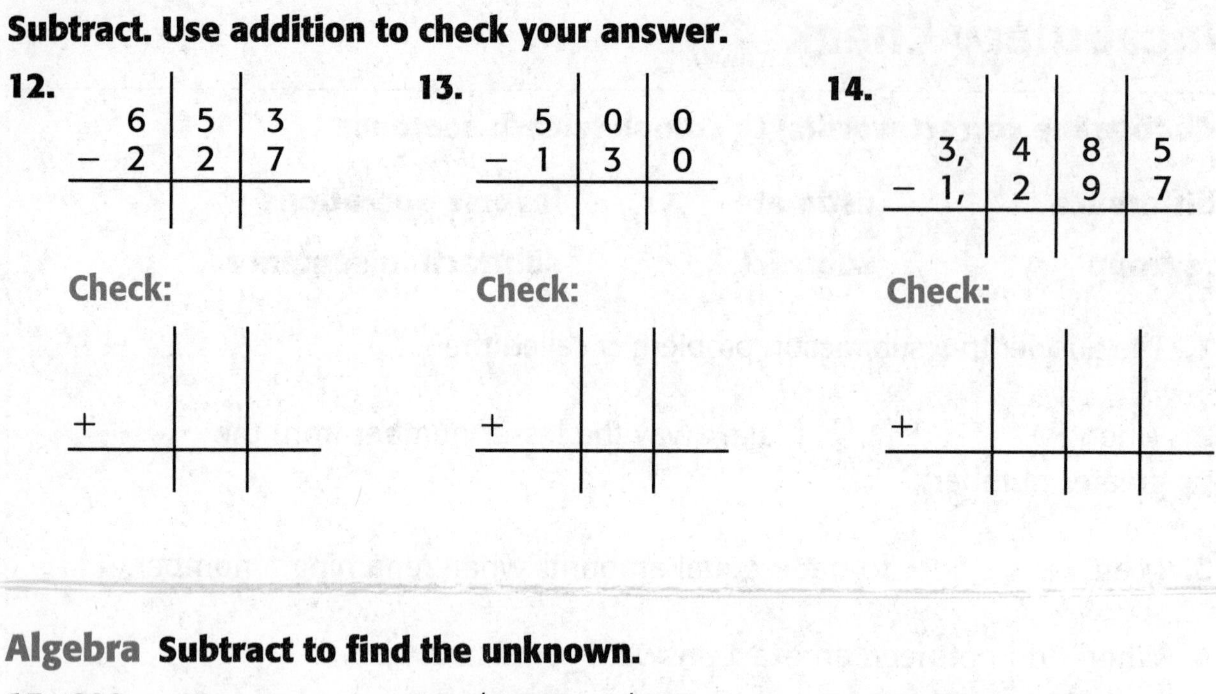

12.

6	5	3
− 2	2	7

Check:

13.

5	0	0
− 1	3	0

Check:

14.

3,	4	8	5
− 1,	2	9	7

Check:

Algebra **Subtract to find the unknown.**

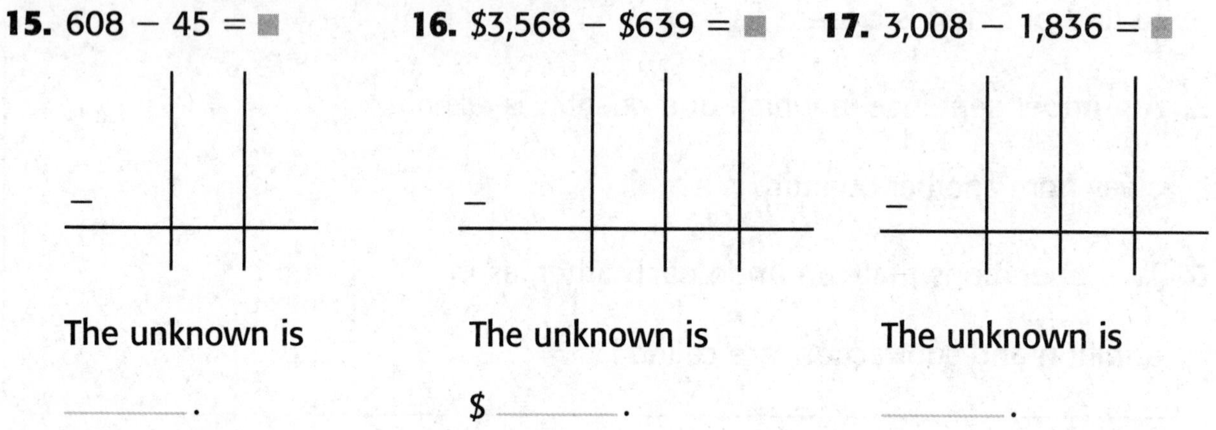

15. 608 − 45 = ■

The unknown is _____.

16. $3,568 − $639 = ■

The unknown is $ _____.

17. 3,008 − 1,836 = ■

The unknown is _____.

Algebra **Use addition to find each unknown.**

18.

5	8	0
− ■	▲	0
4	3	0

■ = _____

▲ = _____

19.

1	■	5
−	4	▲
	8	6

■ = _____

▲ = _____

20.

6,	9	2	0
− ■	1	▲	8
4,	7	2	2

■ = _____

▲ = _____

Name

Real World Problem Solving

My Work!

21. An artist created a piece of art with 675 round glass beads. There were also 179 beads in the shape of a heart. Write a number sentence to find about how many more round beads there were.

22. There are 365 days in one year. There were 173 sunny days this year. Write a number sentence to find the number of days that were not sunny.

23. In which month were fewer raffle tickets sold? Explain.

TV Raffle Ticket Sales		
Day	March	April
Saturday	$3,129	$4,103
Sunday	$3,977	$3,001

Test Practice

24. Students want to make 425 get well cards. So far they have made 165 cards. How many more cards do they need to make?

Ⓐ 240 cards Ⓒ 270 cards

Ⓑ 260 cards Ⓓ 590 cards

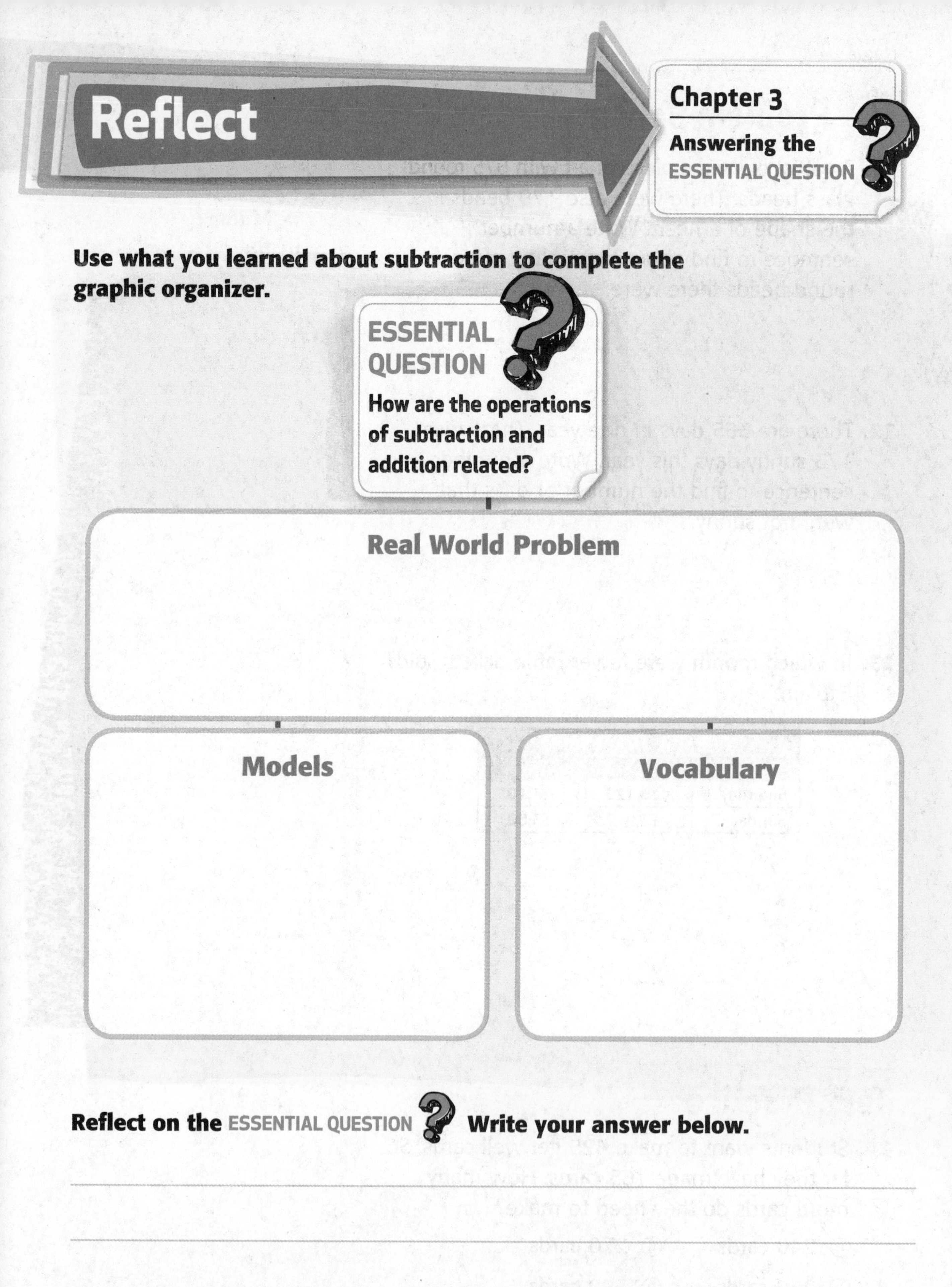

Use what you learned about subtraction to complete the graphic organizer.

Reflect on the ESSENTIAL QUESTION **Write your answer below.**

Chapter

Understand Multiplication

ESSENTIAL QUESTION

What does multiplication mean?

My Favorite Foods

Watch a video!

Watch

MY Common Core State Standards

Operations and Algebraic Thinking

CCSS

3.OA.1 Interpret products of whole numbers, e.g., interpret 5 × 7 as the total number of objects in 5 groups of 7 objects each.

3.OA.3 Use multiplication and division within 100 to solve word problems in situations involving equal groups, arrays, and measurement quantities, e.g., by using drawings and equations with a symbol for the unknown number to represent the problem.

3.OA.5 Apply properties of operations as strategies to multiply and divide.

3.OA.8 Solve two-step word problems using the four operations. Represent these problems using equations with a letter standing for the unknown quantity. Assess the reasonableness of answers using mental computation and estimation strategies including rounding.

Hey, I already know some of these!

Standards for Mathematical PRACTICE

1. Make sense of problems and persevere in solving them.
2. Reason abstractly and quantitatively.
3. Construct viable arguments and critique the reasoning of others.
4. Model with mathematics.
5. Use appropriate tools strategically.
6. Attend to precision.
7. Look for and make use of structure.
8. Look for and express regularity in repeated reasoning.

= focused on in this chapter

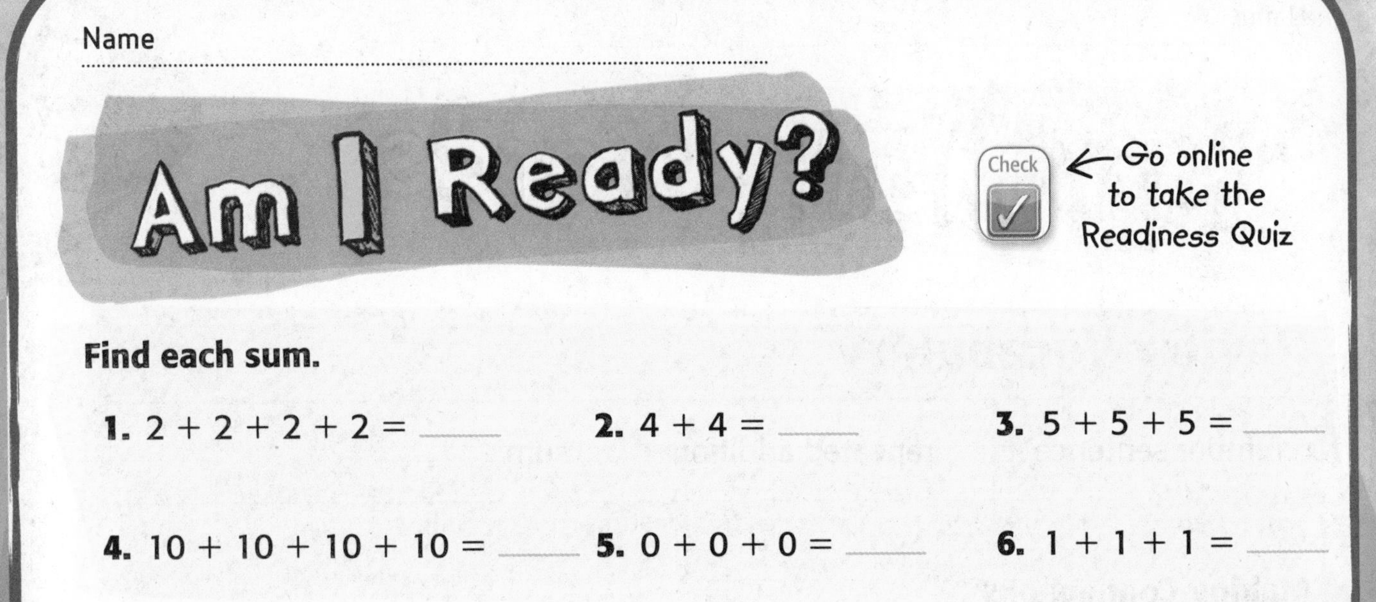

Find each sum.

1. $2 + 2 + 2 + 2 =$ ____

2. $4 + 4 =$ ____

3. $5 + 5 + 5 =$ ____

4. $10 + 10 + 10 + 10 =$ ____

5. $0 + 0 + 0 =$ ____

6. $1 + 1 + 1 =$ ____

Write an addition sentence for each picture.

7. **8.**

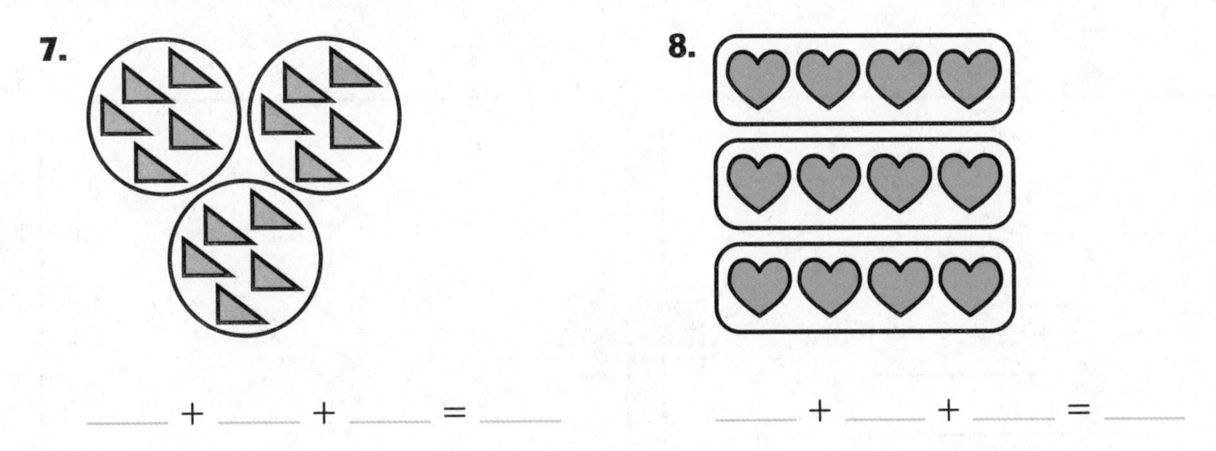

7. ____ + ____ + ____ = ____

8. ____ + ____ + ____ = ____

Solve. Use repeated addition.

9. Larisa has 2 cups with 4 crackers in each cup. How many crackers does she have in all?

____ crackers

10. On Monday and Tuesday, Lance rode his bike around the block 3 times each day. How many times in all did he ride his bike around the block?

____ times

How Did I Do?

Shade the boxes to show the problems you answered correctly.

1	2	3	4	5	6	7	8	9	10

Name

Review Vocabulary

number sentence　　repeated addition　　sum

Making Connections
Use the review words to label each example of addition in the graphic organizer. Use the second column to draw an example of the addition.

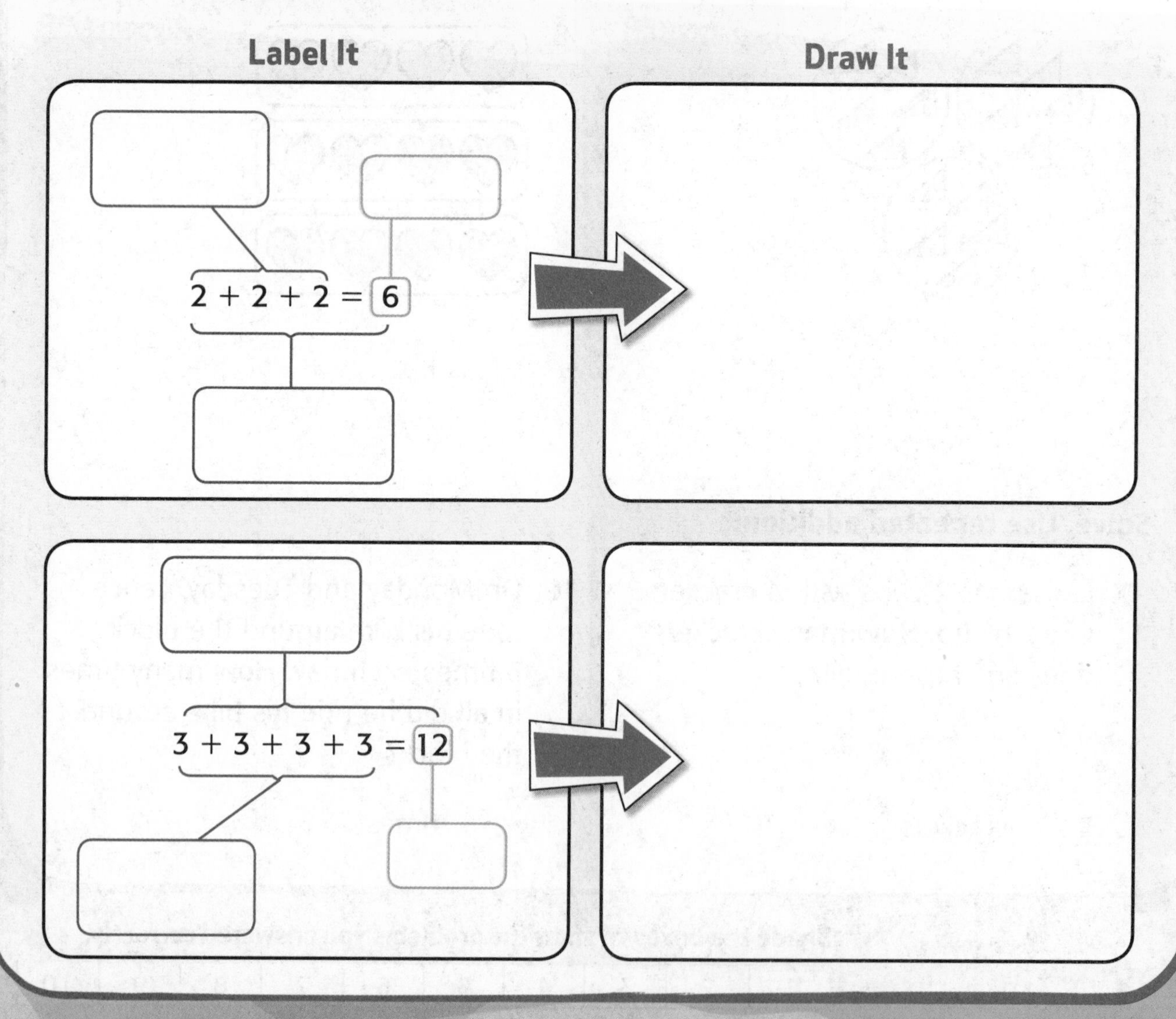

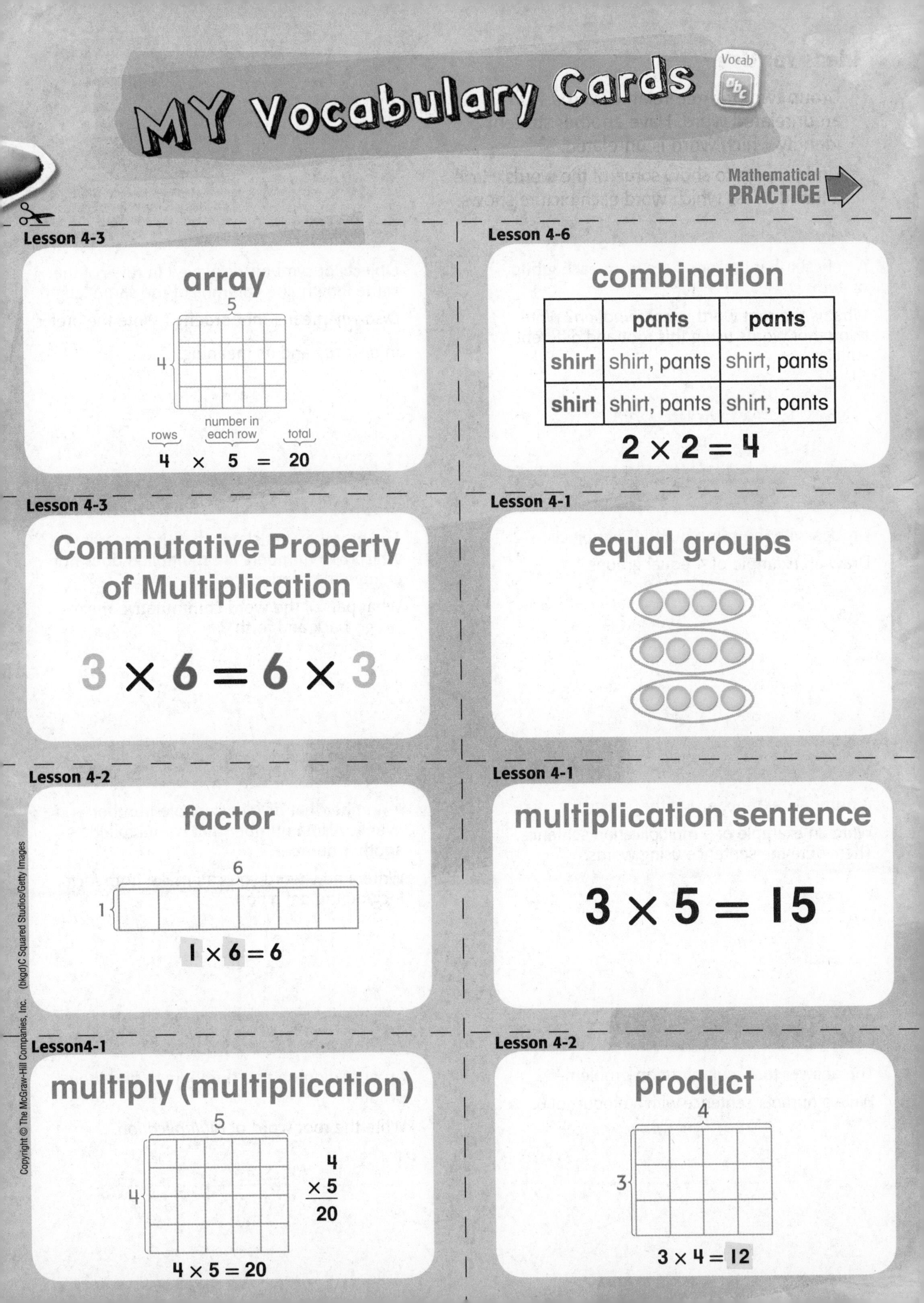

MY Vocabulary Cards

Vocab

Mathematical PRACTICE

Lesson 4-3

array

$4 \times 5 = 20$ (rows × number in each row = total)

Lesson 4-6

combination

	pants	pants
shirt	shirt, pants	shirt, pants
shirt	shirt, pants	shirt, pants

$2 \times 2 = 4$

Lesson 4-3

Commutative Property of Multiplication

$3 \times 6 = 6 \times 3$

Lesson 4-1

equal groups

Lesson 4-2

factor

$1 \times 6 = 6$

Lesson 4-1

multiplication sentence

$3 \times 5 = 15$

Lesson4-1

multiply (multiplication)

$$\begin{array}{r} 4 \\ \times 5 \\ \hline 20 \end{array}$$

$4 \times 5 = 20$

Lesson 4-2

product

$3 \times 4 = 12$

Ideas for Use

- Group two or three related words. Add an unrelated word. Have another student identify which word is unrelated.
- Find pictures to show some of the words. Have a friend guess which word each picture shows.

A new set that has one item from each group of items.

What is the root word in *combination?* Write two other words using this root and different suffixes.

Objects or symbols displayed in rows of the same length and columns of the same length.

Disarray means "not orderly." Write the prefix in *disarray* and its meaning.

Groups with the same number of objects.

Draw an example of 4 equal groups.

The property that states that the order in which two numbers are multiplied does not change the product.

What part of the word commutative means "to go back and forth"?

A number sentence using the × sign.

Write an example of a multiplication sentence. Then write the sentence using words.

A number that divides a whole number evenly. Also a number that is multiplied by another number.

Write a new word you can make from *factor*. Include the definition.

The answer to a multiplication problem.

Write a number sentence with a product of 6.

An operation on two numbers to find their product.

Write the root word of *multiplication.*

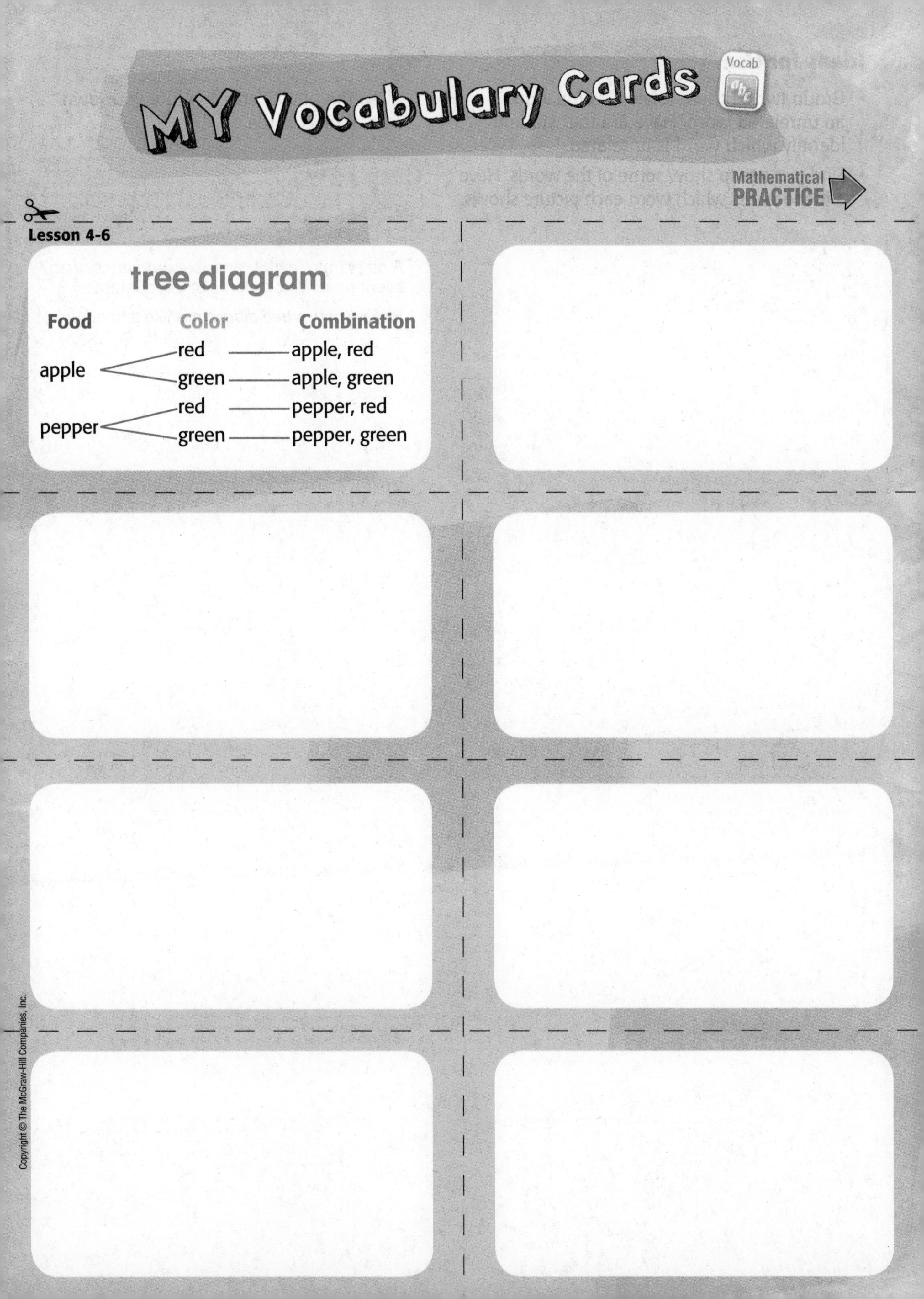

MY Vocabulary Cards

Vocab

Mathematical PRACTICE

Lesson 4-6

tree diagram

Food	Color	Combination
apple	red	apple, red
	green	apple, green
pepper	red	pepper, red
	green	pepper, green

Ideas for Use

- Group two or three related words. Add an unrelated word. Have another student identify which word is unrelated.
- Find pictures to show some of the words. Have a friend guess which word each picture shows.
- Use the blank cards to write your own vocabulary cards.

A diagram of all the possible outcomes of an event or series of events or experiments.

Explain how a *tree diagram* is like a tree.

FOLDABLES® Follow the steps on the back to make your Foldable.

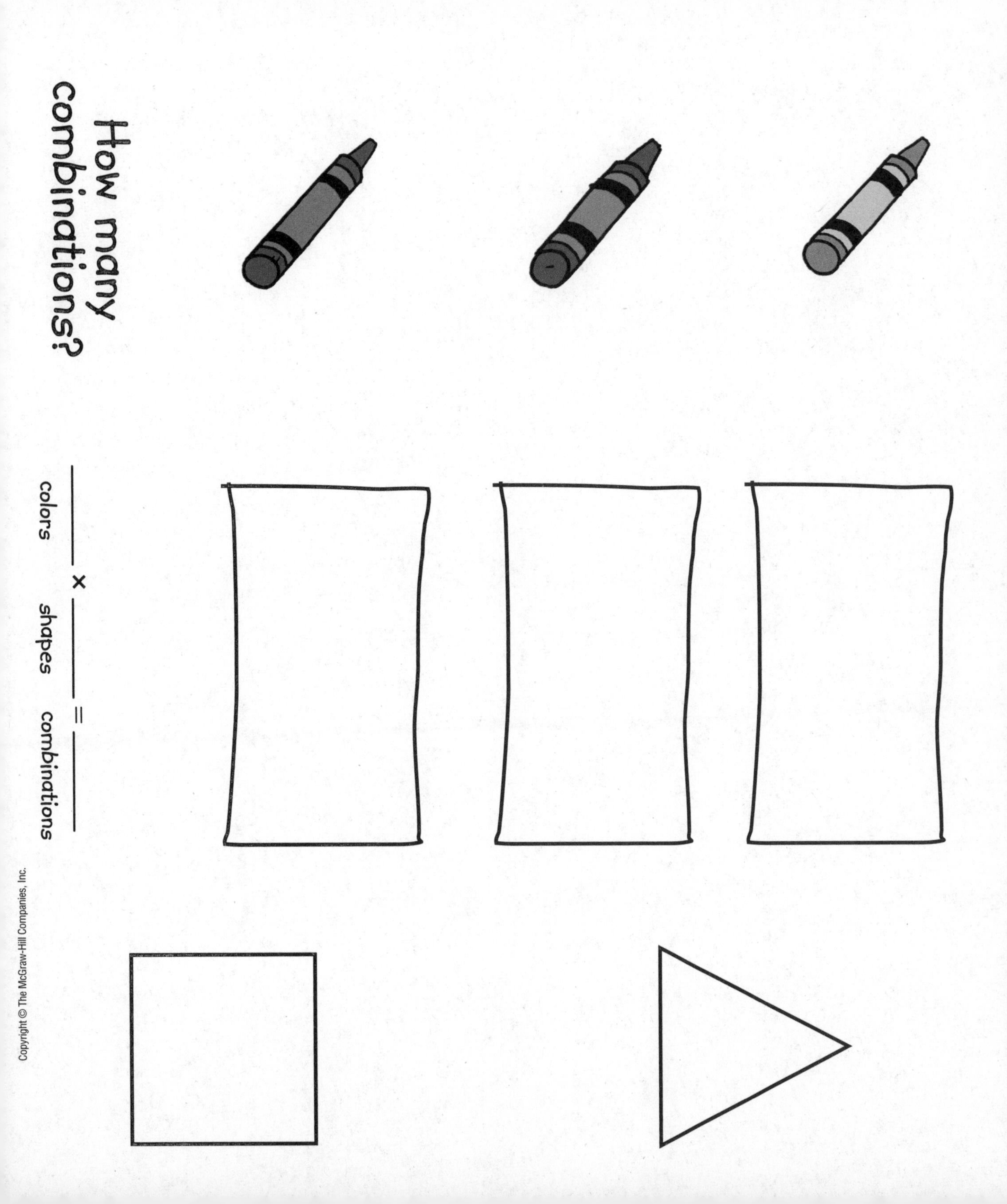

FOLDABLES
Study Organizer

Name ______________________

Hands On
Model Multiplication

Lesson 1

ESSENTIAL QUESTION
What does multiplication mean?

Build It

When you have **equal groups,** you have the same number of objects in each group. Use repeated addition to find the total number of objects.

Find the total in 4 equal groups of 5.

1. Use connecting cubes to show 4 equal groups of 5 cubes. Draw the groups.

   There are ____ groups with ____ in each group.

My Drawing!

2. Write the number of cubes in each group. Use repeated addition to complete the number sentence.

   ____ + ____ + ____ + ____ = ____

   The total in 4 groups of 5 is 20.

3. Record the number of groups, the number in each group, and the total from above.

   Explore other ways to group the 20 cubes equally.

Number of Groups	Number in Each Group	Total
4	5	20
10	2	20

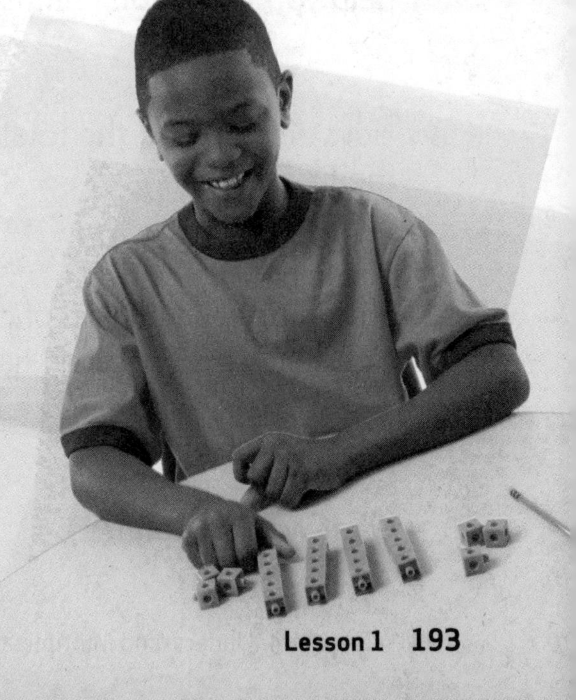

You can also use **multiplication** to find the total number of objects in equal groups. A number sentence with the symbol (×) is called a **multiplication sentence**. It means to **multiply**.

Try It

Shawn helped his mom bake cookies. He served 4 cookies on each plate. There are 2 plates. How many cookies did he serve?

Find the total of 2 plates of 4 cookies.

1. Use counters to model the equal groups. Draw the groups.

My Drawing!

2. Use repeated addition to complete the number sentence.

   ____ + ____ = ____

3. Write a multiplication sentence to show 2 plates of 4 cookies, or 2 groups of 4.

   ____ × ____ = ____

   number of groups, number in each group, total

So, Shawn served ____ cookies.

Talk About It

1. Mathematical PRACTICE 3 **Draw a Conclusion** How can addition help you find a total number of objects in equal groups?

____

2. How did you find the total number of cubes in Step 2 of the first activity?

____

3. Shawn counted a batch of cookies by finding 4 + 4 + 4. How could multiplication have helped him to find the total?

____

What was the total? ____________

Name ______________________

Practice It

Draw a model to find the total number.

4. 6 groups of 2 equals ______

5. 3 groups of 5 equals ______

6. $2 \times 4 =$ ______

7. $1 \times 7 =$ ______

Describe each set of equal groups.

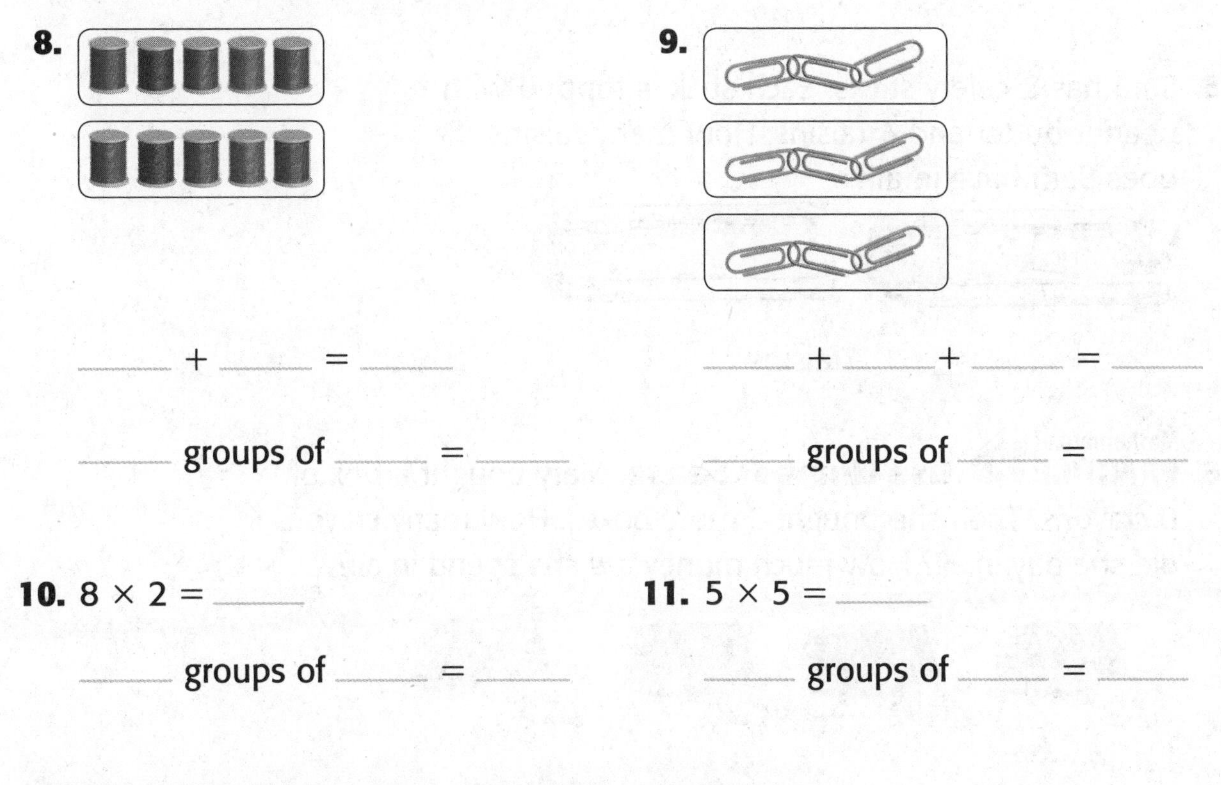

8.

______ + ______ = ______

______ groups of ______ = ______

9.

______ + ______ + ______ = ______

______ groups of ______ = ______

10. $8 \times 2 =$ ______

______ groups of ______ = ______

11. $5 \times 5 =$ ______

______ groups of ______ = ______

Equally group the counters. Draw the equal groups.

12. set of 6 counters

13. set of 18 counters

Apply It

Mathematical PRACTICE 4 Model Math **Draw to complete each model. Then complete each number sentence.**

My Work!

14. Tennis balls come in cans of 3.
How many tennis balls are in 4 cans?

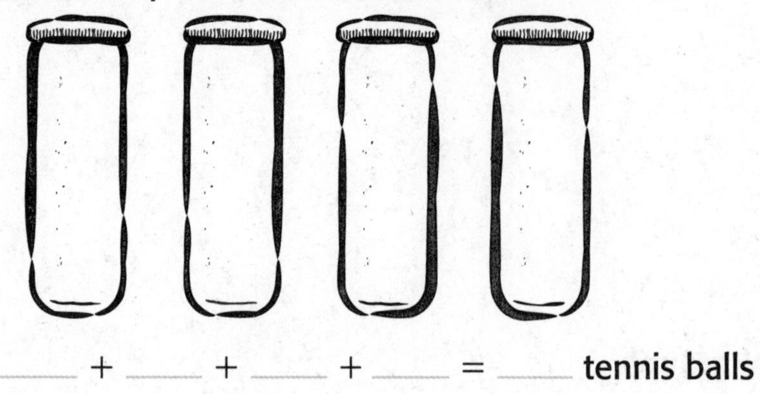

____ + ____ + ____ + ____ = ____ tennis balls

15. Sam has 2 celery sticks. Each stick is topped with peanut butter and 4 raisins. How many raisins does Sam have in all?

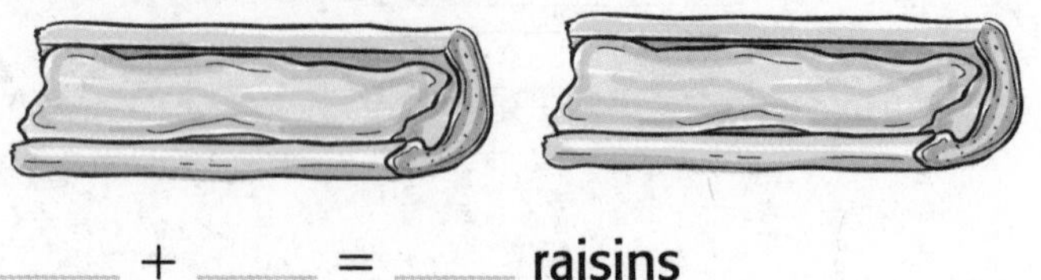

____ + ____ = ____ raisins

16. **Mathematical PRACTICE 2 Use Number Sense** Mary bought a box of 6 crayons. Then she bought 3 more boxes. How many crayons did she buy in all? How much money did she spend in all?

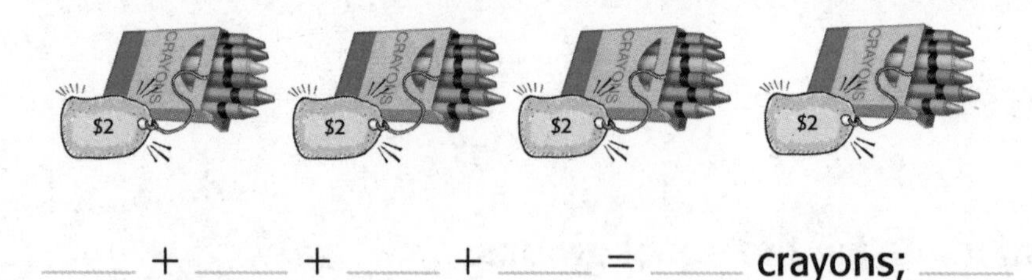

____ + ____ + ____ + ____ = ____ crayons; ____

Write About It

17. What does it mean to multiply?

____

____

Name ______________________

Multiplication as Repeated Addition

Lesson 2

ESSENTIAL QUESTION
What does multiplication mean?

There are many ways to find the total when there are groups of equal objects.

Math in My World

Watch Tutor

Example 1

Gilberto made 4 small pizzas for his party. Each pizza had 5 slices of tomato. How many slices of tomato did Gilberto use to make 4 small pizzas?

Find how many slices of tomato there are in 4 groups of 5.

One Way **Draw a picture.**

1. There are ______ groups. Draw 4 pizzas.
2. There are ______ in each group. Draw 5 slices of tomato on each pizza.
3. Count. There are ______ slices of tomato.

My Drawing!

Another Way **Use repeated addition.**
Write an addition sentence to show the equal groups.

______ + ______ + ______ + ______ = ______

So, ______ groups of ______ is ______.

Gilberto used ______ slices of tomato.

When you find the total of equal groups of objects, you **multiply**. The symbol (×) means to multiply. The numbers multiplied are **factors**. The result is the **product**.

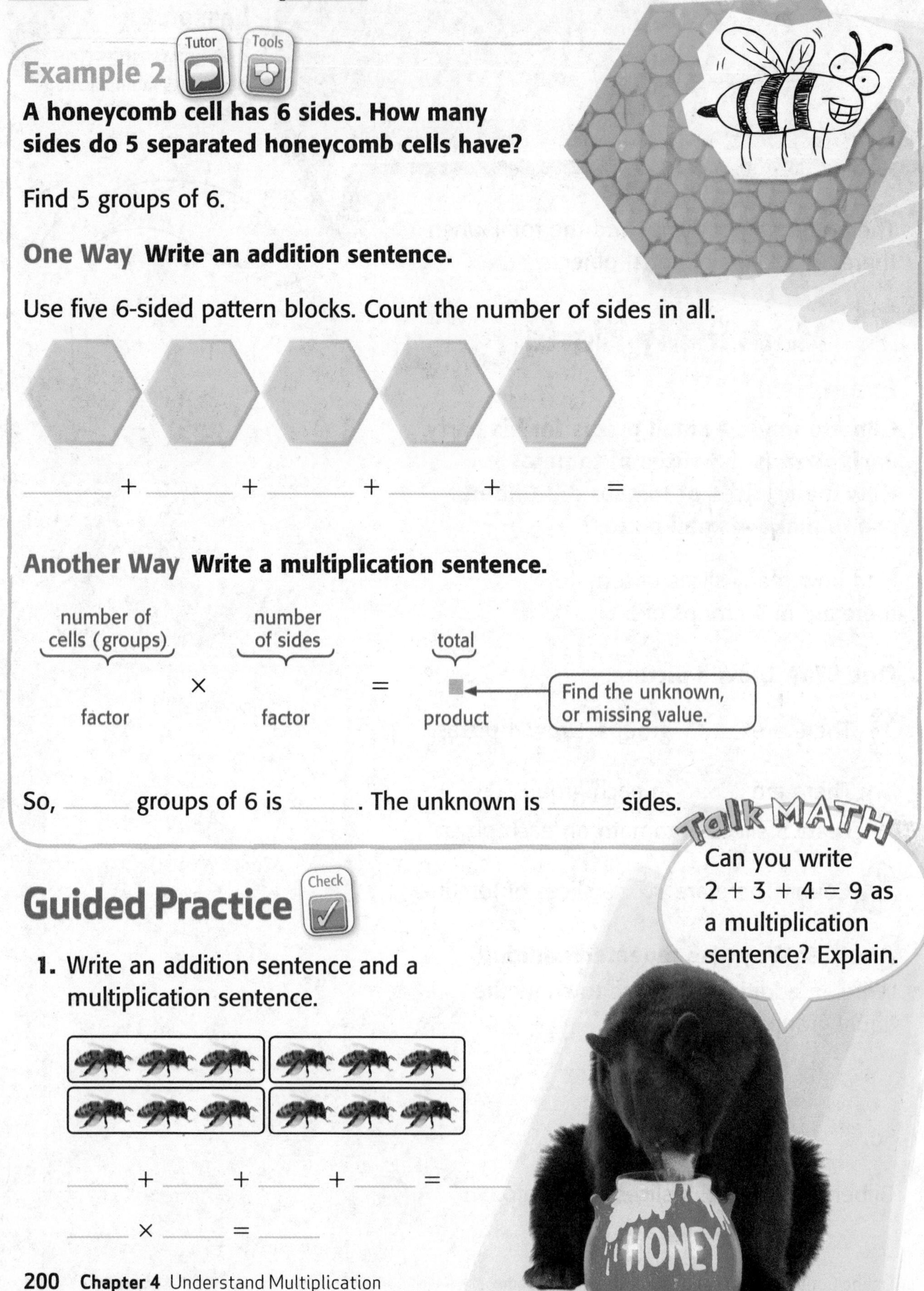

Example 2

A honeycomb cell has 6 sides. How many sides do 5 separated honeycomb cells have?

Find 5 groups of 6.

One Way **Write an addition sentence.**

Use five 6-sided pattern blocks. Count the number of sides in all.

____ + ____ + ____ + ____ + ____ = ____

Another Way **Write a multiplication sentence.**

number of cells (groups)		number of sides		total	
	×		=	■	← Find the unknown, or missing value.
factor		factor		product	

So, ____ groups of 6 is ____. The unknown is ____ sides.

Talk MATH

Can you write $2 + 3 + 4 = 9$ as a multiplication sentence? Explain.

Guided Practice

1. Write an addition sentence and a multiplication sentence.

____ + ____ + ____ + ____ = ____

____ × ____ = ____

Name ....

Independent Practice

Write an addition sentence and a multiplication sentence for each.

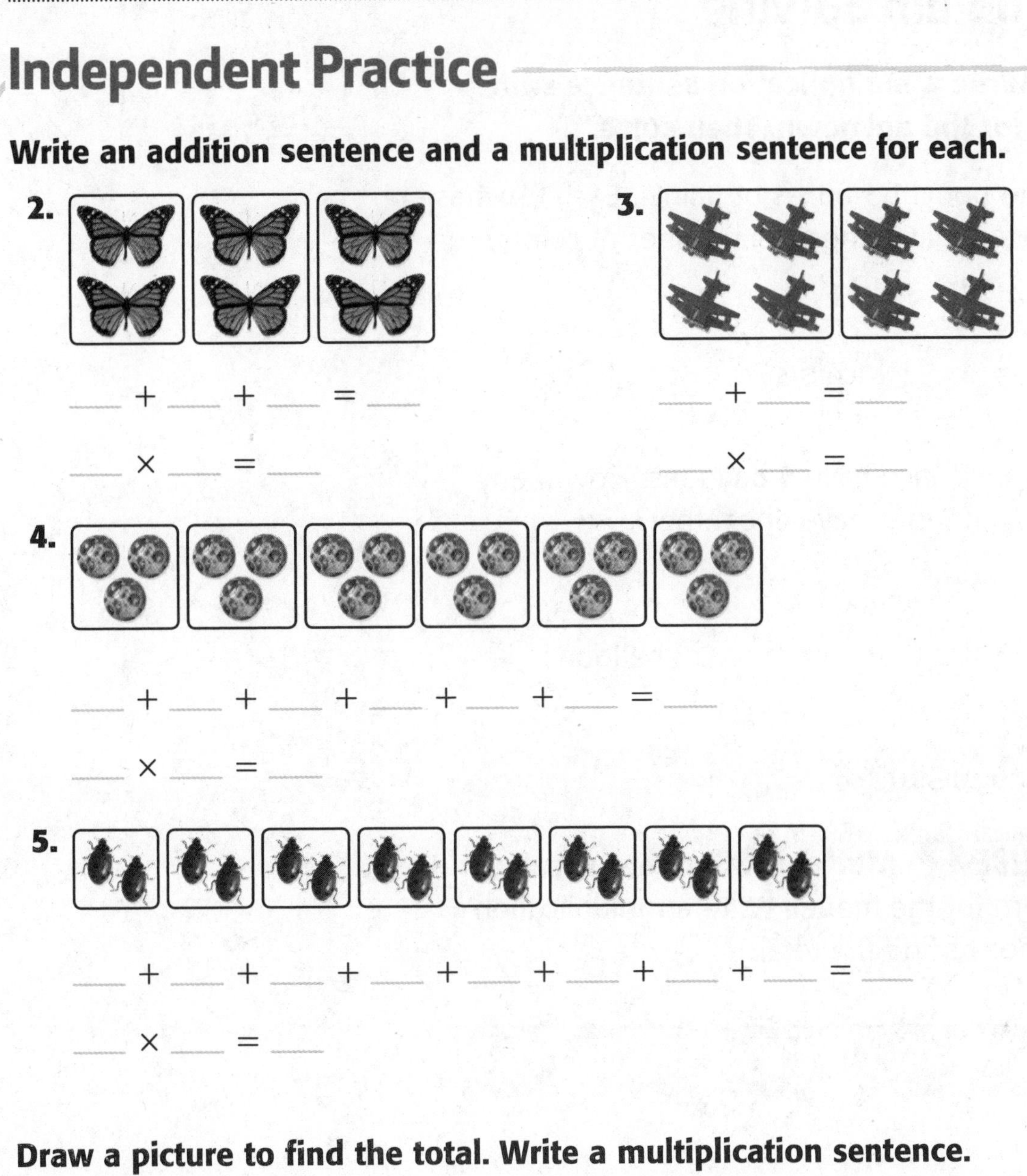

2.

___ + ___ + ___ = ___

___ × ___ = ___

3.

___ + ___ = ___

___ × ___ = ___

4.

___ + ___ + ___ + ___ + ___ + ___ = ___

___ × ___ = ___

5.

___ + ___ + ___ + ___ + ___ + ___ + ___ + ___ = ___

___ × ___ = ___

Draw a picture to find the total. Write a multiplication sentence.

6. 6 groups of 5

___ × ___ = ___

7. 8 groups of 4

___ × ___ = ___

Algebra **Multiply to find the unknown product.**

8. 3 × 5 = ■

The unknown is ___.

9. 5 × 2 = ■

The unknown is ___.

10. 3 × 3 = ■

The unknown is ___.

Problem Solving

Algebra Write a multiplication sentence with a symbol for the unknown. Then solve.

11. Adriano bought 3 boxes of paints. Each box has 8 colors. What is the total number of colors?

$3 \times 8 = \blacksquare$

$3 \times 8 =$ ______ colors

12. Three boys each have 5 balloons. How many balloons do they have altogether?

______ $\times$ ______ $= \blacksquare$

______ $\times$ ______ $=$ ______ balloons

My Work!

HOT Problems

13. Mathematical PRACTICE 4 **Model Math** Write a real-world problem for the model. Write a multiplication sentence to find the total.

____

____

14. Mathematical PRACTICE 2 **Use Number Sense** What is 2 more than 5 groups of 3? ______

15. **Building on the Essential Question** How are multiplication and repeated addition alike?

____

____

Name ....

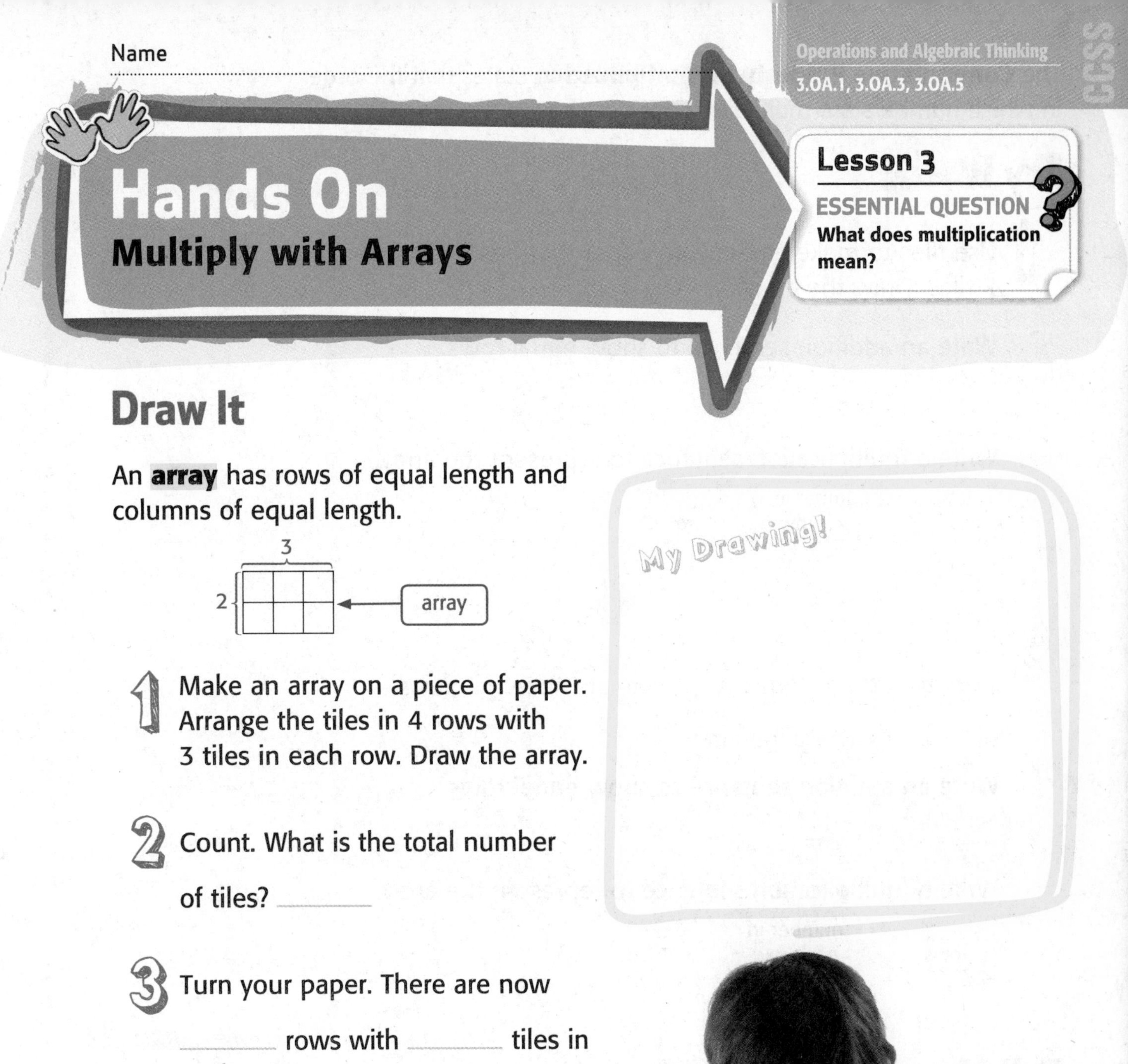

Hands On
Multiply with Arrays

Lesson 3

ESSENTIAL QUESTION
What does multiplication mean?

Draw It

An **array** has rows of equal length and columns of equal length.

My Drawing!

1. Make an array on a piece of paper. Arrange the tiles in 4 rows with 3 tiles in each row. Draw the array.

2. Count. What is the total number of tiles? ______

3. Turn your paper. There are now ______ rows with ______ tiles in each row.

   Draw what the array looks like now.

4. Count. What is the total number of tiles? ______

So, there are the same number of tiles, ______, if you turn the array.

The **Commutative Property of Multiplication** states that the order in which numbers are multiplied does not change the product.

Try It

My Drawing!

1 Use tiles to make an array on paper that has 5 rows of 2 tiles. Draw the array.

Write an addition sentence to show equal rows.

____ + ____ + ____ + ____ + ____ = ____

Write a multiplication sentence to represent the array.

rows		number in each row		total
____	×	____	=	____

2 Turn the array the other way. There are now ____ rows of ____ tiles. Draw the array.

Write an addition sentence to show equal rows.

____ + ____ = ____

Write a multiplication sentence to represent the array.

rows		number in each row		total
____	×	____	=	____

Talk About It

1. Mathematical PRACTICE 2 **Stop and Reflect** What is the connection between repeated addition and an array?

____

2. How can you use an array to model the Commutative Property?

____

3. List 3 everyday objects that are arranged in an array.

____

Name ______________________________

Practice It

Draw an array to find the product.

4. $4 \times 2 =$ ____

5. $3 \times 5 =$ ____

Write an addition sentence and a multiplication sentence to show equal rows.

6. ____ + ____ + ____ + ____ = ____

____ × ____ = ____

7. ____ + ____ + ____ = ____

____ × ____ = ____

8. ____ + ____ = ____

____ × ____ = ____

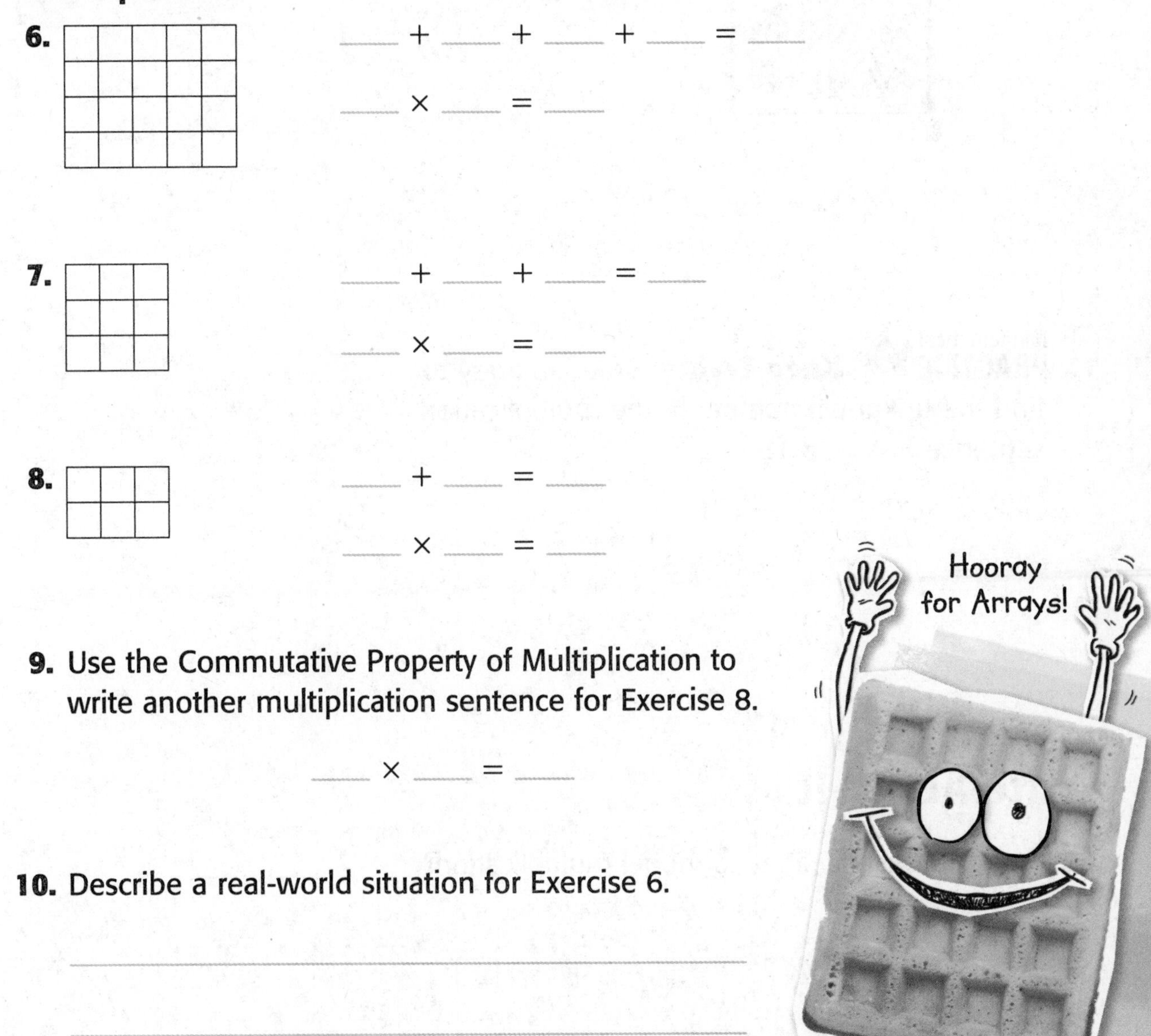

9. Use the Commutative Property of Multiplication to write another multiplication sentence for Exercise 8.

____ × ____ = ____

10. Describe a real-world situation for Exercise 6.

____

____

Apply It

My Work!

11. Marcos has 3 sheets of stickers. Each sheet has 4 stickers on it. Write a multiplication sentence to find how many stickers he has in all.

____ × ____ = ____ stickers

12. Circle the picture that does not represent an array. Explain.

____

____

13. Mathematical PRACTICE 1 **Keep Trying** Draw an array to find the unknown number in the multiplication sentence 6 × ■ = 18.

Write About It

14. How can I use an array to model multiplication?

____

____

Name ....

MY Homework

Lesson 3

Hands On: Multiply with Arrays

Homework Helper

eHelp Need help? connectED.mcgraw-hill.com

James discovered that a sheet of stamps is in an array. The stamps are arranged in 6 equal rows of 3.

Write an addition sentence to show equal rows.

$3 + 3 + 3 + 3 + 3 + 3 = 18$

Write a multiplication sentence to represent the array.

rows, number in each row, total

$6 \times 3 = 18$

James turns the sheet of stamps the other way. There still are 18 stamps. Only now, there are 3 equal rows of 6.

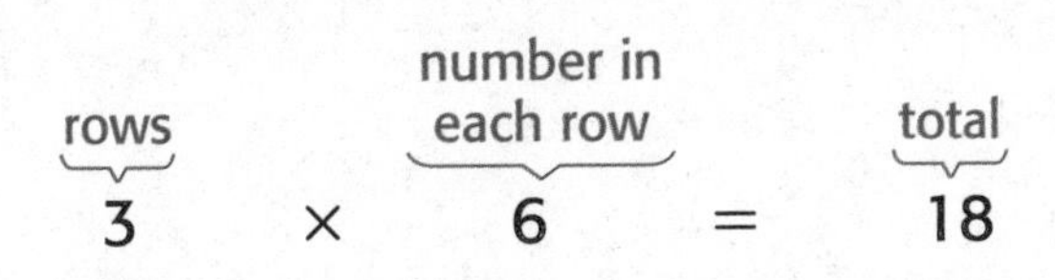

This is the Commutative Property of Multiplication.

Practice

Draw an array to find the product.

1. $5 \times 7 =$ ______

2. $6 \times 5 =$ ______

Write an addition sentence and a multiplication sentence to show equal rows.

3. 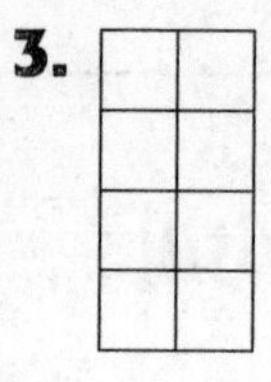

4. 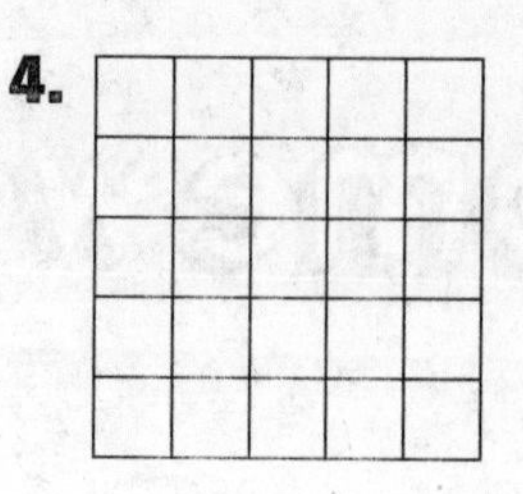

Vocabulary Check

5. Draw two arrays to model $2 \times 3 = 6$. Use the arrays to show the meaning of the Commutative Property of Multiplication.

Problem Solving

6. **Mathematical PRACTICE 4 Model Math** Suki's watercolor set has 3 rows of paint. There are 8 colors in each row. Write a multiplication sentence to find the total number of colors in the set.

7. A checkerboard has 8 rows, with 8 squares in each row. Write a multiplication sentence to find the total number of squares.

My Work!

Name ______________________________

Arrays and Multiplication

Lesson 4

ESSENTIAL QUESTION
What does multiplication mean?

An **array** is a group of objects arranged in equal numbered rows and equal numbered columns. Arrays can help you multiply.

Math in My World

Example 1

Mrs. Roberts baked a batch of bagels. She arranged the bagels in 3 equal rows of 4 bagels each on the cooling rack. How many bagels did she bake?

Find the total number of bagels. Use counters to model the array. Draw the array.

____ rows of ____ = ■

Find the unknown.

My Drawing!

You can use repeated addition or multiplication to find the unknown.

One Way **Add.** ____ + ____ + ____ = ____ ← addition sentence

Another Way **Multiply.** ____ × ____ = ____ ← multiplication sentence

So, ____ rows of ____ is ____ or ____ × ____ = ____.

The unknown is ____. Mrs. Roberts baked 12 bagels.

Example 2

One page of Elsa's photo album is shown. Write two multiplication sentences to find how many photos are on the page.

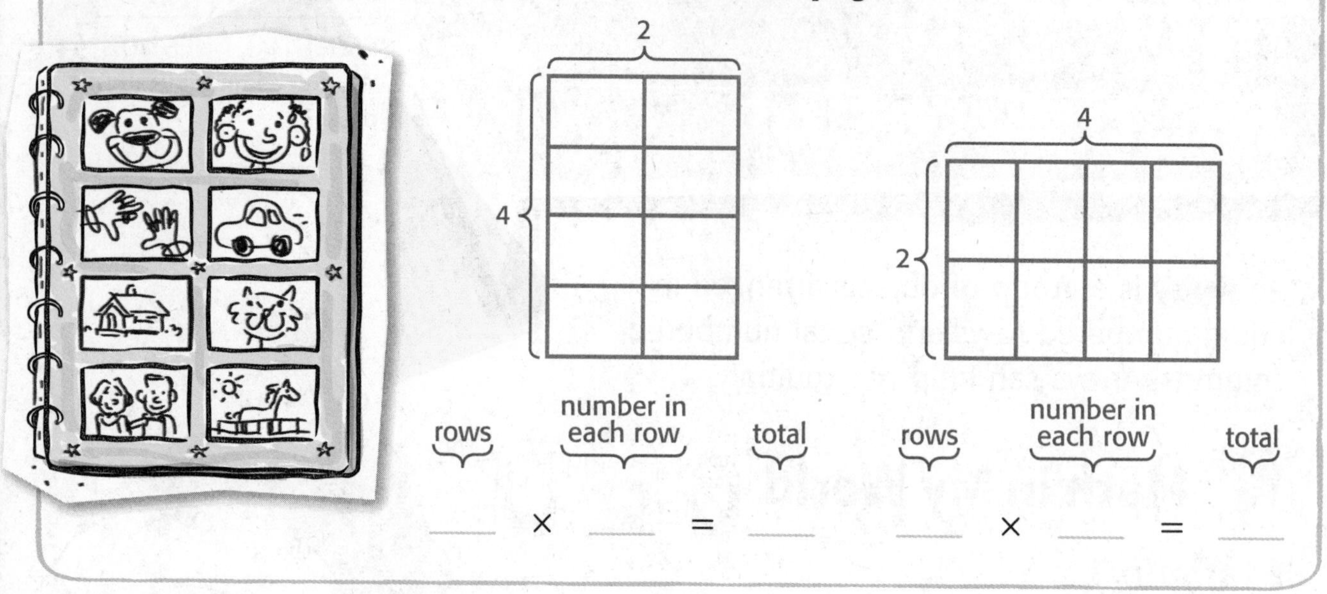

___ × ___ = ___   ___ × ___ = ___

Key Concept Commutative Property

Words	The **Commutative Property of Multiplication** says the order in which numbers are multiplied does not change the product.
Examples	4 × 2 = 8 (factor, factor, product) 2 × 4 = 8 (factor, factor, product)

Guided Practice

Talk MATH What other operation uses the Commutative Property? Explain.

Write an addition sentence and a multiplication sentence to show equal rows.

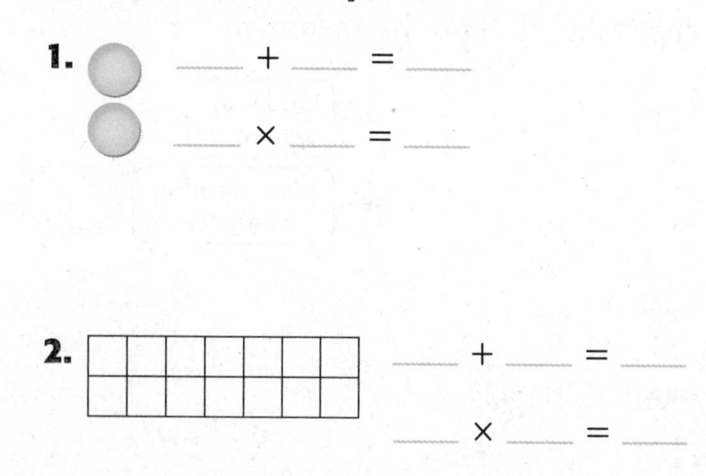

1. ___ + ___ = ___

___ × ___ = ___

2. ___ + ___ = ___

___ × ___ = ___

Name

Independent Practice

Write an addition sentence and a multiplication sentence to show equal rows.

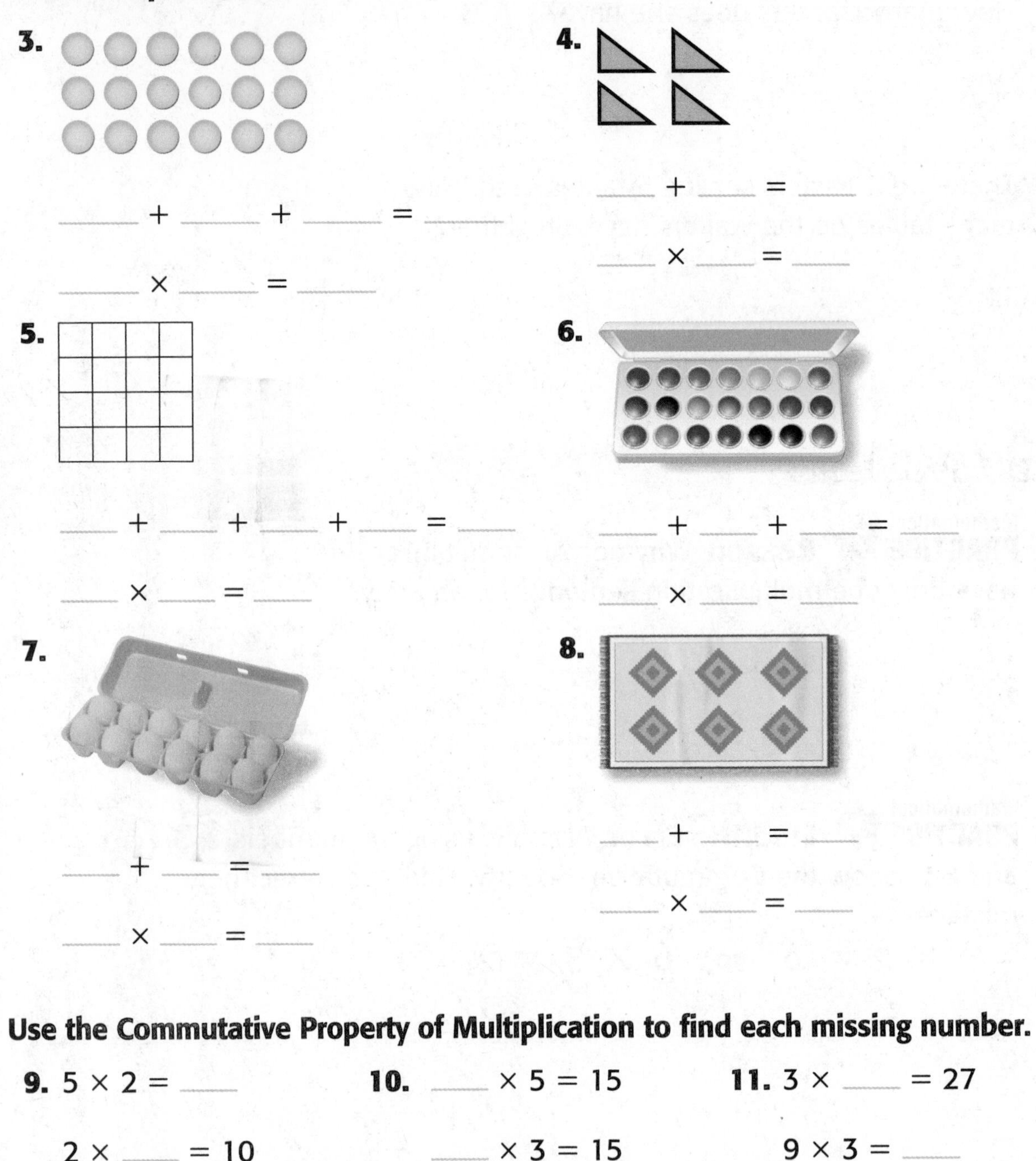

Use the Commutative Property of Multiplication to find each missing number.

9. $5 \times 2 =$ ____

$2 \times$ ____ $= 10$

10. ____ $\times 5 = 15$

____ $\times 3 = 15$

11. $3 \times$ ____ $= 27$

$9 \times 3 =$ ____

12. Hope drew the array at the right. Write a multiplication sentence to represent the model.

____ × ____ = ____

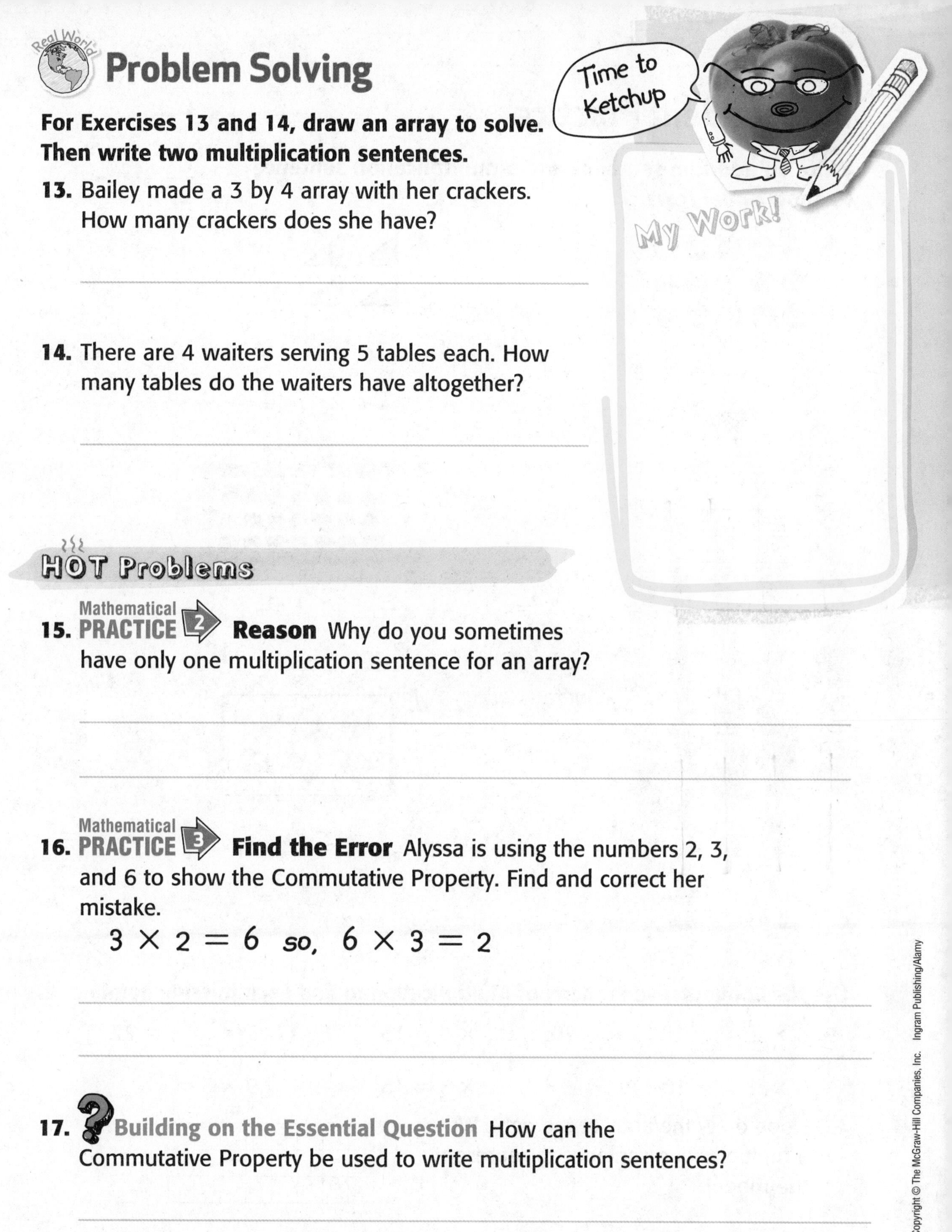

Problem Solving

For Exercises 13 and 14, draw an array to solve. Then write two multiplication sentences.

13. Bailey made a 3 by 4 array with her crackers. How many crackers does she have?

14. There are 4 waiters serving 5 tables each. How many tables do the waiters have altogether?

HOT Problems

15. Mathematical PRACTICE 2 **Reason** Why do you sometimes have only one multiplication sentence for an array?

16. Mathematical PRACTICE 3 **Find the Error** Alyssa is using the numbers 2, 3, and 6 to show the Commutative Property. Find and correct her mistake.

$3 \times 2 = 6$ *so,* $6 \times 3 = 2$

17. **Building on the Essential Question** How can the Commutative Property be used to write multiplication sentences?

Name ....

Real World

Problem-Solving Investigation

STRATEGY: Make a Table

Lesson 5

ESSENTIAL QUESTION
What does multiplication mean?

Learn the Strategy

Watch Tutor

Selma bought 3 shorts and 2 shirts. Laura, bought 4 shorts and 2 shirts. How many different shirt and shorts outfits can each girl make?

I ♥ Shopping!

1 Understand

What facts do you know?

____

What do you need to find?

How many different _______ and _______ outfits they each can make.

2 Plan

Organize the information in a table of columns and rows.

3 Solve

Make a table for each girl. List the possible shirt and shorts outfits.

Selma	Shirt 1	Shirt 2
Shorts A	A1	A2
Shorts B	B1	B2
Shorts C	C1	C2

Laura	Shirt 1	Shirt 2
Shorts A	A1	A2
Shorts B	B1	B2
Shorts C	C1	C2
Shorts D	D1	D2

So, Selma can make _______ outfits, and Laura can make _______.

4 Check

Does your answer make sense? Explain.

____

Practice the Strategy

How many lunches can Malia make if she chooses one main dish and one side dish?

1 Understand

What facts do you know?

____

____

What do you need to find?

____

____

2 Plan

____

3 Solve

4 Check

Does your answer make sense? Explain.

____

Name

Apply the Strategy

Solve each problem by making a table.

1. Trey can choose one type of bread and one type of meat for his sandwich. How many different sandwiches can Trey make?

	Turkey	Chicken
Wheat		
White		

Trey can make ______ sandwiches.

2. The students in Mr. Robb's class are designing a flag. The flag's background can be gold, red, or green. The flag can have either a blue or a purple stripe. Color all the possible flags.

	Blue	Purple
Gold		
Red		
Green		

They can design ______ flags.

My Work!

3. Mathematical PRACTICE 4 **Model Math** Tracy has a picture of her mom, a picture of her dad, and a picture of her dog. She has a black frame and a white frame. What is the question? Solve.

______

______

Review the Strategies

Use any strategy to solve each problem.

- Use the four-step plan.
- Determine reasonable answers.
- Make a table.

4. **Amber has coins in a jar. The sum of the coins is 13¢. What are the possible groups of coins Amber could have?**

5. Solana buys 2 bags of salad mix for $8 and 3 pounds of fresh vegetables for $9. She gives the cashier $20. How much change will she receive?

6. Mathematical PRACTICE 6 **Be Precise** Mr. Grow has 12 tomato plants arranged in 2 rows of 6. List 2 other ways that Mr. Grow could arrange his 12 tomato plants in equal rows. Explain to a classmate how you got your answer.

7. Mathematical PRACTICE 5 **Use Math Tools** One campsite has 3 tents with 5 people in each tent. Another campsite has 3 tents with 4 people in each. How many campers are there in all? Draw an array to solve.

Name ....

Lesson 5
Problem Solving: Make a Table

Homework Helper

eHelp Need help? connectED.mcgraw-hill.com

Jane's new bike can have hand brakes or foot brakes. The bike can be silver, blue, black, or purple. How many possible bikes are there?

1 Understand

There are 2 types of brakes: hand brakes or foot brakes.
There are 4 color choices: silver, blue, black, or purple.

I need to find the number of possible bikes.

2 Plan

Make a table.

	Silver	Blue	Black	Purple
Hand brakes	Hand/ Silver	Hand/Blue	Hand/ Black	Hand/ Purple
Foot brakes	Foot/ Silver	Foot/Blue	Foot/ Black	Foot/ Purple

3 Solve

There are 8 possible bikes.

4 Check

Multiply 2 types of brakes by 4 color choices. $4 \times 2 = 8$

Problem Solving

Real World

1. Solve the problem by making a table.

Claudio will decorate his bedroom. He can choose tan, blue, or gray paint and striped or plaid curtains. How many ways can he decorate his room with different paint and curtains?

	tan, (t)	blue, (b)	gray, (g)
striped, (s)			
plaid, (p)			

Solve each problem by making a table.

2. Jimmy has a number cube labeled 1 through 6, and a penny. How many different ways can the cube and penny land with one roll of the cube and one flip of the penny?

	1	2	3	4	5	6
heads, (h)						
tails, (t)						

____

3. Archie earns $4 each week for doing his chores. How much money will Archie earn in 2 months if there are 4 weeks in each month?

	Week 1	Week 2	Week 3	Week 4
Month 1				
Month 2				

____

4. Mathematical PRACTICE 7 **Identify Structure** Abigail has a green, yellow, and purple shirt to match with either a white, black, or red pair of pants. How many different shirt and pants outfits can she make?

	pants, (w)	pants, (b)	pants, (r)
shirt, (g)			
shirt, (y)			
shirt, (p)			

____

How many outfits would be possible if Abigail had only 2 shirts and 2 pair of pants? Explain.

____

Name ....

Use Multiplication to Find Combinations

Lesson 6

ESSENTIAL QUESTION
What does multiplication mean?

When you make a **combination**, you make a new set that has one item from each group of items.

Math in My World

Example 1

Amos' team has 3 jersey colors: green, red, and yellow. They can wear orange or black shorts. Find all of the jersey and short combinations for the team.

1. Color the first jersey green, the second one red, and the last yellow.

2. Color 1 pair of shorts orange and 1 pair of shorts black below each jersey.

Combinations	1	2		4	5	
Jersey Colors	GREEN	GREEN	RED	RED		
Shorts Colors	ORANGE	BLACK	ORANGE		ORANGE	

Write a multiplication sentence.

_______ × _______ = ■ ← Find the unknown.
jersey colors   shorts colors   combinations

_______ × _______ = _______

So, there are _______ jersey and shorts combinations possible.

Another way to find combinations is to use a tree diagram. A **tree diagram** uses "branches" to show all possible combinations.

Example 2

Tutor

What are all the possible fruit sorbet combinations if you choose one flavor and one fruit to add in?

Complete the tree diagram.

Flavors	Fruits	Combinations
 mango	banana	mango, ______
	berries	mango, berries
	peach	______, peach
 strawberry	banana	______, banana
	berries	______, berries
	peach	______, ______
 vanilla	banana	______, ______
	berries	______, ______
	peach	______, ______

Check Multiply to find the number of possible combinations.

______ flavors × ______ fruits = ______ fruit sorbet combinations

So, there are ______ possible fruit sorbet combinations.

Guided Practice

Check

1. Refer to Example 2. How would the possible number of combinations change if one more flavor was added? Write the multiplication sentence.

______ × ______ = ______

Talk MATH

Explain how a tree diagram helps you find all the possible combinations without repeating any.

Name

Independent Practice

Find all the possible combinations. Write a multiplication sentence to check.

2. Jackie is playing a card game with triangles and circles. Each shape can be blue, red, yellow, or green. How many different cards are there?

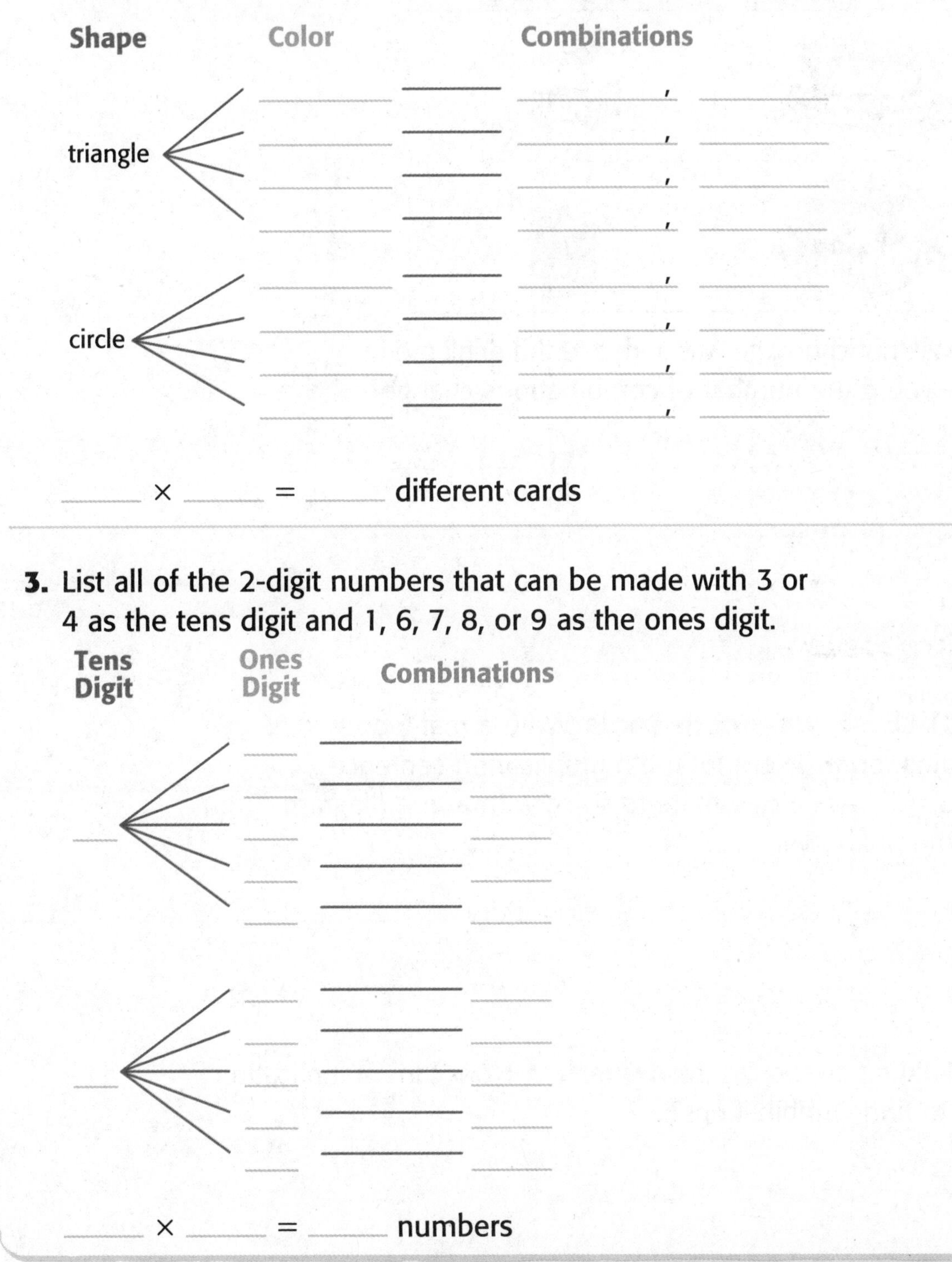

3. List all of the 2-digit numbers that can be made with 3 or 4 as the tens digit and 1, 6, 7, 8, or 9 as the ones digit.

Write a multiplication sentence to solve the problem.

4. Madison needs to choose 1 breakfast item and 1 drink. Find the number of possible combinations.

_____ × _____ = _____ combinations

Breakfast Menu	Drink Menu
	MILK MILK
	Juice

Suppose hot chocolate was added to the drink menu. How would the number of combinations change?

____

____

My Work!

HOT Problems

5. Mathematical PRACTICE 5 **Use Math Tools** Write a real-world combination problem for the multiplication sentence $4 \times 2 = \blacksquare$. Ask a classmate to solve with a tree diagram. Then find the unknown.

____

____

6. **Building on the Essential Question** How can multiplication help to find combinations?

____

____

Name ....

MY Homework

Lesson 6
Use Multiplication to Find Combinations

Homework Helper

Need help? connectED.mcgraw-hill.com

Lucia's three dogs have red, purple, blue, green, and orange collars that they take turns wearing. Find the number of possible dog and collar combinations.

Show all of the possible combinations.

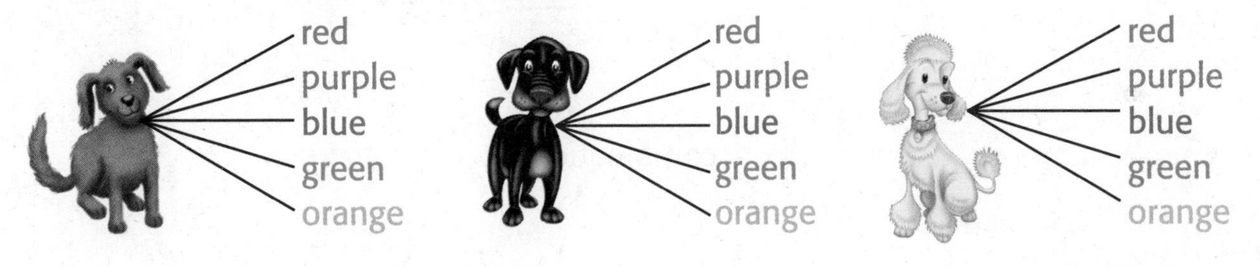

There are 3 dogs and 5 collar colors.
$3 \times 5 = 15$ possible combinations

Practice

1. Diana can take 1 pencil and 1 eraser to school. Her choices are shown. How many different pencil and eraser combinations are there? Complete the tree diagram. Write a multiplication sentence.

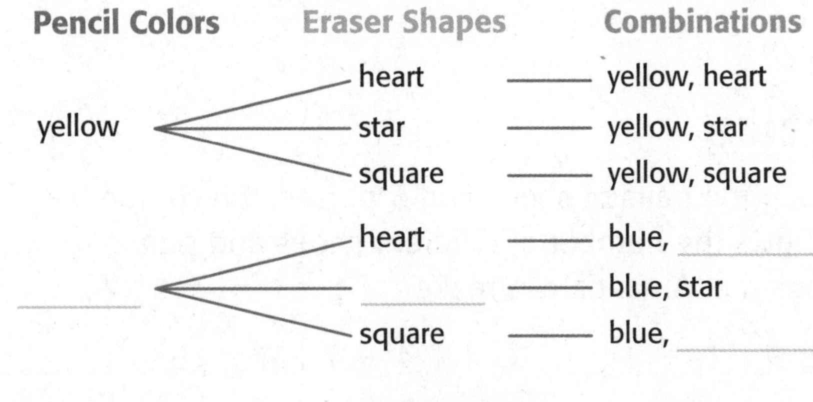

______ × ______ = ______ combinations

2. Mathematical PRACTICE 7 **Identify Structure** For a snack, Randy can choose from peanuts, carrots, or popcorn. He can have water or juice to drink. How many snack and drink combinations are there? Complete the tree diagram. Write a multiplication sentence.

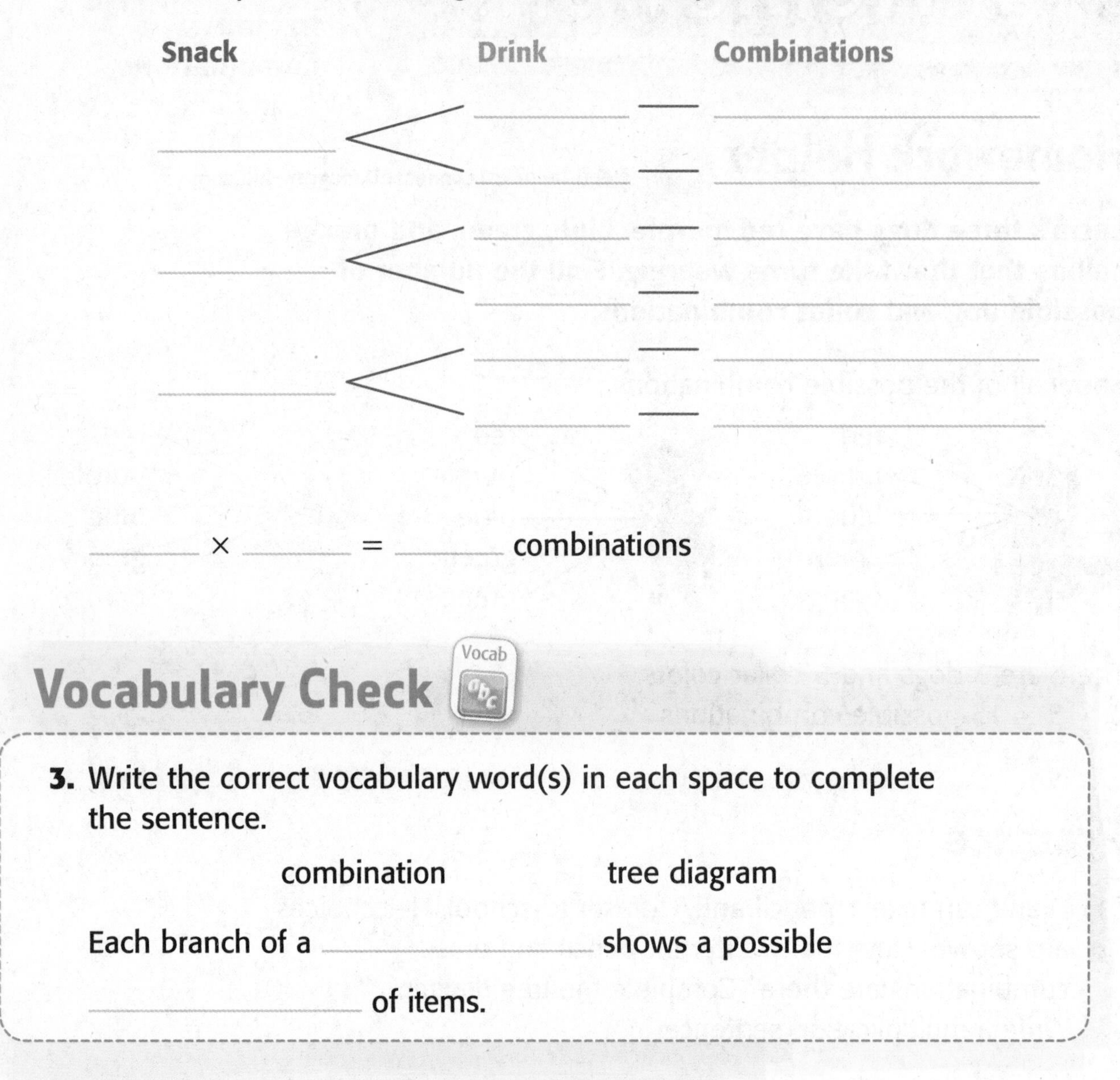

_____ × _____ = _____ combinations

Vocabulary Check

3. Write the correct vocabulary word(s) in each space to complete the sentence.

combination tree diagram

Each branch of a _____ shows a possible _____ of items.

Test Practice

4. Amanda bought 4 pairs of shoes and 5 purses. Which number sentence shows the number of different shoes and purse combinations that Amanda can make?

Ⓐ $4 + 5 = 9$
Ⓑ $5 \times 8 = 40$
Ⓒ $4 + 4 + 4 + 4 = 16$
Ⓓ $4 \times 5 = 20$

Review

Chapter 4

Understand Multiplication

Vocabulary Check

Vocab

Use the word bank to complete each sentence.

array	**combination**	**Commutative Property**
equal groups	**factor**	**multiplication**
multiplication sentence	**multiply**	**product**
tree diagram		

1. An arrangement of objects into rows of equal length and columns of equal length is a(n) ________.

2. The answer to a multiplication problem is the ________.

3. ________ is the operation of two numbers that can be thought of as repeated addition.

4. A number multiplied by another number is a ________.

5. You can put equal groups together to ________.

6. The ________ says the order in which numbers are multiplied does not change the product.

7. A ________ uses "branches" to show all possible combinations.

8. When you make a ________ of items, you make a new set that has one item from each group.

9. When you have ________, you have the same number of objects in each group.

Concept Check

Write an addition and a multiplication sentence to show equal rows.

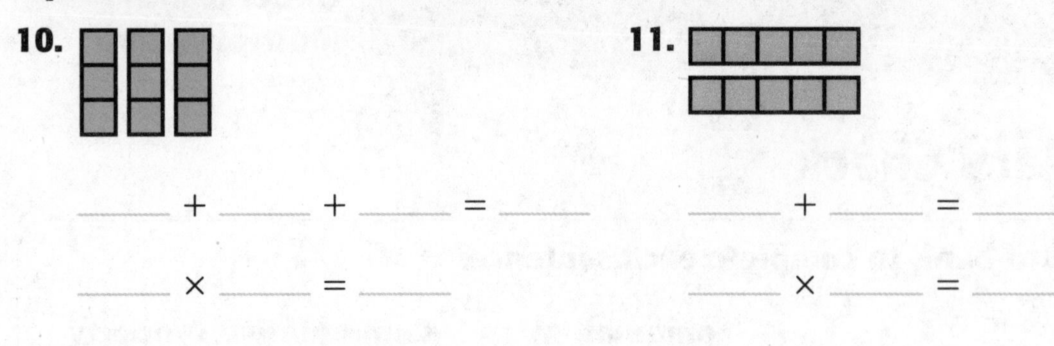

10.

_____ + _____ + _____ = _____

_____ × _____ = _____

11.

_____ + _____ = _____

_____ × _____ = _____

Write two multiplication sentences for each array.

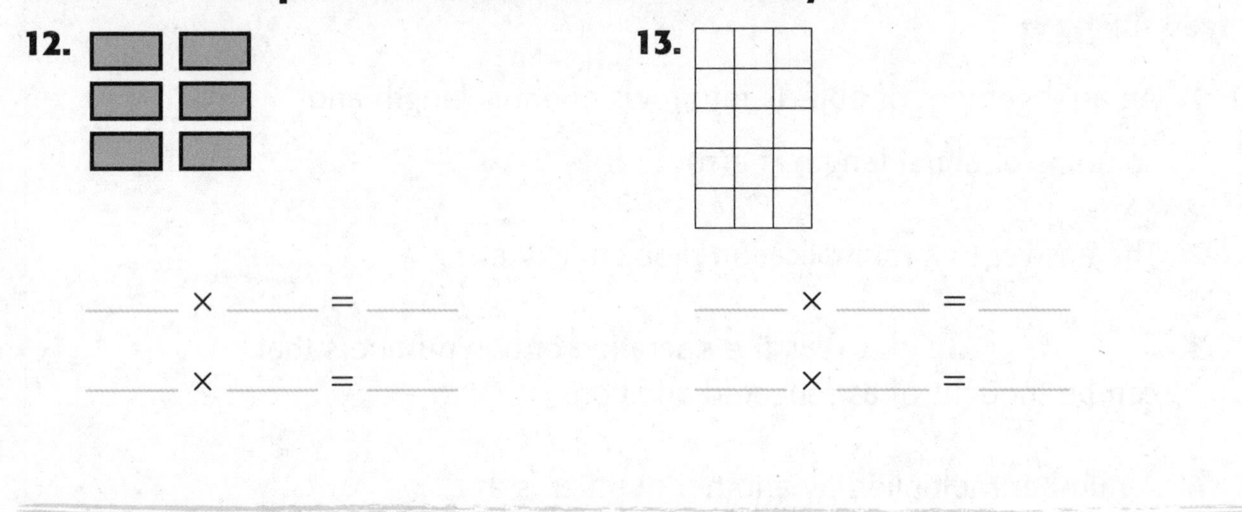

12.

_____ × _____ = _____

_____ × _____ = _____

13.

_____ × _____ = _____

_____ × _____ = _____

14. Find the possible combinations of one yogurt and one topping. Complete the tree diagram. Write a multiplication sentence to check.

Sundaes	
Yogurt	**Toppings**
strawberry	granola
peach	strawberries
vanilla	

_____ × _____ = _____ combinations

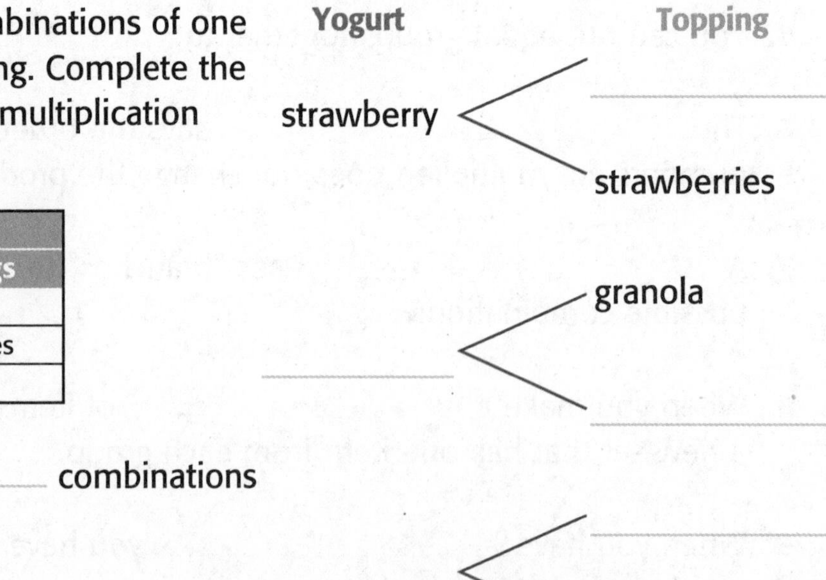

Name

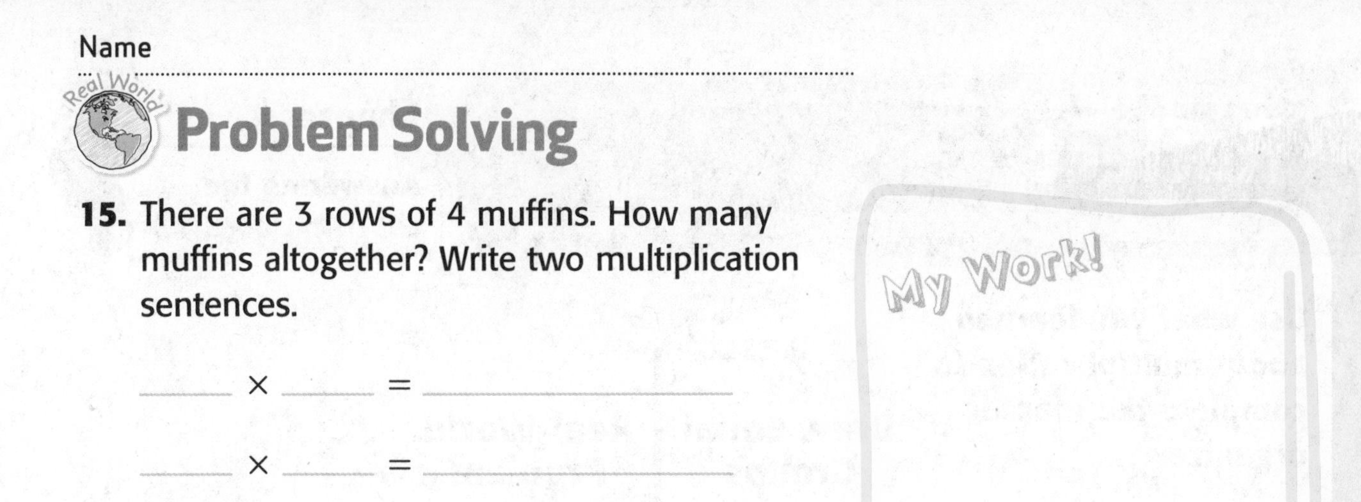

Problem Solving

15. There are 3 rows of 4 muffins. How many muffins altogether? Write two multiplication sentences.

_____ × _____ = __________

_____ × _____ = __________

My Work!

16. Toya finishes reading a book every 3 days. How many days does it take her to read 7 books? Complete the table to solve.

Days	Books
3	1
6	
12	
	5
18	6
21	

So, it takes her _____ days to read 7 books.

Test Practice

17. Timmy downloaded 5 songs each day for five days. How many songs did he download during the 5 days altogether?

Ⓐ 5 songs

Ⓑ 10 songs

Ⓒ 25 songs

Ⓓ 35 songs

Reflect

Chapter 4

Answering the ESSENTIAL QUESTION

Use what you learned about multiplication to complete the graphic organizer.

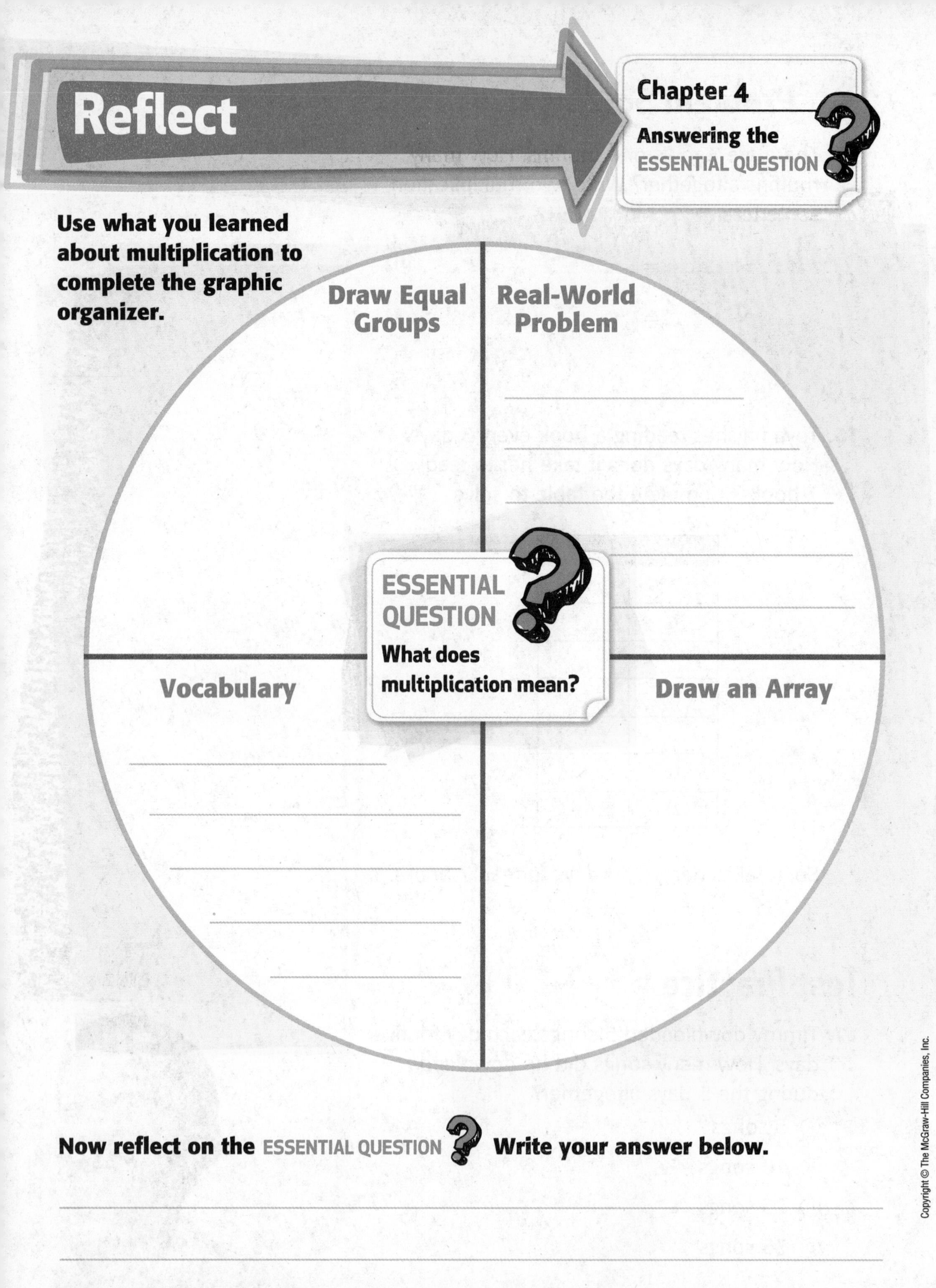

Now reflect on the ESSENTIAL QUESTION Write your answer below.

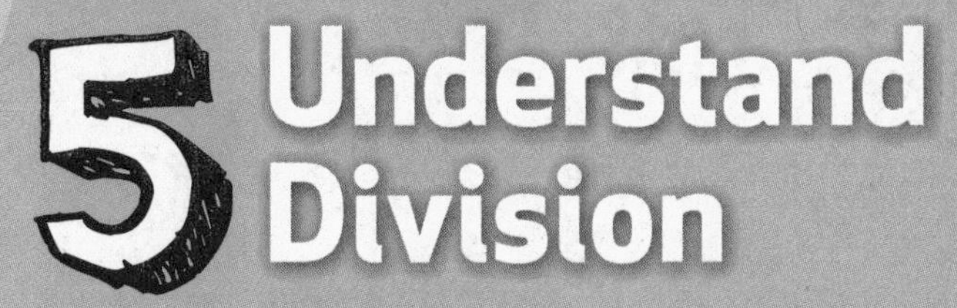

Chapter 5 Understand Division

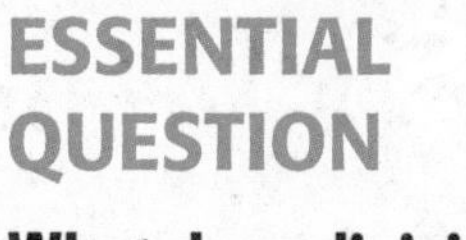

ESSENTIAL QUESTION

What does division mean?

MY Common Core State Standards

Operations and Algebraic Thinking

CCSS

3.OA.2 Interpret whole-number quotients of whole numbers, e.g., interpret $56 \div 8$ as the number of objects in each share when 56 objects are partitioned equally into 8 shares, or as a number of shares when 56 objects are partitioned into equal shares of 8 objects each.

3.OA.4 Determine the unknown whole number in a multiplication or division equation relating three whole numbers.

3.OA.6 Understand division as an unknown-factor problem.

3.OA.7 Fluently multiply and divide within 100, using strategies such as the relationship between multiplication and division (e.g., knowing that $8 \times 5 = 40$, one knows $40 \div 5 = 8$) or properties of operations. By the end of Grade 3, know from memory all products of two one-digit numbers.

Hey, I already know some of these!

Standards for Mathematical PRACTICE

1. Make sense of problems and persevere in solving them.
2. Reason abstractly and quantitatively.
3. Construct viable arguments and critique the reasoning of others.
4. Model with mathematics.
5. Use appropriate tools strategically.
6. Attend to precision.
7. Look for and make use of structure.
8. Look for and express regularity in repeated reasoning.

= focused on in this chapter

Name

Am I Ready?

Check — Go online to take the Readiness Quiz

Write two multiplication sentences for each array.

1.

2.

Identify a pattern. Then find the missing numbers.

3. 30, 25, 20, _____, _____, 5 Pattern: _____

4. 12, _____, 8, _____, 4, 2 Pattern: _____

5. 55, 45, 35, _____, 15, _____ Pattern: _____

Draw the counters in the circles to make equal groups.

6.

7.

8. Colton made 15 party invitations. His brother made 9 invitations. Write a subtraction sentence to find how many more invitations Colton made.

9. Mrs. Jones has 21 pencils. She gives 2 pencils to Carter and 2 pencils to Mandy. Write a subtraction sentence to find how many pencils she has left.

How Did I Do?

Shade the boxes to show the problems you answered correctly.

1	2	3	4	5	6	7	8	9

Name

Review Vocabulary

array equal groups pattern repeated addition

Making Connections

Use the review vocabulary to describe each example in the graphic organizer.

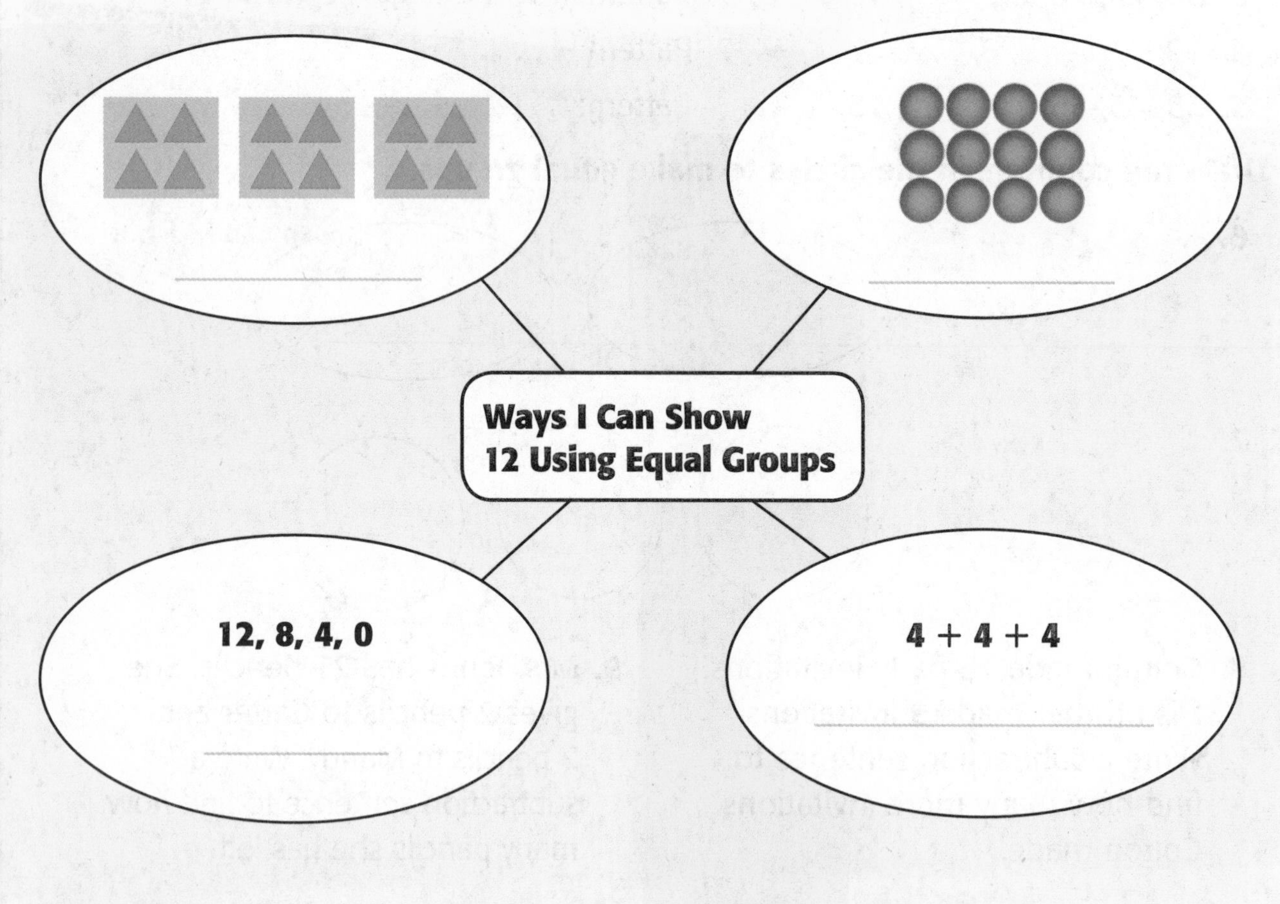

How are the examples similar? How are the examples different?

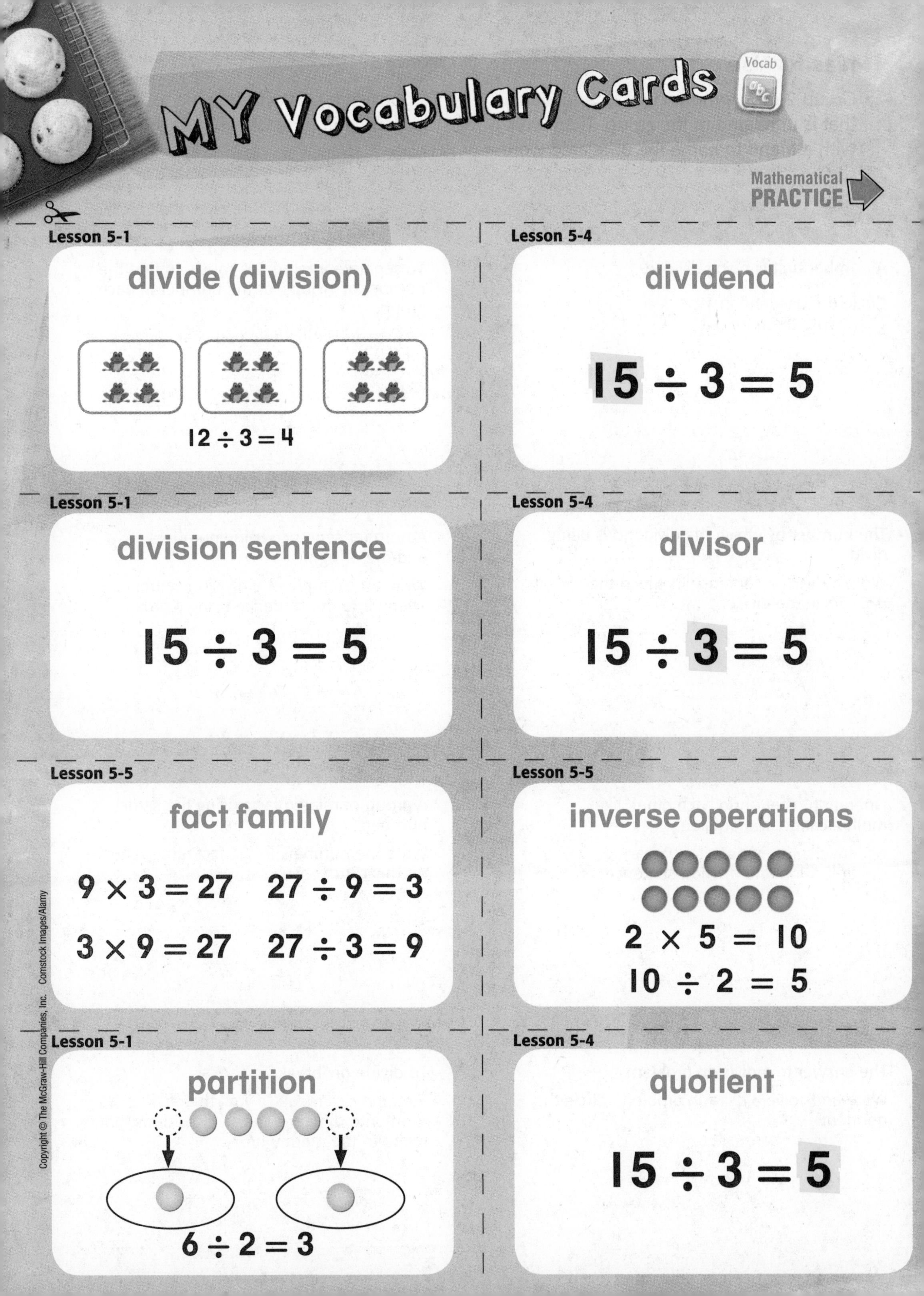

MY Vocabulary Cards

Vocab

Mathematical PRACTICE

Lesson 5-1

divide (division)

$12 \div 3 = 4$

Lesson 5-4

dividend

$15 \div 3 = 5$

Lesson 5-1

division sentence

$15 \div 3 = 5$

Lesson 5-4

divisor

$15 \div 3 = 5$

Lesson 5-5

fact family

$9 \times 3 = 27$ $27 \div 9 = 3$

$3 \times 9 = 27$ $27 \div 3 = 9$

Lesson 5-5

inverse operations

$2 \times 5 = 10$

$10 \div 2 = 5$

Lesson 5-1

partition

$6 \div 2 = 3$

Lesson 5-4

quotient

$15 \div 3 = 5$

Ideas for Use

- Group 2 or 3 common words. Add a word that is unrelated to the group. Then work with a friend to name the unrelated word.
- Design a crossword puzzle. Use the definition for each word as the clues.

A number that is being divided.

Circle the dividend in $12 \div 4 = \blacksquare$.
Then write the quotient.

To separate into equal groups, to find the number of groups, or the number in each group.

How can dividing help you share snacks with friends?

The number by which the dividend is being divided.

Write a division sentence in which the divisor is 5. Circle the divisor.

A number sentence using numbers and the ÷ sign.

Write an example of a division sentence. Then write the sentence using words.

Operations that undo each other, like multiplication and division.

Use inverse operations to write a multiplication and division sentence.

A group of related facts using the same numbers.

Write the numbers in the fact family shown on this card.

The answer to a division problem.

Write and solve a division problem. Circle the quotient.

To divide or "break up."

Partition can mean "a wall that divides a room into different areas." How does that relate to the math word?

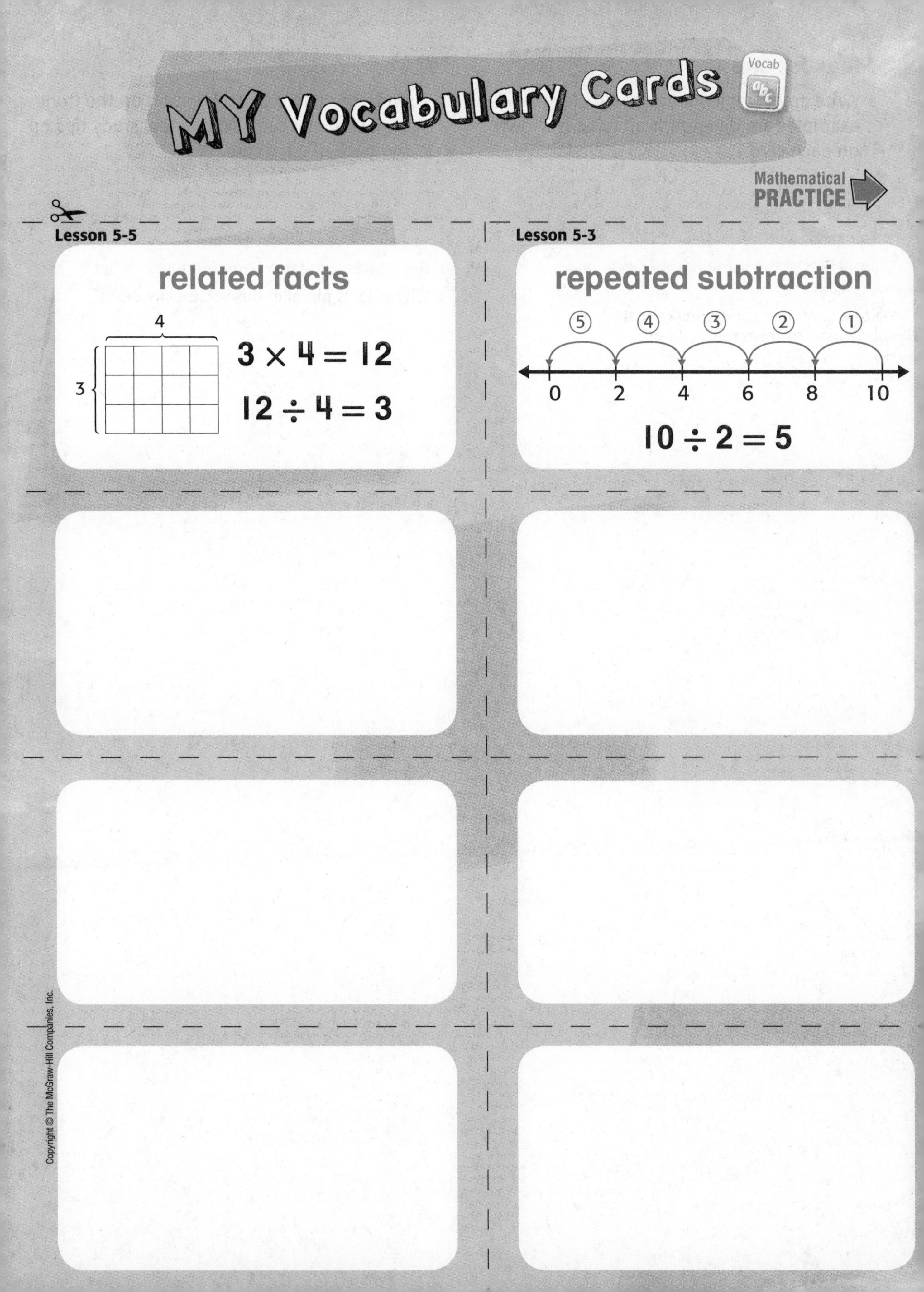
MY Vocabulary Cards
Vocab
abc
Mathematical PRACTICE
Lesson 5-5
related facts
4
3
3 × 4 = 12
12 ÷ 4 = 3
Lesson 5-3
repeated subtraction
5
4
3
2
1
0
2
4
6
8
10
10 ÷ 2 = 5

Ideas for Use

- Write an example for each card. Be sure your examples are different from what is shown on each card.
- Write the name of each lesson on the front of each blank card. Write a few study tips on the back of each card.

Subtraction of the same number over and over again.

Write a sentence comparing repeated subtraction to repeated addition.

Basic facts using the same numbers.

Why is *facts* plural in this vocabulary word?

MY Foldable

FOLDABLES® Follow the steps on the back to make your Foldable.

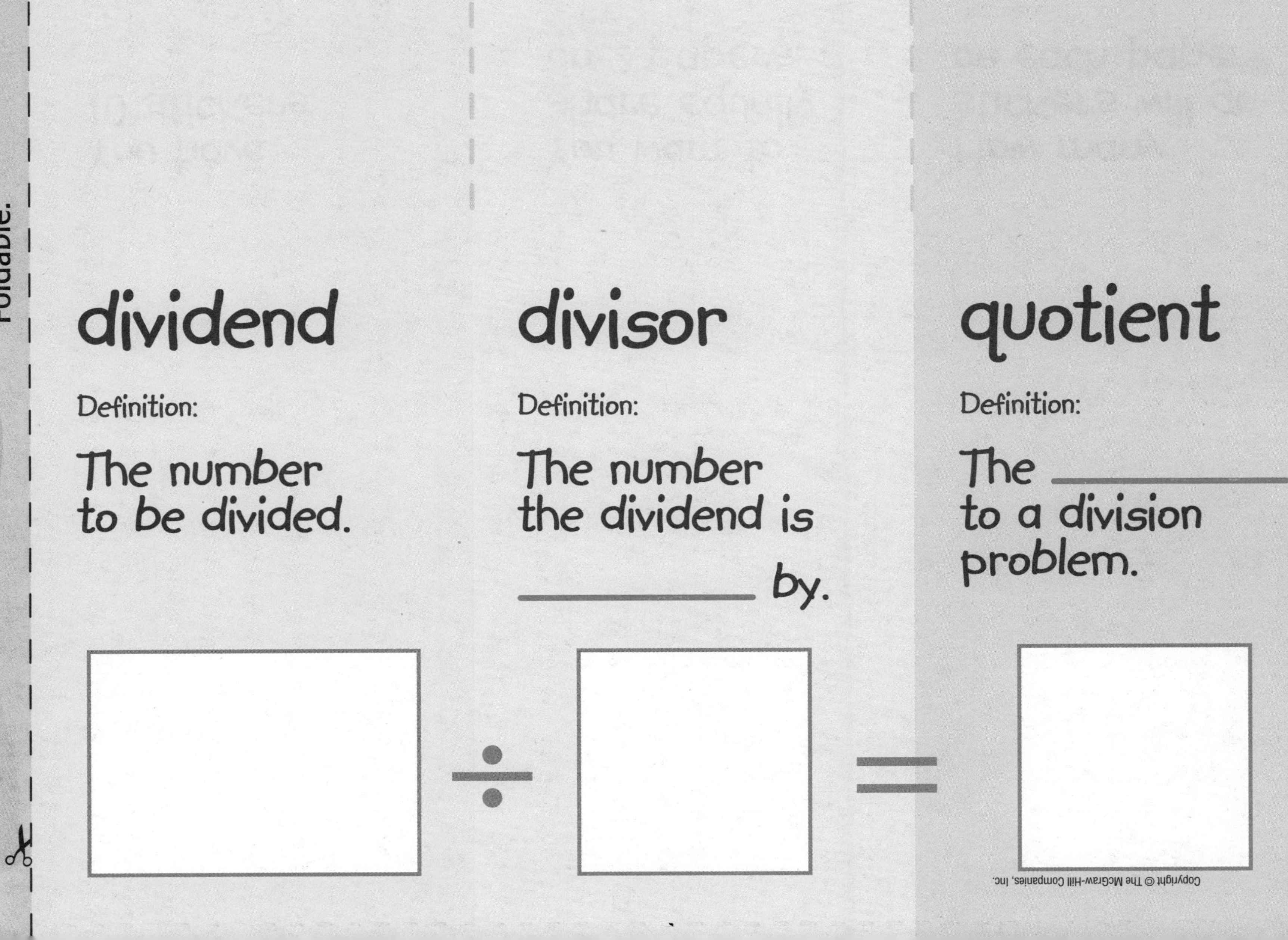

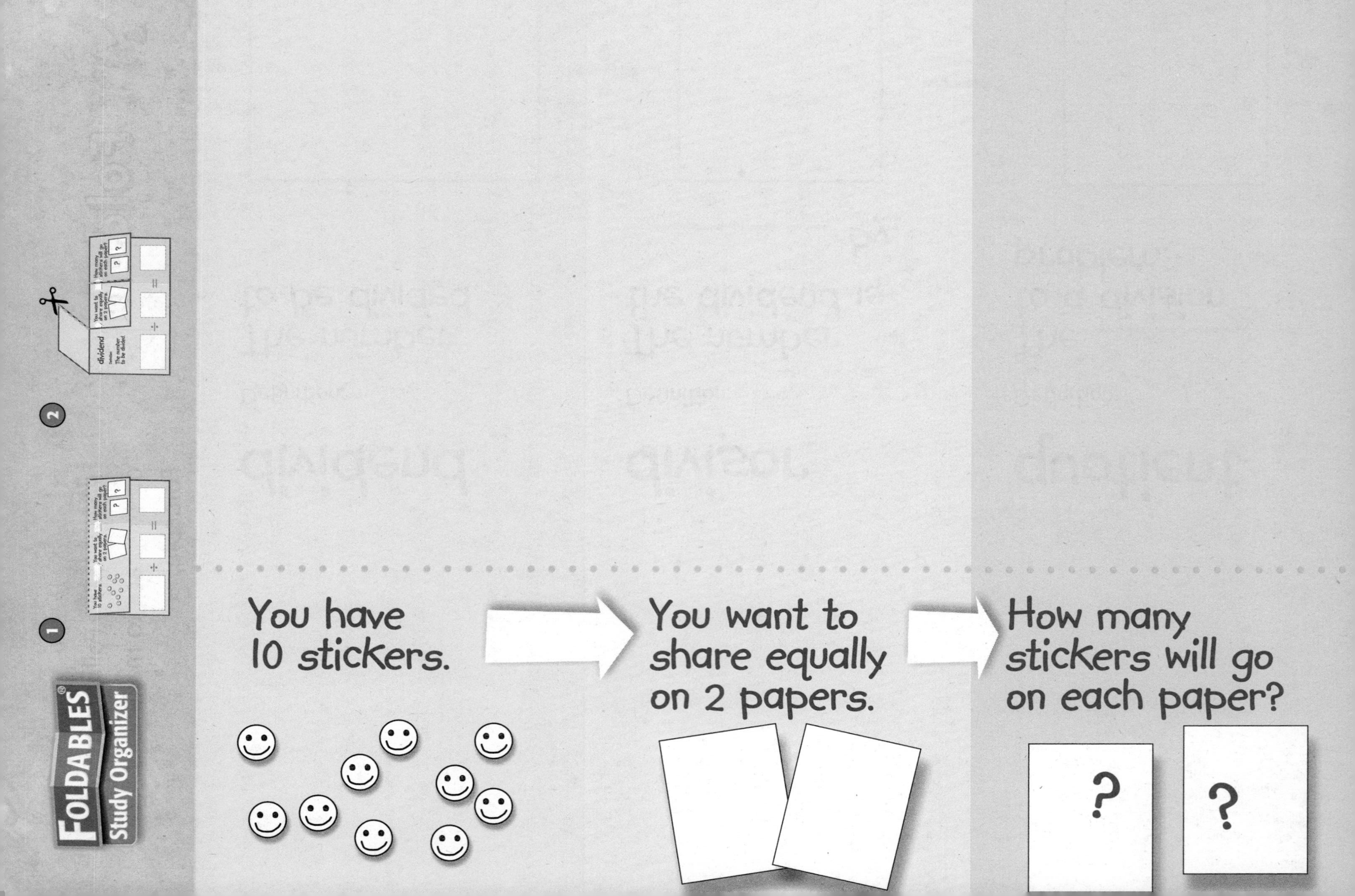
FOLDABLES
Study Organizer
You have 10 stickers.
You want to share equally on 2 papers.
How many stickers will go on each paper?
?
?
dividend
Definition
The number to be divided.
Copyright © The McGraw-Hill Companies, Inc.

Name ....

Hands On
Model Division

Lesson 1

ESSENTIAL QUESTION
What does division mean?

Division is an operation with two numbers. One number tells you how many items you have. The other tells you how many equal shares, or groups, to form or how many to put in each group.

10 ÷ 5 = 2 ← Read ÷ as *divided by.* 10 divided by 5 = 2.

To **divide** means to **partition,** or separate a number into equal groups, to find the number of groups, or find the number in each group.

Build It

Tools

Find how many in each group. Divide 12 counters into 3 equal groups. How many are in each group?

1. Partition one counter at a time into a group until all of the counters are gone.

2. Draw the groups of counters.

3. Write a **division sentence,** or a number sentence that uses division.

12 counters were divided into

_______ groups.

There are _______ counters in each group.

So, 12 ÷ 3 = _______ in each group. ← SAY: Twelve divided by three equals four.

My Drawing!

Try It

Find how many groups. Place 12 counters in groups of 3. How many groups are there?

Make groups of 3 until all the counters are gone. Draw the groups.

My Drawing!

Write a division sentence. 12 counters were

divided into equal groups of ______.

There are ______ groups.

$12 \div 3 =$ ______ groups. ← SAY: Twelve divided by three equals four.

Talk About It

1. Explain how you divided 12 counters into equal groups.

______

______

2. When you divided the counters into groups of 3, how did you find the number of equal groups?

______

______

3. Mathematical PRACTICE 3 **Draw a Conclusion** Explain the difference between the way you partitioned the counters in the first activity to the way you partitioned them in the second activity.

______

______

______

Name

Practice It

4. Partition 8 counters one at a time to find the number of counters in each group. Draw the counters.

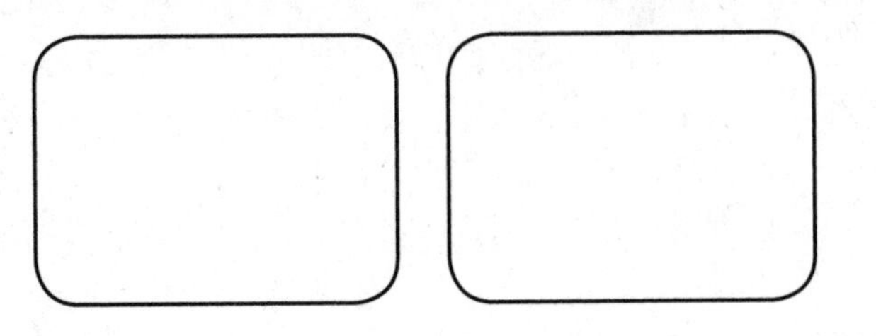

There are _____ counters in each group; 8 ÷ 2 = _____.

5. Circle equal groups of 5 to find the number of equal groups.

There are _____ equal groups; 15 ÷ _____ = 5.

6. Algebra Use counters to find each unknown.

Number of Counters	Number of Equal Groups	Number in Each Group	Division Sentence
9	■	3	9 ÷ ■ = 3
14	2	?	14 ÷ 2 = ?
15	■	5	15 ÷ ■ = 5
6	?	3	6 ÷ ? = 3

7. Choose one division sentence from Exercise 6. Write and solve a real-world problem for that number sentence.

Apply It

Draw a model to solve. Then write a number sentence.

8. A florist needs to make 5 equal-sized bouquets from 25 flowers. How many flowers will be in each bouquet?

My Drawing!

9. Mathematical PRACTICE 4 **Model Math** Mrs. Wilson called the flower shop to place an order for 9 flowers. She wants an equal number of roses, daisies, and tulips. How many of each kind of flower will Mrs. Wilson receive?

10. Mathematical PRACTICE 1 **Make a Plan** Mr. Cutler bought 2 dozen roses to equally arrange in 4 vases. How many roses will he put in each vase? (*Hint:* 1 dozen = 12)

11. Mathematical PRACTICE 2 **Reason** Can 13 counters be partitioned equally into groups of 3? Explain.

Write About It

12. How can I use models to understand division?

Name

Division as Equal Sharing

Lesson 2

ESSENTIAL QUESTION
What does division mean?

One way to **divide** is to find the number in each group. This can be done by equal sharing.

Eat your veggies!

Math in My World

Tools Watch Tutor

Example 1

Nolan feeds 6 carrots equally to 3 rabbits. How many carrots does each rabbit get?

Draw one carrot at a time next to each rabbit until there are no more carrots.

Write a division sentence to represent the problem. A **division sentence** is a number sentence that uses the operation of division.

_____ carrots equally shared by _____ rabbits gives _____ carrots to each rabbit.

$6 \div 3 =$ _____ There are _____ carrots for each rabbit.

You can think of division sentences in two ways.

6 items 3 equal groups 2 items in each group	→ $6 \div 3 = 2$ ←	6 items 2 equal groups 3 items in each group

You can draw an array to help you divide.

Example 2

Fifteen scouts equally shared 3 tents. How many scouts are in each tent? Place one counter (scout) at a time next to each tent until all the counters are gone. Draw a sketch of your counters.

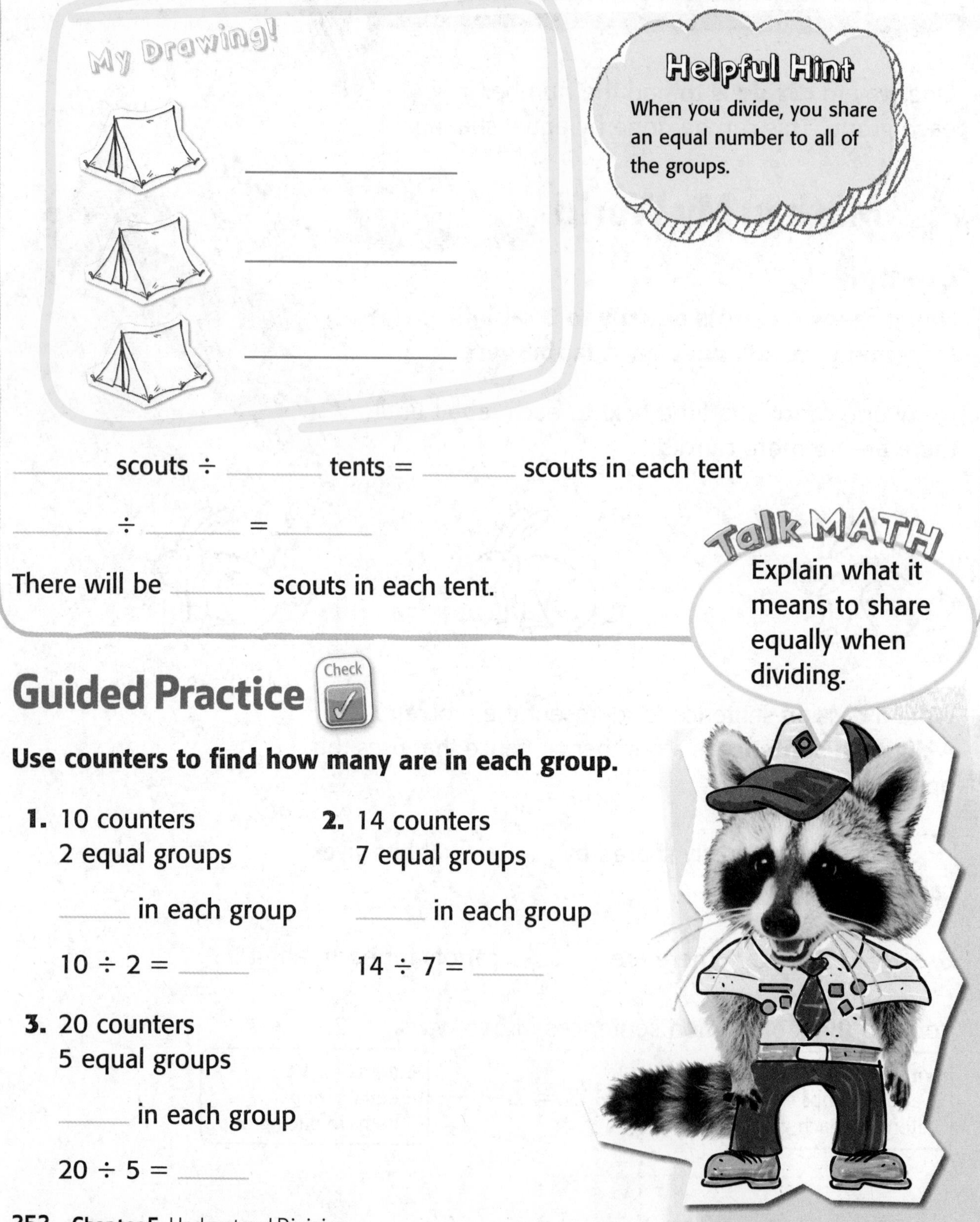

_____ scouts ÷ _____ tents = _____ scouts in each tent

_____ ÷ _____ = _____

There will be _____ scouts in each tent.

Guided Practice

Check

Use counters to find how many are in each group.

1. 10 counters
2 equal groups

_____ in each group

$10 \div 2 =$ _____

2. 14 counters
7 equal groups

_____ in each group

$14 \div 7 =$ _____

3. 20 counters
5 equal groups

_____ in each group

$20 \div 5 =$ _____

Name ....

Independent Practice

Use counters to find how many are in each group.

4. 12 counters
2 equal groups

____ in each group

____ ÷ ____ = ____

5. 16 counters
4 equal groups

____ in each group

____ ÷ ____ = ____

6. 18 counters
6 equal groups

____ in each group

____ ÷ ____ = ____

Use counters to find the number of equal groups.

7. 8 counters

____ equal groups

4 in each group

8 ÷ ____ = 4

8. 21 counters

____ equal groups

7 in each group

21 ÷ ____ = 7

9. 18 counters

____ equal groups

9 in each group

18 ÷ ____ = 9

Use counters to draw an array. Write a division sentence.

10. Draw 9 counters in 3 equal rows.

There are ____ in each row.

____ ÷ ____ = ____

11. Draw 14 counters in 2 equal rows.

There are ____ in each row.

____ ÷ ____ = ____

My Drawing!

Algebra Draw lines to match each division sentence with its correct unknown.

12. 24 ÷ ■ = 3 • 5

13. 30 ÷ 6 = ■ • 7

14. 42 ÷ ■ = 6 • 8

Problem Solving

Draw a picture to solve. Then write a division sentence.

15. Marla has $25. How many hamster wheels can she buy?

16. A seamstress needs 18 feet of fabric. How many yards of fabric does she need? (*Hint:* 1 yard = 3 feet)

17. Mathematical PRACTICE 1 **Plan Your Solution** There are 6 juice boxes in a package. How many packages need to be bought if 24 juice boxes are needed for a picnic? Write a division sentence with a symbol for the unknown. Then solve.

My Drawing!

HOT Problems

18. Mathematical PRACTICE 4 **Model Math** Write a real-world problem that uses the division sentence 12 ÷ 6 = ■. Then find the unknown.

19. **Building on the Essential Question** How is dividing like sharing?

Name

Relate Division and Subtraction

Lesson 3
ESSENTIAL QUESTION
What does division mean?

Math in My World

Tools Watch Tutor

Example 1

A designer makes 15 dresses in equal numbers of red, blue, or yellow. How many dresses of each color are there? Write a division sentence with a symbol for the unknown. Then solve.

15 ÷ 3 = ■ ← unknown

One Way **Use models.**

Draw one counter at a time on each dress until all 15 counters are gone.

There are ____ dresses of each color. The unknown is ____.

So, ____ ÷ ____ = ____.

Another Way **Use a number line.**

You can also divide using **repeated subtraction.** Subtract equ groups of 3 repeatedly until you get to zero.

(5) (4) (3) (2) (1)

0 1 2 3 4 5 6 7 8 9 10 11 12 13 14 15 ← F 15 ÷ 3, start at 15.

15 ÷ 3 = 5

You subtracted groups of three ____ times.

So, 15 ÷ 3 = ____.

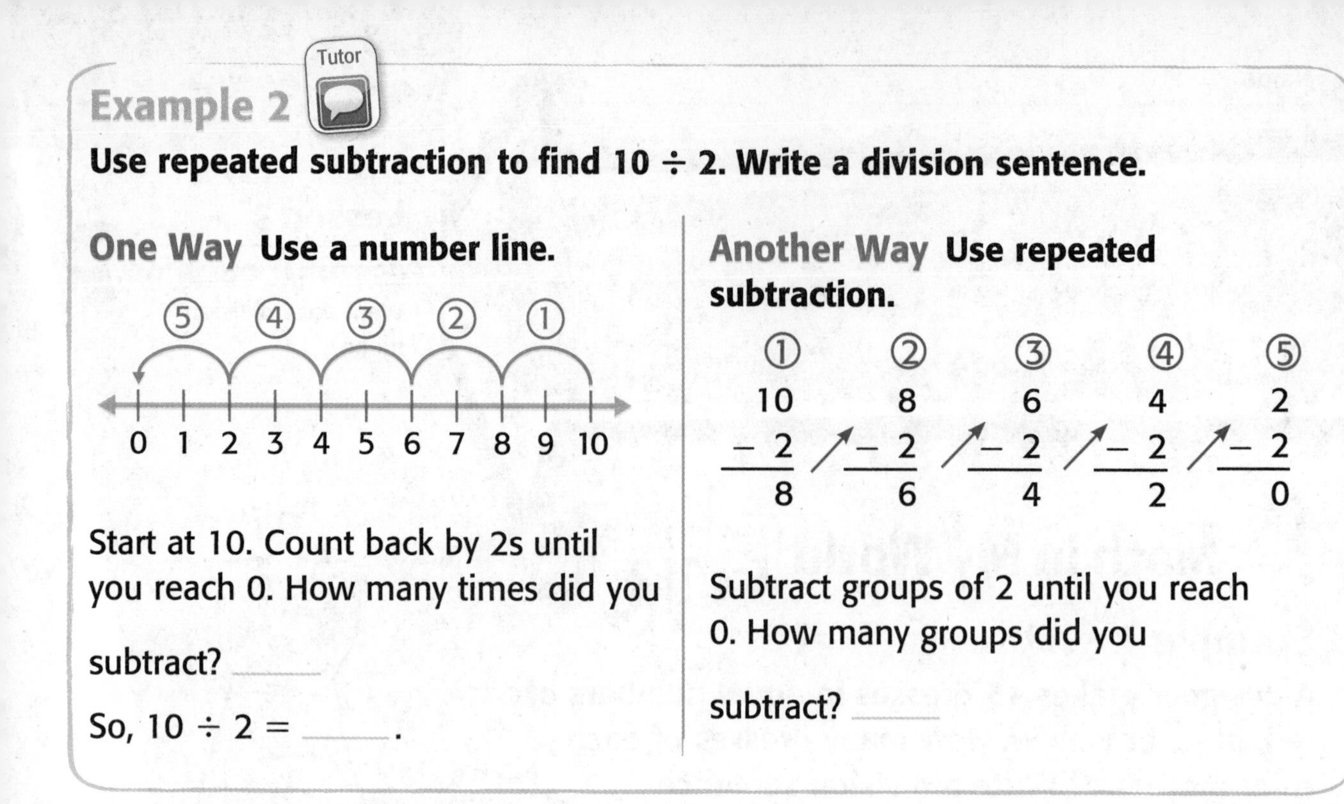

Example 2

Use repeated subtraction to find $10 \div 2$. Write a division sentence.

One Way **Use a number line.**

Start at 10. Count back by 2s until you reach 0. How many times did you subtract? ______

So, $10 \div 2 =$ ______.

Another Way **Use repeated subtraction.**

$10 - 2 = 8$, $8 - 2 = 6$, $6 - 2 = 4$, $4 - 2 = 2$, $2 - 2 = 0$

Subtract groups of 2 until you reach 0. How many groups did you subtract? ______

Guided Practice

Algebra **Write a division sentence with a symbol for the unknown. Then solve.**

1. There are 16 flowers. Each vase has 4 flowers. How many vases are there?

______ $\div 4 =$ ■

There are ______ vases.

2. There are 14 ears. Each dog has 2 ears. How many dogs are there?

$14 \div$ ______ $=$ ■

There are ______ dogs.

Use repeated subtraction to divide.

3.

$12 \div 3 =$ ______

4.

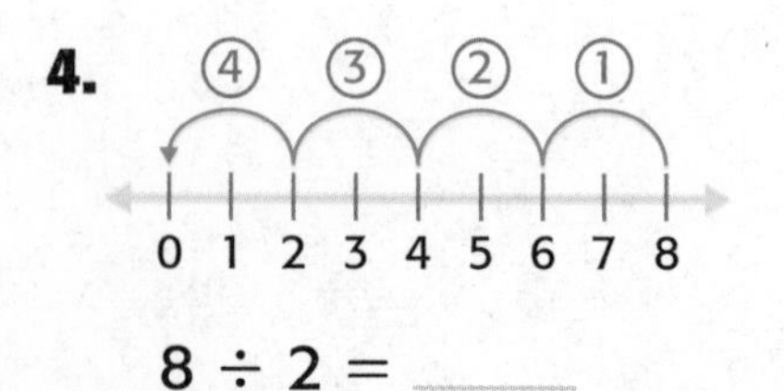

$8 \div 2 =$ ______

Talk MATH

Explain how to use a number line to find $18 \div 9$.

Name

Independent Practice

Algebra Write a division sentence with a symbol for the unknown. Then solve.

5. There are 16 orange slices. Each orange has 8 slices. How many oranges are there?

6. There are 16 miles. Each trip is 2 miles. How many trips are there?

7. There are 25 marbles, with 5 marbles in each bag. How many bags are there?

8. Four friends will share 12 muffins equally. How many muffins will each friend get?

Use repeated subtraction to divide.

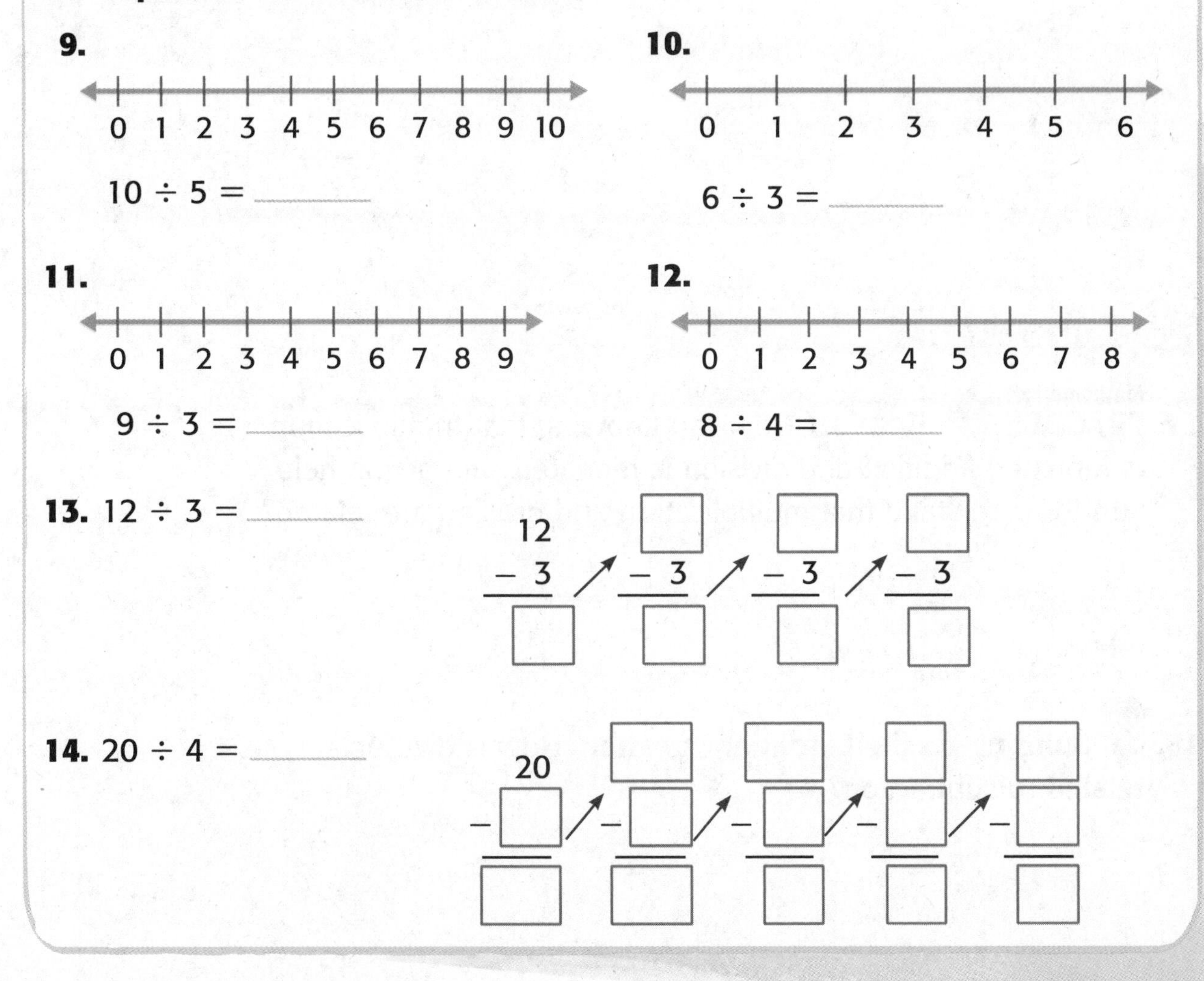

9. $10 \div 5 =$ ______

10. $6 \div 3 =$ ______

11. $9 \div 3 =$ ______

12. $8 \div 4 =$ ______

13. $12 \div 3 =$ ______

14. $20 \div 4 =$ ______

Problem Solving

Chicago's Ferris wheel is 10 stories tall. Each cart can seat up to 6 people for a 7-minute ride.

Write a division sentence with a symbol for the unknown. Then solve.

15. It costs $24 for 4 people to ride the Ferris wheel. How much does each ticket cost?

____

16. Mathematical PRACTICE 2 **Use Symbols** If 30 students from a class wanted to ride, how many carts would they need?

____

My Work!

HOT Problems

17. Mathematical PRACTICE 2 **Reason** How can knowing that multiplication is repeated addition and division is repeated subtraction help you to understand that multiplication and division are related?

____

____

18. **Building on the Essential Question** How is division related to subtraction?

____

____

Check My Progress

Vocabulary Check

Use the word bank to label each definition.

array	division	division sentence
equal groups	repeated subtraction	

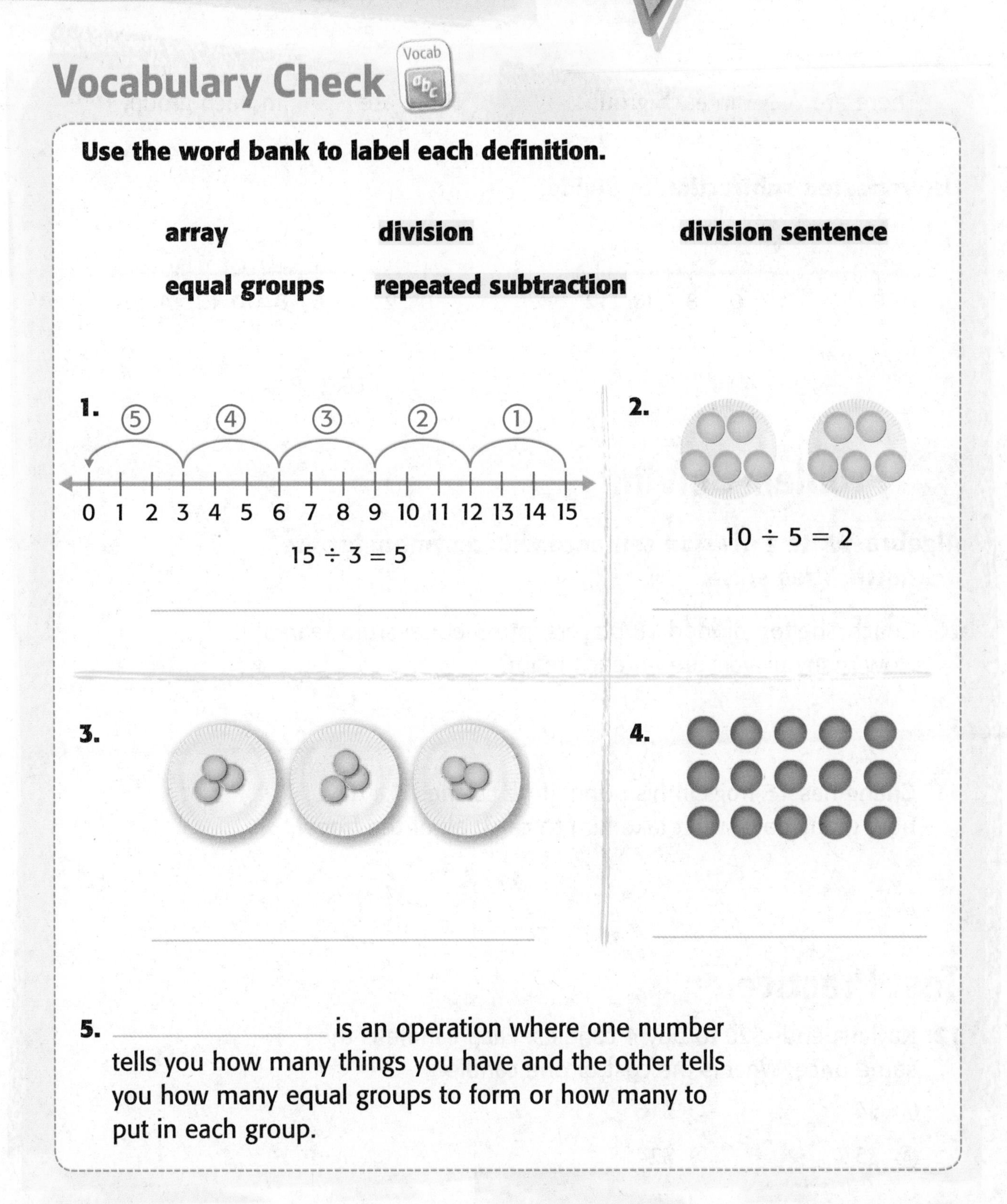

5. ______________ is an operation where one number tells you how many things you have and the other tells you how many equal groups to form or how many to put in each group.

Concept Check

Write a division sentence and divide to find how many are in each group.

6. 12 counters
3 equal groups

____ ÷ ____ = ____

There are ____ in each group.

7. 15 counters
5 equal groups

____ ÷ ____ = ____

There are ____ in each group.

Use repeated subtraction to divide.

8. **9.**

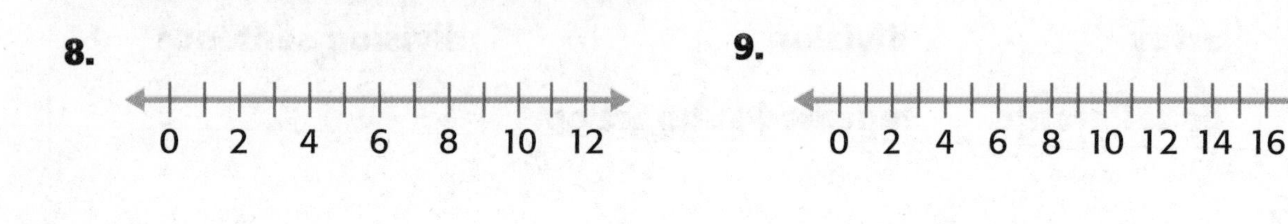

$12 \div 4 =$ ____

$16 \div 8 =$ ____

Problem Solving

Algebra Write a division sentence with a symbol for the unknown. Then solve.

10. Coach Shelton divided 18 players into 3 equal-sized teams. How many players are on each team?

____

11. Chang has 15 frogs in his pond. If he catches 3 a day, how many days will it take him to catch all of the frogs?

____

Test Practice

12. Kayla spends $20 to buy 4 candles. Each candle is the same price. What is the cost of one candle?

Ⓐ $4 Ⓒ $16

Ⓑ $5 Ⓓ $24

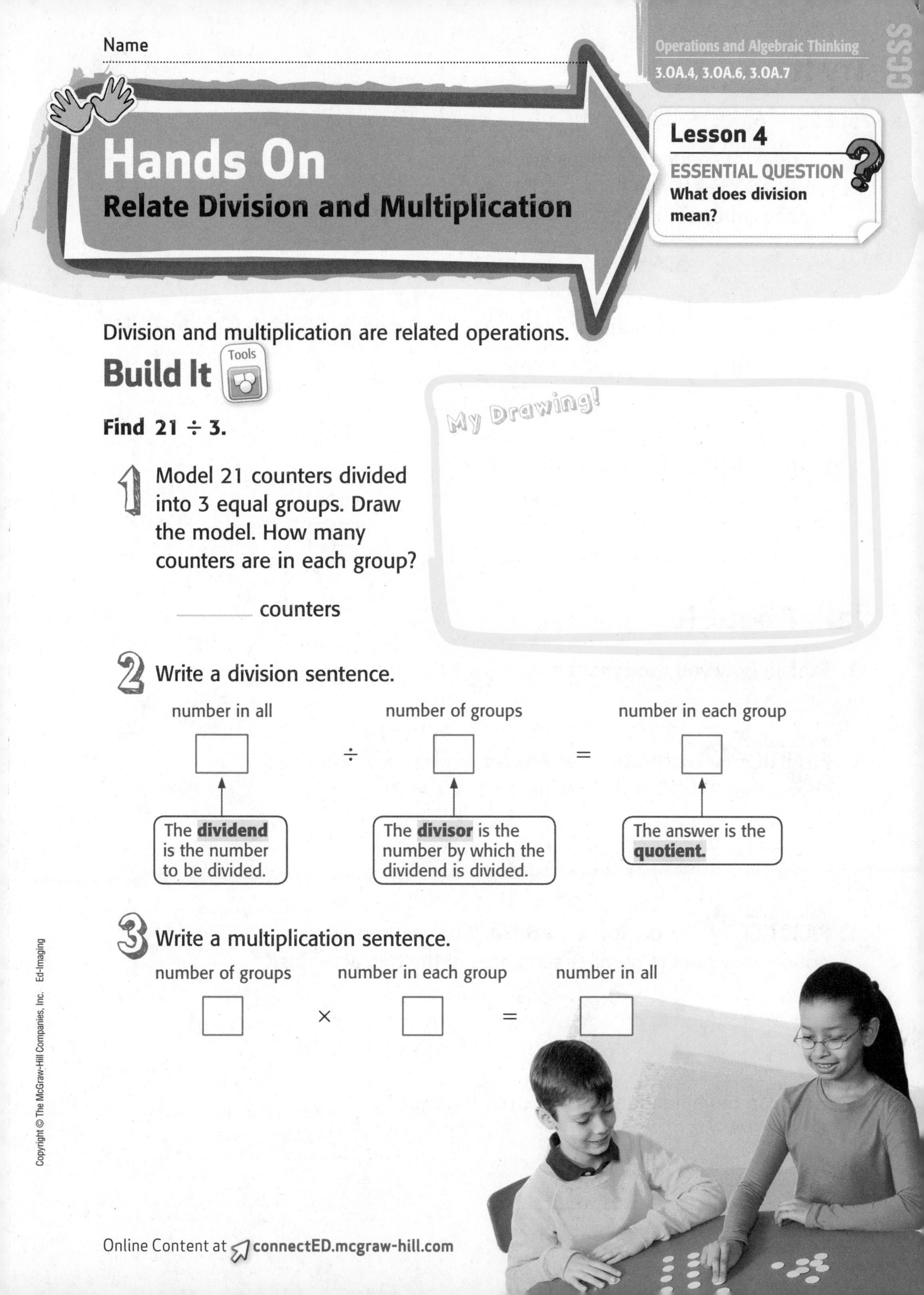

Name

Hands On
Relate Division and Multiplication

Lesson 4

ESSENTIAL QUESTION
What does division mean?

Division and multiplication are related operations.

Build It

Tools

Find 21 ÷ 3.

My Drawing!

1. Model 21 counters divided into 3 equal groups. Draw the model. How many counters are in each group?

   ______ counters

2. Write a division sentence.

number in all		number of groups		number in each group
☐	÷	☐	=	☐
The **dividend** is the number to be divided.		The **divisor** is the number by which the dividend is divided.		The answer is the **quotient.**

3. Write a multiplication sentence.

number of groups		number in each group		number in all
☐	×	☐	=	☐

Try It

Find 20 ÷ 4.

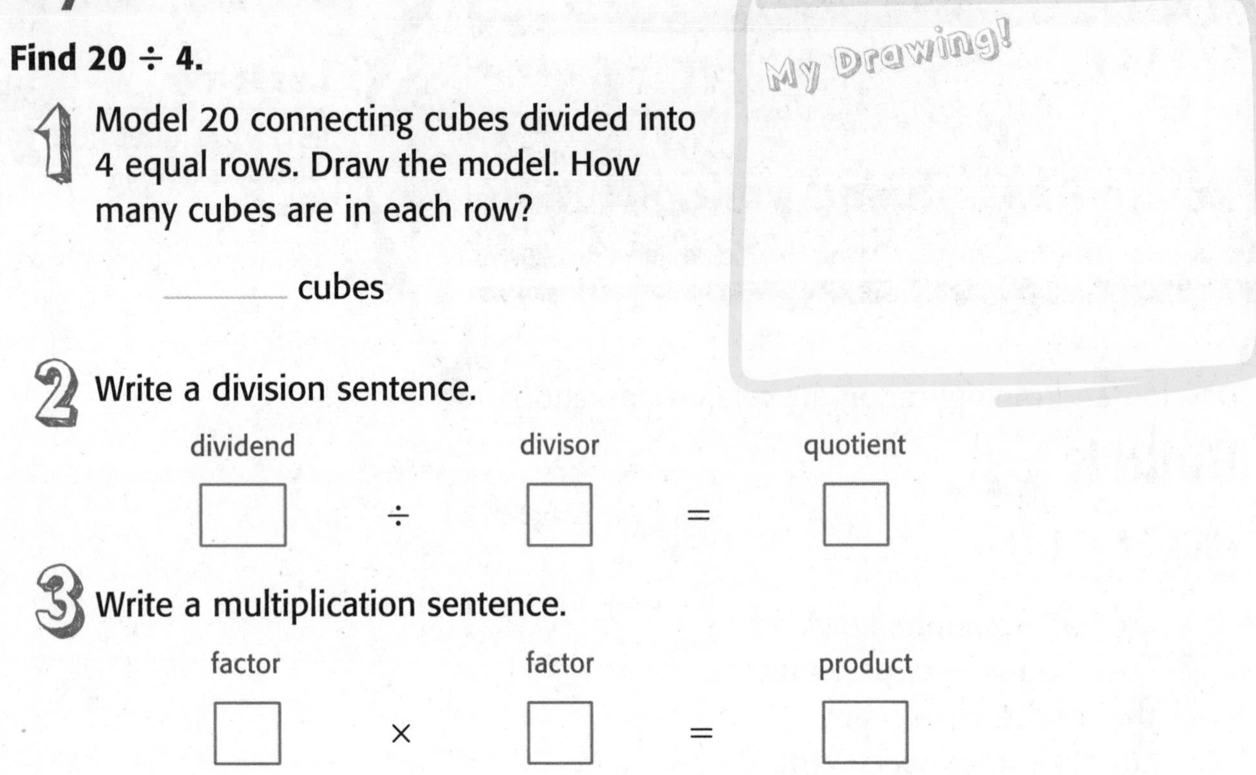

1. Model 20 connecting cubes divided into 4 equal rows. Draw the model. How many cubes are in each row?

   _______ cubes

2. Write a division sentence.

   dividend ☐ ÷ divisor ☐ = quotient ☐

3. Write a multiplication sentence.

   factor ☐ × factor ☐ = product ☐

Talk About It

1. Explain how you used models to show 21 ÷ 3.

____

2. Mathematical PRACTICE 6 **Explain to a Friend** Explain how the array shows that 21 ÷ 3 = 7 is related to 3 × 7 = 21.

____

____

3. Mathematical PRACTICE 8 **Look for a Pattern** What pattern do you notice between the number sentences in the two activities?

____

____

4. How can multiplication facts be used to divide?

____

____

Name

Practice It

Write a related division and multiplication sentence for each.

5.

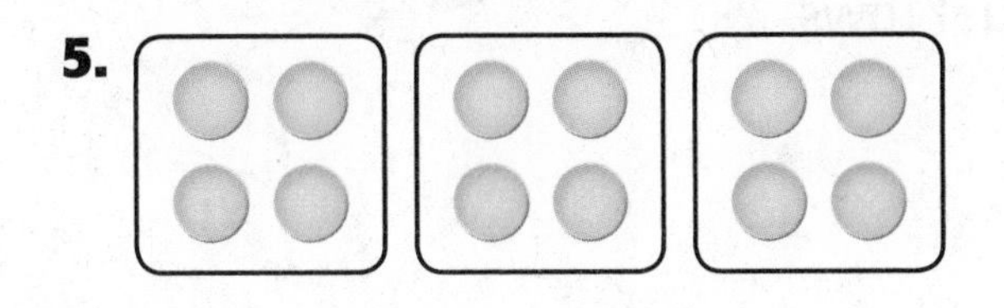

6.

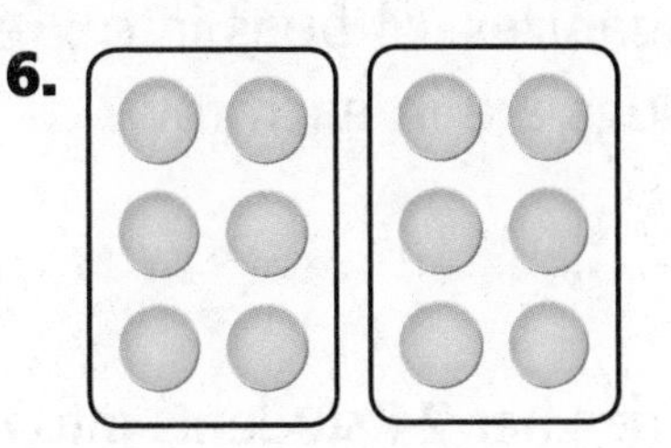

Use connecting cubes to solve. Draw your model. Write a division sentence.

7. Model 10 connecting cubes divided into 2 equal rows. How many cubes are in each row?

8. Model 6 connecting cubes divided into 3 equal rows. How many cubes are in each row?

My Drawing!

My Drawing!

Algebra Use counters to model each problem. Find the unknown. Then write a related multiplication sentence.

9. $12 \div 6 = \blacksquare$

The unknown is ____.

10. $21 \div 7 = \blacksquare$

The unknown is ____.

11. $25 \div 5 = \blacksquare$

The unknown is ____.

Apply It

Draw a model to solve. Then write a division sentence.

My Drawing!

12. A scientist organizes 14 bugs into 2 equal rows. How many bugs are in each row?

____

13. A teacher divides her 24 students into 4 equal activity groups. How many students are in each group?

____

14. Arianna divides 25 gold stars between herself and 4 friends. How many stars does each friend receive?

____

15. Mark checked out 12 books from the library. He reads an equal amount of books each week for 4 weeks. How many books does he read each week?

____

16. Mathematical PRACTICE 1 **Make a Plan** Eli had 20 lemons. He used an equal number in each of 3 pitchers of lemonade. He has 2 lemons left. How many lemons did Eli use to make one pitcher of lemonade?

____

Write About It

17. How are arrays used in both multiplication and division?

____

____

____

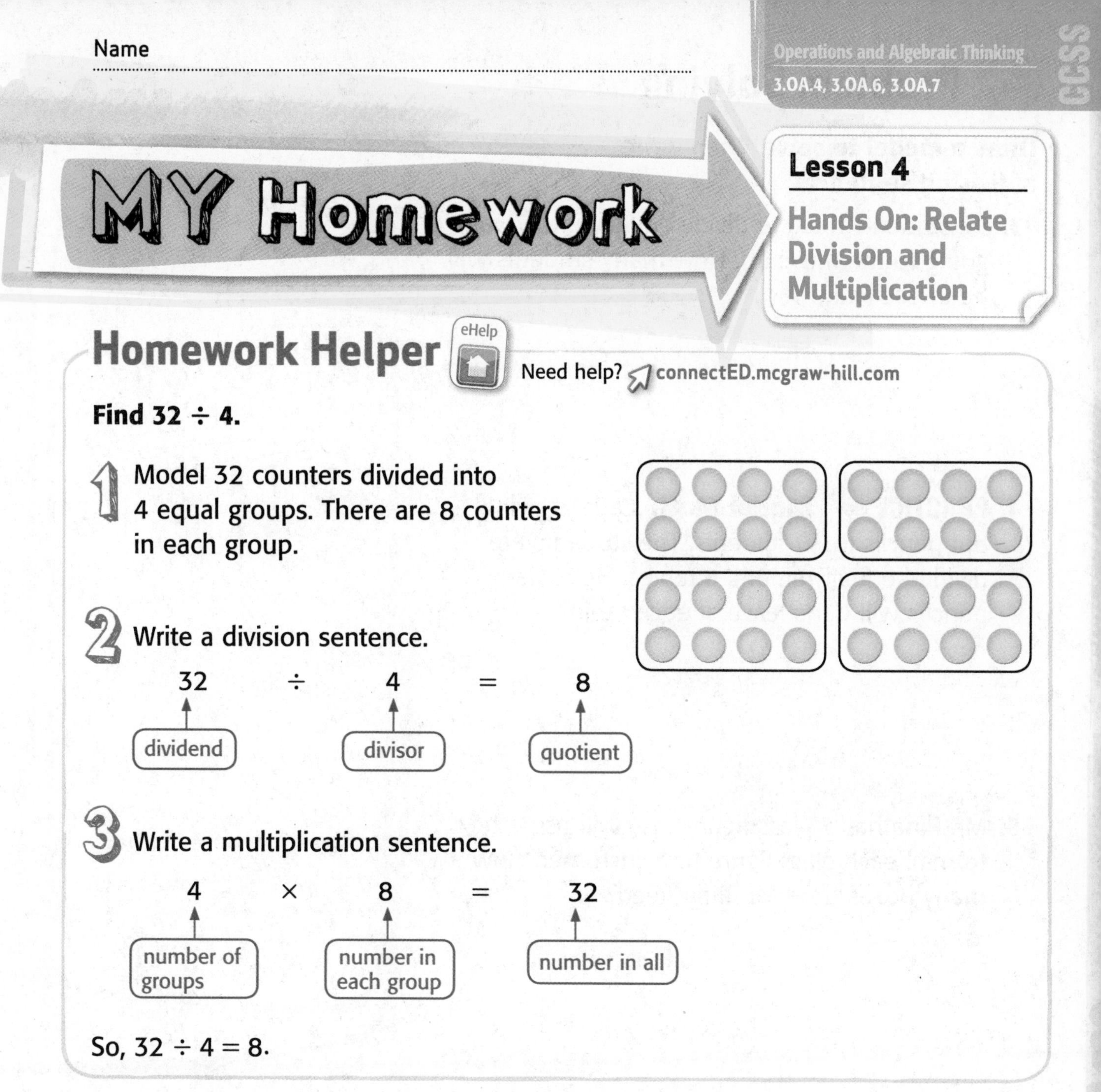

Name

MY Homework

Lesson 4

Hands On: Relate Division and Multiplication

Homework Helper

eHelp Need help? connectED.mcgraw-hill.com

Find 32 ÷ 4.

1 Model 32 counters divided into 4 equal groups. There are 8 counters in each group.

2 Write a division sentence.

$32 \div 4 = 8$

dividend, divisor, quotient

3 Write a multiplication sentence.

$4 \times 8 = 32$

number of groups, number in each group, number in all

So, $32 \div 4 = 8$.

Practice

Write a related division and multiplication sentence for each.

1.

2.

Problem Solving

Draw a model to solve. Then write a division sentence.

3. 42 students need to divide equally into 7 vans going to the museum. How many students will be in each van?

My Drawing!

4. Mathematical PRACTICE 4 **Model Math** Carla is giving out pencils to 30 students. The students are divided equally among 6 tables. How many pencils will Carla leave at each table?

5. Mr. Rina has 7 glass figures. He will use 1 box to mail each glass figure to a customer. How many boxes does Mr. Rina need?

Vocabulary Check

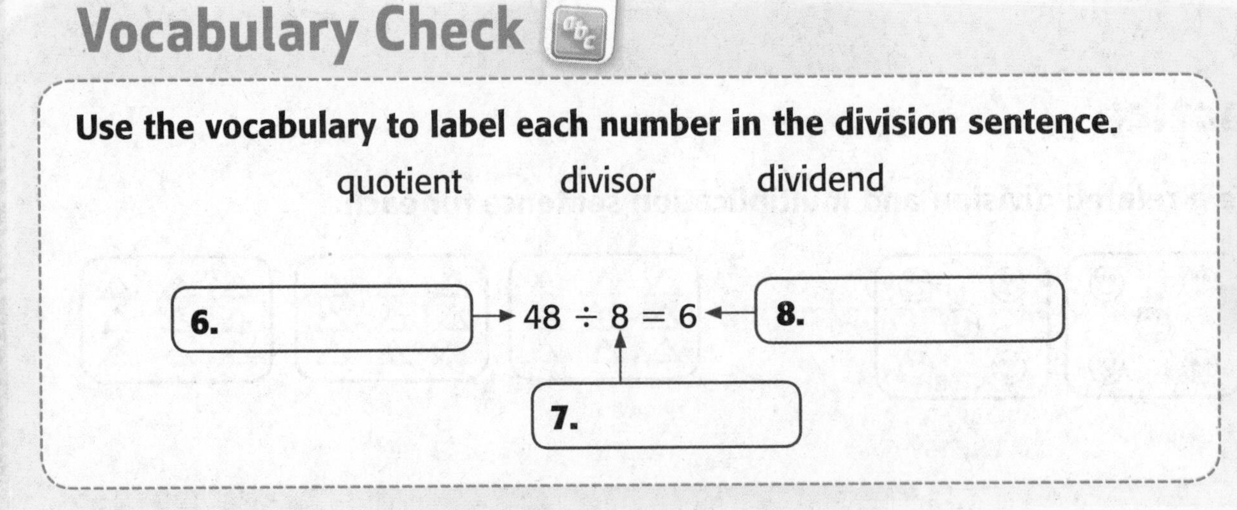

Name

Inverse Operations

Lesson 5

ESSENTIAL QUESTION
What does division mean?

You have learned how division and multiplication are related. Operations that are related are **inverse operations** because they undo each other.

Math in My World

Tools Watch Tutor

Example 1

A baker has made a tray of fresh muffins. Use the array to write a related multiplication and division sentence to find the unknown. How many muffins are there in all?

Multiplication						**Division**				
number of rows		number in each row		number in all		number in all		number of rows		number in each row
3	×	☐	=	■	← unknown →	■	÷	3	=	☐
factor		factor		product		**dividend**		**divisor**		**quotient**

The unknown is ______.

So, there are ______ muffins in all.

The multiplication sentence multiplies 3 by ______ to get 12. The division sentence undoes the multiplication by dividing 12 by 3 to get ______.

A group of **related facts** using the same numbers is a **fact family.** Each fact family follows a pattern by using the same numbers.

Fact Family 3, 4 and 12

$3 \times 4 = 12$
$4 \times 3 = 12$
$12 \div 3 = 4$
$12 \div 4 = 3$

Fact Family 7 and 49

$7 \times 7 = 49$
$49 \div 7 = 7$

Example 2

Tutor

Complete the fact family for the numbers 3, 6, and 18.

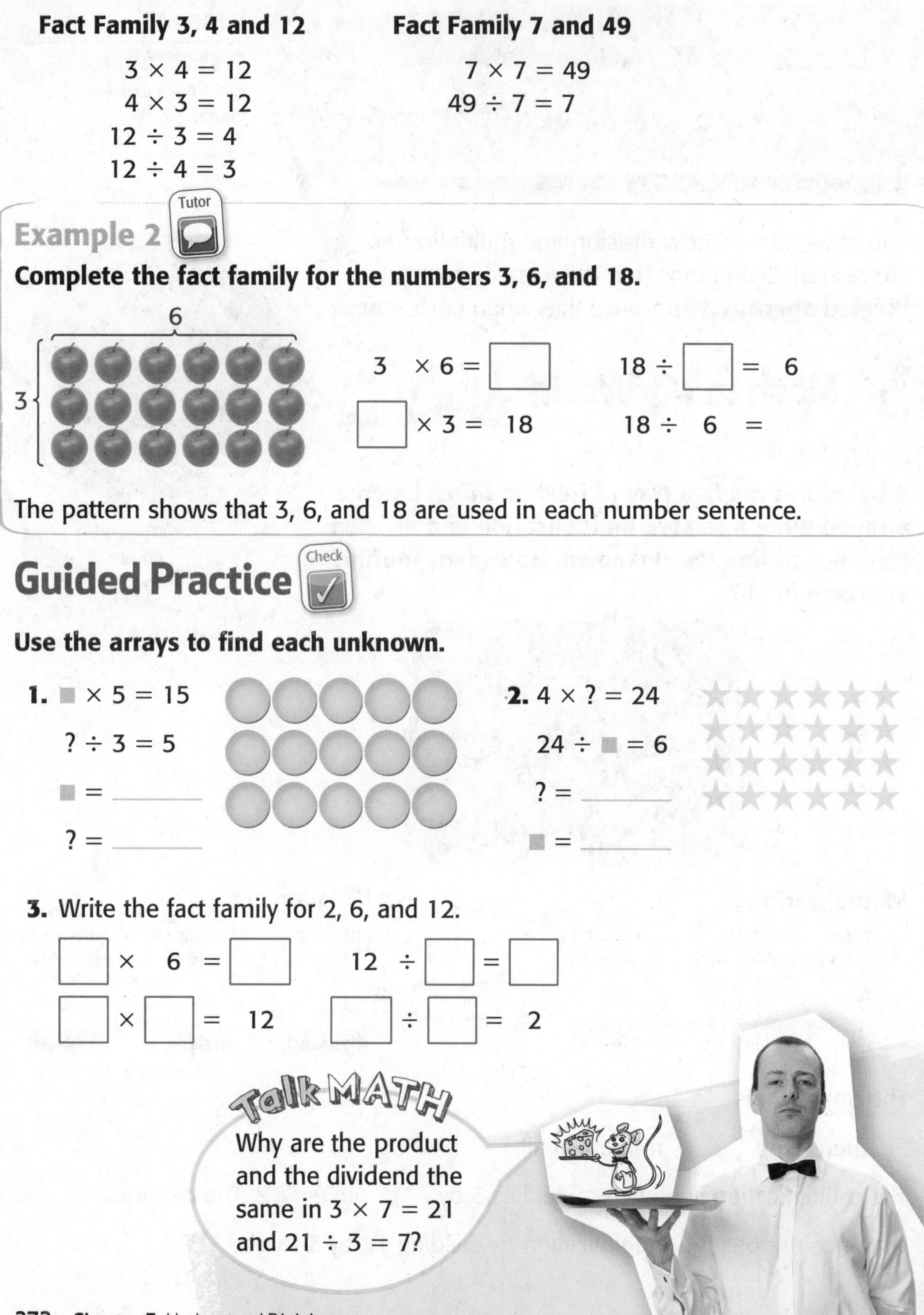

$3 \times 6 = \square$ $\quad$ $18 \div \square = 6$

$\square \times 3 = 18$ $\quad$ $18 \div 6 =$

The pattern shows that 3, 6, and 18 are used in each number sentence.

Guided Practice

Check

Use the arrays to find each unknown.

1. ■ × 5 = 15

? ÷ 3 = 5

■ = ______

? = ______

2. 4 × ? = 24

24 ÷ ■ = 6

? = ______

■ = ______

3. Write the fact family for 2, 6, and 12.

$\square \times 6 = \square$ $\quad$ $12 \div \square = \square$

$\square \times \square = 12$ $\quad$ $\square \div \square = 2$

Talk MATH

Why are the product and the dividend the same in $3 \times 7 = 21$ and $21 \div 3 = 7$?

Name

Independent Practice

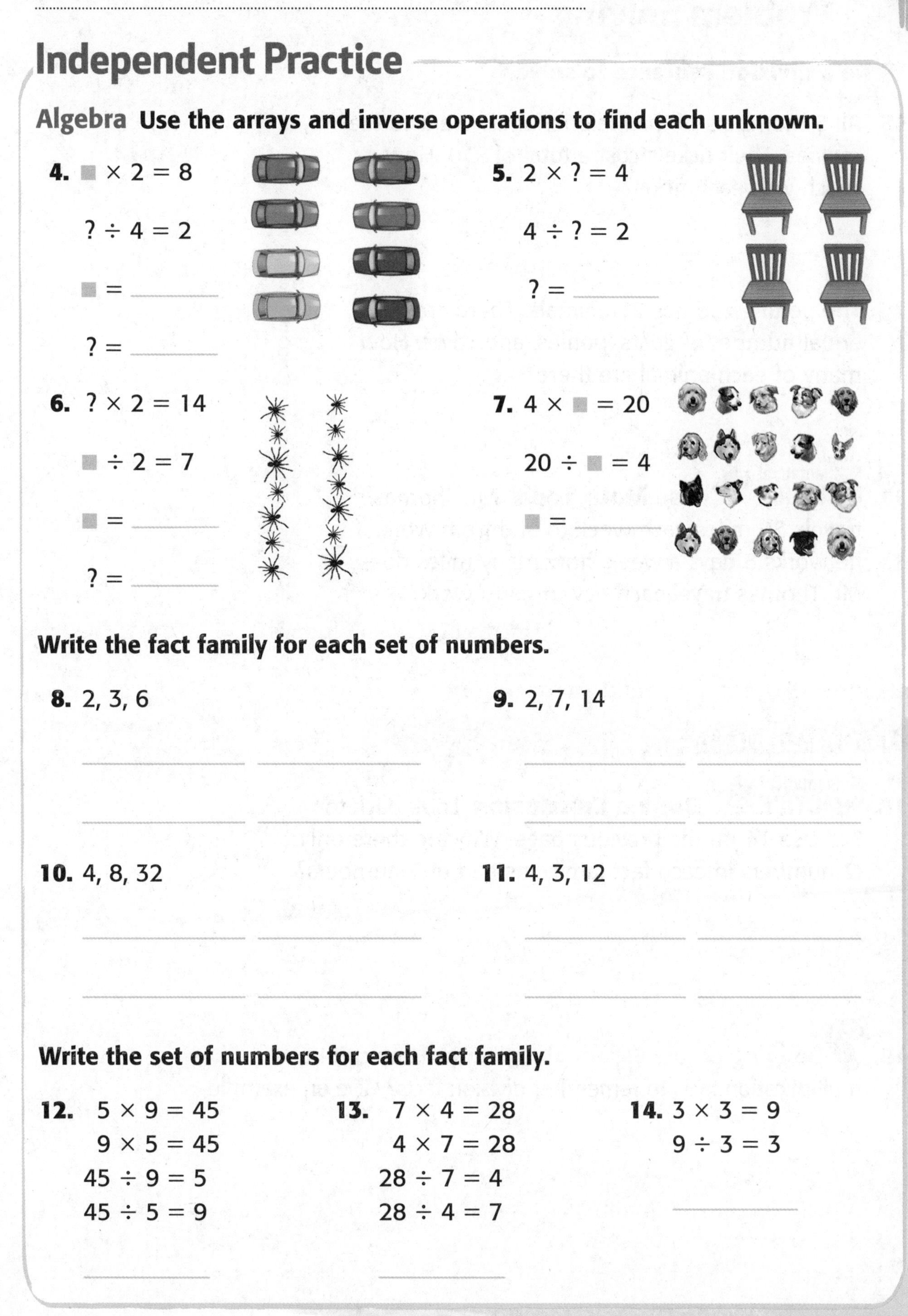

Algebra **Use the arrays and inverse operations to find each unknown.**

4. ■ × 2 = 8

? ÷ 4 = 2

■ = ______

? = ______

5. 2 × ? = 4

4 ÷ ? = 2

? = ______

6. ? × 2 = 14

■ ÷ 2 = 7

■ = ______

? = ______

7. 4 × ■ = 20

20 ÷ ■ = 4

■ = ______

Write the fact family for each set of numbers.

8. 2, 3, 6

9. 2, 7, 14

10. 4, 8, 32

11. 4, 3, 12

Write the set of numbers for each fact family.

12. 5 × 9 = 45
9 × 5 = 45
45 ÷ 9 = 5
45 ÷ 5 = 9

______

13. 7 × 4 = 28
4 × 7 = 28
28 ÷ 7 = 4
28 ÷ 4 = 7

______

14. 3 × 3 = 9
9 ÷ 3 = 3

______

Problem Solving

Write a division sentence to solve.

15. All 5 members of the Malone family went to the movies. Their tickets cost a total of $30. How much was each ticket?

16. The petting zoo has 21 animals. There are an equal number of goats, ponies, and cows. How many of each animal are there?

My Work!

17. Mathematical PRACTICE 5 **Use Math Tools** Mr. Thomas travels 20 miles each week to and from work. If he works 5 days a week, how many miles does Mr. Thomas travel each day to go to work?

HOT Problems

18. Mathematical PRACTICE 3 **Draw a Conclusion** Look back to Exercise 14 on the previous page. Why are there only 2 numbers in each fact family instead of 3 numbers?

19. **Building on the Essential Question** How can I use multiplication facts to remember division facts? Give an example.

Name ____________

Lesson 5
Inverse Operations

Homework Helper

eHelp

Need help? connectED.mcgraw-hill.com

The array represents 27 children lined up in 3 rows. Use the array to find each unknown.

9 × ■ = 27
? ÷ 3 = 9
■ = 3
? = 27

You know 3 rows of 9 = 27.
So, 9 rows of 3 = 27 and 27 ÷ 3 = 9.

Practice

Algebra Use the array to find each unknown.

1. ■ × 4 = 20

? ÷ 5 = 4

■ = ______

? = ______

2. 4 × ■ = 16

? ÷ 4 = 4

■ = ______

? = ______

3. 7 × ■ = 21

? ÷ 7 = 3

■ = ______

? = ______

4. 2 × ■ = 12

? ÷ 2 = 6

■ = ______

? = ______

Write the fact family for each set of numbers.

5. 5, 8, 40

6. 6, 7, 42

Write the set of numbers for each fact family.

7. $4 \times 9 = 36$ $36 \div 4 = 9$

$9 \times 4 = 36$ $36 \div 9 = 4$

8. $2 \times 8 = 16$ $16 \div 2 = 8$

$8 \times 2 = 16$ $16 \div 8 = 2$

Problem Solving

9. Mathematical PRACTICE 4 **Model Math** Tia has \$35 to spend on socks for the family. If each pair of socks costs \$5, how many pairs can she buy? Write a division sentence to solve.

Vocabulary Check

Vocab

Draw a line to connect each vocabulary word with its definition.

10. dividend
11. divisor
12. fact family
13. inverse operations
14. quotient

- the number being divided
- a group of related facts that use the same numbers
- the answer to a division problem
- the number by which a dividend is divided
- operations that undo each other

Test Practice

15. Which pair shows inverse operations?

Ⓐ $2 \times 2 = 4; 4 \div 2 = 2$

Ⓑ $2 \times 2 = 4; 4 - 2 = 2$

Ⓒ $2 \times 2 = 4; 8 \div 4 = 2$

Ⓓ $2 \times 2 = 4; 4 \div 4 = 1$

Name

Problem-Solving Investigation

STRATEGY: Use Models

Lesson 6

ESSENTIAL QUESTION
What does division mean?

Learn the Strategy

Mia has 18 items that need to be split evenly among 3 welcome baskets. How many items will Mia put in each basket?

1 Understand

What facts do you know?

_______ items need to be split evenly among _______ baskets.

What do you need to find?

the number of _______________________

2 Plan

I will make a model to find _______________________.

3 Solve

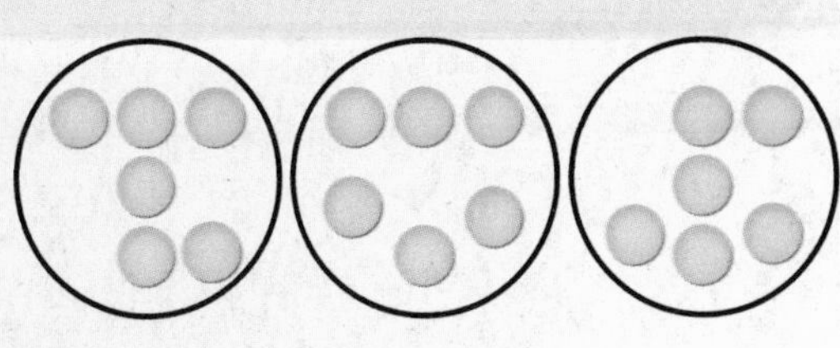

I will use counters to model the problem by placing _____ counter at a time in each group.

The model shows that $18 \div 3 =$ _____.

So, Mia will fill each basket with _____ items.

4 Check

Does your answer make sense? Explain.

____

Practice the Strategy

A veterinarian helped 20 pets from Monday to Friday. She helped an equal number of pets each day. How many pets did she help each day?

1 Understand

What facts do you know?

____

____

What do you need to find?

____

2 Plan

____

3 Solve

4 Check

Does your answer make sense? Explain.

____

____

Name

Apply the Strategy

Solve each problem by using a model.

My Work!

1. Mathematical PRACTICE 5 **Use Math Tools** Jill has 27 blocks. She wants to divide them equally into the bowls shown below. How many blocks will be in each bowl?

2. The owner of an apartment building needs to fix 16 locks in four of his apartments. Each apartment has the same number of locks that needs to be fixed. How many locks in each apartment need to be fixed?

3. A baker used a dozen eggs to make 3 cakes. The recipe called for each cake to have the same number of eggs. How many eggs were used in each cake? (Hint: 1 dozen = 12)

4. There are 13 girls and 11 boys that want to play a game. They need to make 4 teams. How many players will be on each team if each team needs an equal number of players?

Review the Strategies

Use any strategy to solve each problem.

- Determine reasonable answers.
- Use an estimate or exact answer.
- Use models.

My Work!

5. Mathematical PRACTICE 2 **Use Number Sense** Sarah needs 15 pieces of chalk for a project. Each box contains 3 pieces of chalk. How many boxes of chalk will she need to buy?

6. Brooke volunteers to read with young children 5 nights a month. She spends 2 hours each visit. This month, she volunteered one extra night. How many hours did she read with the children this month?

7. Mathematical PRACTICE 4 **Model Math** A chef will make pizzas. He has broccoli, peppers, onions, pepperoni, and sausage. How many types of pizzas can be made with one type of vegetable and one type of meat? Name the combinations.

8. A scientist estimates that a brown bear weighs 700 pounds. It actually weighs 634 pounds. How much more is the estimate than the actual weight?

Name

MY Homework

Lesson 6
Problem Solving: Use Models

Homework Helper

eHelp Need help? connectED.mcgraw-hill.com

Lucy needs 7 craft sticks to make a puzzle. She has 28 craft sticks. How many puzzles can Lucy make? Use a model to solve.

1 Understand

Lucy has 28 craft sticks. She needs 7 sticks to make a puzzle. Find how many puzzles Lucy can make.

2 Plan

Divide 28 craft sticks into equal groups of 7.

3 Solve

There are 4 equal groups of 7 craft sticks.
The model shows that $28 \div 7 = 4$.
So, Lucy can make 4 puzzles.

4 Check

Use multiplication to check. $4 \times 7 = 28$
So, the answer is correct.

Problem Solving

1. Brandon spent $20 on school supplies. He bought five different items that each cost the same amount. How much did each item cost? Use a model to solve.

   Each item cost ______.

Solve each problem by using a model.

2. Mathematical PRACTICE 5 **Use Math Tools** Alice planted 6 tomato plants, 4 bean plants, and 2 pepper plants. Each row had 6 plants. How many rows did Alice plant?

3. At the circus, there are 18 clowns. The clowns drive around in little cars. If there are 3 clowns in each car, how many cars are there?

4. Mr. and Mrs. Carson took Sarah, Brent, and Joanie to see a movie. They paid $50 in all. The Carsons spent $15 on snacks. How much did each ticket cost?

5. Mrs. Glover had 25 rare coins. She divided them evenly among her 5 grandchildren. How many coins did each grandchild get?

6. A singer performed 9 songs at a recital. She had 3 weeks to practice. How many songs did she practice each week if she practiced an equal number of songs each week?

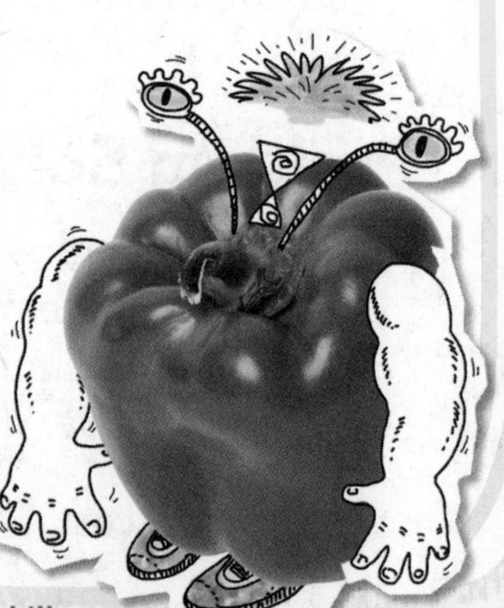

Review

Chapter 5
Understand Division

Vocabulary Check

Vocab

Use the word bank to complete each sentence.

array	**divide**	**fact family**	**inverse operations**
partition	**related facts**	**repeated subtraction**	

1. ______________ are a set of basic facts using the same three numbers.
2. An arrangement of objects into equal rows and equal columns is a(n) ______________.
3. A way to divide by sharing one object at a time until all the objects are gone is to ______________.
4. To ______________ means to separate a number into equal groups, to find the number of groups, or find the number in each group.
5. ______________ is a way to subtract the same number over and over again until you reach 0.
6. Operations that are related are ______________ because they undo each other.
7. $3 \times 5 = 15$, $5 \times 3 = 15$, $15 \div 5 = 3$, and $15 \div 3 = 5$ are the facts in the 3, 5, 15 ______________.
8. Write a **division sentence** in the space below. Label the **dividend**, **divisor**, and **quotient**.

Concept Check

Use counters to find how many are in each group.

9. 14 counters
2 equal groups

_____ ÷ _____ = _____

_____ in each group

10. 25 counters
5 equal groups

_____ ÷ _____ = _____

_____ in each group

Use repeated subtraction to divide.

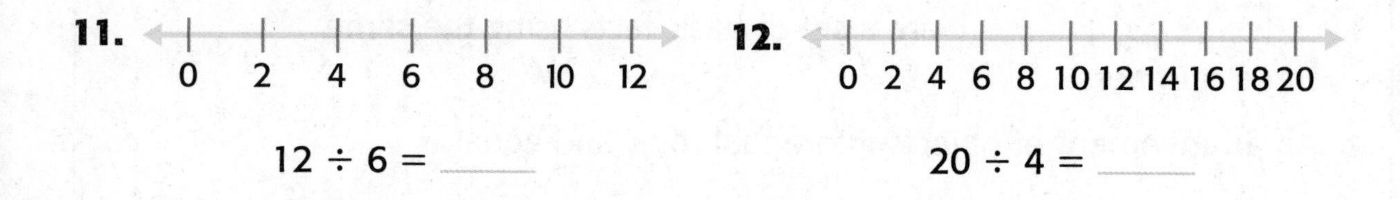

11. $12 \div 6 =$ _____

12. $20 \div 4 =$ _____

Write a related division and multiplication sentence for each.

13. **14.**

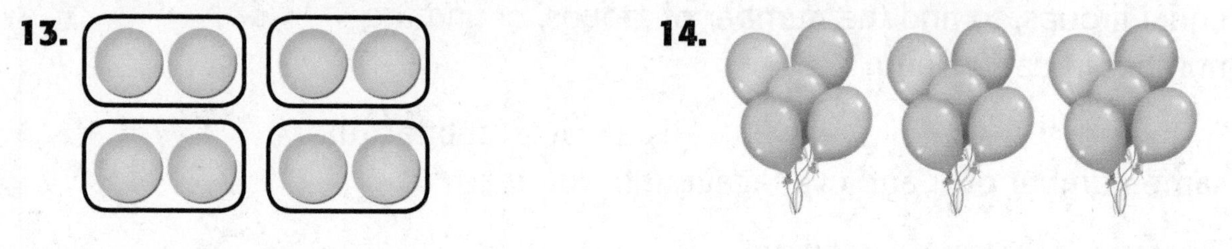

_______________ _______________

Write the fact family for each set of numbers.

15. 4, 7, 28

_____ _____

_____ _____

16. 3, 9, 27

_____ _____

_____ _____

Name

Problem Solving

My Work!

17. Brandon's dentist gave him 12 toothbrushes. Brandon wants to share them equally among himself and his 2 friends. How many toothbrushes will each person get? Write a division sentence.

18. A teacher has 24 pencils. She keeps 4 and shares the others equally with 5 students. How many pencils did each student get?

19. Circle the number sentence that does not belong. Explain. Then write the missing number sentence.

$3 \times 6 = 18$ $\quad$ $18 \div 2 = 9$

$18 \div 6 = 3$ $\quad$ $6 \times 3 = 18$

Test Practice

20. Harper saved $30 from mowing lawns April through September. She saved an equal amount each month. How much money did Harper save each month?

Ⓐ $5 Ⓒ $8

Ⓑ $6 Ⓓ $10

Reflect

Chapter 5

Answering the ESSENTIAL QUESTION

Use what you learned about division to complete the graphic organizer.

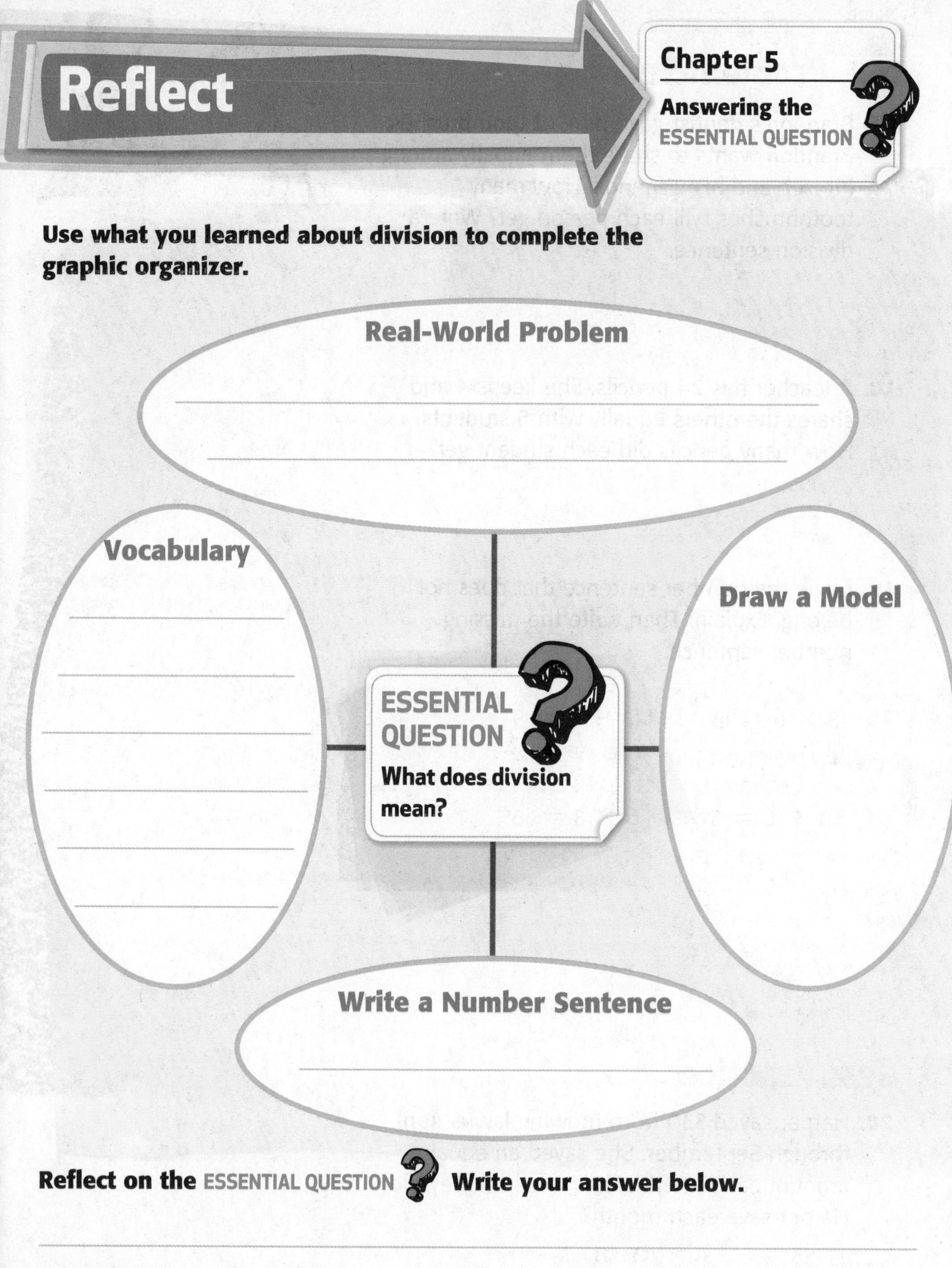

Reflect on the ESSENTIAL QUESTION Write your answer below.

Chapter

6 Multiplication and Division Patterns

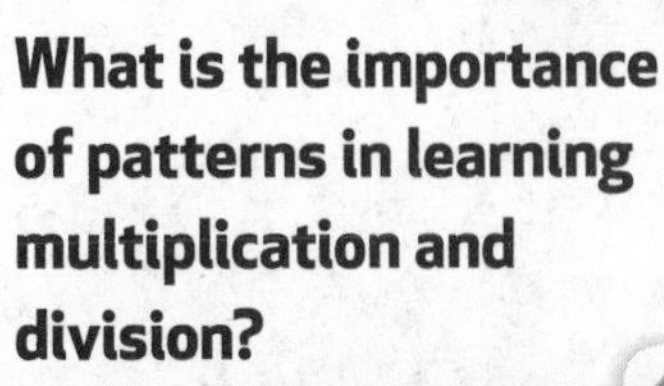

ESSENTIAL QUESTION

What is the importance of patterns in learning multiplication and division?

Watch a video!

Watch

MY Common Core State Standards

Operations and Algebraic Thinking

CCSS

3.OA.1 Interpret products of whole numbers, e.g., interpret 5 × 7 as the total number of objects in 5 groups of 7 objects each.

3.OA.2 Interpret whole-number quotients of whole numbers, e.g., interpret 56 ÷ 8 as the number of objects in each share when 56 objects are partitioned equally into 8 shares, or as a number of shares when 56 objects are partitioned into equal shares of 8 objects each.

3.OA.3 Use multiplication and division within 100 to solve word problems in situations involving equal groups, arrays, and measurement quantities, e.g., by using drawings and equations with a symbol for the unknown number to represent the problem.

3.OA.4 Determine the unknown whole number in a multiplication or division equation relating three whole numbers.

3.OA.5 Apply properties of operations as strategies to multiply and divide.

3.OA.6 Understand division as an unknown-factor problem.

3.OA.7 Fluently multiply and divide within 100, using strategies such as the relationship between multiplication and division (e.g., knowing that 8 × 5 = 40, one knows 40 ÷ 5 = 8) or properties of operations. By the end of Grade 3, know from memory all products of two one-digit numbers.

3.OA.9 Identify arithmetic patterns (including patterns in the addition table or multiplication table), and explain them using properties of operations.

Number and Operations in Base Ten *This chapter also addresses this standard:*

3.NBT.3 Multiply one-digit whole numbers by multiples of 10 in the range 10-90 (e.g., 9 × 80, 5 × 60) using strategies based on place value and properties of operations.

Standards for Mathematical PRACTICE

1. Make sense of problems and persevere in solving them.
2. Reason abstractly and quantitatively.
3. Construct viable arguments and critique the reasoning of others.
4. Model with mathematics.
5. Use appropriate tools strategically.
6. Attend to precision.
7. Look for and make use of structure.
8. Look for and express regularity in repeated reasoning.

= focused on in this chapter

Name

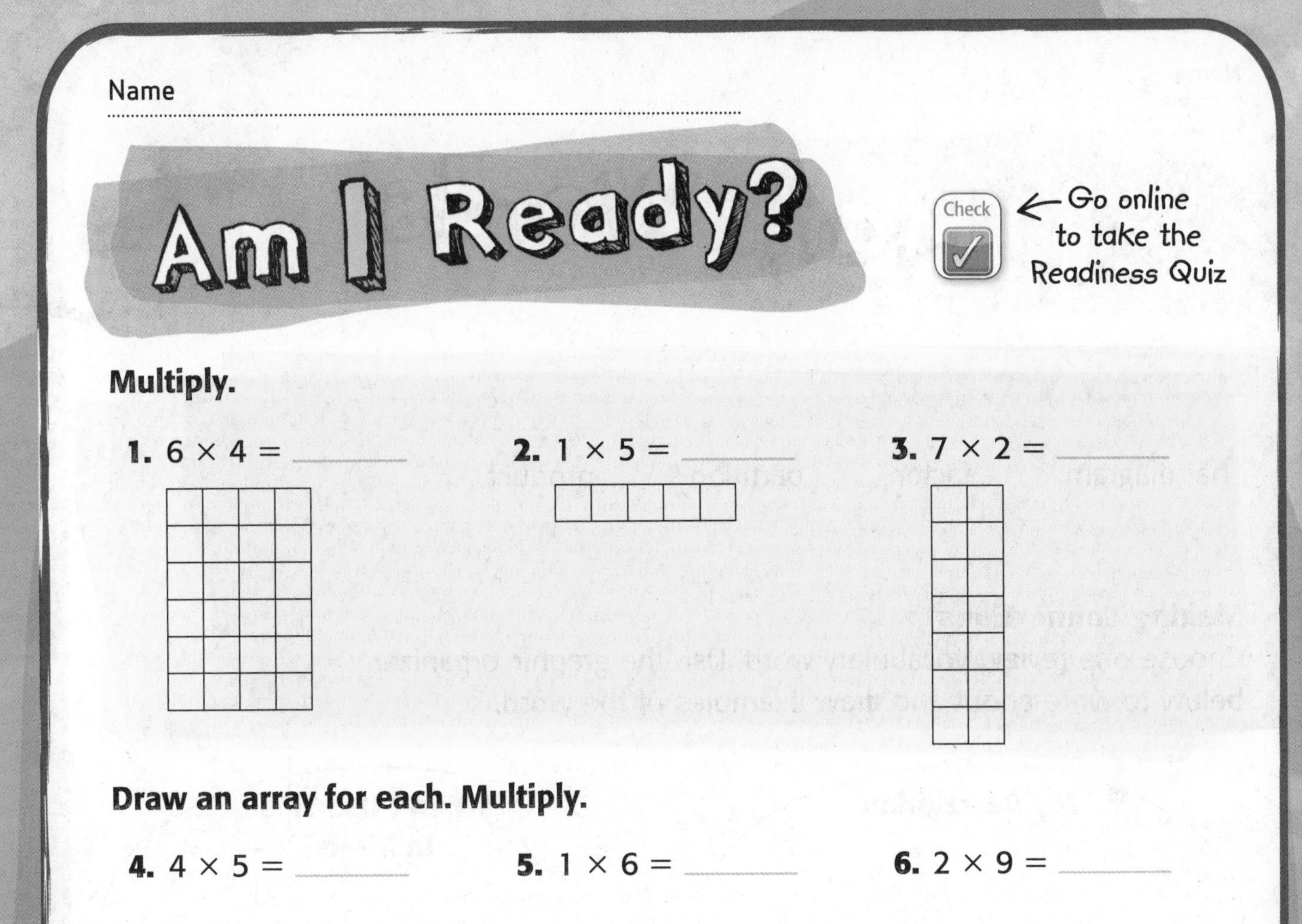

Multiply.

1. $6 \times 4 =$ ______

2. $1 \times 5 =$ ______

3. $7 \times 2 =$ ______

Draw an array for each. Multiply.

4. $4 \times 5 =$ ______

5. $1 \times 6 =$ ______

6. $2 \times 9 =$ ______

Identify a pattern. Then find the missing numbers.

7. ______, ______, 30, 25, 20, 15

The pattern is ______.

8. ______, ______, 16, 14, 12, 10

The pattern is ______.

9. Louis has 2 quarters. Yellow whistles cost 5¢ each. Louis wants to buy 8 whistles. Does he have enough money? Explain.

______

______

10. Nine trees lined each side of a street. Some trees were cut down leaving a total of 7 trees. How many trees were cut down?

______

Shade the boxes to show the problems you answered correctly.

How Did I Do?

1	2	3	4	5	6	7	8	9	10

Name

Review Vocabulary

bar diagram   factor   partition   product

Making Connections

Choose one review vocabulary word. Use the graphic organizer below to write about and draw examples of the word.

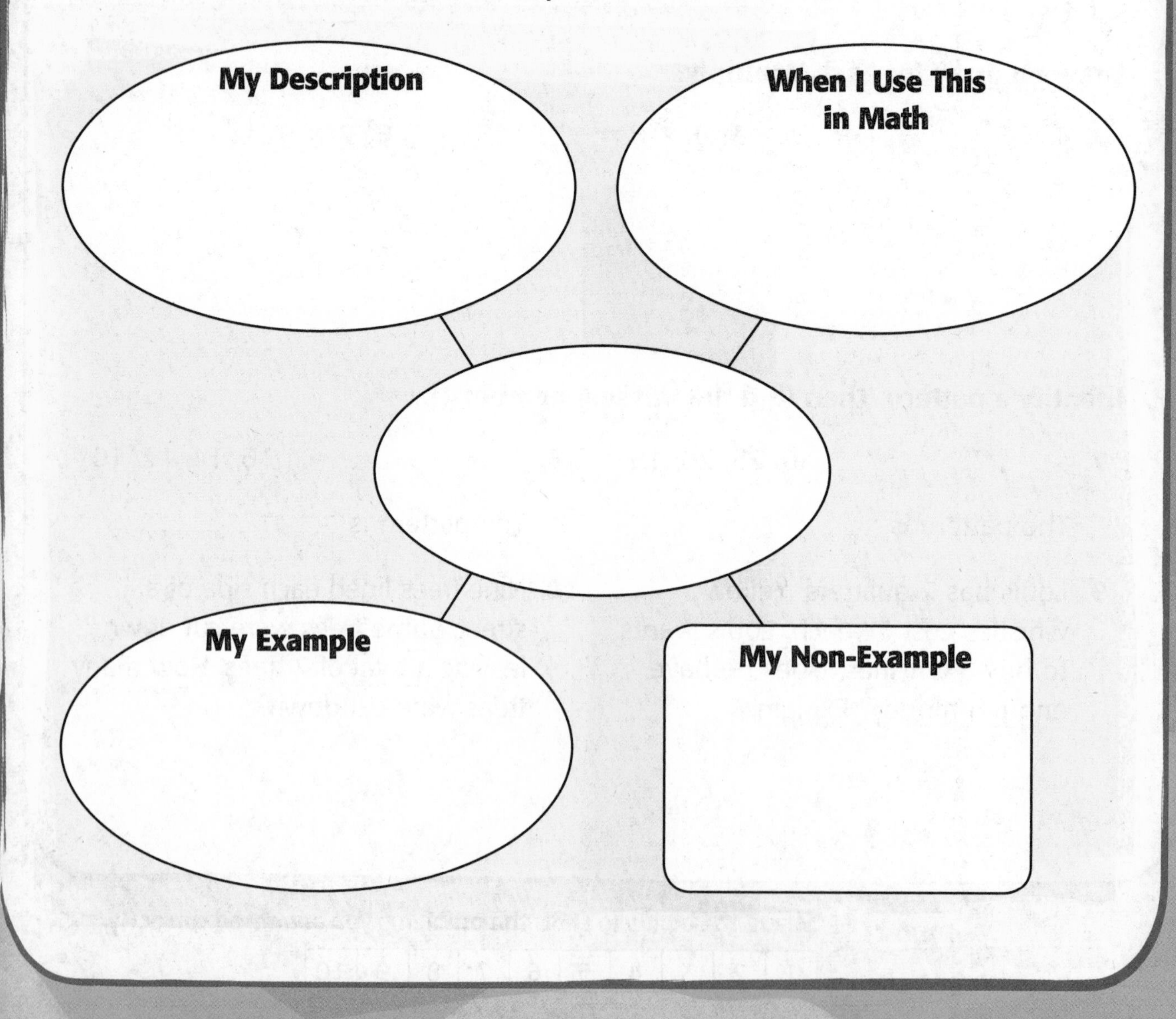

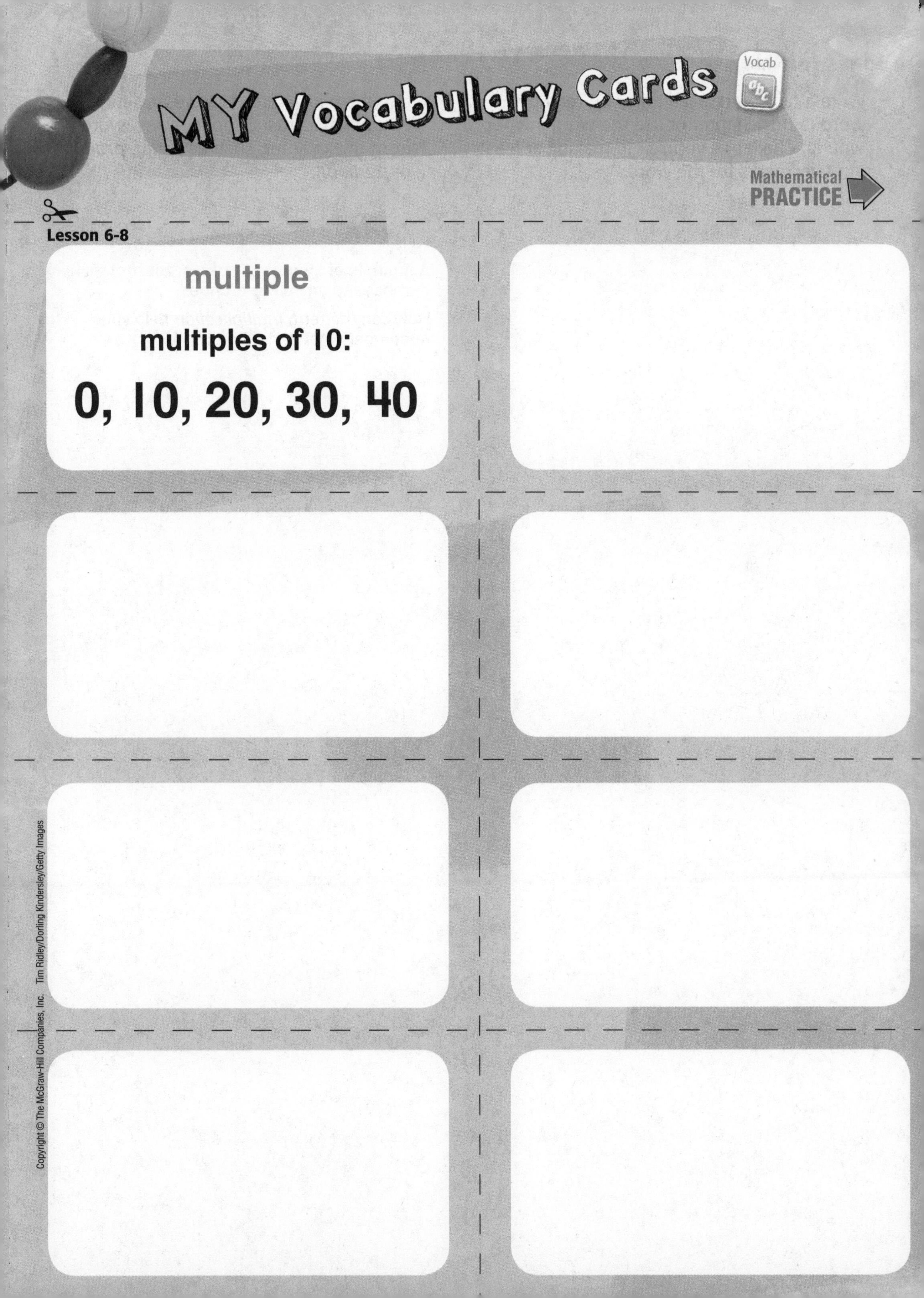

MY Vocabulary Cards

Vocab abc

Mathematical PRACTICE

Lesson 6-8

multiple

multiples of 10:

0, 10, 20, 30, 40

Ideas for Use

- Write a tally mark each time you read the word in this chapter or use the word in your writing. Challenge yourself to making at least ten tally marks for the word.
- Use the blank cards to write review vocabulary cards. Choose review words from this chapter, such as *factor, product,* or *partition.*

A multiple of a number is the product of that number and any other number.

How can the term *multiplication* help you remember what a multiple is?

MY Foldable

FOLDABLES® Follow the steps on the back to make your Foldable.

1

2

3

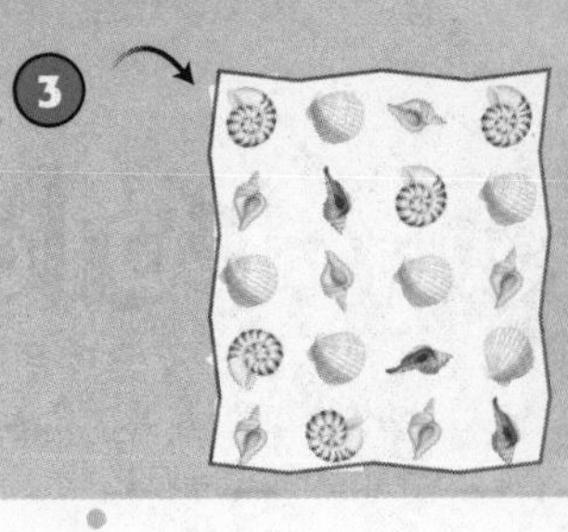

Name ______________________________

Patterns in the Multiplication Table

Lesson 1

ESSENTIAL QUESTION
What is the importance of patterns in learning multiplication and division?

Patterns in the multiplication table can help you remember products and find unknown factors.

Math in My World

Watch Tutor

Example 1

Enrique noticed he could find the product of two factors in the multiplication table. What is the product of 2 × 3?

The **black numbers** in the table are products. The column and row of **blue numbers** are factors.

×	0	1	2	3	4	5	6	7	8	9	10
0	0	0	0	0	0	0	0	0	0	0	0
1	0	1	2	3	4	5	6	7	8	9	10
2	0	2	4	6	8	10	12	14	16	18	20
3	0	3	6	9	12	15	18	21	24	27	30
4	0	4	8	12	16	20	24	28	32	36	40
5	0	5	10	15	20	25	30	35	40	45	50
6	0	6	12	18	24	30	36	42	48	54	60
7	0	7	14	21	28	35	42	49	56	63	70
8	0	8	16	24	32	40	48	56	64	72	80
9	0	9	18	27	36	45	54	63	72	81	90
10	0	10	20	30	40	50	60	70	80	90	100

1. Look at the two circled factors. Follow the numbers across and down until they meet. This is the product. Complete the number sentence.

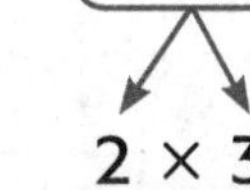

factors

2 × 3 = ________ ← product

2. Draw a triangle around the product in the multiplication table that has the same factors. Follow left and above to its factors. Draw a triangle around each factor. Complete the number sentence.

________ × ________ = 6

The two number sentences are examples of the

____________ Property of Multiplication.

Example 2

Enrique found a pattern when he multiplied 4 by any factor.

Use a yellow crayon to finish Enrique's pattern. Write the numbers.

0, 4, 8, 12, _____, _____, _____,

_____, _____, _____, _____

Circle whether the product of 4 and any number is even or odd.

even   odd

The product of 4 and 5 is 20. Write this product as the sum of two equal numbers.

_____ + _____ = 20

×	0	1	2	3	4	5	6	7	8	9	10
0	0	0	0	0	0	0	0	0	0	0	0
1	0	1	2	3	4	5	6	7	8	9	10
2	0	2	4	6	8	10	12	14	16	18	20
3	0	3	6	9	12	15	18	21	24	27	30
4	0	4	8	12	16	20	24	28	32	36	40
5	0	5	10	15	20	25	30	35	40	45	50
6	0	6	12	18	24	30	36	42	48	54	60
7	0	7	14	21	28	35	42	49	56	63	70
8	0	8	16	24	32	40	48	56	64	72	80
9	0	9	18	27	36	45	54	63	72	81	90
10	0	10	20	30	40	50	60	70	80	90	100

Example 3

Use a blue crayon to color the products with a factor of 3. What do you notice about these products?

The list of products with a factor of _____ increase by _____. It is as if you are counting by 3s.

Guided Practice

1. Use an orange crayon to color the products with a factor of 5. What do you notice about the products in this row and column?

   The products with a factor of

   _____ end in _____ or _____.

2. Use a purple crayon to color the products with a factor of 10. What do you notice about the products in this row and column?

   The products with a factor of

   _____ end in _____.

×	0	1	2	3	4	5	6	7	8	9	10
0	0	0	0	0	0	0	0	0	0	0	0
1	0	1	2	3	4	5	6	7	8	9	10
2	0	2	4	6	8	10	12	14	16	18	20
3	0	3	6	9	12	15	18	21	24	27	30
4	0	4	8	12	16	20	24	28	32	36	40
5	0	5	10	15	20	25	30	35	40	45	50
6	0	6	12	18	24	30	36	42	48	54	60
7	0	7	14	21	28	35	42	49	56	63	70
8	0	8	16	24	32	40	48	56	64	72	80
9	0	9	18	27	36	45	54	63	72	81	90
10	0	10	20	30	40	50	60	70	80	90	100

Name ....

CCSS

Independent Practice

3. Shade a row of numbers **blue** that show the products with a factor of 2. What do you notice about the products in this row?

The 2s products end in

_______, _______, _______,

_______, or _______.

Are all the products in this row even or odd?

____

×	0	1	2	3	4	5	6	7	8	9	10
0	0	0	0	0	0	0	0	0	0	0	0
1	0	1	2	3	4	5	6	7	8	9	10
2	0	2	4	6	8	10	12	14	16	18	20
3	0	3	6	9	12	15	18	21	24	27	30
4	0	4	8	12	16	20	24	28	32	36	40
5	0	5	10	15	20	25	30	35	40	45	50
6	0	6	12	18	24	30	36	42	48	54	60
7	0	7	14	21	28	35	42	49	56	63	70
8	0	8	16	24	32	40	48	56	64	72	80
9	0	9	18	27	36	45	54	63	72	81	90
10	0	10	20	30	40	50	60	70	80	90	100

4. Shade a column of numbers **green** that show the products with a factor of 3. Describe the pattern of even and odd products.

____

5. Shade a row of numbers **yellow** that show the products with a factor of 1. What do you notice about this row?

____

6. Look at the product shaded **gray**. Circle the two factors that make this product. Complete the number sentence.

$4 \times$ _______ $= 36$

Draw a triangle around the product that has the same factors. Draw a triangle around each factor. Complete the number sentence.

$9 \times$ _______ $= 36$

The two number sentences show the _______________ Property of Multiplication.

Problem Solving

7. Layne packed 3 toy cars in each of 4 cases. Circle the factors and shade the product to find how many toy cars Layne packed.

×	0	1	2	3	4	5	6	7	8	9	10
0	0	0	0	0	0	0	0	0	0	0	0
1	0	1	2	3	4	5	6	7	8	9	10
2	0	2	4	6	8	10	12	14	16	18	20
3	0	3	6	9	12	15	18	21	24	27	30
4	0	4	8	12	16	20	24	28	32	36	40
5	0	5	10	15	20	25	30	35	40	45	50

____

Joey packed 4 toy cars in each of 3 cases. Circle the other two factors and shade the product to find how many toy cars Joey packed.

____

8. Write the two number sentences that show the ways each boy packed the toy cars in Exercise 7.

____

Which property is this an example of?

_______________ Property of _______________

HOT Problems

9. Mathematical PRACTICE 7 **Identify Structure** Write a real-world problem for which you can use the multiplication table and the Commutative Property of Multiplication to solve. Then solve.

____

____

____

10. **Building on the Essential Question** How can a multiplication table help you multiply?

____

____

Name ______________________________

MY Homework

Lesson 1
Patterns in the Multiplication Table

Homework Helper

eHelp

Need help? connectED.mcgraw-hill.com

Find the product of 3 × 4.

1. Find 3 in the far left column.
2. Find 4 in the row along the top.
3. Follow the numbers across and down until they meet. This is the product.

×	0	1	2	3	4	5	6	7	8	9	10
0	0	0	0	0	0	0	0	0	0	0	0
1	0	1	2	3	4	5	6	7	8	9	10
2	0	2	4	6	8	10	12	14	16	18	20
3	0	3	6	9	12	15	18	21	24	27	30
4	0	4	8	12	16	20	24	28	32	36	40
5	0	5	10	15	20	25	30	35	40	45	50
6	0	6	12	18	24	30	36	42	48	54	60
7	0	7	14	21	28	35	42	49	56	63	70
8	0	8	16	24	32	40	48	56	64	72	80
9	0	9	18	27	36	45	54	63	72	81	90
10	0	10	20	30	40	50	60	70	80	90	100

factors

3 × 4 = 12 ← product

The Commutative Property tells you that you can change the order of the factors without changing the product.

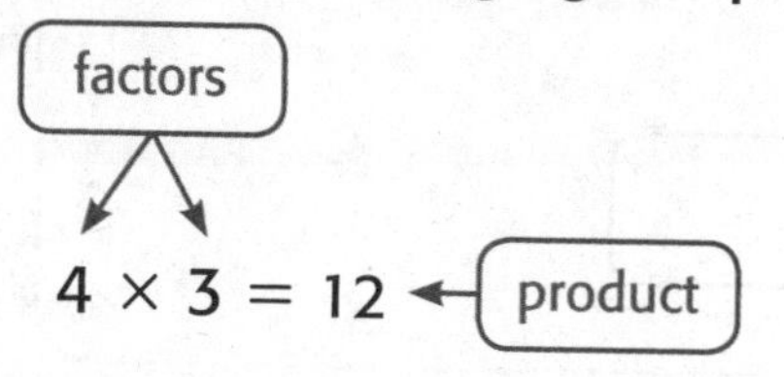

Practice

1. Look at the products with a factor of 5. What pattern do you see? The products with a factor of 5 end in ______ or ______.

2. Look at the products with a factor of 0. What do you notice? The products with a factor of 0 end in ______.

3. Find 10 × 5. Circle the factors and the product. Write the product.

____

×	0	1	2	3	4	5	6	7	8	9	10
0	0	0	0	0	0	0	0	0	0	0	0
1	0	1	2	3	4	5	6	7	8	9	10
2	0	2	4	6	8	10	12	14	16	18	20
3	0	3	6	9	12	15	18	21	24	27	30
4	0	4	8	12	16	20	24	28	32	36	40
5	0	5	10	15	20	25	30	35	40	45	50
6	0	6	12	18	24	30	36	42	48	54	60
7	0	7	14	21	28	35	42	49	56	63	70
8	0	8	16	24	32	40	48	56	64	72	80
9	0	9	18	27	36	45	54	63	72	81	90
10	0	10	20	30	40	50	60	70	80	90	100

4. Shade a row of numbers yellow to show the products with a factor of 10. What do you notice about this row?

The products with a factor of 10 end in ________.

Problem Solving

5. Mathematical PRACTICE 4 **Model Math** Mason has 1 notebook for science and 1 notebook for reading. He put 9 stickers on each notebook. How many stickers did Mason use in all? Write two multiplication sentences.

____

Vocabulary Check

Vocab

6. Label each with the correct word.

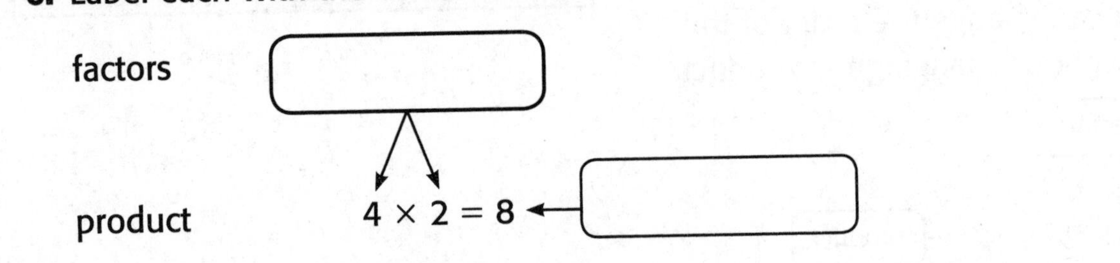

Test Practice

7. Which property states the order in which two numbers are multiplied does not change the product?

Ⓐ Associative Property of Addition

Ⓑ Commutative Property of Multiplication

Ⓒ Inverse Operations

Ⓓ Identity Property of Addition

Name ....

Multiply by 2

Lesson 2

ESSENTIAL QUESTION
What is the importance of patterns in learning multiplication and division?

Math in My World

Tools Watch Tutor

Group Project!

Example 1

The students in an art class are working on a project. How many students are there in the art class if there are 8 groups of 2?

Find 8 groups of 2.

Write 8 groups of 2 as 8 × 2.

One Way Use an array.
Draw an array with 8 rows and 2 columns.

Another Way Draw a picture.
Draw 8 equal groups of 2.

My Drawing!

Write an addition sentence and multiplication sentence.

___ + ___ + ___ + ___ + ___ + ___ + ___ + ___ = ___ ___ × ___ = ___

So, 8 × 2 = ______. There are ______ students in the art class.

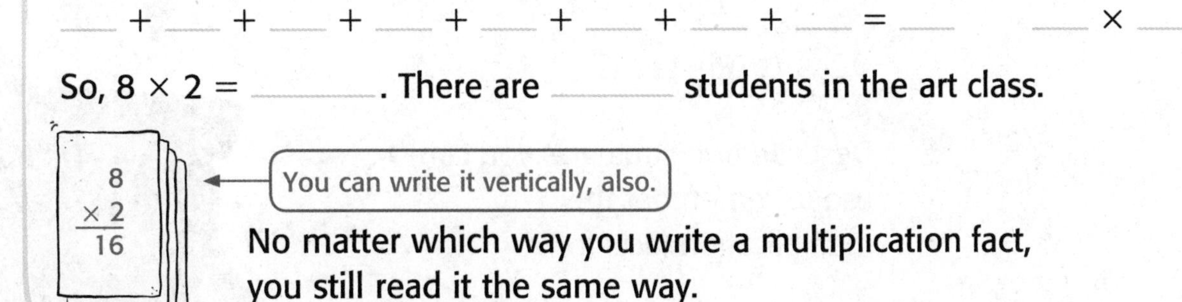

$$\begin{array}{r} 8 \\ \times\ 2 \\ \hline 16 \end{array}$$

You can write it vertically, also.

No matter which way you write a multiplication fact, you still read it the same way.

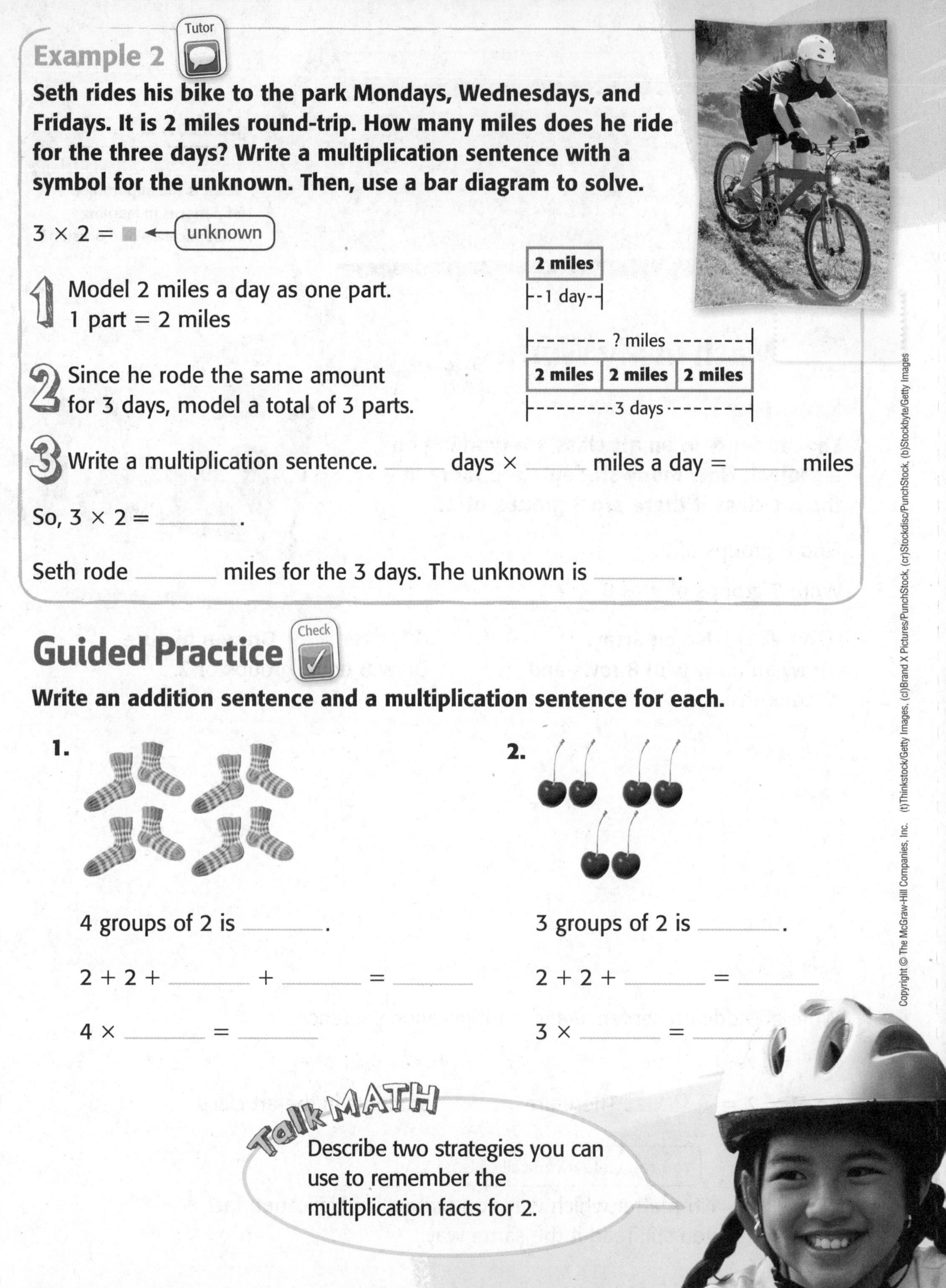

Example 2

Tutor

Seth rides his bike to the park Mondays, Wednesdays, and Fridays. It is 2 miles round-trip. How many miles does he ride for the three days? Write a multiplication sentence with a symbol for the unknown. Then, use a bar diagram to solve.

$3 \times 2 = \blacksquare$ ← unknown

1 Model 2 miles a day as one part.
1 part = 2 miles

2 Since he rode the same amount for 3 days, model a total of 3 parts.

3 Write a multiplication sentence. _____ days × _____ miles a day = _____ miles

So, $3 \times 2 =$ _____.

Seth rode _____ miles for the 3 days. The unknown is _____.

Guided Practice

Check

Write an addition sentence and a multiplication sentence for each.

1.

4 groups of 2 is _____.

2 + 2 + _____ + _____ = _____

4 × _____ = _____

2.

3 groups of 2 is _____.

2 + 2 + _____ = _____

3 × _____ = _____

Talk MATH

Describe two strategies you can use to remember the multiplication facts for 2.

Lesson 2

Multiply by 2

Homework Helper

eHelp

Need help? connectED.mcgraw-hill.com

Helen buys 2 bunches of bananas. There are 10 bananas in each bunch. How many bananas does Helen buy in all?

Find 2 × 10.

This can be written vertically, also.

Use an array to model 2 groups of 10.

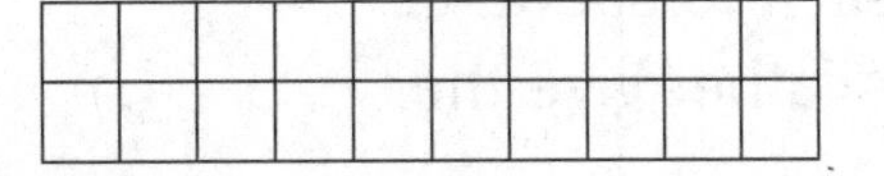

You can write an addition sentence to represent the models.

10 + 10 = 20

OR

You can write a multiplication sentence to represent the models.

2 × 10 = 20

So, Helen bought 20 bananas in all.

Practice

Write an addition sentence and a multiplication sentence.

1.

3 groups of 2 is ______.

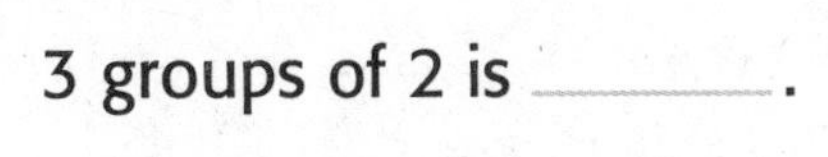

2 + ______ + ______ = ______

______ × 2 = ______

2.

4 groups of 2 is ______.

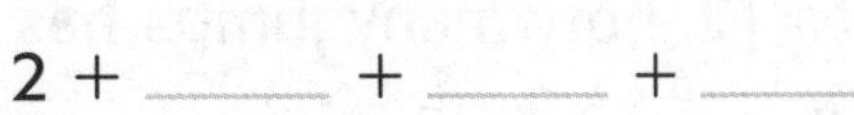

2 + ______ + ______ + ______ = ______

______ × 2 = ______

Draw an array for each. Then write a multiplication sentence.

3. 7 rows of 2

4. 2 rows of 5

______ × ______ = ______

______ × ______ = ______

Problem Solving

Mathematical PRACTICE 2 **Use Algebra** **Write a multiplication sentence with a symbol for the unknown. Then solve.**

5. Franklin's father gave him and his sister $8 each to spend at the movies. How much money did Franklin's father give the children altogether?

____

6. There are 7 people in the Watson family. They all keep their gloves in 1 box in the closet. If each person has a pair of gloves, how many gloves are in the box?

____

Vocabulary Check

7. Write or draw the meaning of a bar diagram.

____

____

Test Practice

8. James is jumping on a pogo stick. He is counting by twos. If he counts to 12, how many jumps has he made?

Ⓐ 2 jumps
Ⓑ 4 jumps
Ⓒ 6 jumps
Ⓓ 10 jumps

Name ....

Divide by 2

Lesson 3

ESSENTIAL QUESTION
What is the importance of patterns in learning multiplication and division?

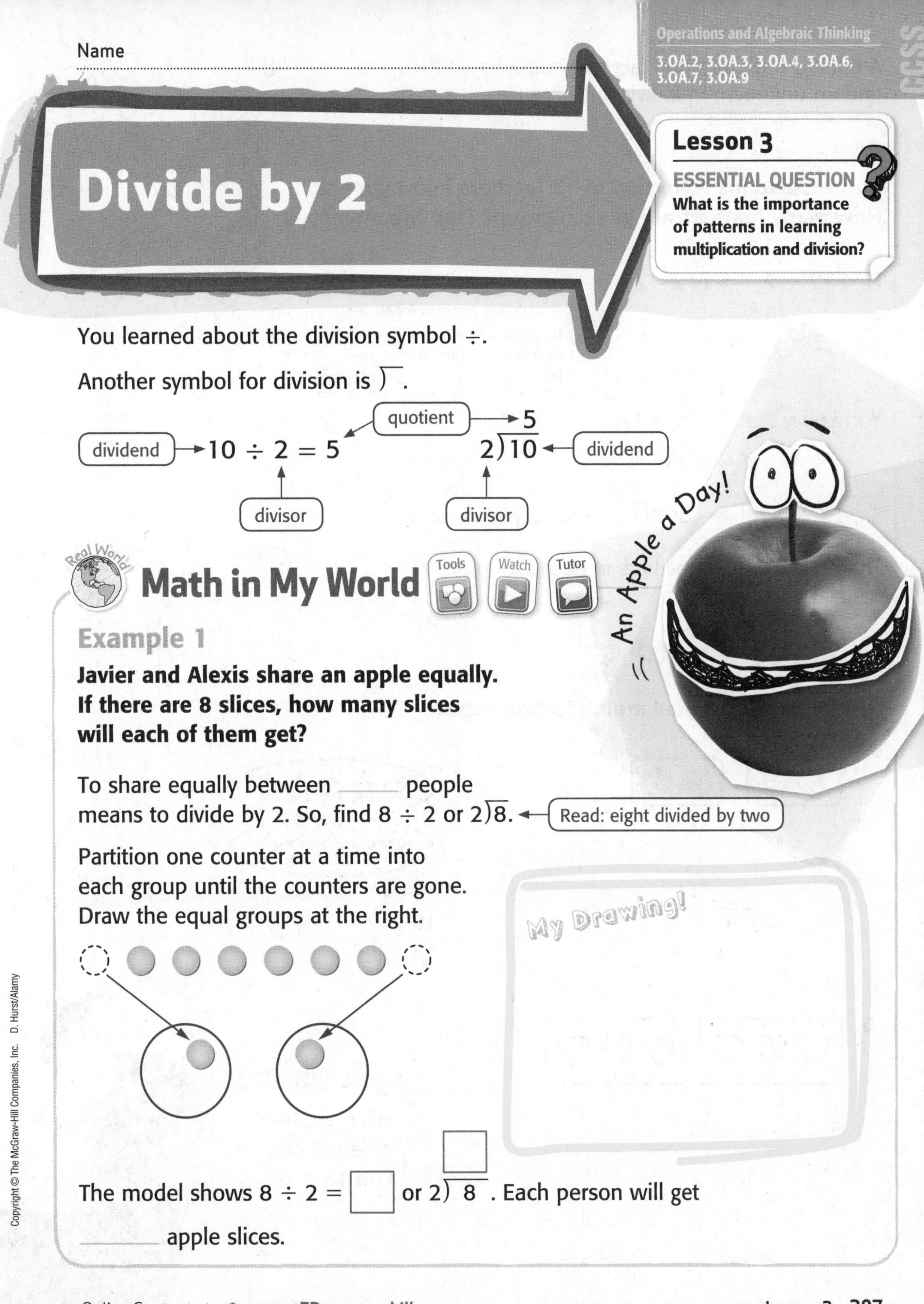

You learned about the division symbol ÷.

Another symbol for division is $\overline{)\ \ }$.

dividend → $10 \div 2 = 5$ ← quotient (divisor: 2)

quotient → $\begin{array}{r}5\\ 2\overline{)10}\end{array}$ ← dividend (divisor: 2)

Math in My World

Tools Watch Tutor

Example 1

Javier and Alexis share an apple equally. If there are 8 slices, how many slices will each of them get?

To share equally between ______ people means to divide by 2. So, find $8 \div 2$ or $2\overline{)8}$. ← Read: eight divided by two

Partition one counter at a time into each group until the counters are gone. Draw the equal groups at the right.

The model shows $8 \div 2 =$ ☐ or $2\overline{)\ 8}$ (quotient ☐). Each person will get ______ apple slices.

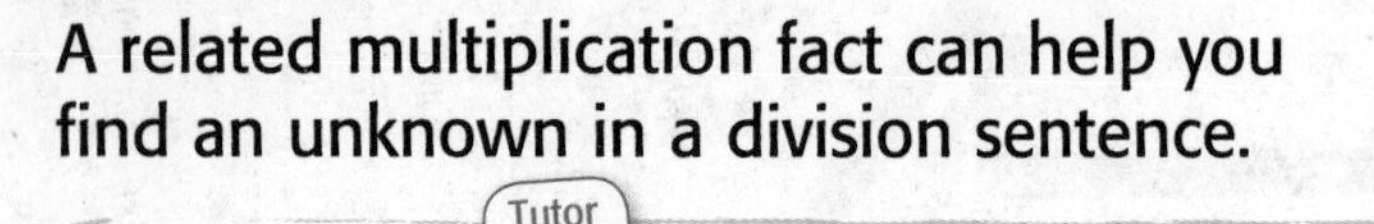

A related multiplication fact can help you find an unknown in a division sentence.

Example 2

Max divided his collection of 12 feathers into 2 groups. How many feathers are in each group? Find the unknown.

Find $12 \div 2 = ■$ or $2\overline{)12}$.

$12 \div 2 = ■ \rightarrow 2 \times ■ = 12$ ← A division sentence can be thought of as a multiplication sentence in which you are looking for an unknown factor.

Tee-hee-hee!!

You know that $2 \times 6 = 12$.

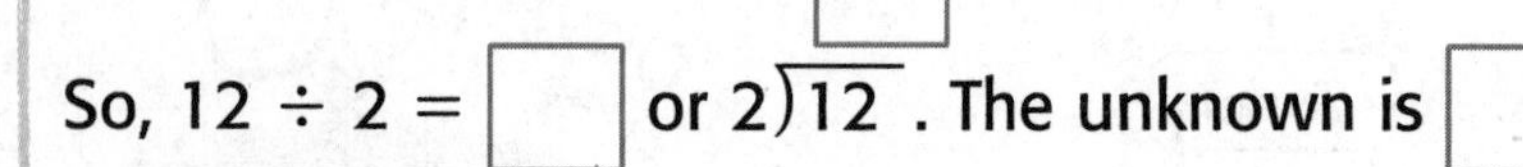

So, $12 \div 2 =$ ☐ or $2\overline{)12}$ ☐. The unknown is ☐.

There are ______ feathers in each group.

Guided Practice

Divide. Write a related multiplication fact.

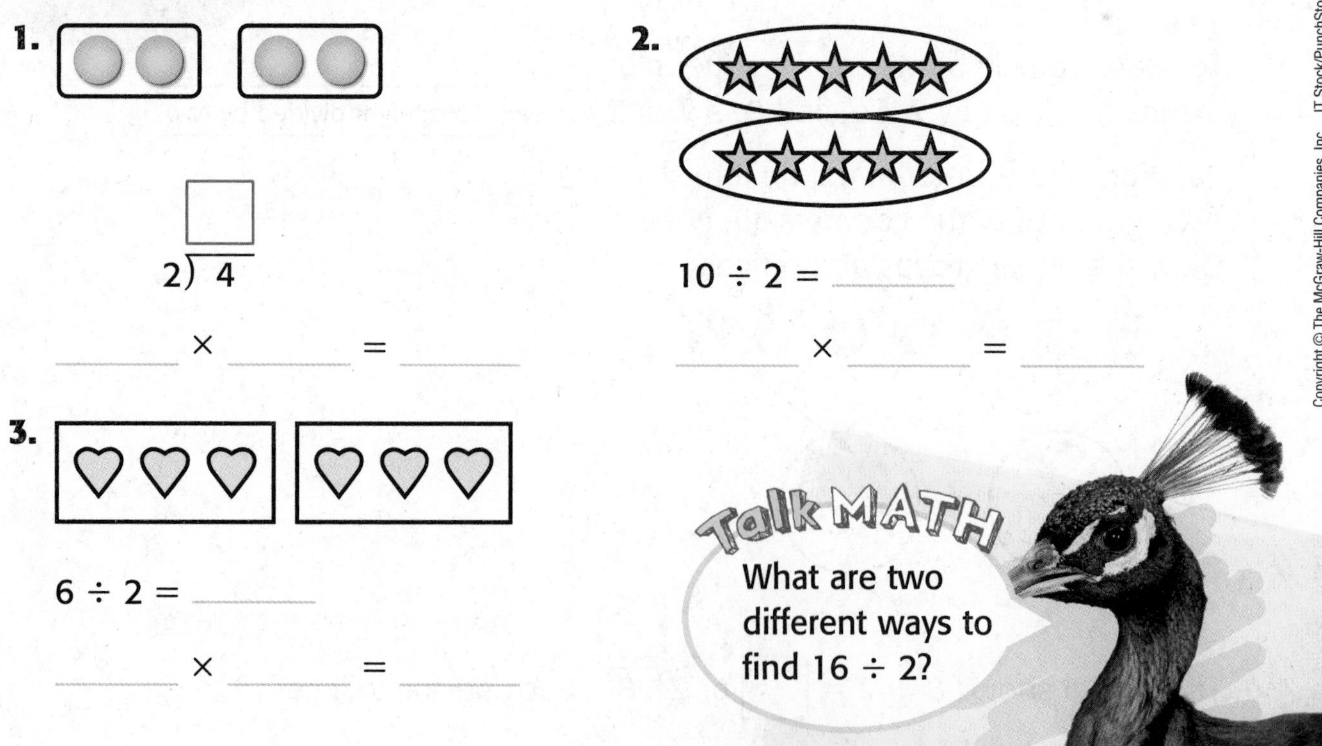

1. $2\overline{)4}$ ☐

______ × ______ = ______

2. $10 \div 2 =$ ______

______ × ______ = ______

3. $6 \div 2 =$ ______

______ × ______ = ______

Talk MATH

What are two different ways to find $16 \div 2$?

Name ....

CCSS

Independent Practice

Divide. Write a related multiplication fact.

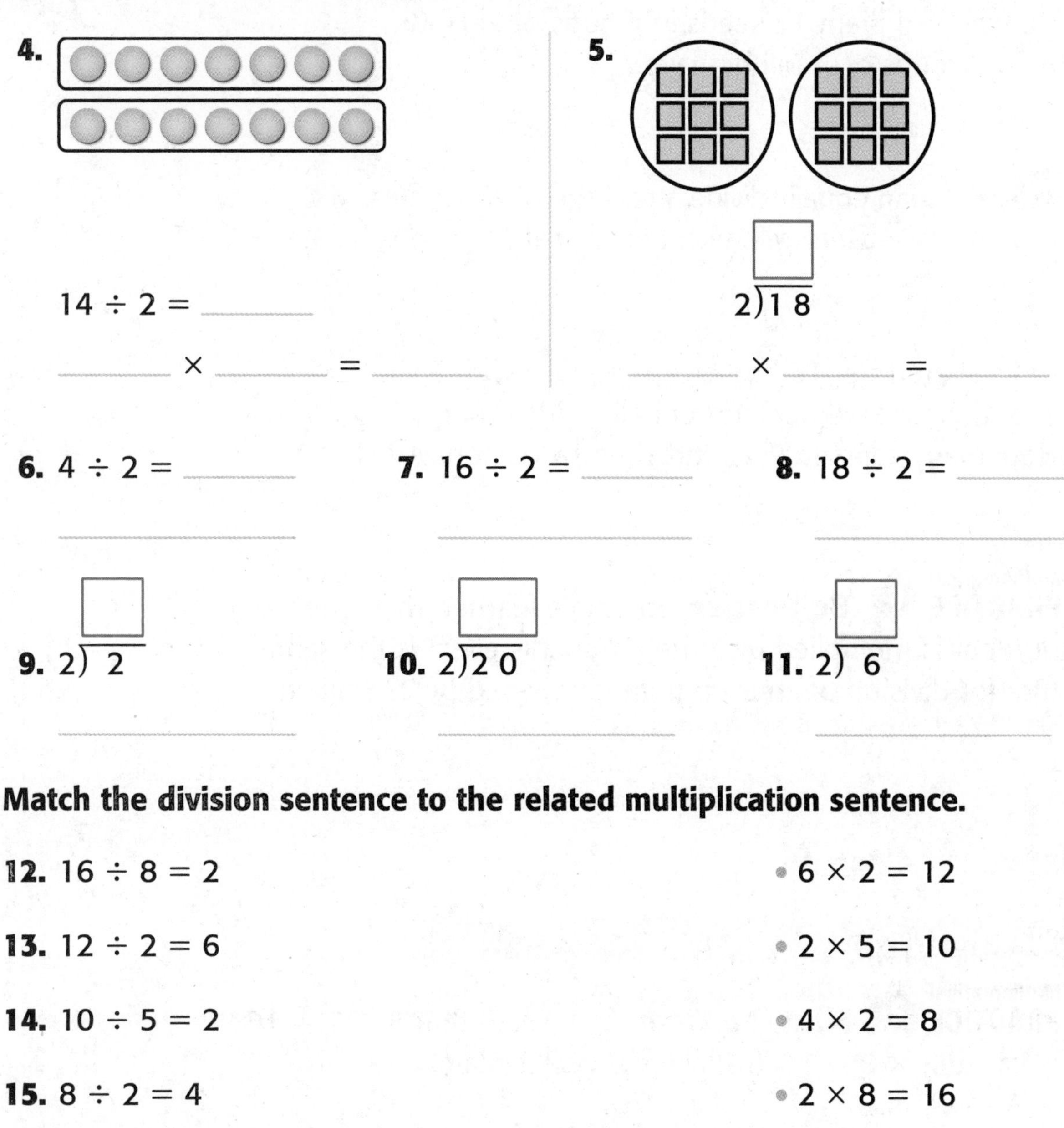

4. $14 \div 2 =$ ______

______ $\times$ ______ $=$ ______

5. $2\overline{)18}$ ☐

______ $\times$ ______ $=$ ______

6. $4 \div 2 =$ ______

______

7. $16 \div 2 =$ ______

______

8. $18 \div 2 =$ ______

______

9. $2\overline{)\ 2}$ ☐

______

10. $2\overline{)20}$ ☐

______

11. $2\overline{)\ 6}$ ☐

______

Match the division sentence to the related multiplication sentence.

12. $16 \div 8 = 2$ • $6 \times 2 = 12$

13. $12 \div 2 = 6$ • $2 \times 5 = 10$

14. $10 \div 5 = 2$ • $4 \times 2 = 8$

15. $8 \div 2 = 4$ • $2 \times 8 = 16$

Algebra Find the unknown. Then write a related multiplication sentence.

16. $12 \div 6 =$ ■

The unknown is ______.

______

17. $14 \div$ ■ $= 2$

The unknown is ______.

______

18. ■ $\div 2 = 3$

The unknown is ______.

______

Problem Solving

Algebra Write a division sentence with a symbol for the unknown for Exercises 19–20. Then solve.

My Work!

19. Damian will plant 12 seeds in groups of 2. How many groups of 2 will he have?

20. Kyle and Alan equally divide a package of 14 erasers. How many erasers will each person get?

21. Lydia shared her 16 bottle caps equally with Pilar. Pilar then shared her caps equally with Timothy. How many caps do Pilar and Timothy each have?

22. Mathematical PRACTICE 6 **Be Precise** You have learned that when any number is multiplied by 2 the product is even. Is the same true for division of an even number divided by 2? Explain.

HOT Problems

23. Mathematical PRACTICE 3 **Find the Error** Blake says that $8 \div 2 = 16$ because $2 \times 8 = 16$. Is Blake correct? Explain.

24. **Building on the Essential Question** How does the relationship between division and multiplication help you find the unknown?

Name ....

Lesson 3
Divide by 2

Homework Helper

eHelp Need help? connectED.mcgraw-hill.com

The school van can carry 12 passengers. There are 2 passengers to a seat. How many seats are in the van?

Find $12 \div 2$, or $2\overline{)12}$.

Partition 12 counters between 2 groups until there are none left.

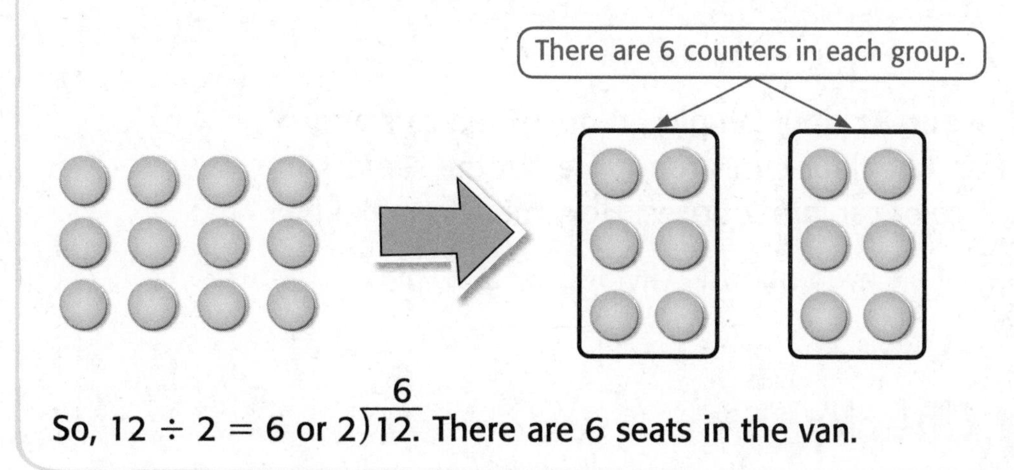

So, $12 \div 2 = 6$ or $2\overline{)12}$ = 6. There are 6 seats in the van.

Practice

Divide. Write a related multiplication fact.

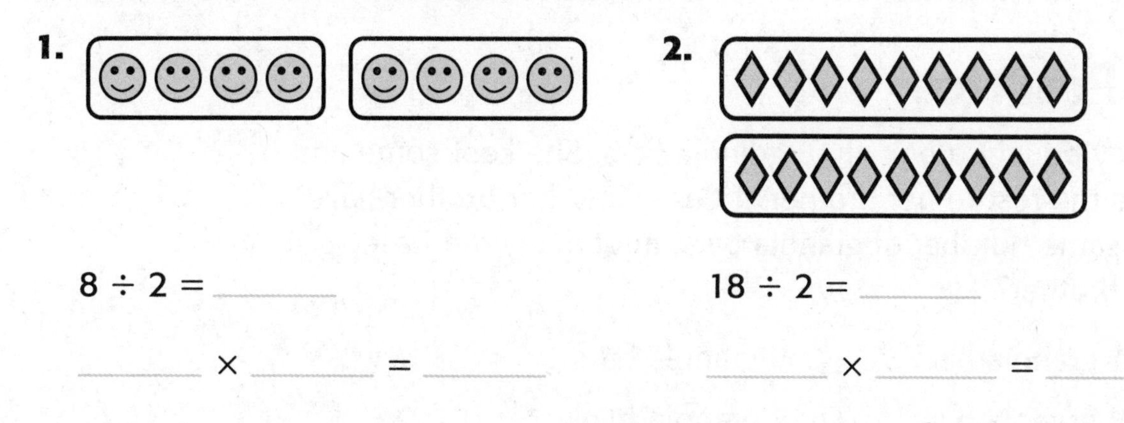

1. $8 \div 2 =$ ______

______ × ______ = ______

2. $18 \div 2 =$ ______

______ × ______ = ______

Divide. Write a related multiplication fact.

3. $20 \div 2 =$ ______ **4.** $6 \div 2 =$ ______ **5.** $12 \div 2 =$ ______

______ ______ ______

6. $2\overline{)\,8}$ **7.** $2\overline{)14}$ **8.** $2\overline{)\,4}$

______ ______ ______

Real World Problem Solving

9. Algebra Britt spent $12 equally at 2 stores. How much did she spend at each store? Write a number sentence with a symbol for the unknown. Then solve.

______

10. Mathematical PRACTICE 1 **Keep Trying** Ian picked up 16 red cars and 12 black cars from the floor of his room. He put the same number of each color car into 2 boxes. How many cars did he put in each box?

______

Vocabulary Check

11. Write or draw a definition of the word partition.

______

______

Test Practice

12. Casey bought a box of 18 granola bars. She kept some and gave the rest to her brother. If Casey and her brother have the same number of granola bars, how many did Casey give her brother?

Ⓐ 1 granola bar Ⓒ 9 granola bars

Ⓑ 8 granola bars Ⓓ 7 granola bars

Name

Multiply by 5

Lesson 4

ESSENTIAL QUESTION
What is the importance of patterns in learning multiplication and division?

You can use patterns to multiply by 5. Multiplying by a number is the same as skip counting by that number.

I'm multiplying!

Math in My World

Tools Watch Tutor

Example 1

Leandro has 7 nickels. How much money does he have?

One nickel equals 5¢. Skip count by fives to find 7 × 5¢.

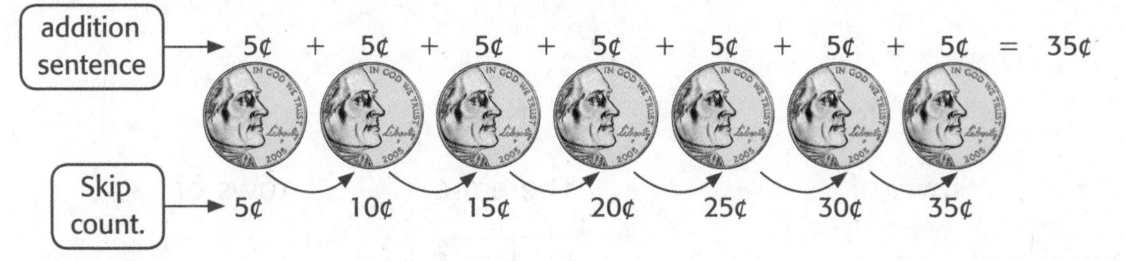

7 nickels is ____ ¢. 7 × 5¢ = ____ ¢

So, Leandro has ____ ¢.

Notice the pattern in the products.

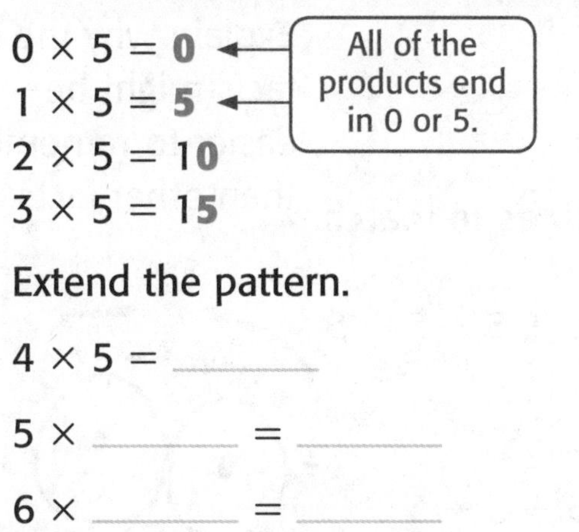

Helpful Hint
When you multiply by 5, the product will always end in 0 or 5.

Extend the pattern.

4 × 5 = ____

5 × ____ = ____

6 × ____ = ____

7 × ____ = ____

Example 2

Tutor

A watermelon patch has 6 rows of watermelons. Each row has 5 watermelons. How many watermelons are in the farmer's patch? Write a multiplication sentence with a symbol for the unknown.

Line Up!

$6 \times 5 = \blacksquare$ ← unknown

1 Draw an array with 6 rows.

2 Use the Commutative Property to draw another array with 5 rows.

My Drawing!

There are ______ rows of ______.

So, $6 \times 5 =$ ______.

The unknown is ______.

There are ______ rows of ______.

So, $5 \times 6 =$ ______.

The unknown is ______.

There are ______ watermelons in the farmer's patch.

Talk MATH

Explain why the 5s facts might be easier to remember than other facts.

Guided Practice

Check

Skip count by fives to find each product. Draw lines to match.

1. $4 \times 5 =$ ☐ • $5 + 5 + 5 + 5 + 5 + 5 + 5 + 5$
2. $3 \times 5 =$ ☐ • $5 + 5 + 5 + 5$
3. $8 \times 5 =$ ☐ • $5 + 5 + 5 + 5 + 5 + 5 + 5$
4. $7 \times 5 =$ ☐ • $5 + 5 + 5$

Name

Independent Practice

Write an addition sentence to help find each product.

5. $2 \times 5 =$ ____

$5 +$ ____ $=$ ____

6. $3 \times 5 =$ ____

$5 +$ ____ $+$ ____ $=$ ____

7. $7 \times 5 =$ ____

________________ $=$ ____

8. $8 \times 5 =$ ____

________________ $=$ ____

9. $5 \times 5 =$ ____

________________ $=$ ____

10. $9 \times 5 =$ ____

________________ $=$ ____

Draw an array for each. Then write a multiplication sentence.

11. 7 rows of 5

$7 \times$ ____ $=$ ____

12. 3 rows of 5

____ $\times$ ____ $=$ ____

13. 4 rows of 5

____ $\times$ ____ $=$ ____

Algebra **Find each unknown. Use the Commutative Property.**

14. $■ \times 6 = 30$
$6 \times ■ = 30$

The unknown is ____.

15. $5 \times ■ = 10$
$■ \times 5 = 10$

The unknown is ____.

16. $9 \times 5 = ■$
$5 \times 9 = ■$

The unknown is ____.

Real World! Problem Solving

17. Kai, Lakita, and Maxwell collected acorns. If each gets 5 acorns, how many acorns did they collect? Explain.

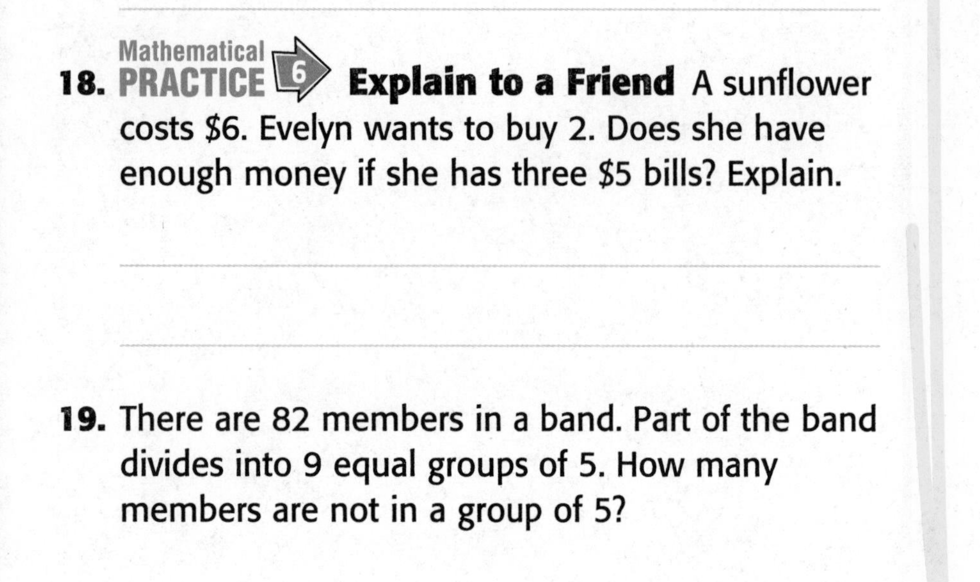

My Work!

18. Mathematical PRACTICE 6 **Explain to a Friend** A sunflower costs $6. Evelyn wants to buy 2. Does she have enough money if she has three $5 bills? Explain.

19. There are 82 members in a band. Part of the band divides into 9 equal groups of 5. How many members are not in a group of 5?

HOT Problems

20. Mathematical PRACTICE 2 **Reason** Circle the strategy that will not help you find 6×5. Explain.

skip counting	rounding
make an array	draw a picture

21. **Building on the Essential Question** What do you notice about all the products of 5? Use the multiplication table if needed.

Lesson 4

Multiply by 5

Homework Helper

eHelp

Need help? connectED.mcgraw-hill.com

There are 6 students. Each student donates $5 to a school fundraiser. How much money did the students donate in all?

Find 6 × $5.

One Way **Skip count by fives.**

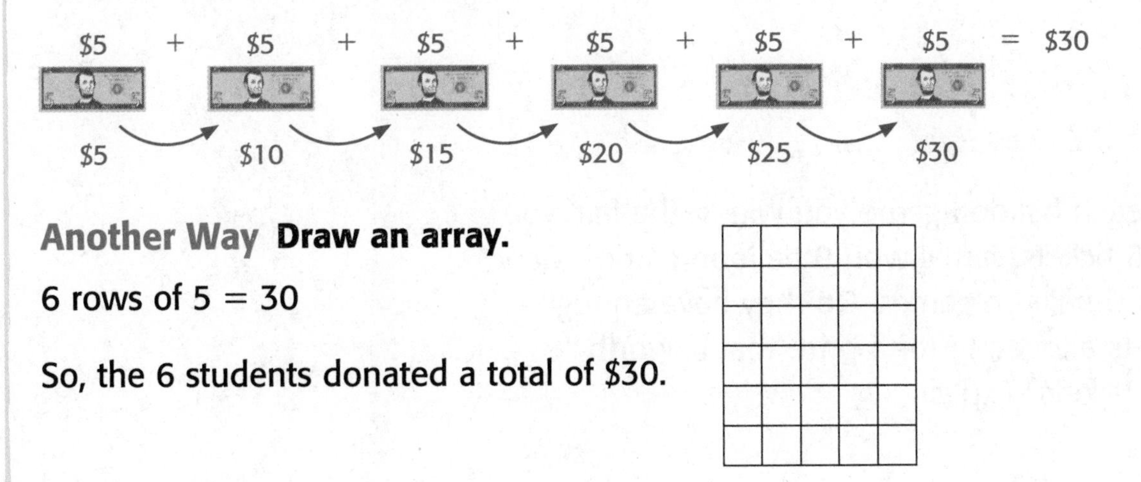

Another Way **Draw an array.**

6 rows of 5 = 30

So, the 6 students donated a total of $30.

Practice

Write an addition sentence to help find each product.

1. 3 × 5 = ______

____

2. 8 × 5 = ______

____

3. 5 × 5 = ______

____

Write a multiplication sentence for each array.

4. 1 row of 5

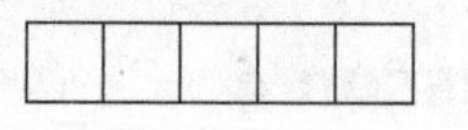

5. 5 rows of 4

6. 5 rows of 9

Problem Solving

7. Each pair of tennis shoes costs \$25. If Andrea has four \$5-bills, does she have enough money to buy one pair? Write a number sentence. Then solve.

My Work!

8. For each balloon game you win at the fair, you get 5 tickets. Jamal won 9 balloon games. Gary won 6 balloon games. Do they have enough tickets altogether for a prize that is worth 100 tickets? Explain.

9. Mathematical PRACTICE 1 **Make Sense of Problems** For a craft, each student will need 5 rubber bands. There are 8 students. Rubber bands come in bags of 9. How many bags will be needed? How many rubber bands will be left over?

Test Practice

10. Shawn has 4 nickels. How many walnuts can he buy if he spends all 4 nickels?

Nuts for Sale
Peanuts -- 3¢ each
Walnuts -- 5¢ each
Chestnuts -- 10¢ each

Ⓐ 1 walnut
Ⓑ 4 walnuts
Ⓒ 5 walnuts
Ⓓ 20 walnuts

Name ....

Divide by 5

Lesson 5

ESSENTIAL QUESTION
What is the importance of patterns in learning multiplication and division?

Use what you know about patterns and multiplying by 5 to divide by 5.

Math in My World

Tools Watch Tutor

Example 1

A group of 5 friends sold a total of 20 glasses of lemonade. They each sold the same number of glasses. How many glasses of lemonade did they each sell?

Find $20 \div 5$.

One Way Use counters and partition.

Partition 20 counters into 5 equal groups. Draw the equal groups.

My Drawing!

There are _____ counters in each group.

$20 \div 5 =$ _____

So, they each sold _____ glasses of lemonade.

Another Way Use repeated subtraction.

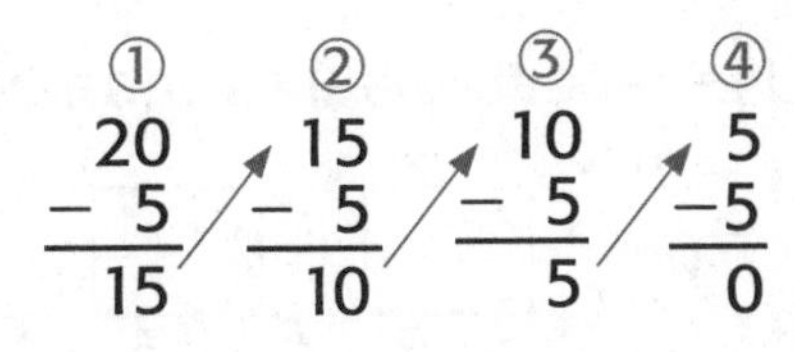

Subtract groups of 5 until you reach 0.

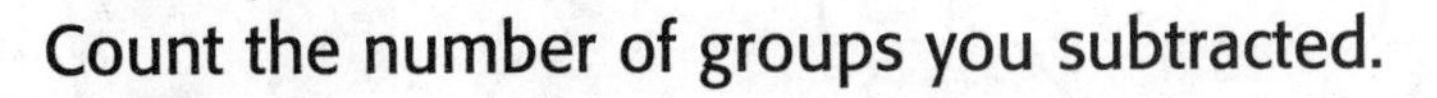

Count the number of groups you subtracted.

Groups of _____ were subtracted _____ times.

There are _____ groups. So, $20 \div 5 =$ _____.

Think of division as an unknown factor problem. Use a related multiplication fact.

Helpful Hint
Nickels can be used to represent the number 5.

Example 2

The school store is selling pencils for 5¢ each. If Corey has 45¢, how many pencils can he buy with his money?

Find the unknown in 45¢ ÷ 5¢ = ■ or $5¢\overline{)45¢}$ (quotient ■).

Draw an array. Then use the inverse operation to find the unknown.

My Drawing!

Think ■ × 5 = 45 (■ = unknown factor)

You know that ______ × 5 = 45.

So, 45¢ ÷ 5¢ = ☐ or $5¢\overline{)45¢}$ (quotient ☐).

The unknown is ______. Corey can buy ______ pencils.

Talk MATH
How can you tell if a number is divisible by 5?

Guided Practice

Use counters to find the number of equal groups or how many are in each group.

1. 35 counters
5 equal groups

______ in each group

35 ÷ 5 = ______

2. 10 counters
5 equal groups

______ in each group

10 ÷ 5 = ______

3. Use repeated subtraction to find 30 ÷ 5.

30	☐	☐	☐	☐	☐
− 5	− 5	− 5	− 5	− 5	− 5
☐	☐	☐	☐	☐	☐

30 ÷ 5 = ______

Name ....

CCSS

Independent Practice

Use counters to find the number of equal groups or how many are in each group.

4. 15 counters
5 equal groups
_____ in each group
15 ÷ 5 = _____

5. 10 counters
_____ equal groups
5 in each group
10 ÷ _____ = 5

6. 25 counters
5 equal groups
_____ in each group
25 ÷ 5 = _____

Use repeated subtraction to divide.

7. 10 ÷ 5 = _____

8. 5 ÷ 1 = _____

Algebra Draw an array and use the inverse operation to find each unknown.

9. ■ × 5 = 20
? ÷ 4 = 5
■ = _____
? = _____

10. 5 × ■ = 40
40 ÷ ? = 8
■ = _____
? = _____

Use the recipe for Buttermilk Corn Bread. Find how much of each ingredient is needed to make 1 loaf of corn bread.

11. cornmeal _____

12. flour _____

13. eggs _____

14. vanilla extract _____

Buttermilk Corn Bread

10 cups cornmeal	3 cups butter
5 cups flour	8 cups buttermilk
1 cup sugar	5 tsp vanilla extract
5 Tbsp baking powder	15 eggs
4 tsp salt	2 tsp baking soda

Makes: 5 loaves

Real World

Problem Solving

Mathematical PRACTICE 2 Use Algebra Write a division sentence with a symbol for the unknown. Then solve.

My Work!

15. Rose had a 30-inch piece of ribbon. She divided the ribbon into 5 equal pieces. How many inches long is each piece?

____

16. Garrison collected 45 flags. He displays them in his room in 5 equal rows. How many flags does Garrison have in each row?

____

HOT Problems

17. Mathematical PRACTICE 1 **Keep Trying** Addison got 40 points on yesterday's 10-question math quiz. Each question is worth 5 points and there is no partial credit. How many questions did she miss?

____

18. Mathematical PRACTICE 2 **Stop and Reflect** Circle the division sentence that does not belong. Explain your reasoning.

$20 \div 2 = 10$	$30 \div 5 = 6$
$30 \div 6 = 5$	$35 \div 5 = 7$

____

19. **Building on the Essential Question** How can an array help you solve a related multiplication and division problem?

____

____

____

Name ....

Lesson 5
Divide by 5

Homework Helper

eHelp Need help? connectED.mcgraw-hill.com

Rudy spent \$30 for 5 car models. Each model costs the same amount. How much did each car model cost?

Find \$30 ÷ 5, or 5)\$30.

One Way **Use counters and partition.**
Partition 30 counters equally among 5 groups until there are none left.

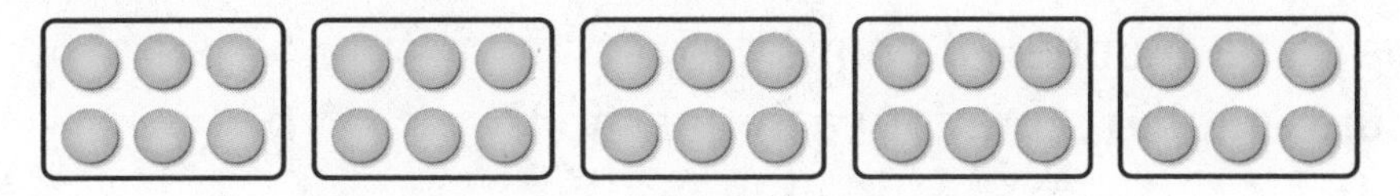

There are 5 equal groups of 6.

Another Way **Use repeated subtraction.**
Subtract 5 until you get to 0. Count the number of times you subtracted.

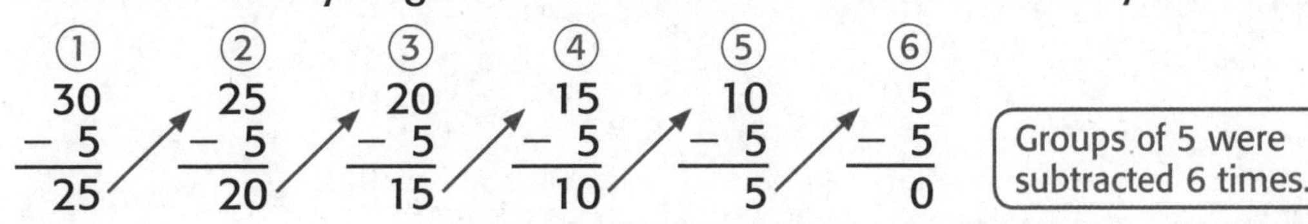

Since \$30 ÷ 5 = \$6, each model cost \$6.

Practice

Partition to find the number of equal groups or how many are in each group.

1. 45 counters
5 equal groups
______ in each group

2. 5 counters
______ equal groups
1 in each group

3. 20 counters
5 equal groups
______ in each group

4. 50 counters
______ equal groups
5 in each group

5. Algebra Draw an array and use the inverse operation to find the unknown.

■ × 5 = 15

? ÷ 3 = 5

■ = ______

? = ______

My Work!

Problem Solving

Write a division sentence with a symbol for the unknown for Exercises 6 and 7. Then solve.

6. Antonio scored 40 points on his math test. There were 5 questions on the test, and each was worth the same number of points. How many points did Antonio score for each question?

______

7. Lunch costs $5. Marcus has $35. How many days can he buy lunch?

______

8. Mathematical PRACTICE 4 **Model Math** Today 25 girls and 20 boys rode their bikes to school. Each bike rack at school holds 5 bikes. How many bike racks were filled?

______

Test Practice

9. Which number sentence represents this repeated subtraction exercise?

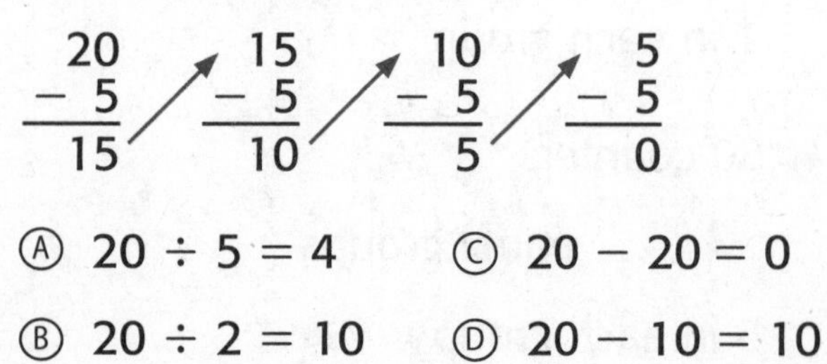

Ⓐ 20 ÷ 5 = 4

Ⓑ 20 ÷ 2 = 10

Ⓒ 20 − 20 = 0

Ⓓ 20 − 10 = 10

Check My Progress

Vocabulary Check

Label each with the correct word(s).

bar diagram **factors** **partition** **product**

1.

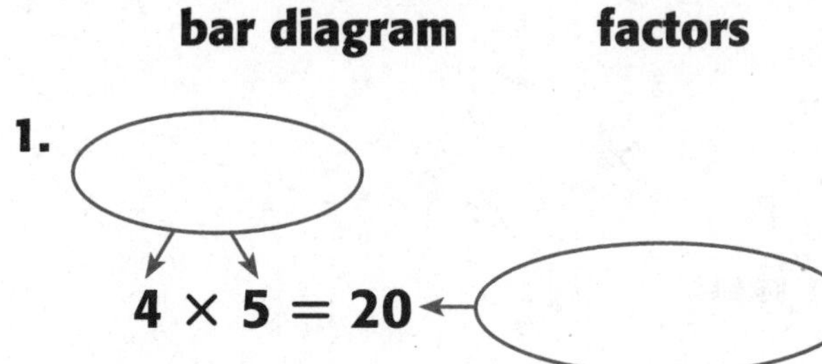

2.

? miles		
2 miles	2 miles	2 miles
3 days		

____

3.

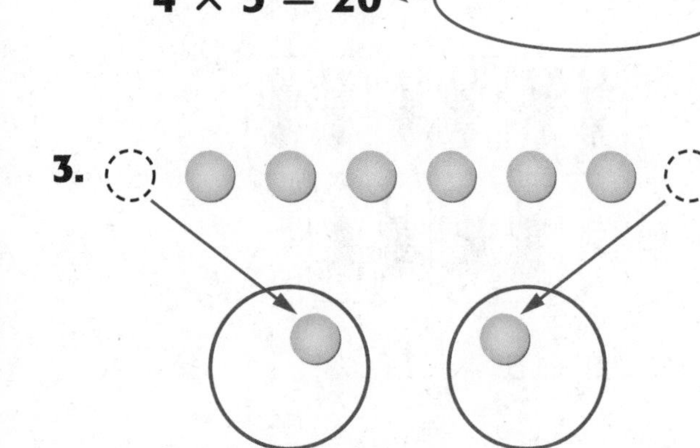

_________ one counter at a time into each group until the counters are gone.

Concept Check

4. Shade the product of the two circled factors. Complete the number sentence.

$6 \times 4 =$ _______

5. Draw a triangle around the product that has the same factors. Write the number sentence that shows the Commutative Property of Multiplication.

____

×	0	1	2	3	4	5	6	7	8	9	10
0	0	0	0	0	0	0	0	0	0	0	0
1	0	1	2	3	4	5	6	7	8	9	10
2	0	2	4	6	8	10	12	14	16	18	20
3	0	3	6	9	12	15	18	21	24	27	30
4	0	4	8	12	16	20	24	28	32	36	40
5	0	5	10	15	20	25	30	35	40	45	50
6	0	6	12	18	24	30	36	42	48	54	60
7	0	7	14	21	28	35	42	49	56	63	70
8	0	8	16	24	32	40	48	56	64	72	80
9	0	9	18	27	36	45	54	63	72	81	90
10	0	10	20	30	40	50	60	70	80	90	100

Write an addition sentence and a multiplication sentence for each.

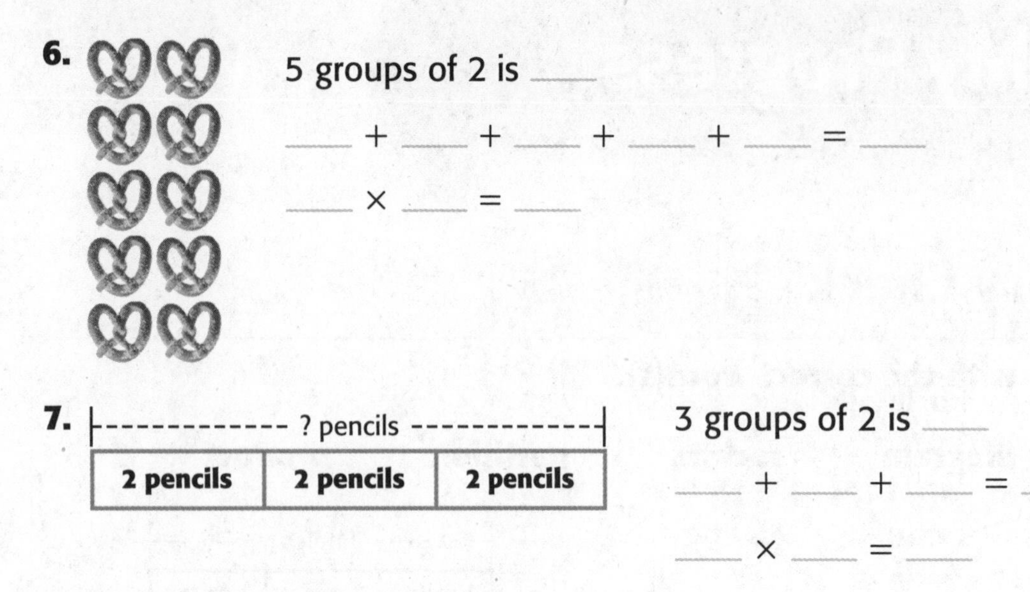

6. 5 groups of 2 is ____

____ + ____ + ____ + ____ + ____ = ____

____ × ____ = ____

7. ? pencils

2 pencils	2 pencils	2 pencils

3 groups of 2 is ____

____ + ____ + ____ = ____

____ × ____ = ____

Divide. Write a related multiplication fact.

8. 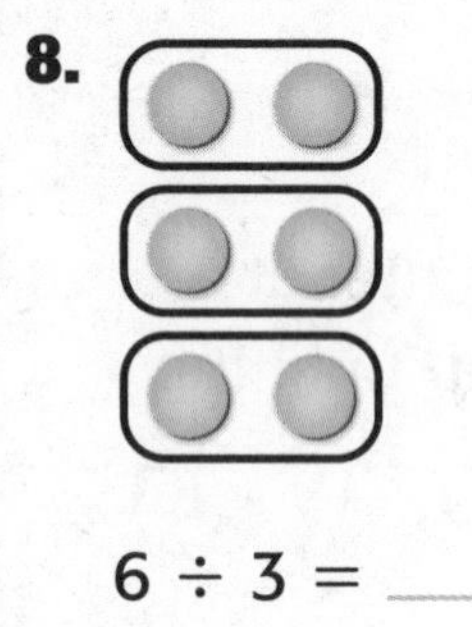

$6 \div 3 =$ ____

____

9. 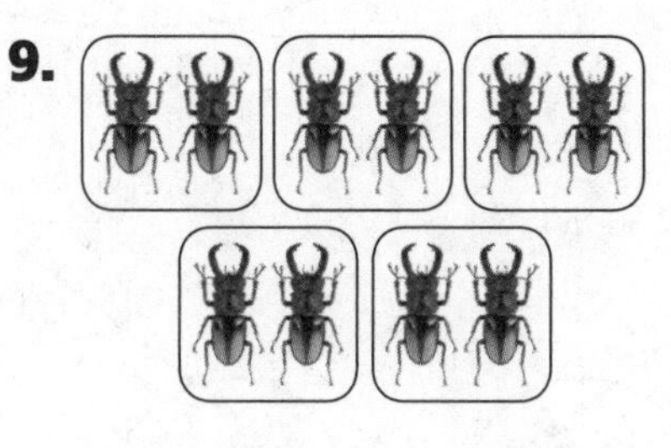

$10 \div 5 =$ ____

____

Problem Solving

10. A postal worker makes 8 trips to deliver some packages. She carries 2 packages at a time. How many packages are delivered?

____

Test Practice

11. Five times as many students bought lunch than packed lunch. Three students packed lunch. Which of the following could be used to find how many students bought lunch?

Ⓐ $5 - 3$ Ⓑ 5×3 Ⓒ $5 + 3$ Ⓓ $5 \div 3$

Name ....

Real World

Problem-Solving Investigation

STRATEGY: Look for a Pattern

Lesson 6

ESSENTIAL QUESTION
What is the importance of patterns in learning multiplication and division?

Learn the Strategy

Tools Watch Tutor

In the first row of her tile pattern, Christina uses 2 tiles. She uses 4 tiles in the second row, 8 tiles in the third row, and 16 tiles in the fourth row. If she continues the pattern, how many tiles will be in the sixth row?

1 Understand

What facts do you know?

There will be ______ tiles in the first row, ______ in the second row, ______ in the third row, and ______ tiles in the fourth row.

What do you need to find?

The number of tiles that will be in row ______.

2 Plan

I will make a table for the information. Then I will look for a pattern.

3 Solve

1st	2nd	3rd	4th	5th	6th
2	4	8	16		

+ 2 + 4 + 8 + 16 + 32

Put the information in a table. Look for a pattern. The numbers double. Now I can continue the pattern. There will be ____ tiles in the sixth row.

4 Check

Does your answer make sense? Explain.

____

Practice the Strategy

Jacy mows lawns every other day. He earns $5 the first day. After that, he earns $1 more than the day before. If he starts mowing on the first day of the month, how much money will he earn the 9th day of the month?

1 Understand

What facts do you know?

____

____

What do you need to find?

____

2 Plan

____

____

3 Solve

4 Check

Does your answer make sense? Explain.

____

Name

Apply the Strategy

My Work!

Solve each problem by looking for a pattern.

1. A collection of bears is shown. If there are 3 more rows, how many bears are there in all? Identify the pattern.

Row	1	2	3	4	5	6	7
Bears							

2. Mathematical PRACTICE 8 **Look for a Pattern** Yutaka is planting 15 flowers. He uses a pattern of 1 daisy and then 2 tulips. If the pattern continues, how many tulips will he use? Explain.

Daisies	Tulips	Total
1	2	
2	4	
3		

Review the Strategies

Use any strategy to solve each problem.

- Use an estimate or exact answer.
- Make a table.
- Look for a pattern.
- Use models.

For Exercises 3–5, use the sign shown.

HEALTH SHACK'S SNACKS	
Sunflower seeds	10¢ per package
Dried fruit	10 pieces for 50¢
Juice	20¢ each
Yogurt	2 for 80¢

My Work!

3. Julius spent 70¢ on sunflower seeds. How many packages did he buy?

____

4. How much did Neila pay for 1 yogurt?

____

5. How much would it cost to buy 1 of everything, including 1 piece of dried fruit?

____

6. Orlon collected 40 comic books. He keeps 10 comic books for himself and divides the rest equally among his 5 friends. How many comic books does each friend get?

____

7. Mathematical PRACTICE 5 **Use Math Tools** The amount of light given off by a light bulb is measured in lumens. Each of the 2 light bulbs in Hudson's room gives 1,585 lumens of light. Together, about how many lumens do both light bulbs give off?

____

Name ....

MY Homework

Lesson 6
Problem Solving: Look for a Pattern

Homework Helper

eHelp Need help? connectED.mcgraw-hill.com

At the arcade, Kelly began with 48 tokens. She gave 24 to Kuri. Then she gave 12 to Tonya. If this pattern continues, how many tokens will Kelly give away next?

1 **Understand**
What facts do you know?
Kelly began with 48 tokens.
She gave away 24 tokens, then 12 tokens.

What do you need to find?
how many tokens Kelly will give away next

2 **Plan**
I will look for a pattern.

3 **Solve**
The pattern is 48, 24, 12, . . .
Each number is half as much as the one before it.
The pattern is divide by 2.
$12 \div 2 = 6$
So, Kelly will give away 6 tokens next.

4 **Check**
Does the answer make sense?
Half of 12 is 6. The answer makes sense.

Real World!

Problem Solving

Solve each problem by looking for a pattern.

1. Adam is lining up his toy train cars. If he continues this color pattern, what color will the 18th car be?

____

1

2. Marissa delivers newspapers on Highview Drive. The first house number is 950, the next is 940, and the third is 930. If the pattern continues, what will the next house number be?

____

3. Kyle is training for a bike race. He rides 5 miles one day, 10 miles the next day, and 15 miles the third day. If Kyle repeats this schedule, what is the total distance he will have ridden after 5 days?

____

4. The Hornets basketball team won their first game by 18 points, their second game by 15 points, and their third game by 12 points. If the pattern continues, by how many points will they win their fifth game?

____

5. Darcy wears brown pants to work one day, blue pants the next day, and a skirt on the third day. If this pattern continues every three days, what will she wear to work on the seventh day?

____

My Work!

Name ....

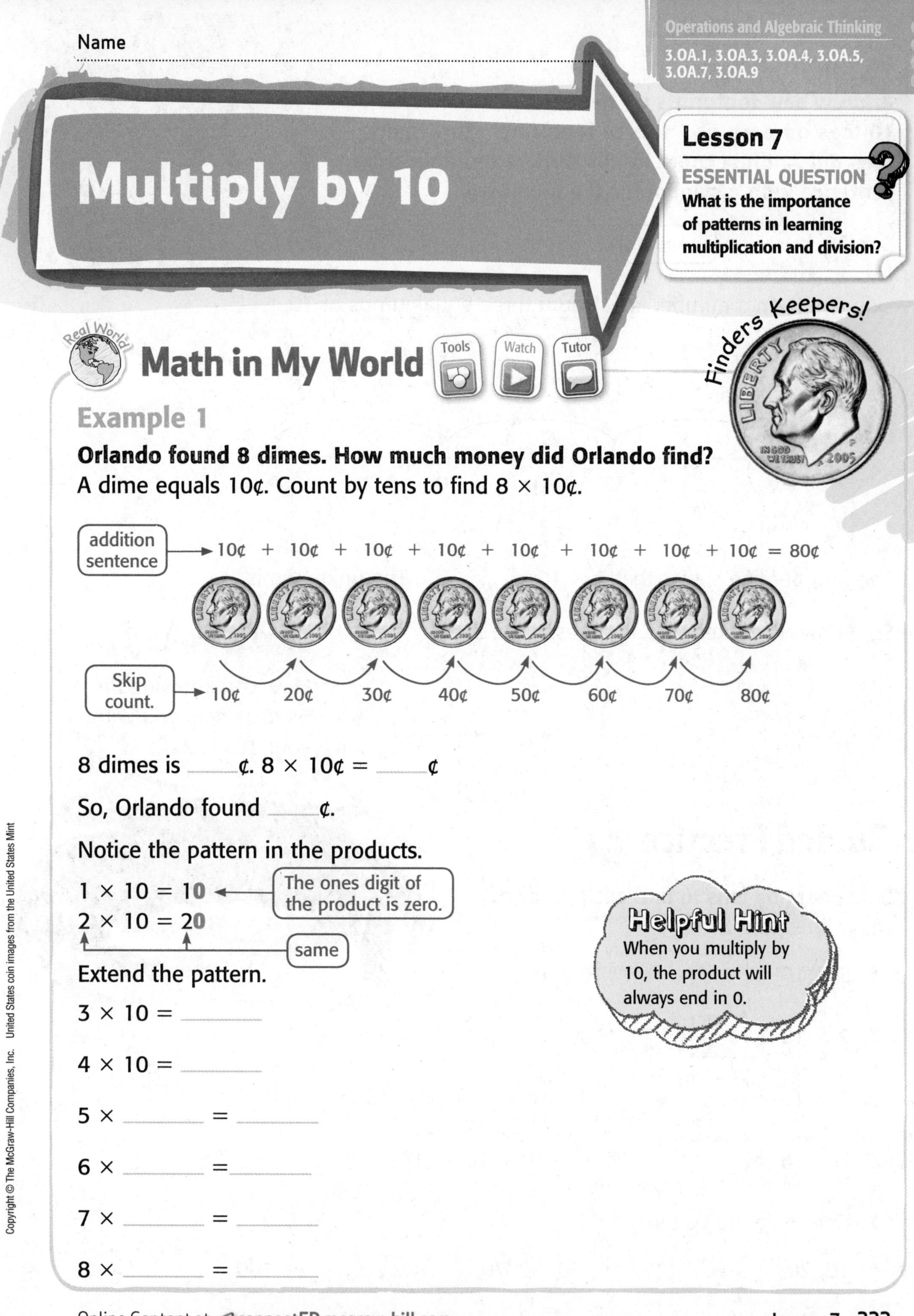

Multiply by 10

Lesson 7

ESSENTIAL QUESTION
What is the importance of patterns in learning multiplication and division?

Math in My World

Tools Watch Tutor

Example 1

Orlando found 8 dimes. How much money did Orlando find?
A dime equals 10¢. Count by tens to find 8 × 10¢.

8 dimes is _____¢. 8 × 10¢ = _____¢

So, Orlando found _____¢.

Notice the pattern in the products.

1 × 10 = 10 ← The ones digit of the product is zero.
2 × 10 = 20

same

Extend the pattern.

3 × 10 = _____

4 × 10 = _____

5 × _____ = _____

6 × _____ = _____

7 × _____ = _____

8 × _____ = _____

Helpful Hint
When you multiply by 10, the product will always end in 0.

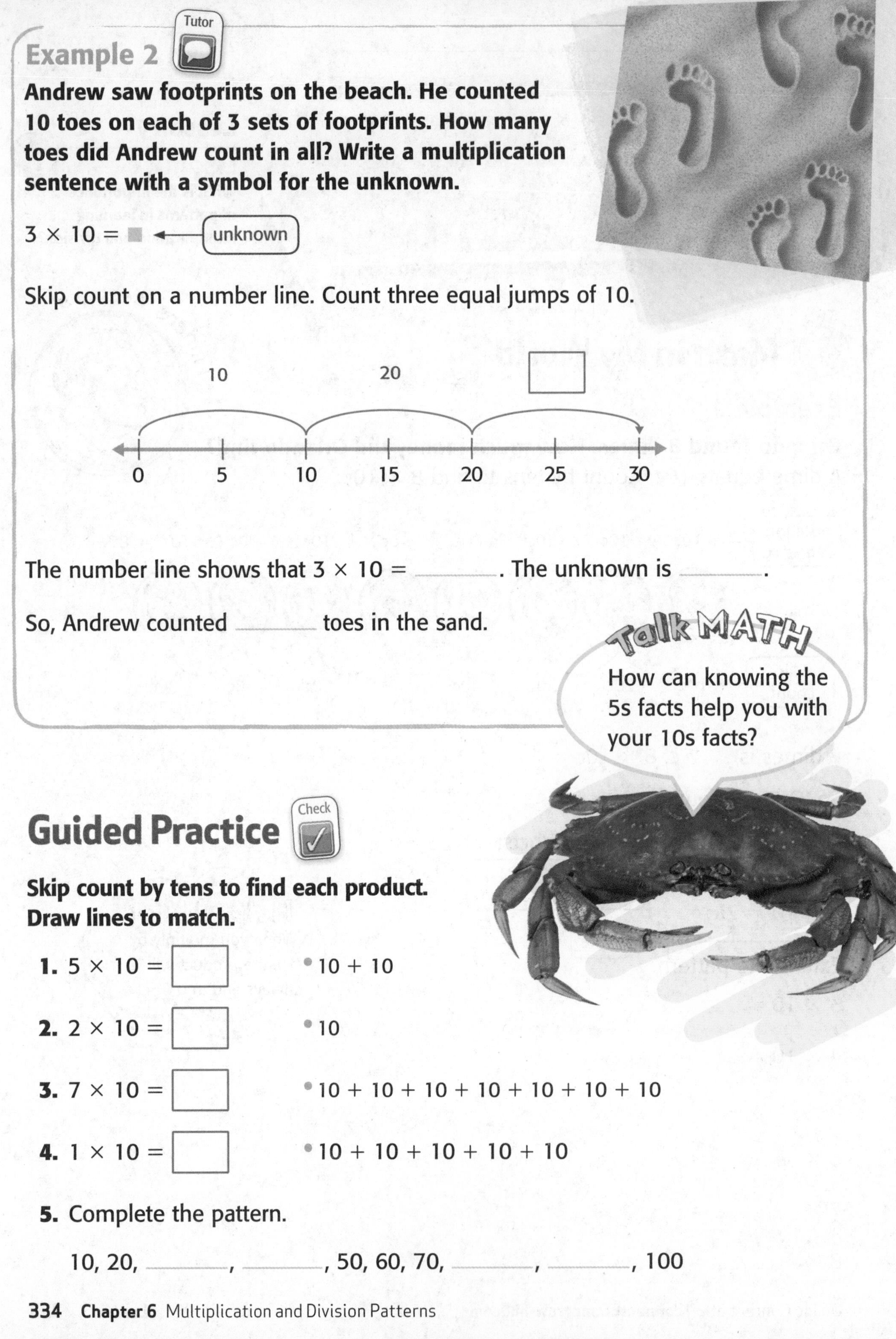

Example 2

Tutor

Andrew saw footprints on the beach. He counted 10 toes on each of 3 sets of footprints. How many toes did Andrew count in all? Write a multiplication sentence with a symbol for the unknown.

$3 \times 10 = \blacksquare$ ← unknown

Skip count on a number line. Count three equal jumps of 10.

The number line shows that $3 \times 10 =$ ______. The unknown is ______.

So, Andrew counted ______ toes in the sand.

Talk MATH

How can knowing the 5s facts help you with your 10s facts?

Guided Practice

Check

Skip count by tens to find each product. Draw lines to match.

1. $5 \times 10 =$ ☐ • $10 + 10$

2. $2 \times 10 =$ ☐ • 10

3. $7 \times 10 =$ ☐ • $10 + 10 + 10 + 10 + 10 + 10 + 10$

4. $1 \times 10 =$ ☐ • $10 + 10 + 10 + 10 + 10$

5. Complete the pattern.

10, 20, ______, ______, 50, 60, 70, ______, ______, 100

Name

Independent Practice

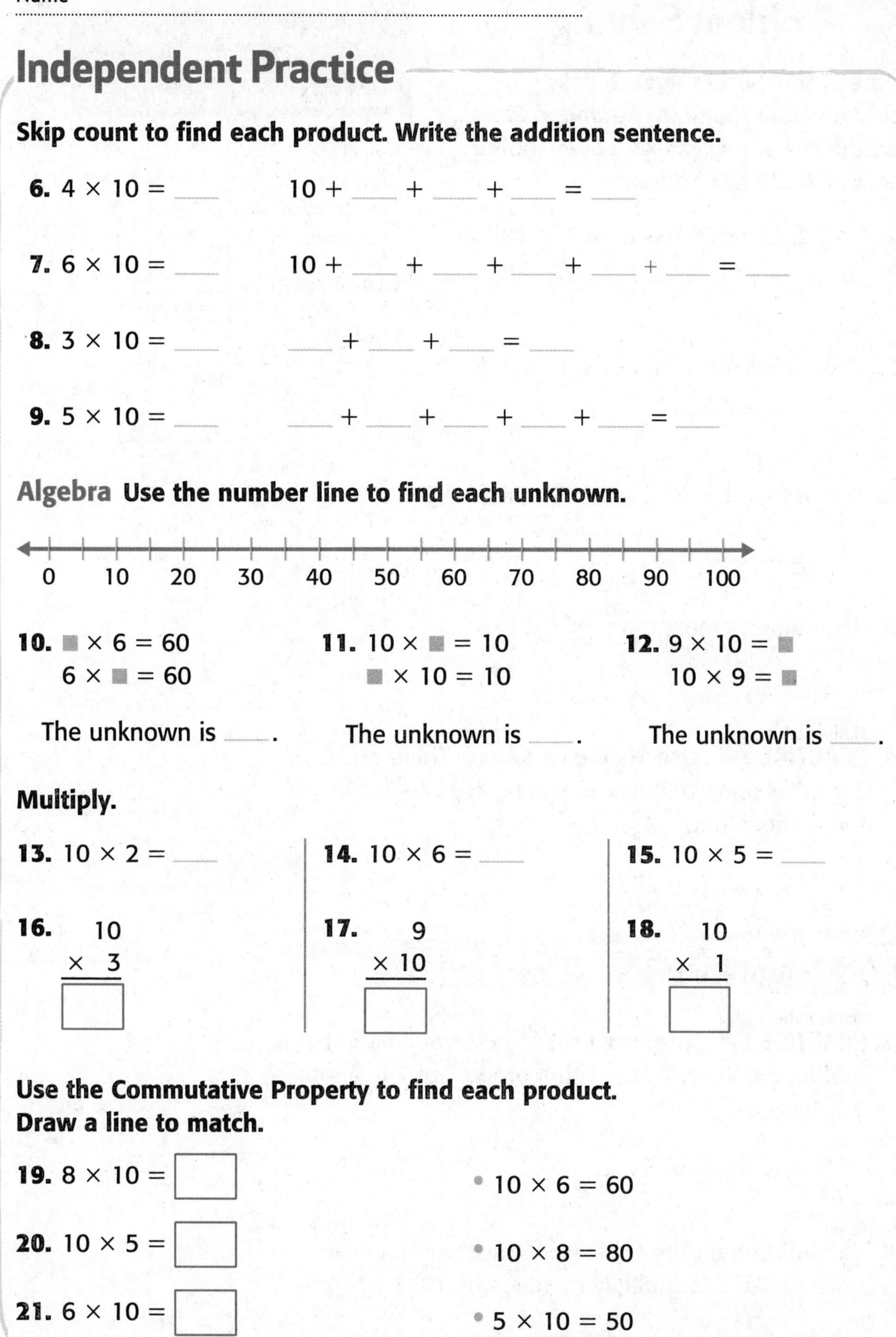

Skip count to find each product. Write the addition sentence.

6. $4 \times 10 =$ ____ $10 +$ ____ $+$ ____ $+$ ____ $=$ ____

7. $6 \times 10 =$ ____ $10 +$ ____ $+$ ____ $+$ ____ $+$ ____ $+$ ____ $=$ ____

8. $3 \times 10 =$ ____ ____ $+$ ____ $+$ ____ $=$ ____

9. $5 \times 10 =$ ____ ____ $+$ ____ $+$ ____ $+$ ____ $+$ ____ $=$ ____

Algebra **Use the number line to find each unknown.**

0 10 20 30 40 50 60 70 80 90 100

10. $\blacksquare \times 6 = 60$
$6 \times \blacksquare = 60$

The unknown is ____.

11. $10 \times \blacksquare = 10$
$\blacksquare \times 10 = 10$

The unknown is ____.

12. $9 \times 10 = \blacksquare$
$10 \times 9 = \blacksquare$

The unknown is ____.

Multiply.

13. $10 \times 2 =$ ____

14. $10 \times 6 =$ ____

15. $10 \times 5 =$ ____

16. $\begin{array}{r} 10 \\ \times\ 3 \\ \hline \square \end{array}$

17. $\begin{array}{r} 9 \\ \times\ 10 \\ \hline \square \end{array}$

18. $\begin{array}{r} 10 \\ \times\ 1 \\ \hline \square \end{array}$

Use the Commutative Property to find each product. Draw a line to match.

19. $8 \times 10 = \square$ • $10 \times 6 = 60$

20. $10 \times 5 = \square$ • $10 \times 8 = 80$

21. $6 \times 10 = \square$ • $5 \times 10 = 50$

Problem Solving

Some of the world's largest glass sculptures are found in the United States. Use the clues in Exercises 22–25 to find the length of each sculpture.

Worlds's Largest Glass Sculptures	
Sculpture Name	**Length (feet)**
Fiori di Como, NV	?
Chihuly Tower, OK	?
Cobalt Blue Chandelier, WA	?
River Blue, CT	?

22. Fiori di Como: 5 less than 7×10

23. Chihuly Tower: 5 more than 10×5

24. Cobalt Blue Chandelier: 9 more than 2×10

25. River Blue: 4 more than 10×1

26. Mathematical PRACTICE 2 **Use Number Sense** There are 5 giraffes and 10 birds. How many legs are there altogether?

My Work!

HOT Problems

27. Mathematical PRACTICE 2 **Reason** Explain how you know that a multiplication sentence with a product of 25 cannot be a 10s fact.

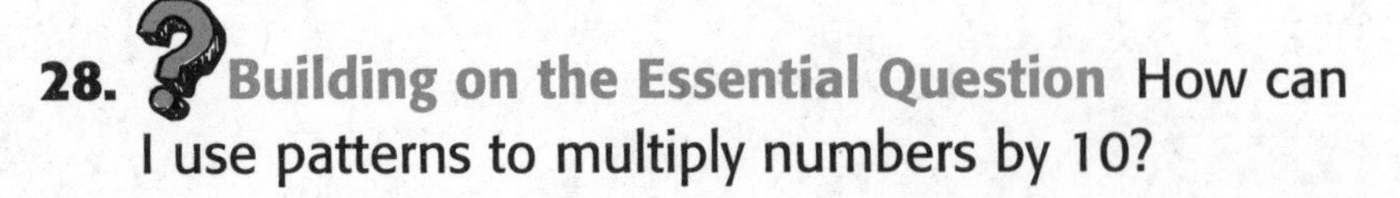

28. **Building on the Essential Question** How can I use patterns to multiply numbers by 10?

Name ....

Lesson 7
Multiply by 10

Homework Helper

eHelp Need help? connectED.mcgraw-hill.com

There are 8 players on the tennis team. Each family contributes $10 toward a gift for the coach. What is the total amount collected for the coach's gift?

Find 8 × $10.

Skip count by tens.

$10 + $10 + $10 + $10 + $10 + $10 + $10 + $10 = $80

$10 $20 $30 $40 $50 $60 $70 $80

So, the total amount collected from 8 families was $80.

Practice

Skip count by tens to find each product. Write the addition sentence.

1. 5 × 10 = ______

____

2. 2 × 10 = ______

____

3. 7 × 10 = ______

____

4. 3 × 10 = ______

____

Algebra **Use the number line to find each unknown.**

0 5 10 15 20 25 30 35 40 45 50

5. ■ × 4 = 40

4 × ■ = 40

The unknown is ______.

6. 10 × ■ = 20

■ × 10 = 20

The unknown is ______.

7. 10 × ■ = 50

■ × 10 = 50

The unknown is ______.

Problem Solving

For Exercises 8–9, write a multiplication sentence to solve.

8. Fiona's class went on a field trip to the art museum. The class rode in vans with 10 people in each van. How many people went on the field trip if they took 4 full vans?

____

9. Mathematical PRACTICE 5 **Use Math Tools** During the football game, Carlos ran with the ball 3 times. Each time, he ran 10 yards. How many yards did Carlos run altogether?

____

10. Each time Allison goes to the recycling center, she takes 10 bags of cans. She will go twice this month, 3 times next month, and once the following month. How many bags of cans will Allison take to the recycling center in these three months?

____

Test Practice

11. Byron has 70 pennies. He stacks them in groups of 10. How many stacks of pennies can Byron make?

Ⓐ 7 stacks

Ⓑ 8 stacks

Ⓒ 9 stacks

Ⓓ 10 stacks

Name ....

Multiples of 10

Lesson 8

ESSENTIAL QUESTION
What is the importance of patterns in learning multiplication and division?

The product of a given number, such as 10, and any other number is a **multiple.** You can use a basic fact and patterns of zeros to mentally find multiples of 10.

Be our guest!

Math in My World

Tools Watch Tutor

Example 1

A new hotel has 3 floors. There are 20 rooms on each floor. What is the total number of rooms in the hotel?

Find 3×20. ← 20 is a multiple of 10, since $2 \times 10 = 20$

One Way **Use a basic fact and patterns.**

$3 \times 2 = 6$ ← basic fact

$3 \times 20 =$ _____ ← $3 \times 2 = 6$, so $3 \times 20 = 60$

Another Way **Use place value.**

Think of 3×20 as 3×2 tens.

Use base-ten blocks to model 3 equal groups of 2 tens. Draw your model at the right.

3×2 tens = _____ tens

So, $3 \times 20 =$ _____.

← 6 tens = 60

My Drawing!

Check for Reasonableness

Use repeated addition.

20 + _____ + _____ = _____

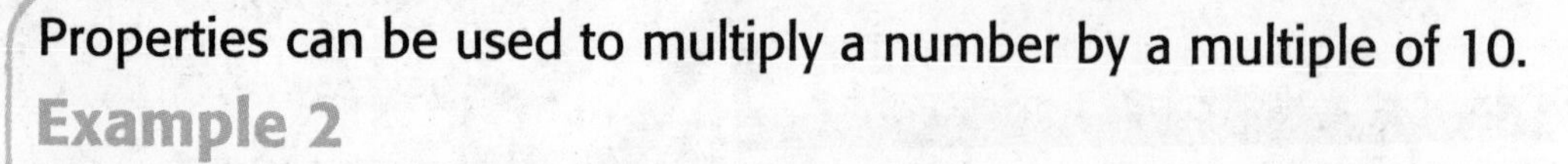

Properties can be used to multiply a number by a multiple of 10.

Example 2

Ellie is buying 2 bags of beads to add to her bead collection. Each bag has 40 beads. How many beads is Ellie buying?

Find 2×40.

$2 \times 40 = 2 \times (4 \times 10)$ — Write 40 as 4×10.

$= (2 \times 4) \times 10$ — Find 2×4 first.

$=$ ____ $\times 10$ — Multiply.

$=$ ____

So, Ellie is buying ____ beads.

Helpful Hint
The way in which numbers are grouped does not change their product.

Example 3

Find the unknown in $4 \times 50 = \blacksquare$.

$4 \times 50 = 200$ — 4×5 tens $= 20$ tens

Sometimes the basic fact has a zero. Keep that zero, then add the other zero.

So, $4 \times 50 =$ ____. The unknown is ____.

Talk MATH

Find the product of 3×20 and 2×30. What do you notice about the products? Is this an example of the Commutative Property of Multiplication? Explain.

Guided Practice

Multiply. Use place value.

1. $2 \times 20 = 2 \times$ ____ tens

$=$ ____ tens

So, $2 \times 20 =$ ____.

2. $5 \times 60 = 5 \times$ ____ tens

$=$ ____ tens

So, $5 \times 60 =$ ____.

Name

Independent Practice

Multiply. Use a basic fact.

3. $5 \times 5 =$ ____

So, $5 \times 50 =$ ____.

4. $6 \times 2 =$ ____

So, $6 \times 20 =$ ____.

5. $5 \times 7 =$ ____

So, $5 \times 70 =$ ____.

Multiply. Use place value.

6. $5 \times 20 =$

____ $\times$ ____ tens = ____ tens

So, $5 \times 20 =$ ____.

7. $2 \times 70 =$

____ $\times$ ____ tens = ____ tens

So, $2 \times 70 =$ ____.

8. $8 \times 50 =$

____ $\times$ ____ tens = ____ tens

So, $8 \times 50 =$ ____.

9. $2 \times 80 =$

____ $\times$ ____ tens = ____ tens

So, $2 \times 80 =$ ____.

Multiply to find each product. Draw lines to match.

10. $2 \times 90 =$ ____

11. $5 \times 40 =$ ____

12. $5 \times 90 =$ ____

• $5 \times (4 \times 10) = (5 \times 4) \times 10$
$= 20 \times 10$
$=$ ____

• $5 \times (9 \times 10) = (5 \times 9) \times 10$
$= 45 \times 10$
$=$ ____

• $2 \times (9 \times 10) = (2 \times 9) \times 10$
$= 18 \times 10$
$=$ ____

Algebra Find each unknown.

13. $2 \times$ ■ $= 100$

The unknown is ____.

14. $2 \times$ ■ $= 60$

The unknown is ____.

15. $6 \times 50 =$ ■

The unknown is ____.

Problem Solving

Write a multiplication sentence with a symbol for the unknown for Exercises 16–17. Then solve.

16. Mathematical PRACTICE 2 **Use Algebra** Demont's card album has 20 pages, and 6 trading cards are on each page. How many cards are there in all?

17. There are 90 houses with 10 windows each. How many windows are there in all?

18. Carlita collected 2 boxes of teddy bears. Each box holds 20 bears. She sells each bear for \$2. How much money did she earn?

My Work!

HOT Problems

19. Mathematical PRACTICE 4 **Model Math** Write a multiplication sentence that uses a multiple of 10 and has a product of 120.

20. Mathematical PRACTICE 8 **Look for a Pattern** Describe the pattern you see when multiplying 5×30.

What is the product of 5×300?

21. **Building on the Essential Question** How do basic facts and patterns help me multiply a number by a multiple of 10?

Name ______________________

Operations and Algebraic Thinking
3.OA.1, 3.OA.3, 3.OA.4, 3.OA.5, 3.OA.7, 3.OA.9, 3.NBT.3

CCSS

MY Homework

Lesson 8
Multiples of 10

Homework Helper

eHelp

Need help? connectED.mcgraw-hill.com

There are 3 shelves in the cabinet. Each shelf holds 40 cans. How many cans will fit in the cabinet?

You need to find 3×40.

One Way Use a basic fact and patterns.

$3 \times 4 = 12$ ← basic fact

$3 \times 40 = 120$ ← pattern

Another Way Use place value.

Use base-ten blocks to model 3 groups of 4 tens.

3×4 tens = 12 tens; 12 tens = 120. ← Use repeated addition to check: $40 + 40 + 40 = 120$

So, $3 \times 40 = 120$.

So, 120 cans will fit in the cabinet.

Practice

Multiply. Use place value.

1. $2 \times 40 =$

$2 \times$ ____ tens = ____ tens

So, $2 \times 40 =$ ____.

2. $5 \times 60 =$

$5 \times$ ____ tens = ____ tens

So, $5 \times 60 =$ ____.

3. $5 \times 30 =$

$5 \times$ ____ tens = ____ tens

So, $5 \times 30 =$ ____.

4. $10 \times 20 =$

$10 \times$ ____ tens = ____ tens

So, $10 \times 20 =$ ____.

Multiply. Use a basic fact.

5. $10 \times 3 =$ ______

So, $10 \times 30 =$ ______

6. $2 \times 9 =$ ______

So, $2 \times 90 =$ ______

7. $2 \times 8 =$ ______

So, $2 \times 80 =$ ______

8. $5 \times 5 =$ ______

So, $5 \times 50 =$ ______

Real World

Problem Solving

Write a multiplication sentence to solve.

9. Harlan has 5 antique watches. Each watch has a value of \$90. How much are Harlan's watches worth in all?

______

10. Mathematical PRACTICE 1 **Keep Trying** Trey uses 40 nails to put up the frame around each window. There are 5 windows in the bedroom. How many nails will Trey use in the bedroom?

______

11. Chloe uses 80 candy wrappers to make a paper necklace. She is making necklaces for herself and 9 friends. How many candy wrappers will Chloe need?

______

Vocabulary Check

Vocab

12. Circle the number sentence that shows 20 is a multiple of 2.

$2 \times 10 = 20$ $2 + 10 = 12$

$2 \times 5 = 10$ $10 \div 2 = 5$

Test Practice

13. Which is equal to 52 tens?

Ⓐ 52,010   Ⓒ 5,200

Ⓑ 5,210   Ⓓ 520

Name ________________

Divide by 10

Lesson 9

ESSENTIAL QUESTION
What is the importance of patterns in learning multiplication and division?

Let's have a party!

Math in My World

Tools Watch Tutor

Example 1

The third grade class needs 50 juice bars. How many boxes of juice bars will they need if there are 10 bars in each box?

Find 50 ÷ 10.

One Way **Use a number line.**

Start at 50 and count back by 10s.

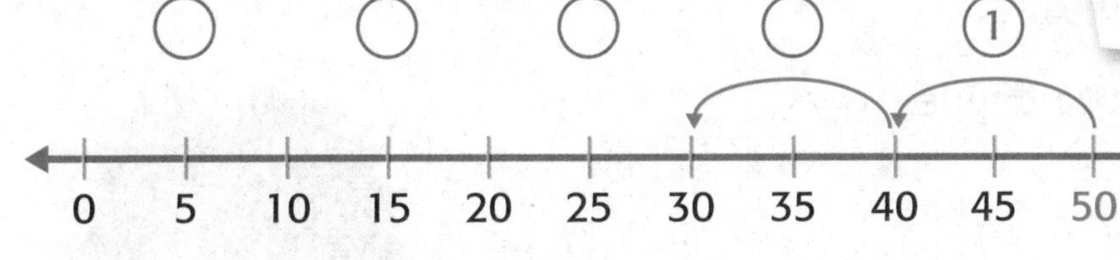

Groups of 10 were counted back ______ times.

Another Way **Use repeated subtraction.**

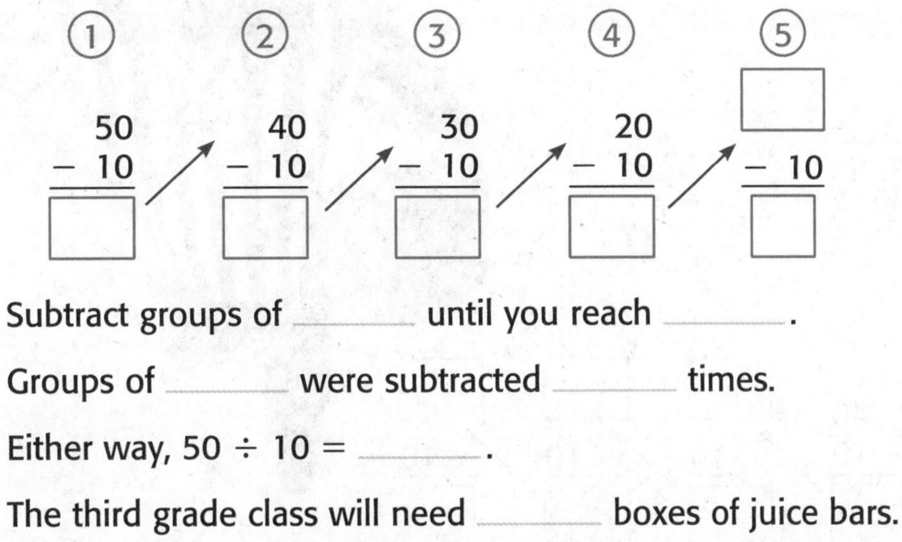

Subtract groups of ______ until you reach ______.

Groups of ______ were subtracted ______ times.

Either way, 50 ÷ 10 = ______.

The third grade class will need ______ boxes of juice bars.

Think of division as an unknown factor problem. Use a related multiplication fact.

Example 2

A quarterback threw the football for a total of 70 yards. Each time he threw the ball, the team gained 10 yards. How many throws did the quarterback make? Write a division sentence with a symbol for the unknown.

Find $70 \div 10 = \blacksquare$.

You know that $10 \times$ ______ $= 70$. The unknown factor is ______.

Since division and multiplication are inverse operations,

$70 \div 10 =$ ______

The unknown is ______.

The quarterback made ______ throws.

Talk MATH

When you divide by 10, what do you notice about the quotient and the dividend?

Guided Practice

Check

Use repeated subtraction to divide.

1. $90 \div 10 =$ ______

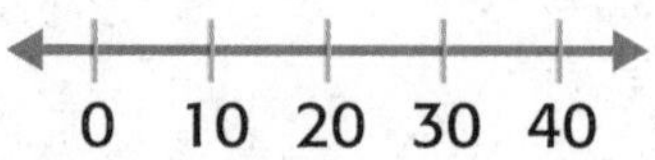

2. $40 \div 10 =$ ______

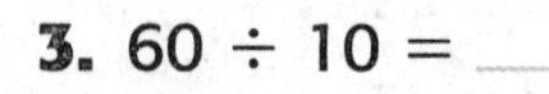

3. $60 \div 10 =$ ______

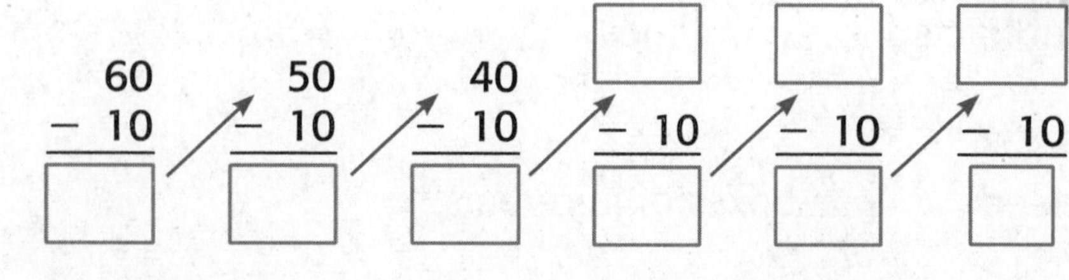

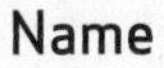

Name ....

CCSS

Independent Practice

Use repeated subtraction to divide.

4. $20 \div 10 =$ ______

5. $10 \div 10 =$ ______

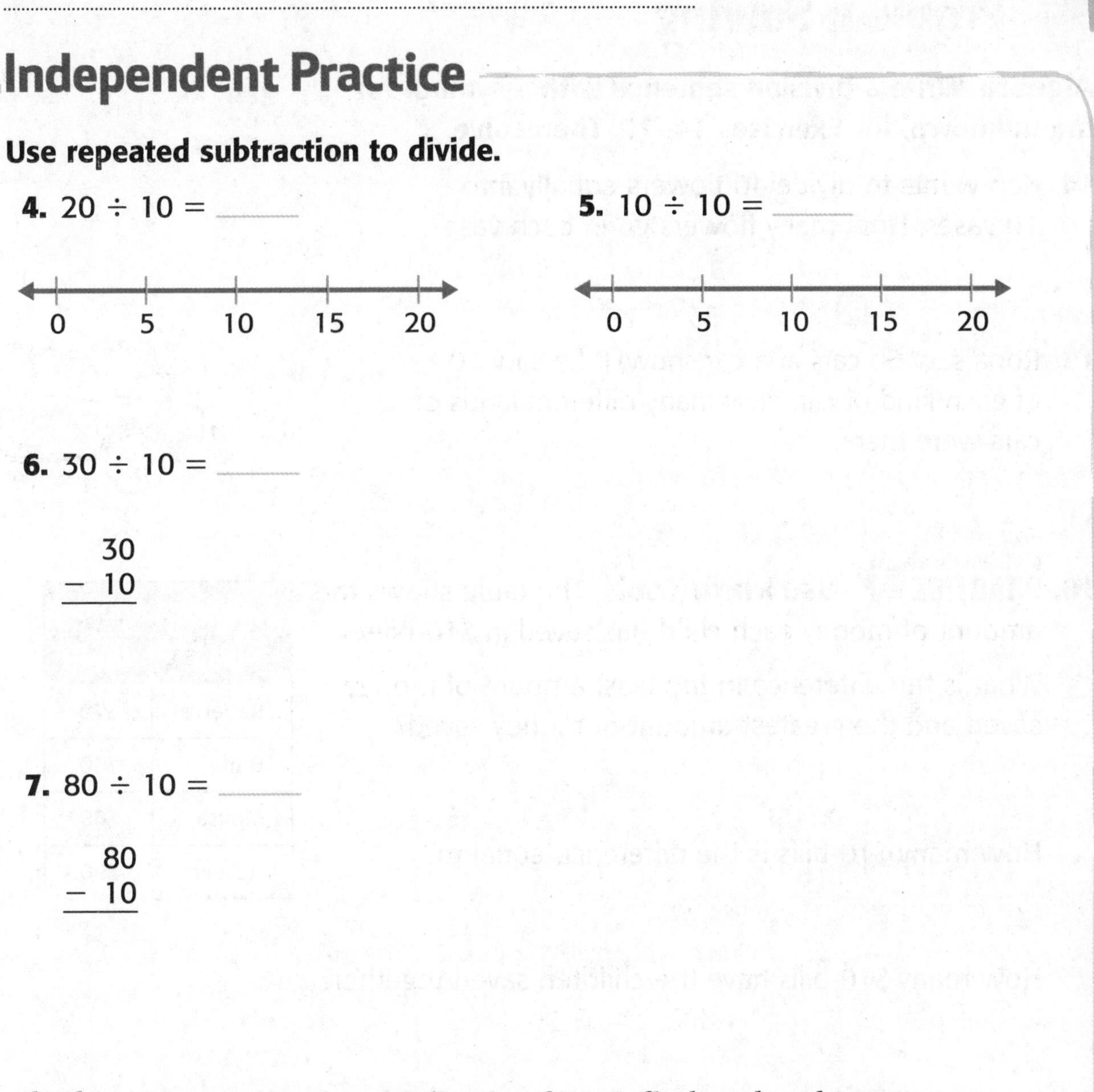

6. $30 \div 10 =$ ______

$$\begin{array}{r} 30 \\ -\ 10 \\ \hline \end{array}$$

7. $80 \div 10 =$ ______

$$\begin{array}{r} 80 \\ -\ 10 \\ \hline \end{array}$$

Algebra Use a related multiplication fact to find each unknown.

8. $50 \div 10 = ■$

$10 \times$ ______ $= 50$

The unknown is ______.

9. $70 \div ■ = 7$

______ $\times 7 = 70$

The unknown is ______.

10. $90 \div 10 = ■$

$10 \times$ ______ $= 90$

The unknown is ______.

11. $60 \div ■ = 6$

______ $\times 6 = 60$

The unknown is ______.

12. $100 \div 10 = ■$

$10 \times$ ______ $= 100$

The unknown is ______.

13. $■ \div 10 = 4$

$10 \times 4 =$ ______

The unknown is ______.

Problem Solving

Algebra Write a division sentence with a symbol for the unknown, for Exercises 14–15. Then solve.

14. Ken wants to divide 40 flowers equally into 10 vases. How many flowers go in each vase?

15. Rona saw 60 cars at a car show. If he saw 10 of each kind of car, how many different kinds of cars were there?

16. Mathematical PRACTICE 5 **Use Math Tools** The table shows the amount of money each child has saved in $10-bills.

What is the difference in the least amount of money saved and the greatest amount of money saved?

How many $10-bills is the difference equal to?

How many $10-bills have the children saved together?

Savings Accounts	
Name	**Saved**
Rebecca	$70
Bret	$30
Monsa	$80
Hakeem	$90

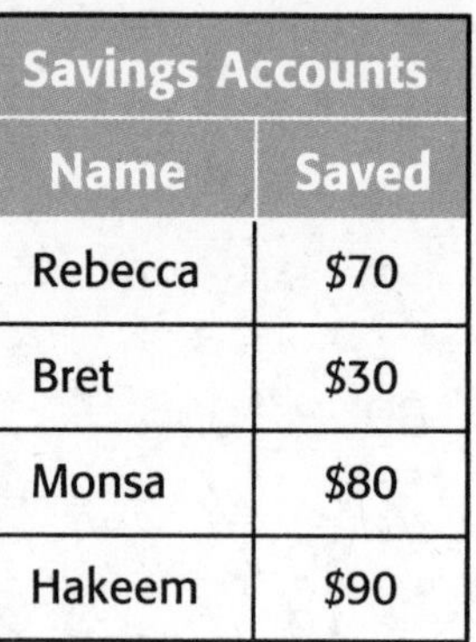

HOT Problems

17. Mathematical PRACTICE 2 **Use Number Sense** Use the numerals 0, 7, and 8 to write two 2-digit numbers that can each be divided by 10. The numerals can be used more than one time.

18. **Building on the Essential Question** How can skip counting by 10s help you find the quotient of 10s facts?

Name ....

Lesson 9

Divide by 10

Homework Helper

eHelp Need help? connectED.mcgraw-hill.com

Ms. Mickle's classroom has 30 desks with 10 desks in each row. How many rows of desks are there?

Find $30 \div 10$.

Subtract groups of 10 until you reach 0.

One Way **Use a number line.**

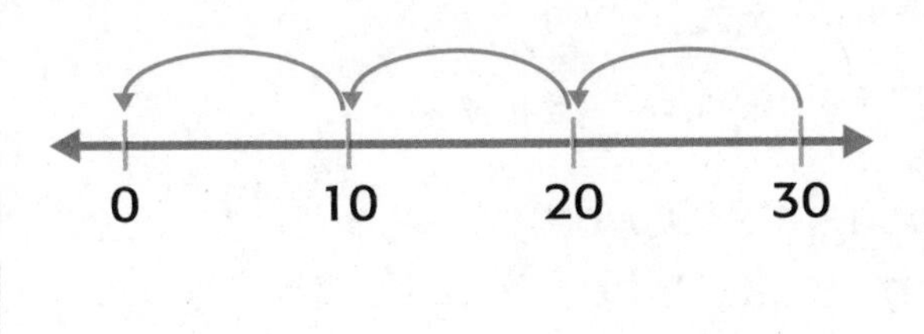

Another Way **Use repeated subtraction.**

① $30 - 10 = 20$ → ② $20 - 10 = 10$ → ③ $10 - 10 = 0$

3 groups of 10 were subtracted and you know that $10 \times 3 = 30$.

So, $30 \div 10 = 3$. There are 3 rows of desks.

Practice

Use repeated subtraction to divide.

1. $70 \div 10 =$ ______

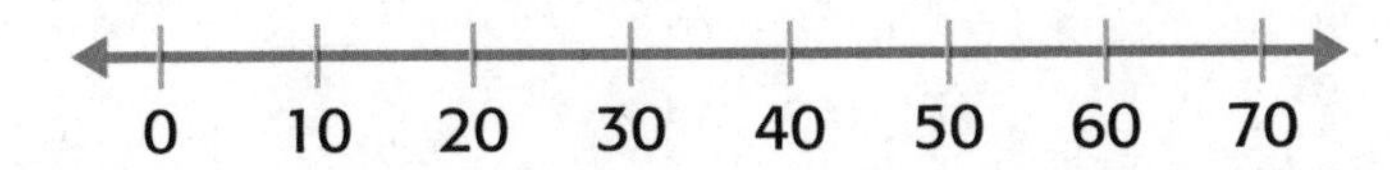

2. $60 \div 10 =$ ______

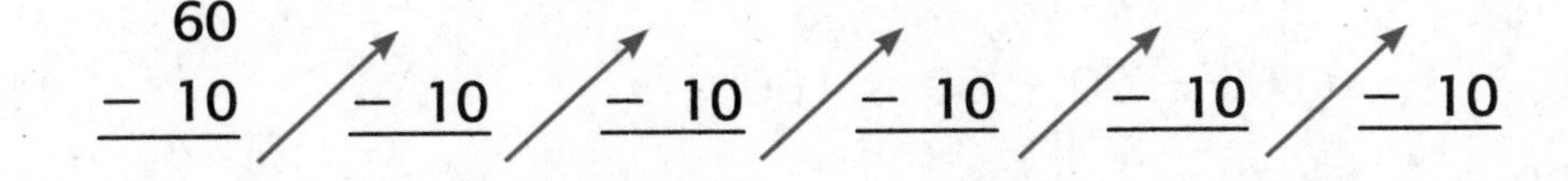

Algebra Use a related multiplication fact to find each unknown.

3. 80 ÷ 10 = ■

10 × ______ = 80

The unknown is ______.

4. ■ ÷ 10 = 3

10 × 3 = ______

The unknown is ______.

5. ■ ÷ 10 = 10

10 × 10 = ______

The unknown is ______.

6. 20 ÷ 10 = ■

10 × ______ = 20

The unknown is ______.

Problem Solving

7. Morgan has 90 cents in her pocket. All of the change is in dimes. How many dimes does Morgan have in all?

______

8. Ricky spent \$90 at the supermarket. He bought \$30 worth of fruit. He spent the rest of the money on steaks. If he bought 10 steaks and they each cost the same amount, what was the price of each steak?

______

9. Mathematical PRACTICE 1 **Make Sense of Problems** Annie bought a bag of 80 mini-carrots. She eats 5 carrots with lunch each day and eats another 5 each night as a snack. In how many days will the bag of carrots be gone?

______

Test Practice

10. Bill has a collection of 60 books he wants to donate to the library. Which number sentence shows how Bill can divide the books equally as he packs them in boxes?

Ⓐ 60 ÷ 6 = 10

Ⓑ 60 − 10 = 50

Ⓒ 60 + 60 + 60 = 180

Ⓓ 60 × 1 = 60

Name ....

Fluency Practice

Mathematical PRACTICE 6

Multiply.

1. $2 \times 9 =$ ____ **2.** $5 \times 3 =$ ____ **3.** $2 \times 4 =$ ____ **4.** $10 \times 6 =$ ____

5. $2 \times 3 =$ ____ **6.** $2 \times 5 =$ ____ **7.** $2 \times 2 =$ ____ **8.** $5 \times 1 =$ ____

9. $5 \times 4 =$ ____ **10.** $2 \times 6 =$ ____ **11.** $2 \times 7 =$ ____ **12.** $10 \times 2 =$ ____

13. $\begin{array}{r} 10 \\ \times\ 3 \\ \hline \end{array}$ **14.** $\begin{array}{r} 5 \\ \times\ 6 \\ \hline \end{array}$ **15.** $\begin{array}{r} 2 \\ \times\ 8 \\ \hline \end{array}$ **16.** $\begin{array}{r} 10 \\ \times\ 4 \\ \hline \end{array}$

17. $\begin{array}{r} 5 \\ \times\ 7 \\ \hline \end{array}$ **18.** $\begin{array}{r} 5 \\ \times\ 5 \\ \hline \end{array}$ **19.** $\begin{array}{r} 10 \\ \times\ 6 \\ \hline \end{array}$ **20.** $\begin{array}{r} 5 \\ \times\ 8 \\ \hline \end{array}$

21. $\begin{array}{r} 2 \\ \times\ 1 \\ \hline \end{array}$ **22.** $\begin{array}{r} 5 \\ \times\ 2 \\ \hline \end{array}$ **23.** $\begin{array}{r} 5 \\ \times\ 9 \\ \hline \end{array}$ **24.** $\begin{array}{r} 10 \\ \times\ 5 \\ \hline \end{array}$

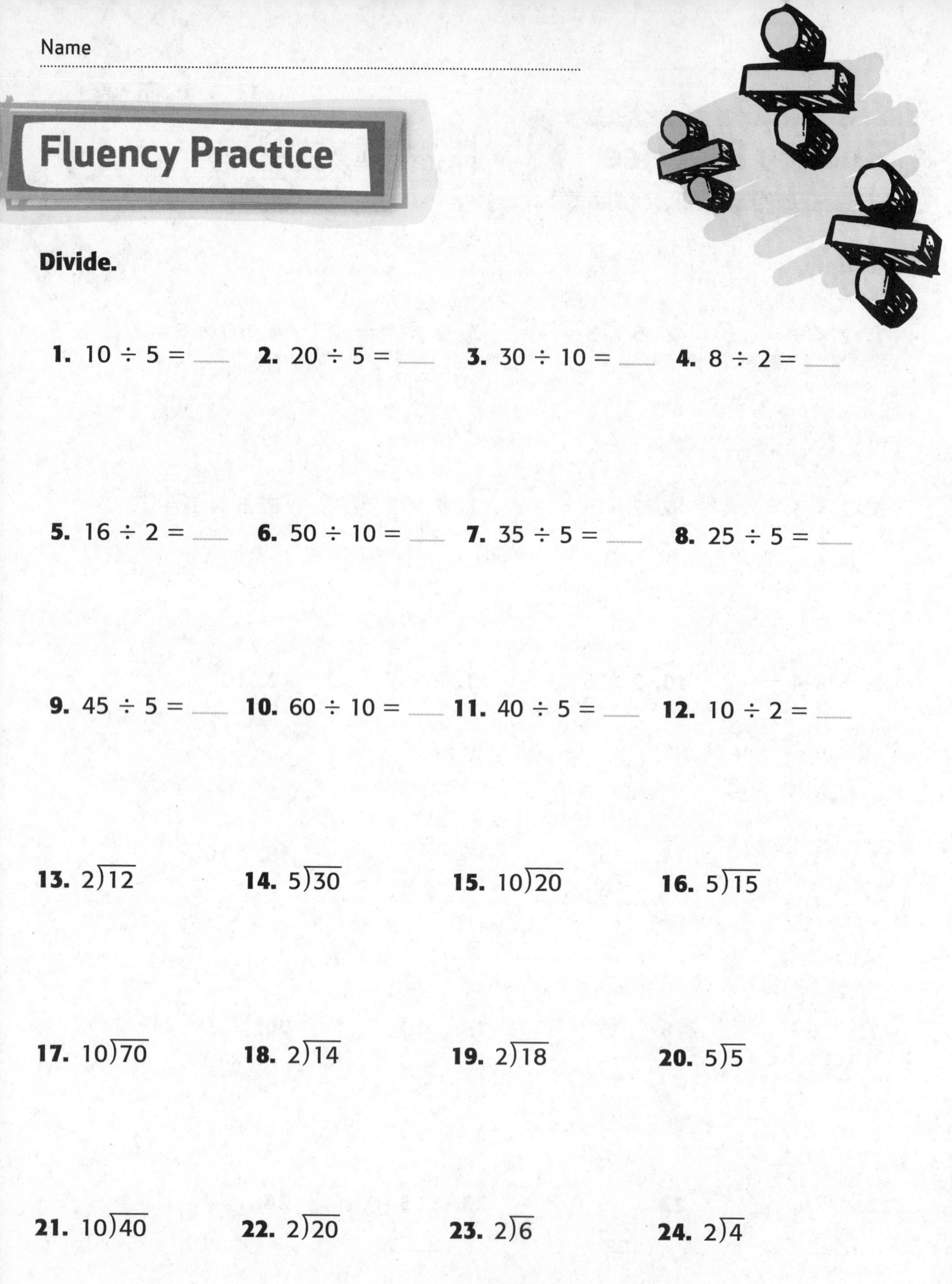

Name ....

Fluency Practice

Divide.

1. $10 \div 5 =$ ___ **2.** $20 \div 5 =$ ___ **3.** $30 \div 10 =$ ___ **4.** $8 \div 2 =$ ___

5. $16 \div 2 =$ ___ **6.** $50 \div 10 =$ ___ **7.** $35 \div 5 =$ ___ **8.** $25 \div 5 =$ ___

9. $45 \div 5 =$ ___ **10.** $60 \div 10 =$ ___ **11.** $40 \div 5 =$ ___ **12.** $10 \div 2 =$ ___

13. $2\overline{)12}$ **14.** $5\overline{)30}$ **15.** $10\overline{)20}$ **16.** $5\overline{)15}$

17. $10\overline{)70}$ **18.** $2\overline{)14}$ **19.** $2\overline{)18}$ **20.** $5\overline{)5}$

21. $10\overline{)40}$ **22.** $2\overline{)20}$ **23.** $2\overline{)6}$ **24.** $2\overline{)4}$

Review

Chapter 6

Multiplication and Division Patterns

Vocabulary Check

Use the clues and word bank below to complete the crossword puzzle.

bar diagram	**factor**	**multiple**
partition	**product**	

Across

1. The answer to a multiplication problem.
2. A drawing that helps to organize your information.
3. The product of a given number and any other whole number.
4. A number that is multiplied by another number.

Down

1. To separate a number of objects into equal groups.

Concept Check

Write an addition sentence and a multiplication sentence. Then draw an array.

5. 2 rows of 3 is ______.

______ + ______ = ______

______ × ______ = ______

6. 5 rows of 2 is ______.

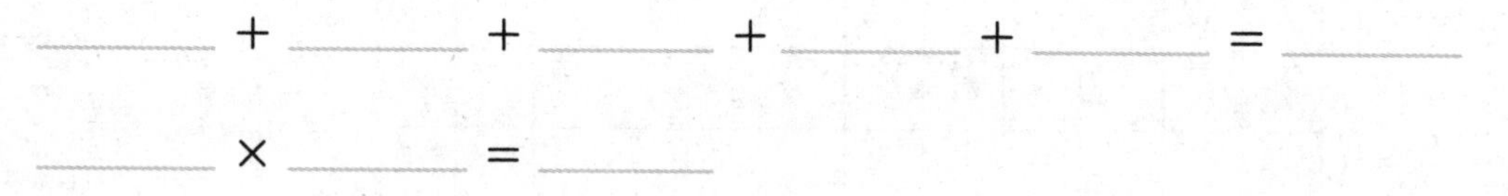

______ + ______ + ______ + ______ + ______ = ______

______ × ______ = ______

Multiply.

7. 7 × 10 = ______

8. 6 × 5 = ______

9. 1 × 5 = ______

10. 2 × 10 = ______

11. 9 × 10 = ______

12. 4 × 5 = ______

Algebra **Use a related multiplication fact to find each unknown.**

13. 30 ÷ 10 = ■

______ × 10 = 30

The unknown is ______.

14. 60 ÷ ■ = 6

______ × 6 = 60

The unknown is ______.

15. 40 ÷ 5 = ■

______ × 5 = 40

The unknown is ______.

Find each product or quotient.

16. 5 ÷ 1 = ______

17. 50 ÷ 10 = ______

18. 9 × 5 = ______

19. 30 × 2 = ______

20. 2 × 80 = ______

21. 5 × 70 = ______

Name

Problem Solving

Algebra Write a number sentence with a symbol for the unknown for Exercises 22 and 23. Then solve.

22. Lucia's classroom has tables that are a total of 25 feet long. If there are 5 tables, how long is each table?

23. Julian bought 8 books this year. He gets a free book every time he buys 1. How many books altogether did he get this year?

24. Robert wrote this division sentence.

$$20 \div 2 = 10$$

Write a number sentence that he could use to check his work.

25. The table below shows the cost of movie tickets.

Movie Tickets					
Input	1	2	3	4	5
Output	\$8	\$16	\$24	?	?

What is the cost for 5 movie tickets?

Test Practice

26. When Xavier buys his lunch, he saves the nickel he gets as change. How much money will Xavier have if he buys his lunch 6 times?

Ⓐ 5¢ Ⓒ 25¢

Ⓑ 6¢ Ⓓ 30¢

My Work!

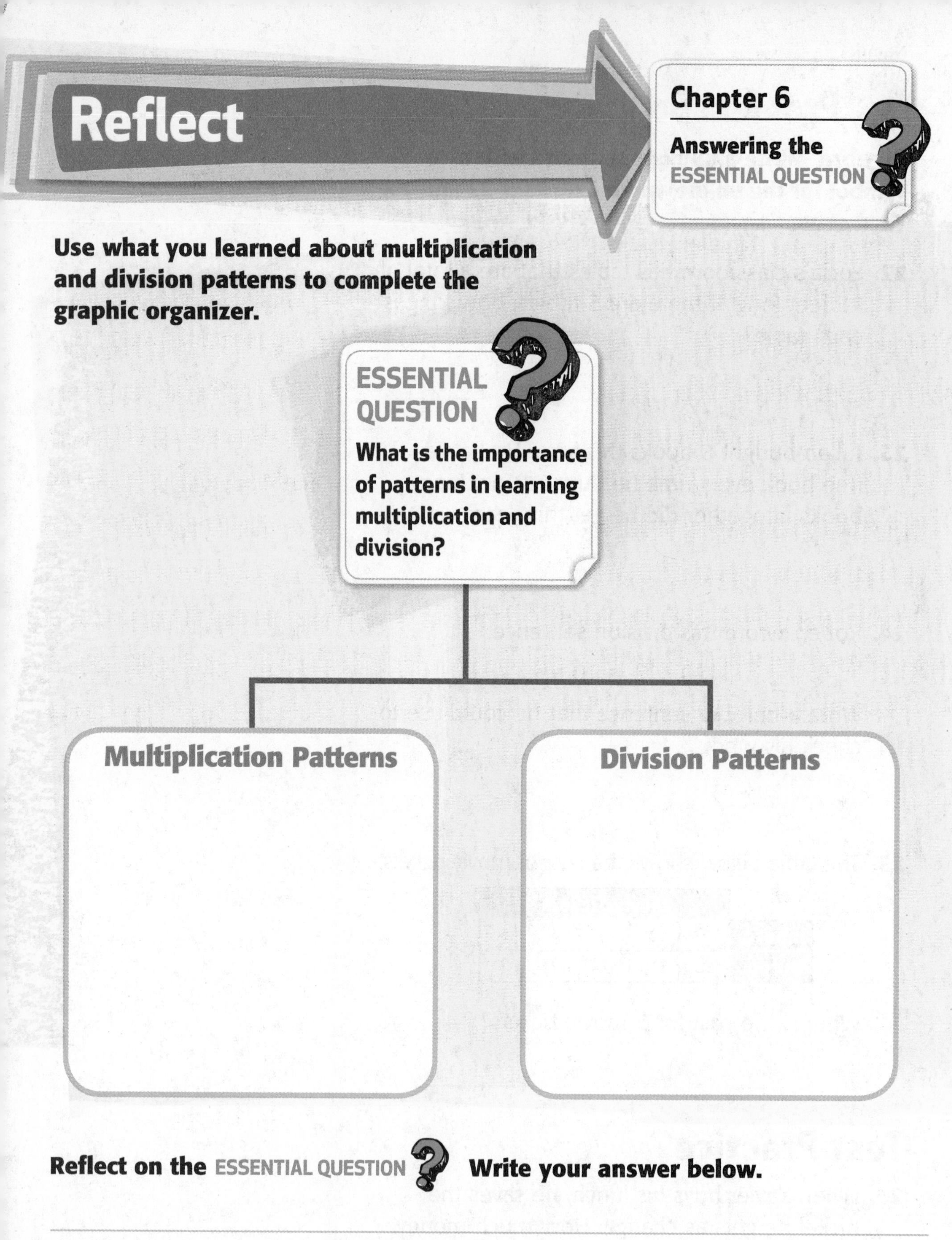

Reflect on the ESSENTIAL QUESTION **Write your answer below.**

Chapter

7 Multiplication and Division

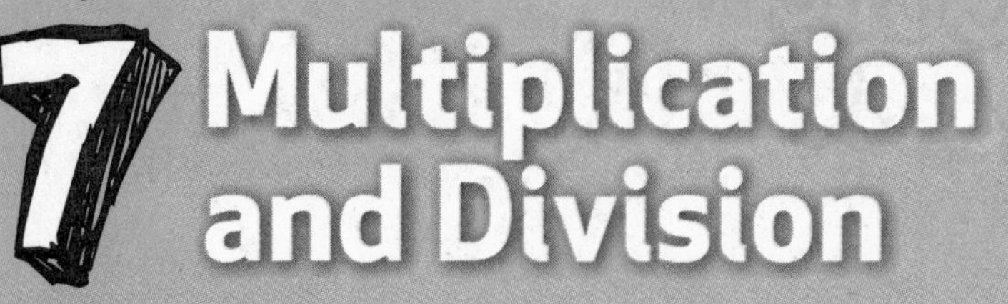

ESSENTIAL QUESTION

What strategies can be used to learn multiplication and division facts?

My Fun Friends

Watch a video!

Watch

MY Common Core State Standards

CCSS

Operations and Algebraic Thinking

CCSS

3.OA.1 Interpret products of whole numbers, e.g., interpret 5×7 as the total number of objects in 5 groups of 7 objects each.

3.OA.2 Interpret whole-number quotients of whole numbers, e.g., interpret $56 \div 8$ as the number of objects in each share when 56 objects are partitioned equally into 8 shares, or as a number of shares when 56 objects are partitioned into equal shares of 8 objects each.

3.OA.3 Use multiplication and division within 100 to solve word problems in situations involving equal groups, arrays, and measurement quantities, e.g., by using drawings and equations with a symbol for the unknown number to represent the problem.

3.OA.4 Determine the unknown whole number in a multiplication or division equation relating three whole numbers.

3.OA.5 Apply properties of operations as strategies to multiply and divide.

3.OA.6 Understand division as an unknown-factor problem.

3.OA.7 Fluently multiply and divide within 100, using strategies such as the relationship between multiplication and division (e.g., knowing that $8 \times 5 = 40$, one knows $40 \div 5 = 8$) or properties of operations. By the end of Grade 3, know from memory all products of two one-digit numbers.

3.OA.9 Identify arithmetic patterns (including patterns in the addition table or multiplication table), and explain them using properties of operations.

Cool! This is what I'm going to be doing!

Standards for Mathematical PRACTICE

1. Make sense of problems and persevere in solving them.
2. Reason abstractly and quantitatively.
3. Construct viable arguments and critique the reasoning of others.
4. Model with mathematics.
5. Use appropriate tools strategically.
6. Attend to precision.
7. Look for and make use of structure.
8. Look for and express regularity in repeated reasoning.

= focused on in this chapter

Name

Am I Ready?

Check — Go online to take the Readiness Quiz

Tell whether the groups in each pair are equal.

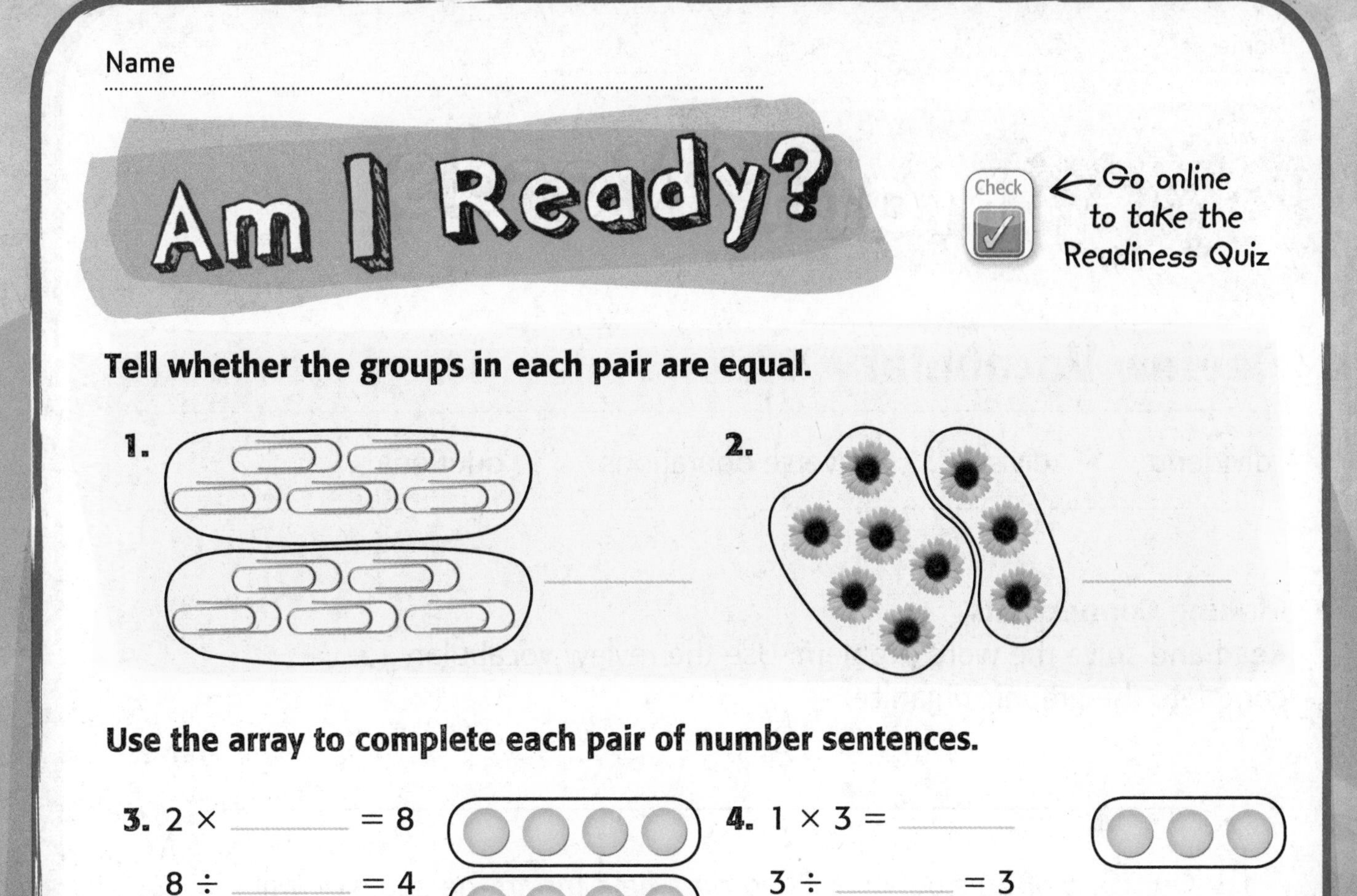

1. ______

2. ______

Use the array to complete each pair of number sentences.

3. $2 \times$ ______ $= 8$
   $8 \div$ ______ $= 4$

4. $1 \times 3 =$ ______
   $3 \div$ ______ $= 3$

Draw lines to match the division sentence to the related multiplication sentence.

5. $6 \div 2 = 3$ • $5 \times 3 = 15$
6. $15 \div 3 = 5$ • $8 \times 2 = 16$
7. $20 \div 4 = 5$ • $4 \times 5 = 20$
8. $16 \div 2 = 8$ • $3 \times 2 = 6$

9. **Algebra** Mrs. June wants to divide 30 folders equally among 10 students. How many folders will each student get? Write a division sentence with a symbol for the unknown. Then solve.

______

Shade the boxes to show the problems you answered correctly.

How Did I Do?

1	2	3	4	5	6	7	8	9

Name

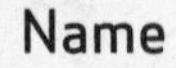

Review Vocabulary

dividend   divisor   inverse operations   quotient

Making Connections

Read and solve the word problem. Use the review vocabulary to complete the graphic organizer.

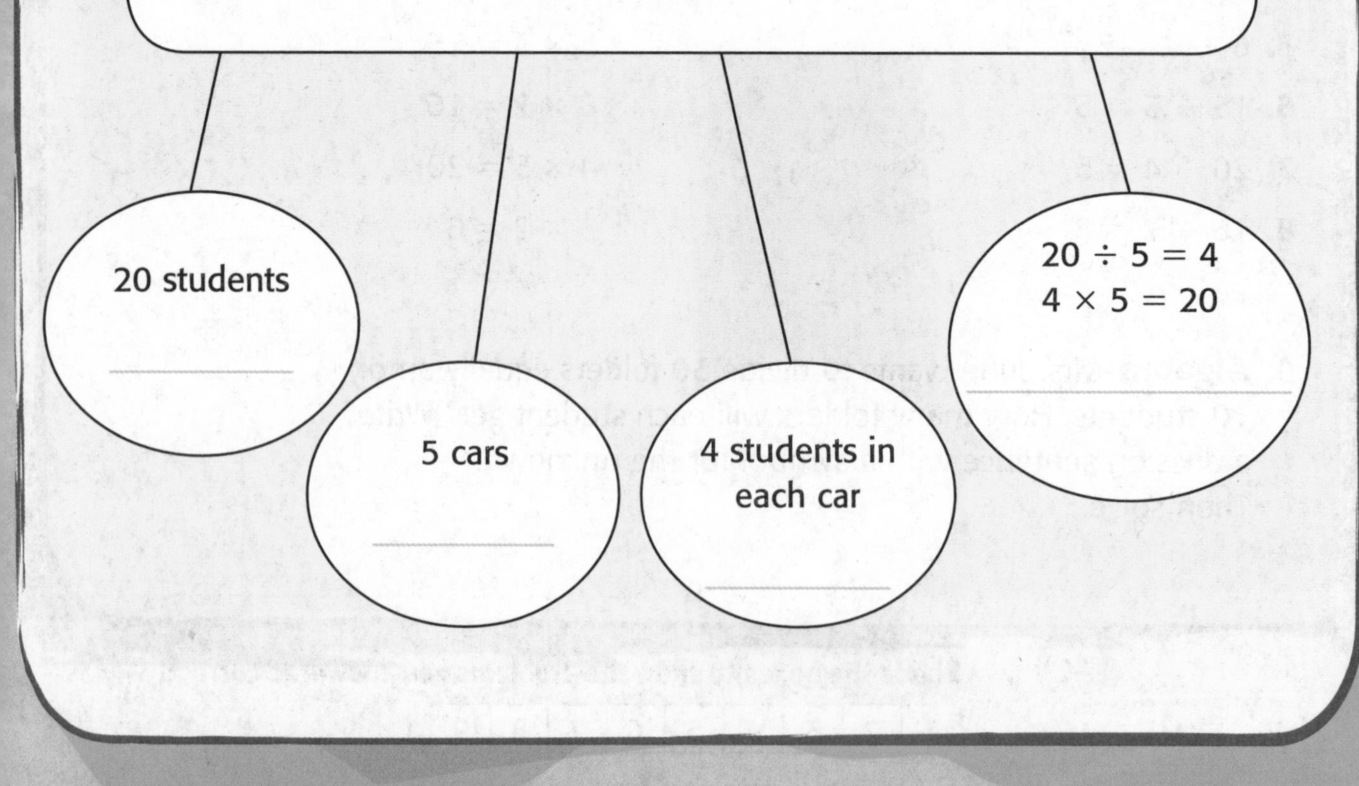

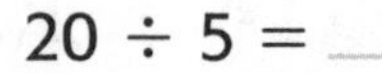

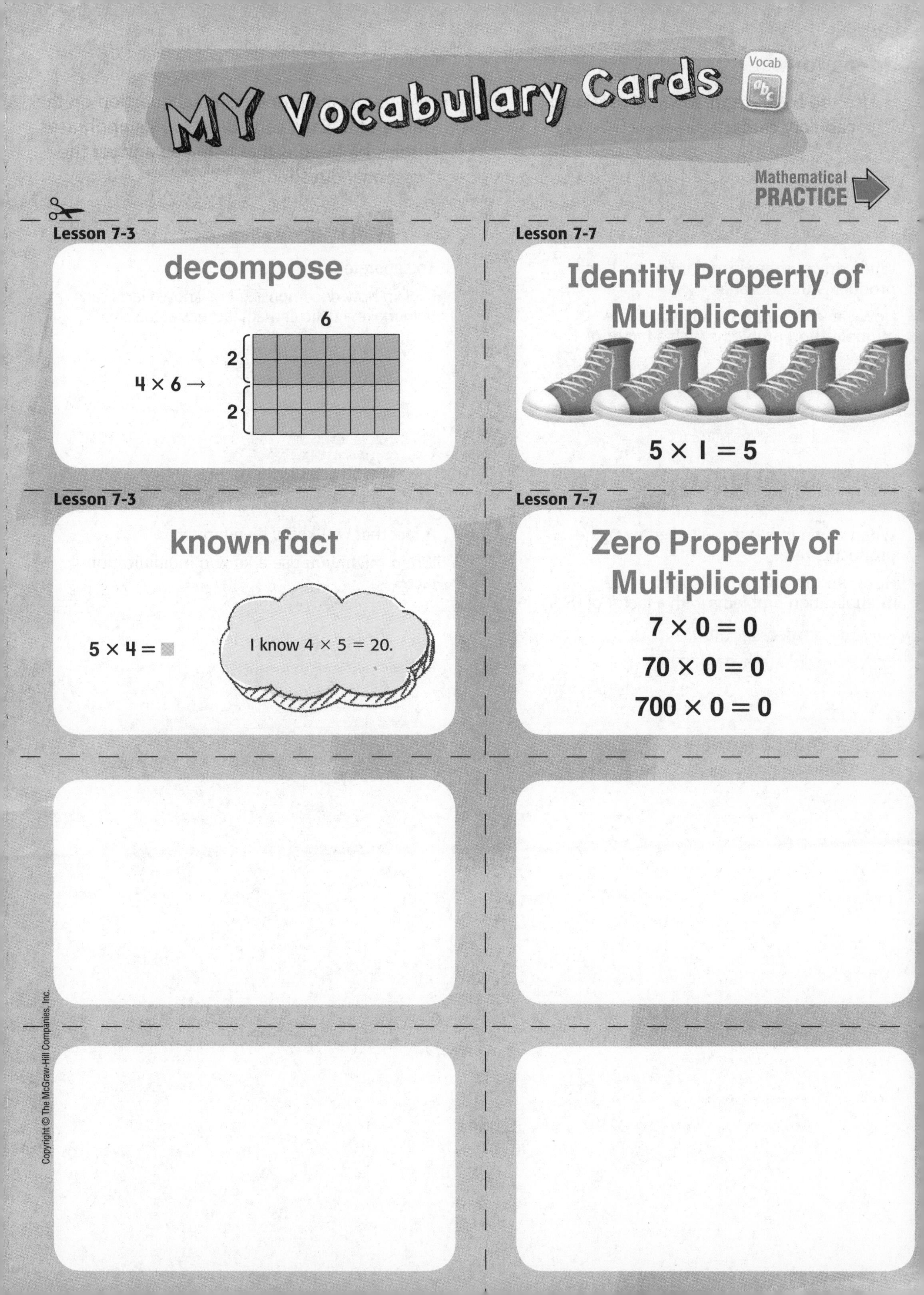

MY Vocabulary Cards

Vocab

Mathematical PRACTICE

Lesson 7-3

decompose

$4 \times 6 \rightarrow$

Lesson 7-7

Identity Property of Multiplication

$5 \times 1 = 5$

Lesson 7-3

known fact

$5 \times 4 = \blacksquare$

Lesson 7-7

Zero Property of Multiplication

$7 \times 0 = 0$

$70 \times 0 = 0$

$700 \times 0 = 0$

Ideas for Use

- Use the blank cards to write your own vocabulary cards.
- Write this chapter's essential question on the front of a blank card. Write words or phrases from the lessons that help you answer the essential question.

When any number is multiplied by 1, the product is that number.

How can this property help you solve multiplication problems with a factor of 1?

To separate into parts.

Explain how decomposing into known facts can help make a difficult math fact easier to solve.

When any number is multiplied by 0, the product is zero.

How can this property help you solve multiplication problems with a factor of 0?

A fact that you know by memory.

When might you use a known multiplication fact?

Follow the steps on the back to make your Foldable.

Divide 24 by 3

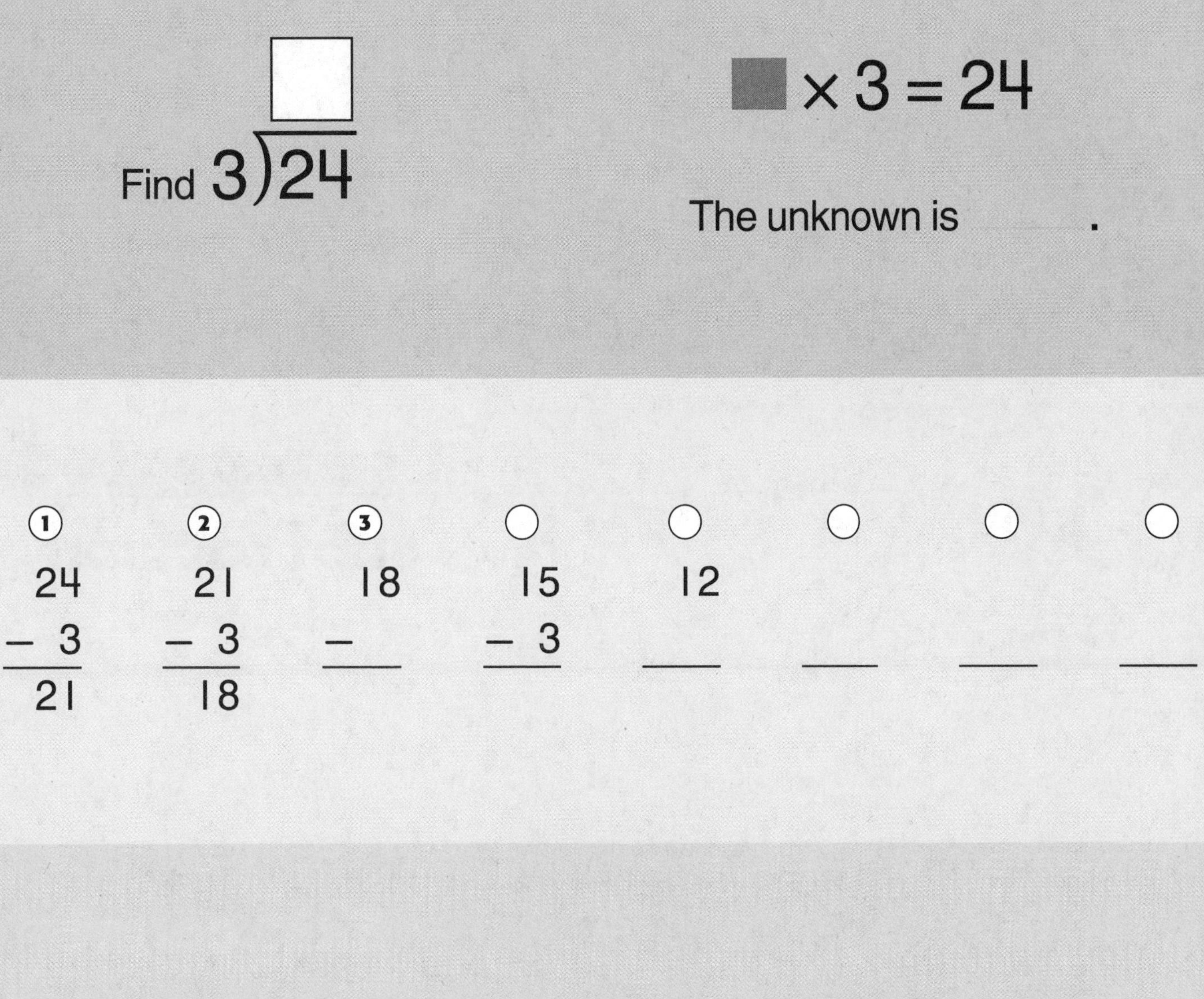

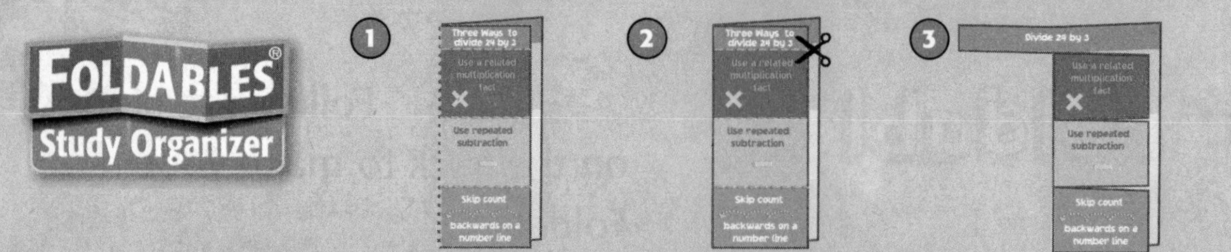

Three Ways to divide 24 by 3

Use a related multiplication fact

×

Use repeated subtraction

−

Skip count backwards on a number line

Name ______________________

Multiply by 3

Lesson 1

ESSENTIAL QUESTION
What strategies can be used to learn multiplication and division facts?

Math in My World

Tools Watch Tutor

Example 1

There are 3 dogs. Each dog buried 4 bones in a yard. How many bones are buried in the yard?

Find 3×4.

Write it like this, also. $\begin{array}{r} 3 \\ \times\ 4 \\ \hline \end{array}$

One Way **Use an array.**

Find 3 rows of 4 bones.
The array shows that $3 \times 4 =$ ______.

There are ______ bones buried in the yard.

Use the Commutative Property to write another multiplication sentence for this array.

______ × ______ = ______

Another Way **Use a number line.**

Skip count to find 3 groups of 4.

① ② ③
0 1 2 3 4 5 6 7 8 9 10 11 12

The number line shows ______ jumps of 4 is ______.

So, $3 \times 4 =$ ______. There are ______ bones buried in the yard.

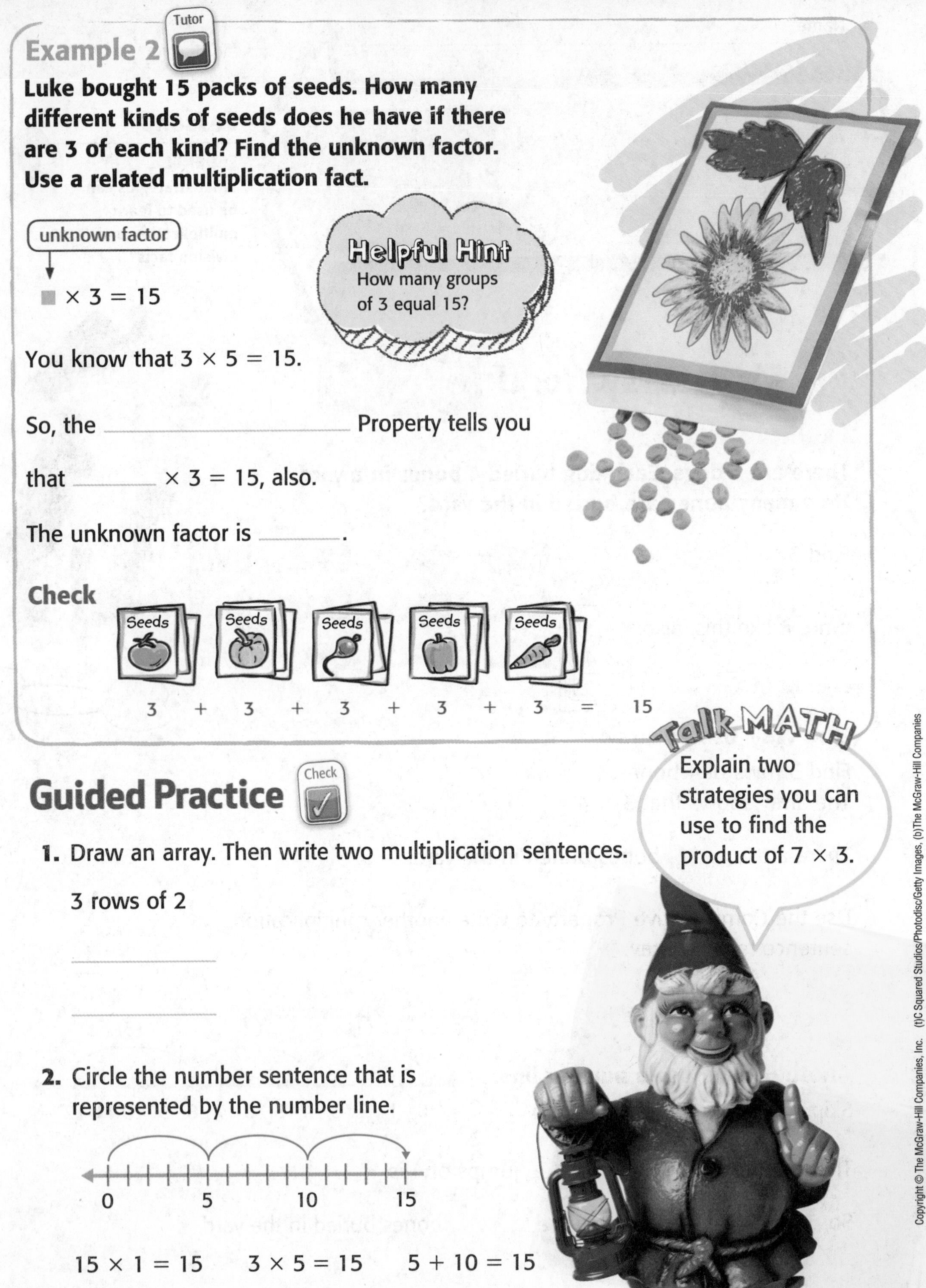

Example 2

Luke bought 15 packs of seeds. How many different kinds of seeds does he have if there are 3 of each kind? Find the unknown factor. Use a related multiplication fact.

unknown factor

■ × 3 = 15

Helpful Hint
How many groups of 3 equal 15?

You know that 3 × 5 = 15.

So, the ____________ Property tells you

that ________ × 3 = 15, also.

The unknown factor is ________.

Check

3 + 3 + 3 + 3 + 3 = 15

Talk MATH
Explain two strategies you can use to find the product of 7 × 3.

Guided Practice

Check

1. Draw an array. Then write two multiplication sentences.

   3 rows of 2

   ____

   ____

2. Circle the number sentence that is represented by the number line.

   0 5 10 15

   15 × 1 = 15 3 × 5 = 15 5 + 10 = 15

Name

Independent Practice

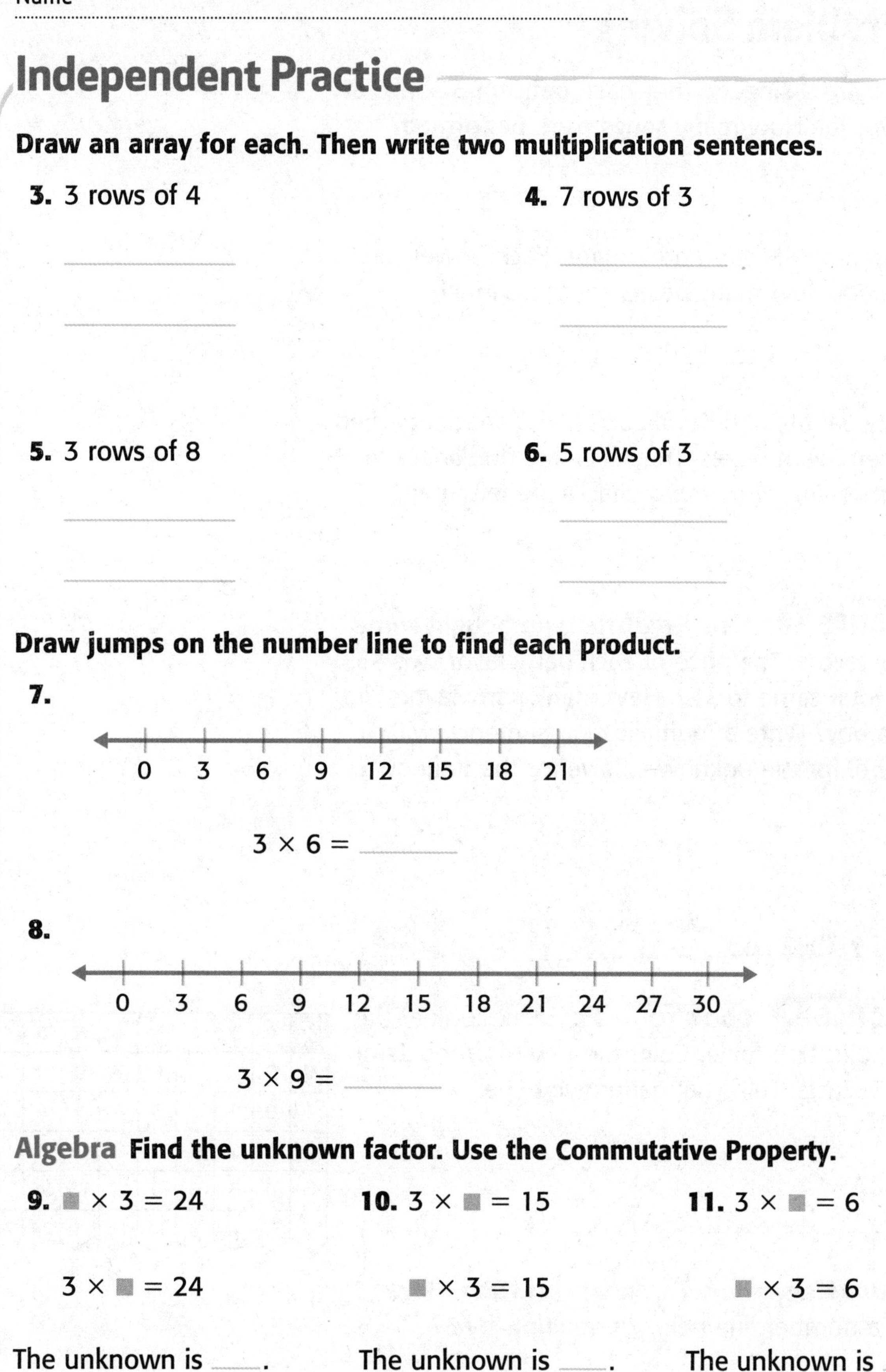

Draw an array for each. Then write two multiplication sentences.

3. 3 rows of 4

4. 7 rows of 3

5. 3 rows of 8

6. 5 rows of 3

Draw jumps on the number line to find each product.

7.

$3 \times 6 =$ ____

8.

$3 \times 9 =$ ____

Algebra **Find the unknown factor. Use the Commutative Property.**

9. ■ × 3 = 24

3 × ■ = 24

The unknown is ____.

10. 3 × ■ = 15

■ × 3 = 15

The unknown is ____.

11. 3 × ■ = 6

■ × 3 = 6

The unknown is ____.

Problem Solving

12. There are 9 singers. They each perform 3 songs at the recital. How many songs were performed?

13. There are 7 daisies and 7 tulips. Each flower has 3 petals. How many petals are there in all?

14. Henry, Jaime, and Kayla each had 3 snacks packed in their lunch boxes. They each ate one snack in the morning. How many snacks are left in all?

15. Mathematical PRACTICE 2 **Use Algebra** Lynn bought some party favors. The price of each party favor was $3. Her total came to $12. How many party favors did Lynn buy? Write a multiplication sentence with a symbol for the unknown. Solve for the unknown.

My Work!

HOT Problems

16. Mathematical PRACTICE 8 **Look for a Pattern** Look at the multiplication table. Color the row of products of the 3s facts. Tell what pattern you see.

×	0	1	2	3	4	5	6	7
0	0	0	0	0	0	0	0	0
1	0	1	2	3	4	5	6	7
2	0	2	4	6	8	10	12	14
3	0	3	6	9	12	15	18	21
4	0	4	8	12	16	20	24	28

17. **Building on the Essential Question** How can a number line help you multiply by 3?

Name

Lesson 1
Multiply by 3

Homework Helper

eHelp Need help? connectED.mcgraw-hill.com

Tyra has 3 posters on each of 3 walls in her bedroom. How many posters does Tyra have in her room? Find 3 × 3.

One Way **Use an array to model 3 rows of 3.**

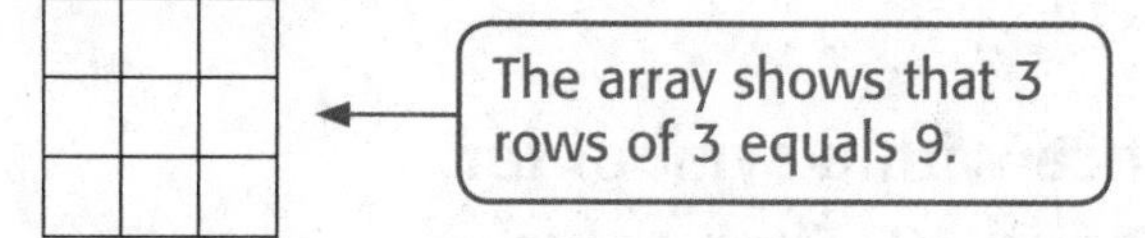

Another Way **Use a number line.**

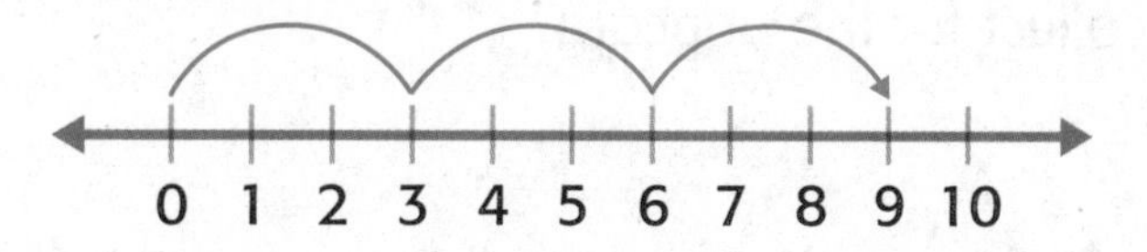

The number line shows that 3 jumps of 3 = 9.

So, 3 × 3 = 9. You can also write it like this. →

$$\begin{array}{r} 3 \\ \times\ 3 \\ \hline 9 \end{array}$$

Tyra has 9 posters in her room.

Practice

Draw an array for each. Then write two multiplication sentences.

1. 3 rows of 8

2. 6 rows of 3

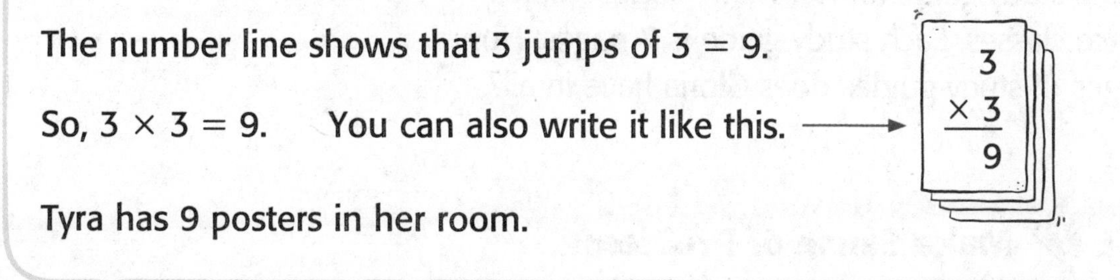

Multiply. Use the number line to skip count if needed.

0 1 2 3 4 5 6 7 8 9 10 11 12 13 14 15 16 17 18 19 20 21 22 23 24 25

3. $5 \times 3 =$ ______

4. $8 \times 3 =$ ______

5. $7 \times 3 =$ ______

6. $4 \times 3 =$ ______

Algebra Find the unknown factor. Use the Commutative Property.

7. $\blacksquare \times 3 = 30$

$3 \times \blacksquare = 30$

The unknown is ______.

8. $3 \times \blacksquare = 18$

$\blacksquare \times 3 = 18$

The unknown is ______.

Problem Solving

Write a multiplication sentence with a symbol for the unknown for Exercises 9 and 10. Then solve.

9. A box of popcorn costs $3 at the baseball game. The vendor sells 5 boxes to people in row 22. How much money did the vendor collect for the popcorn?

______

10. Gloria has a study guide for her math, social studies, and science classes. Each study guide is 7 pages. How many pages of study guides does Gloria have in all?

______

11. Mathematical PRACTICE 1 **Make Sense of Problems** Meredith feeds her 3 dogs twice a day. How many times does she feed the dogs in 3 days?

______

My Work!

Test Practice

12. There are 3 rows of cars in the parking lot. Each row has 5 cars. How many cars are in the parking lot?

Ⓐ 18 cars

Ⓑ 15 cars

Ⓒ 12 cars

Ⓓ 9 cars

Name ______________________

Divide by 3

Lesson 2

ESSENTIAL QUESTION
What strategies can be used to learn multiplication and division facts?

Math in My World

Watch Tutor

Example 1

Max, Maria, and Tani have 24 markers in all. Each person has the same number of markers. How many markers does each person have?

Find the unknown quotient. $3\overline{)24}$ ■ ← unknown quotient

One Way **Use the multiplication table.**

1. Locate row 3. Circle the divisor.
2. Follow row 3 to 24. Circle the dividend.
3. Move straight up the column to 8. Circle the quotient.

The unknown quotient is ______.

×	1	2	3	4	5	6	7	8
1	1	2	3	4	5	6	7	8
2	2	4	6	8	10	12	14	16
3	3	6	9	12	15	18	21	24
4	4	8	12	16	20	24	28	32
5	5	10	15	20	25	30	35	40
6	6	12	18	24	30	36	42	48
7	7	14	21	28	35	42	49	56
8	8	16	24	32	40	48	56	64

Another Way **Use a related fact.**

Find $24 \div 3$ by thinking of a related multiplication fact.

Find the unknown factor. → ■ $\times 3 = 24$

THINK What times 3 equals 24? → ______ $\times 3 = 24$

The unknown factor is ______.

So, $3\overline{)24}$ (answer box) or $24 \div 3 =$ ______. The unknown is ______.

Each person has ______ markers.

Example 2

Tools Tutor

To travel to the beach, Angela and her 5 friends will divide up equally into 3 cars. How many friends will be in each car?

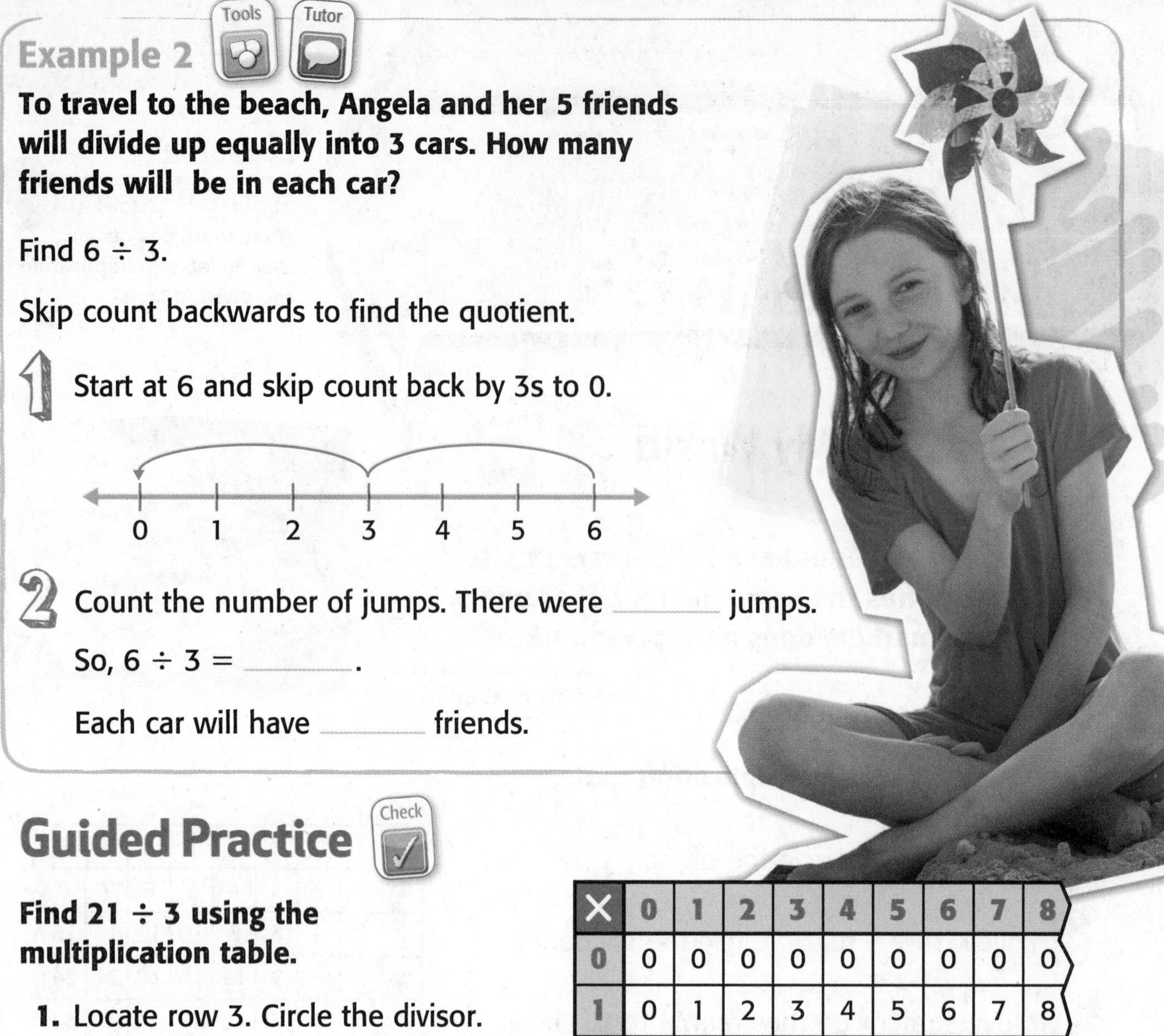

Find $6 \div 3$.

Skip count backwards to find the quotient.

1. Start at 6 and skip count back by 3s to 0.

0 1 2 3 4 5 6

2. Count the number of jumps. There were ______ jumps.

   So, $6 \div 3 =$ ______.

   Each car will have ______ friends.

Guided Practice

Check

Find $21 \div 3$ using the multiplication table.

×	0	1	2	3	4	5	6	7	8
0	0	0	0	0	0	0	0	0	0
1	0	1	2	3	4	5	6	7	8
2	0	2	4	6	8	10	12	14	16
3	0	3	6	9	12	15	18	21	24
4	0	4	8	12	16	20	24	28	32
5	0	5	10	15	20	25	30	35	40

1. Locate row 3. Circle the divisor.
2. Follow row 3 to 21. Circle the dividend.
3. Circle the quotient to solve.

   $21 \div 3 =$ ______

4. Skip count to find the quotient.

   $12 \div 3 =$ ______

0 1 2 3 4 5 6 7 8 9 10 11 12

Talk MATH

Look back at the circled numbers on the multiplication table. Write the four related facts for the 3 numbers.

Name ....

CCSS

Independent Practice

Skip count backwards to find the quotient.

5.

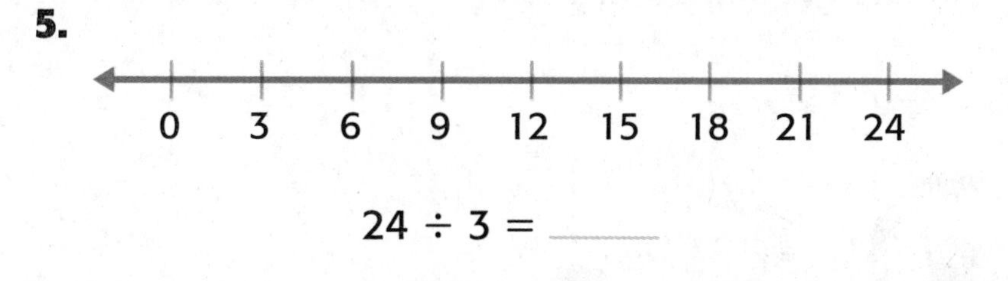

$24 \div 3 =$ ______

Algebra Use a related multiplication fact to find the unknown.

6. $15 \div 3 = \blacksquare$

$3 \times$ ______ $= 15$

The unknown is ______.

7. $3\overline{)27}$ = ■

$3 \times$ ______ $= 27$

The unknown is ______.

8. $3\overline{)21}$ = ■

$3 \times$ ______ $= 21$

The unknown is ______.

Algebra Find each unknown to solve the puzzle. Write the letter corresponding to each quotient on the line above each exercise number.

9. $3 \div 3 = \blacksquare$ ______

10. $9 \div \blacksquare = 3$ ______

11. $15 \div \blacksquare = 3$ ______

12. $27 \div 9 = \blacksquare$ ______

13. $\blacksquare \div 3 = 8$ ______

14. $6 \div 2 = \blacksquare$ ______

15. $\blacksquare \div 3 = 4$ ______

16. $18 \div 3 = \blacksquare$ ______

3 i	15 p	5 v
6 n	8 n	24 s
1 d	12 o	4 t

Exercise: ____ ____ ____ ____ ____ ____ ____ ____
9 10 11 12 13 14 15 16

Problem Solving

17. Mathematical PRACTICE 4 **Model Math** A soccer coach buys 3 equally-priced soccer balls for $21. What is the price for each ball? Write a division sentence. Then label each price tag.

18. Stanley is on a 3-day hike. He will hike a total of 18 miles. If he hikes the same number of miles each day, how many miles will he hike the first day?

19. Mathematical PRACTICE 1 **Make a Plan** Jessica arranged 27 stickers in 3 equal rows. Then she gave 1 row of stickers to Dane and 3 stickers to Kim. How many stickers does Jessica have left?

My Work!

HOT Problems

20. Mathematical PRACTICE 7 **Identify Structure** There are 24 bananas to be divided equally among 3 monkeys. How many bananas will each monkey receive?

Rewrite the story using a related multiplication fact. Solve.

21.  **Building on the Essential Question** Besides using models, how else can I find $18 \div 3$?

Name ....

MY Homework

Lesson 2
Divide by 3

Homework Helper

eHelp

Need help? connectED.mcgraw-hill.com

Chuck and his 2 brothers read 12 books about the solar system. Each boy read the same number of books. How many books did each boy read? Find the unknown quotient.

You need to find 12 ÷ 3, or $3\overline{)12}$.

Use the multiplication table.

1 Find row 3. Circle the divisor.

2 Move across row 3 to 12. Circle the dividend.

3 Move straight up the column to 4. Circle the quotient.

×	0	1	2	3	(4)	5	6
0	0	0	0	0	0	0	0
1	0	1	2	3	4	5	6
2	0	2	4	6	8	10	12
(3)	0	3	6	9	(12)	15	18
4	0	4	8	12	16	20	24
5	0	5	10	15	20	25	30

So, 12 ÷ 3 = 4. Each boy read 4 books.

Practice

Algebra Use a related multiplication fact to find the unknown.

1. 30 ÷ 3 = ■

3 × ______ = 30

The unknown is ______.

2. 18 ÷ 3 = ■

3 × ______ = 18

The unknown is ______.

3. 15 ÷ 3 = ■

3 × ______ = 15

The unknown is ______.

4. 21 ÷ 3 = ■

3 × ______ = 21

The unknown is ______.

Use the multiplication table to divide.

5. 24 ÷ 3 = ______

6. 9 ÷ 3 = ______

7. 27 ÷ 3 = ______

8. 3 ÷ 3 = ______

9. 18 ÷ 3 = ______

×	0	1	2	3	4	5	6	7	8	9	10
0	0	0	0	0	0	0	0	0	0	0	0
1	0	1	2	3	4	5	6	7	8	9	10
2	0	2	4	6	8	10	12	14	16	18	20
3	0	3	6	9	12	15	18	21	24	27	30
4	0	4	8	12	16	20	24	28	32	36	40
5	0	5	10	15	20	25	30	35	40	45	50
6	0	6	12	18	24	30	36	42	48	54	60
7	0	7	14	21	28	35	42	49	56	63	70
8	0	8	16	24	32	40	48	56	64	72	80
9	0	9	18	27	36	45	54	63	72	81	90
10	0	10	20	30	40	50	60	70	80	90	100

Real World

Problem Solving

10. Alana mailed 6 letters in 3 different mailboxes. She put the same number of letters in each mailbox. How many letters did Alana put in each mailbox?

______

My Work!

11. Mathematical PRACTICE 2 **Use Number Sense** Ms. Banks divides 18 basketballs evenly among 3 bags. For class, she takes 2 basketballs from each bag. How many basketballs are left in one of the bags?

______

Test Practice

12. Elyse served herself and 2 friends 24 ounces of juice. She poured the same amount of juice in each glass. How many ounces of juice were in each glass?

Ⓐ 22 ounces
Ⓑ 12 ounces
Ⓒ 8 ounces
Ⓓ 6 ounces

Name ....

Hands On
Double a Known Fact

Lesson 3

ESSENTIAL QUESTION
What strategies can be used to learn multiplication and division facts?

A **known fact** is a fact you have memorized. You can use a known multiplication fact to solve a multiplication fact you do not know.

Build It

Find 4 × 6.
Decompose, or separate, the number 4 into equal addends of 2 + 2.

My Drawing!

1 Model the known fact, 2 × 6.

Use counters to make an array. Show 2 rows of 6. Draw your array.

Write the number sentence.

_____ × _____ = _____

2 Double the known fact.

Make one more 2 × 6 array. Draw your array. Write the number sentence for this array.

_____ × _____ = _____

Add the two products together.

12 + _____ = _____

3 Find 4 × 6.

Push the two arrays together. Write the new number sentence.

_____ × _____ = _____

2 × 6 plus 2 × 6 = _____ So, 4 × 6 = _____

Try It

Find 6 × 5.

Decompose 6 into two equal addends. 6 = 3 + ______

Model the known fact, 3 × 5 two times.

Draw two 3 × 5 arrays.
Write the product on each array.

2 Double the known fact 3 × 5.

Draw the two arrays pushed together.

Add the products.

15 + ______ = ______

3 Find 6 × 5.

Write the multiplication sentence.

______ × ______ = ______

3 × 5 plus 3 × 5 = ______. So, 6 × 5 = ______.

My Drawing!

Talk About It

1. Why can you double the product of 2 × 6 to find 4 × 6?

____

____

2. What doubles fact would 3 × 6 help you find? ______

3. Give an example of doubling a known fact. Explain.

____

____

4. Draw two arrays that you could put together to find 4 × 5. Label the arrays with a number sentence.

Name ____________________

Practice It

Use counters to model a known fact that will help you find the first product. Draw the model two times.

5. $4 \times 3 =$ ____

Known fact: $2 \times 3 =$ ____

Double the product: $6 + 6 =$ ____

6. $4 \times 4 =$ ____

Known fact: $2 \times 4 =$ ____

Double the product: $8 + 8 =$ ____

7. $7 \times 4 =$ ____

Known fact: ____

Double the product: ____

8. $6 \times 7 =$ ____

Known fact: ____

Double the product: ____

Use counters to double the known fact. Write the product it helps you find.

9. $3 \times 8 =$ ____

$3 \times 8 =$ ____

____ $\times$ ____ $=$ ____

10. $3 \times 6 =$ ____

$3 \times 6 =$ ____

____ $\times$ ____ $=$ ____

Apply It

11. Megan and Paul each have a tray of cookies. Each tray has 2 rows of 6 cookies. How many cookies do they have if they put their trays together?

What known fact did you double?

What fact is found by doubling the known fact?

My Work!

12. Mathematical PRACTICE 4 **Model Math** Two heart buttons have 8 holes. Jeffery doubles the number of heart buttons shown. How many holes will there be now?

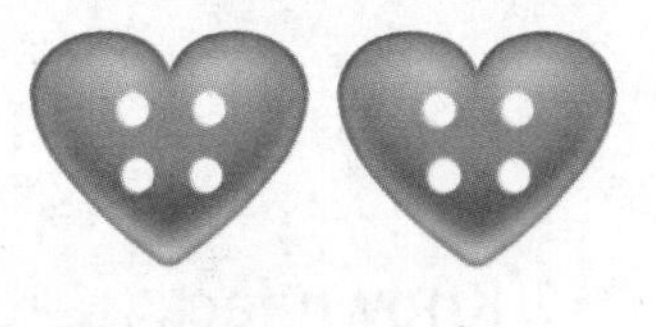

What known fact did you double?

What fact is found by doubling the known fact?

13. Mathematical PRACTICE 3 **Justify Conclusions** Can you double a known fact to find 7×6? Explain.

Write About It

14. When is doubling a known fact useful?

Name ....

Lesson 3

Hands On: Double a Known Fact

Homework Helper

eHelp Need help? connectED.mcgraw-hill.com

Find 6 × 5.

Decompose, or separate, the factor 6 into two equal addends of 3 + 3. Then you can double the known fact, 3 × 5.

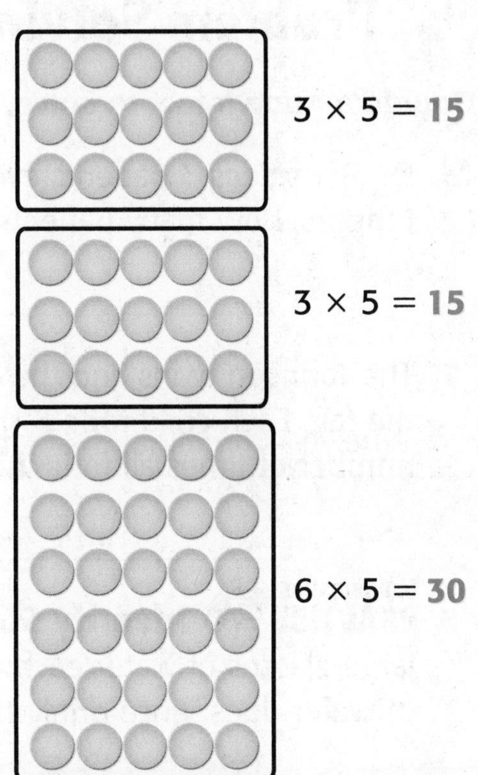

1 Model the known fact, 3 × 5.

Use counters to make an array that shows 3 rows of 5.

2 Double the known fact.

Make another array that shows 3 rows of 5.

3 Find 6 × 5.

Push the two arrays together into 6 rows of 5.

Add the products from the two arrays:

15 + 15 = 30

The combined arrays show 6 × 5 = 30.

Practice

1. Draw counters to model a known fact that will help you find 4 × 5. Draw the model two times.

4 × 5 = ______

Known fact: 2 × 5 = ______

Double the product: 10 + 10 = ______

Double the known fact. Write the product it helps you find.

2. $3 \times 7 =$ ______

$3 \times 7 =$ ______

______ $\times$ ______ $=$ ______

3. $3 \times 3 =$ ______

$3 \times 3 =$ ______

______ $\times$ ______ $=$ ______

4. $2 \times 6 =$ ______

$2 \times 6 =$ ______

______ $\times$ ______ $=$ ______

5. $2 \times 9 =$ ______

$2 \times 9 =$ ______

______ $\times$ ______ $=$ ______

Problem Solving

Double a known fact to solve.

My Work!

6. Dr. Berry sees 3 patients every hour. If she works 8 hours, how many patients does she see?

______

7. The Johnson twins and the Clayton twins went to the fair. Each child rode 5 rides. What is the total number of times all four children went on a ride?

______

8. Mathematical PRACTICE 7 **Identify Structure** Vince drinks 4 large glasses of water each day. How many glasses of water does Vince drink in 7 days?

______

Vocabulary Check

Vocab abc

Choose the correct word(s) to complete each sentence.

known fact　　　decompose

9. One way to ______ the number 8 is to write it as 4 + 4.

10. A fact that you have memorized is a ______.

Name

Multiply by 4

Lesson 4

ESSENTIAL QUESTION
What strategies can be used to learn multiplication and division facts?

Math in My World

Watch Tutor

Example 1

A box is packed with 4 rows of oranges. Each row has 9 oranges. How many oranges are in the box?

Catching some rays!

Find 4×9.

Decompose 4 into equal addends of $2 + 2$.

Use the **known fact** of 2×9 and double it.

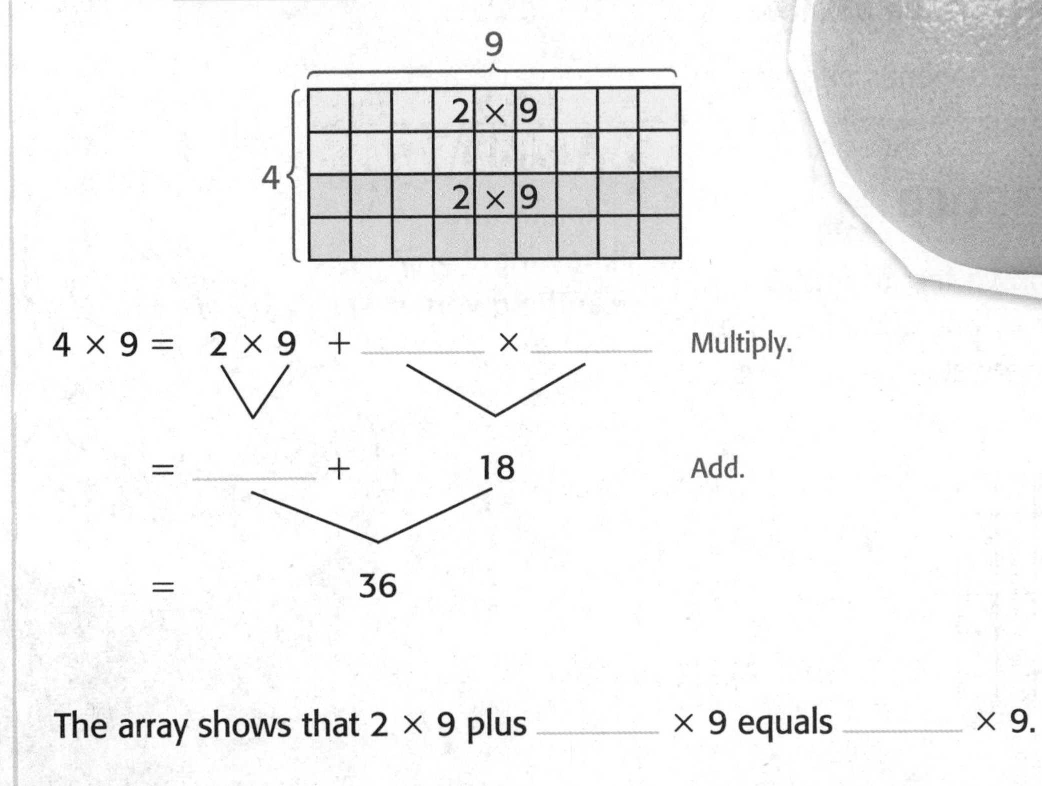

$4 \times 9 = 2 \times 9 + \_\_\_\_ \times \_\_\_\_$ Multiply.

$= \_\_\_\_ + 18$ Add.

$= 36$

The array shows that 2×9 plus ____ $\times 9$ equals ____ $\times 9$.

So, $4 \times 9 =$ ____. There are ____ oranges in the box.

Example 2

There are 3 bunches of bananas. Each bunch has 4 bananas. How many bananas are there altogether? Write a multiplication sentence with a symbol for the unknown. Then solve.

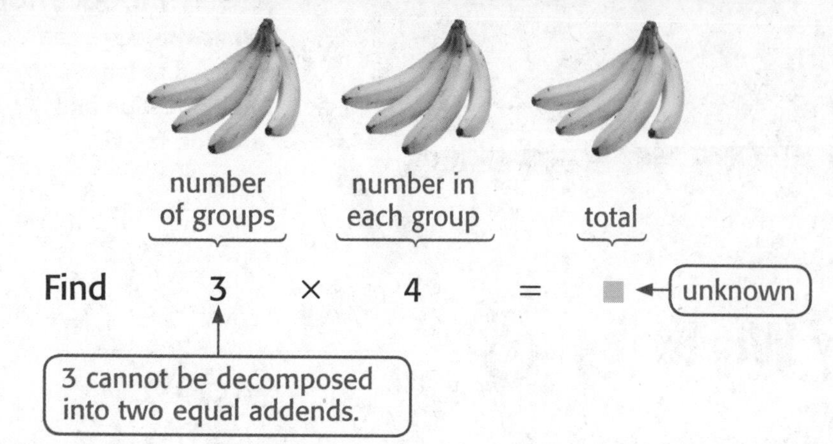

Find $3 \times 4 = ■$ ← unknown

3 cannot be decomposed into two equal addends.

Solve by decomposing the factor 4 into two equal addends of 2. Use the known fact 3×2 plus 3×2 to find the unknown.

3×2 plus _____ × _____ = ■

6 + _____ = _____

So, $3 \times 4 =$ _____. The unknown is _____.

There are _____ bananas altogether.

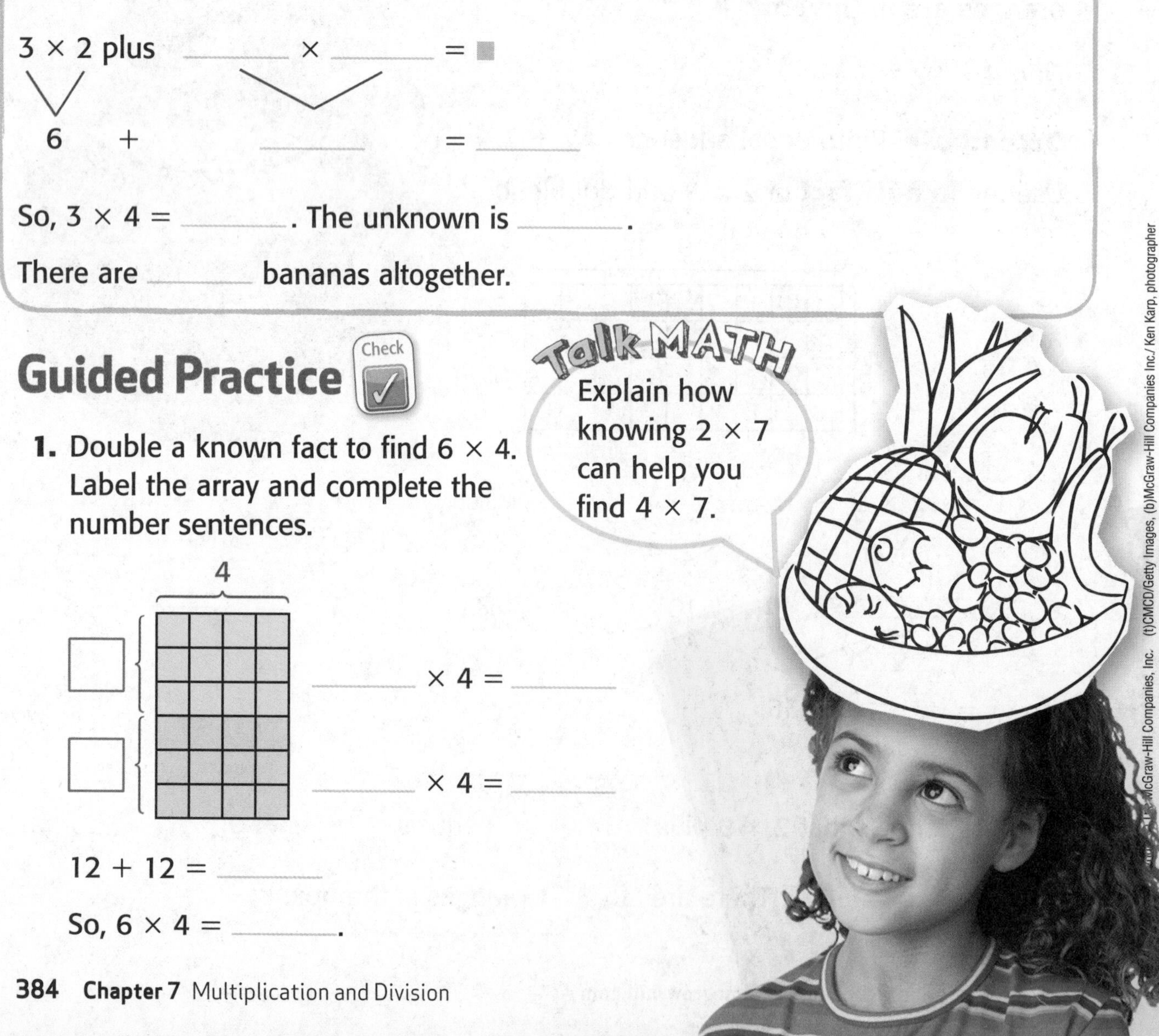

Guided Practice

Check

Talk MATH

Explain how knowing 2×7 can help you find 4×7.

1. Double a known fact to find 6×4. Label the array and complete the number sentences.

4

☐ _____ × 4 = _____

☐ _____ × 4 = _____

$12 + 12 =$ _____

So, $6 \times 4 =$ _____.

Name

Independent Practice

Double a known fact to find each product. Draw and label an array.

2. 8 × 4 = ______

3. 5 × 4 = ______

4. 4 × 6 = ______

5. 7 × 4 = ______

Algebra Find each unknown. Double a known fact.

6. 7 × 4 = ■

The unknown is ______.

7. 9 × 4 = ■

The unknown is ______.

8. $\begin{array}{r} 4 \\ \times\ 4 \\ \hline ■ \end{array}$

The unknown is ______.

9. $\begin{array}{r} 4 \\ \times\ 10 \\ \hline ■ \end{array}$

The unknown is ______.

Problem Solving

10. The library offers 4 activities on Saturday. Each activity table has room for 10 children. How many children can take part in the activities? Write a multiplication sentence to solve.

11. Mathematical PRACTICE 2 **Reason** Mitch bought 4 bottles of suntan lotion for $6 each. Later the lotion went on sale for $4 each. How many more bottles could he have bought for the same amount of money if he waited for the sale?

HOT Problems

12. Mathematical PRACTICE 2 **Use Number Sense** Look at the multiplication table. Circle the two numbers that represent the product of 4 and 10. Write this product as a sum of two equal addends.

×	0	1	2	3	4	5	6	7	8	9	10
0	0	0	0	0	0	0	0	0	0	0	0
1	0	1	2	3	4	5	6	7	8	9	10
2	0	2	4	6	8	10	12	14	16	18	20
3	0	3	6	9	12	15	18	21	24	27	30
4	0	4	8	12	16	20	24	28	32	36	40
5	0	5	10	15	20	25	30	35	40	45	50
6	0	6	12	18	24	30	36	42	48	54	60
7	0	7	14	21	28	35	42	49	56	63	70
8	0	8	16	24	32	40	48	56	64	72	80
9	0	9	18	27	36	45	54	63	72	81	90
10	0	10	20	30	40	50	60	70	80	90	100

Can the product of 4 and any number always be written as a sum of two equal addends? Explain.

13. **Building on the Essential Question** What is one strategy I could use to multiply by 4?

Lesson 4
Multiply by 4

Homework Helper

eHelp Need help? connectED.mcgraw-hill.com

Find 4 × 7.

Decompose 4 into equal addends of 2 + 2.

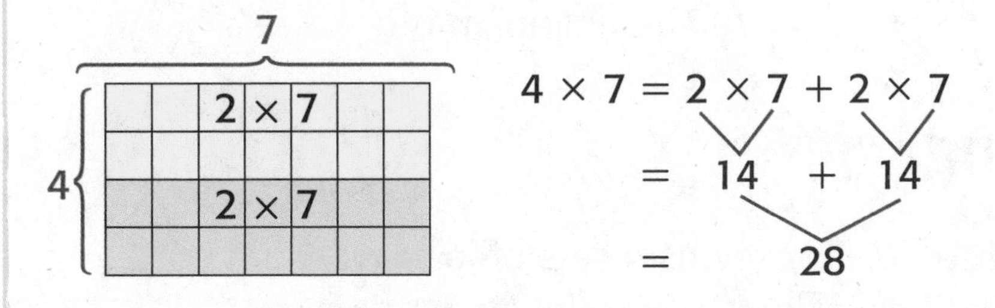

$$4 \times 7 = 2 \times 7 + 2 \times 7$$
$$= 14 + 14$$
$$= 28$$

The array shows that 4 groups of 7 equals
2 groups of 7 plus 2 groups of 7.

So, 4 × 7 = 28.

Practice

Double a known fact to find the product. Label the array and complete the number sentences.

1. 4 × 5

2. 3 × 4

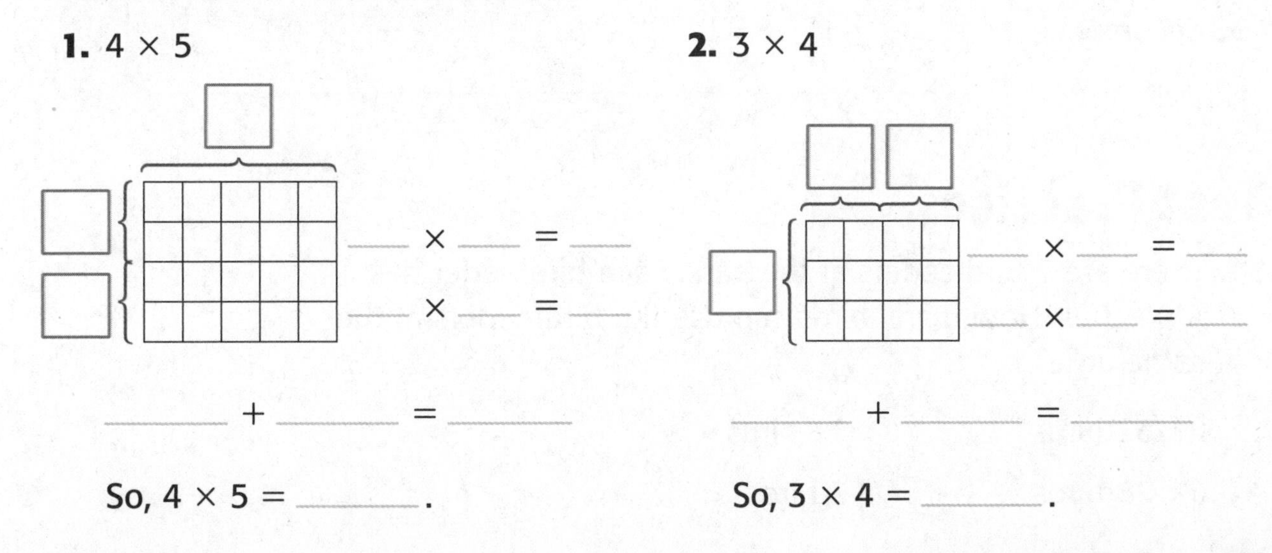

1. ___ × ___ = ___
___ × ___ = ___
___ + ___ = ___
So, 4 × 5 = ___.

2. ___ × ___ = ___
___ × ___ = ___
___ + ___ = ___
So, 3 × 4 = ___.

Double a known fact to find each product. Draw and label an array.

3. $6 \times 4 =$ ______

4. $4 \times 8 =$ ______

Algebra **Find each unknown. Double a known fact.**

5. $9 \times 4 = \blacksquare$

The unknown is ______.

6. $4 \times 4 = \blacksquare$

The unknown is ______.

Real World

Problem Solving

7. Mathematical PRACTICE 4 **Model Math** Melissa owns 4 sets of jewelry. Each set contains 2 earrings, 1 necklace, 1 bracelet, and 1 ring. How many pieces of jewelry does Melissa have altogether? Write a multiplication sentence.

______

Vocabulary Check

Vocab

Write a definition for the vocabulary word(s).

8. known fact ______

9. decompose ______

Test Practice

10. There are 7 birdfeeders at the park. Each birdfeeder has 4 perches. How many birds can use the birdfeeders at the same time?

Ⓐ 32 birds

Ⓑ 28 birds

Ⓒ 11 birds

Ⓓ 3 birds

Name ____________________

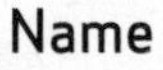

Divide by 4

Lesson 5

ESSENTIAL QUESTION
What strategies can be used to learn multiplication and division facts?

Math in My World

Tools Watch Tutor

Example 1

The distance around a square window in Peter's house is 12 feet. What is the length of each side?

Find 12 ÷ 4.

One Way **Use models.**

Start with 12 counters. Circle 4 equal groups.

There are ______ counters in each group. 12 ÷ 4 = ______

So, the length of each side of the window is ______ feet.

Another Way **Use repeated subtraction.**

Subtract groups of 4 until you reach 0.

Count the number of times you subtracted.

Groups of ______ were subtracted ______ times.

So, 12 ÷ 4 = ______.

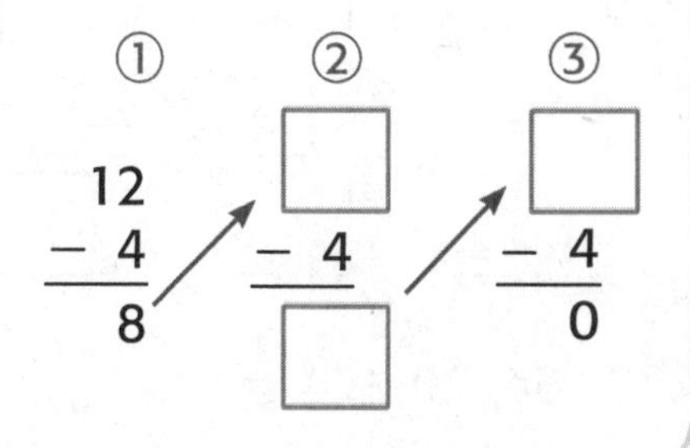

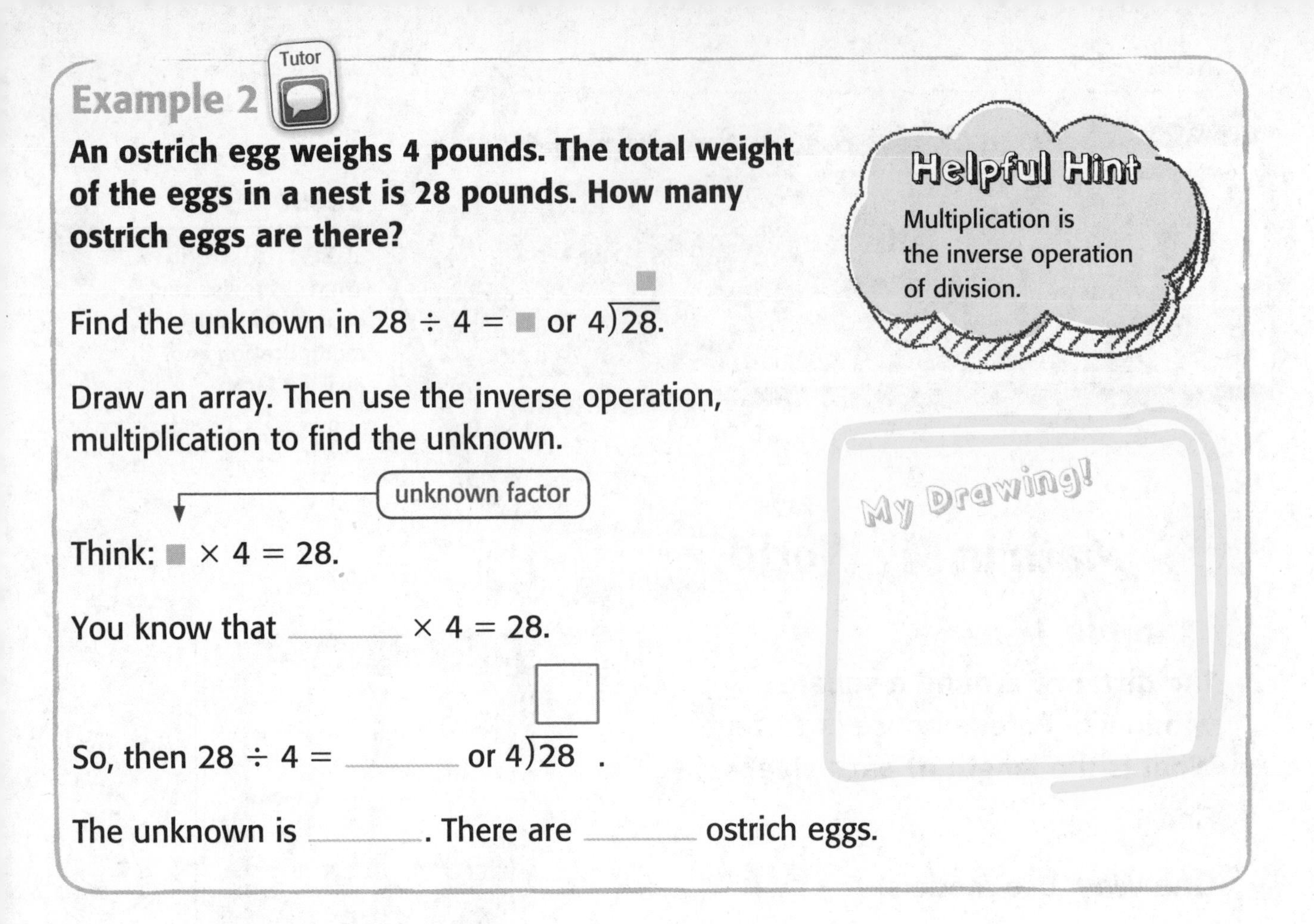

Example 2

Tutor

An ostrich egg weighs 4 pounds. The total weight of the eggs in a nest is 28 pounds. How many ostrich eggs are there?

Find the unknown in $28 \div 4 = \blacksquare$ or $4\overline{)28}$ (with ■ above the 28).

Draw an array. Then use the inverse operation, multiplication to find the unknown.

unknown factor

Think: $\blacksquare \times 4 = 28$.

You know that _____ $\times 4 = 28$.

So, then $28 \div 4 =$ _____ or $4\overline{)28}$ (with a box above the 28).

The unknown is _____. There are _____ ostrich eggs.

Helpful Hint

Multiplication is the inverse operation of division.

My Drawing!

Guided Practice

Check

Use counters to find how many are in each group.

1. 8 counters
4 equal groups

_____ in each group

So, $8 \div 4 =$ _____.

2. 24 counters
4 equal groups

_____ in each group

So, $24 \div 4 =$ _____.

Talk MATH

Without dividing, how do you know that the quotient of $12 \div 3$ is greater than the quotient of $12 \div 4$?

3. Use repeated subtraction to find $20 \div 4$.

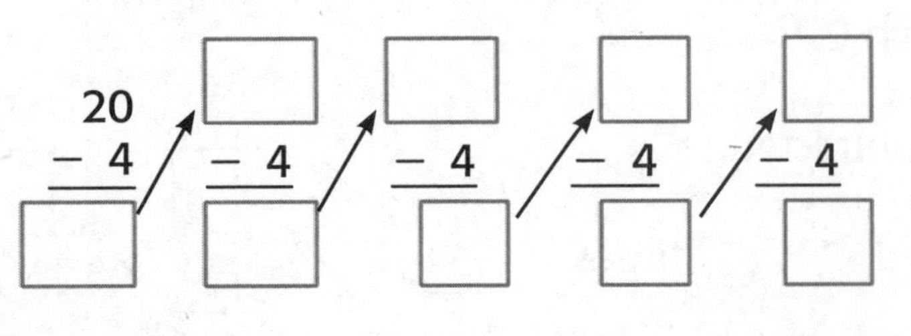

So, $20 \div 4 =$ _____.

Name ....

CCSS

Independent Practice

Use counters to find the number of equal groups or how many are in each group.

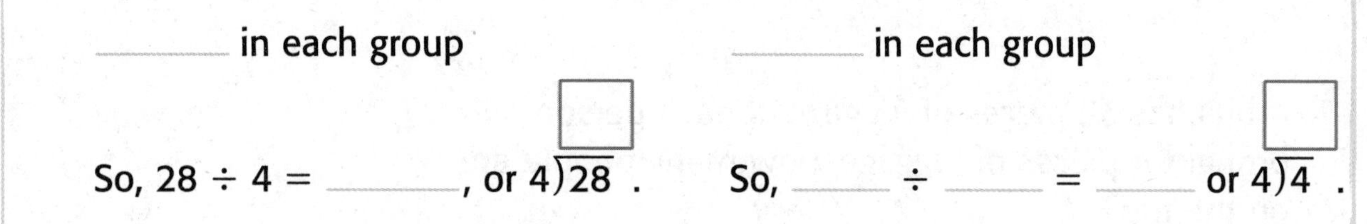

4. 28 counters
4 equal groups

_____ in each group

So, $28 \div 4 =$ _____, or $4\overline{)28}$.

5. 4 counters
4 equal groups

_____ in each group

So, _____ ÷ _____ = _____ or $4\overline{)4}$.

Use repeated subtraction to divide.

6. $8 \div 4 =$ _____

7. $16 \div 4 =$ _____

Algebra **Draw an array and use the inverse operation to find each unknown.**

8. $? \times 6 = 24$

■ $\div 4 = 6$

? = _____

■ = _____

9. $8 \times ? = 32$

$32 \div 4 =$ ■

? = _____

■ = _____

My Drawing!

Problem Solving

Algebra Write a division sentence with a symbol for the unknown in Exercises 10 and 11. Solve.

My Work!

10. Gwen, Clark, Elvio, and Trent will be on vacation for 20 days. They are dividing the planning equally. How many days will Clark have to plan?

11. A bus has 32 pieces of luggage. If each person brought 4 pieces of luggage, how many people are on the trip?

12. Mathematical PRACTICE 2 **Reason** It costs $40 for 4 friends to ride go-carts for 1 hour. How much does it cost 1 person to ride for 2 hours?

HOT Problems

13. Mathematical PRACTICE 3 **Find the Error** Kendra wrote the two number sentences below to help her find $12 \div 4$. Explain and correct her mistake.

$4 + 8 = 12$
So, $12 \div 4 = 8$.

14. **Building on the Essential Question** How can an array help me divide?

Name ....

MY Homework

Lesson 5
Divide by 4

Homework Helper

eHelp

Need help? connectED.mcgraw-hill.com

The Dolan family bought a pack of 20 juice boxes. There are 4 people in the Dolan family. If they divide the juice boxes evenly, how many will each family member get?

Find the unknown in $20 \div 4 = \blacksquare$, or $4\overline{)20}$.

Use multiplication to find the unknown.

Think: $\blacksquare \times 4 = 20$

You know that $5 \times 4 = 20$.

So, $20 \div 4 = 5$, or $4\overline{)20}$ with quotient 5.

The unknown is 5. Each family member will get 5 juice boxes.

Practice

Use counters to find how many are in each group.

1. 8 counters
4 equal groups

_______ in each group

So, $8 \div 4 =$ _______.

2. 40 counters
4 equal groups

_______ in each group

So, $40 \div 4 =$ _______.

3. 28 counters
4 equal groups

_______ in each group

So, $28 \div 4 =$ _______.

4. 12 counters
4 equal groups

_______ in each group

So, $12 \div 4 =$ _______.

Use repeated subtraction to divide.

5. $24 \div 4 =$ ______

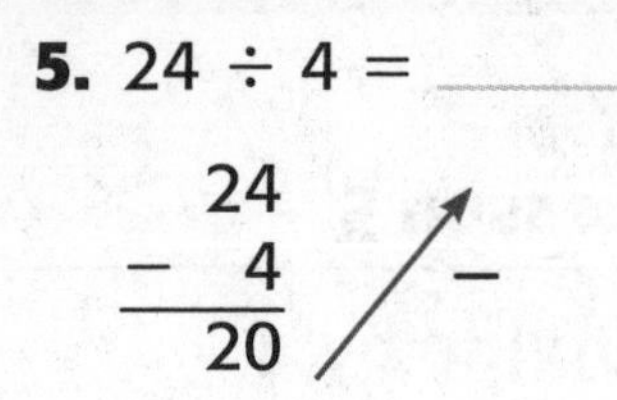

6. $16 \div 4 =$ ______

Algebra **Use the inverse operation to find each unknown.**

7. ______ $\times 2 = 8$

______ $\div 4 = 2$

8. $9 \times$ ______ $= 36$

$36 \div 4 =$ ______

Problem Solving

Write a division sentence to solve.

9. A boat rental shop has enough boats for 28 people to ride. Each boat seats 4 people. How many boats does the shop have?

______

10. Mathematical PRACTICE 6 **Explain to a Friend** Ollie and 3 friends evenly shared 24 marbles. Kaitlyn and 2 friends evenly shared 18 marbles. Did Ollie or Kaitlyn's friends get more marbles? Explain.

______

Test Practice

11. Each minute, 4 gallons of water flow into a bathtub. How many minutes does it take for the bathtub to have 32 gallons of water?

Ⓐ 6 minutes
Ⓑ 7 minutes
Ⓒ 8 minutes
Ⓓ 9 minutes

Check My Progress

Vocabulary Check

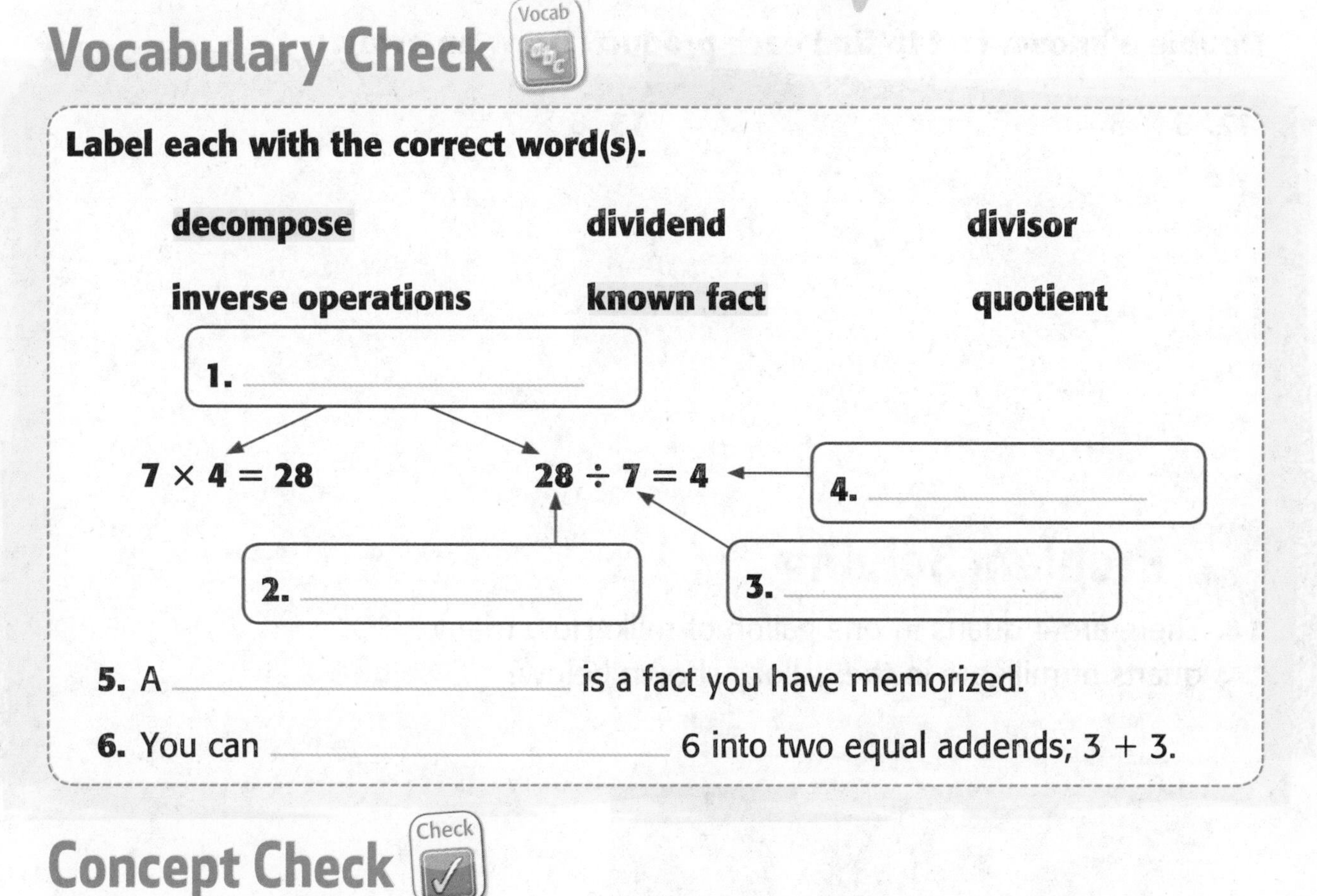

Label each with the correct word(s).

decompose	**dividend**	**divisor**
inverse operations	**known fact**	**quotient**

1. ____________

$7 \times 4 = 28$ $\quad$ $28 \div 7 = 4$

2. ____________

3. ____________

4. ____________

5. A ____________ is a fact you have memorized.

6. You can ____________ 6 into two equal addends; 3 + 3.

Concept Check

Draw an array for each. Then write two multiplication sentences.

7. 3 rows of 6

____

____

8. 4 rows of 6

____

____

9. Draw jumps on the number line to find $27 \div 3$.

$27 \div 3 =$ ______

0 3 6 9 12 15 18 21 24 27 30

Algebra Use a related multiplication fact to find the unknown.

10. $21 \div 3 = \blacksquare$

$\blacksquare \times 3 = 21$

The unknown is ______.

11. $32 \div 4 = \blacksquare$

$4 \times \blacksquare = 32$

The unknown is ______.

Double a known fact to find each product. Draw an array.

12. $6 \times 5 =$ ______

13. $6 \times 7 =$ ______

Problem Solving

14. There are 4 quarts in one gallon of milk. How many quarts of milk are in the gallons shown below?

______

My Work!

Test Practice

15. Aisha picked 27 apples. She placed an equal number of apples in 3 bags. How many apples did she place in each bag?

Ⓐ 8 apples
Ⓑ 9 apples
Ⓒ 24 apples
Ⓓ 30 apples

Name ....

Real World

Problem-Solving Investigation

STRATEGY: Extra or Missing Information

Lesson 6

ESSENTIAL QUESTION
What strategies can be used to learn multiplication and division facts?

Learn the Strategy

The school's hayride starts at 6 P.M. There are 4 wagons that can hold 9 children each. Half of the children going are girls. What is the total number of children that can ride on the wagons?

1 Understand

What facts do you know?

The hayride starts at ________ P.M.

There are ________ wagons that hold ________ children each.

Half of the children are girls.

What do you need to find?

the number of ____________ that can ride on the 4 wagons

2 Plan

Decide what facts are important.

the number of ____________

the number of ____________ each wagon holds

- The hayride starts at 6 P.M.
- Half of the children are girls.

3 Solve

4 (wagons) × 9 (children) = ________ (total children)

4 Check

Does your answer make sense? Explain.

____

Practice the Strategy

An animal shelter has 23 cats and 14 dogs. How much would it cost to adopt 1 cat and 1 dog? Tell what facts are extra or missing.

$35

$40

1 Understand

What facts do you know?

What do you need to find?

2 Plan

3 Solve

4 Check

Does your answer make sense? Explain.

Name

Apply the Strategy

Determine if there is extra or missing information to solve each problem. Then solve if possible.

My Work!

1. Mathematical PRACTICE 3 **Draw a Conclusion** Mrs. Friedman had 2 boxes of chalk. She bought 4 more boxes with 10 pieces each. She paid $2 per box. How much did she spend on the 4 boxes of chalk?

2. Bert bought 4 of each of the items below. How much change did he get back?

Item	Cost
Pencils	$2
Paper	$1
Binder	$3

3. Mathematical PRACTICE 5 **Use Math Tools** Alejandra is 58 inches tall. Her sister is in the first grade and is 10 inches shorter than Alejandra. What known fact when doubled, is equal to the height of Alejandra's sister?

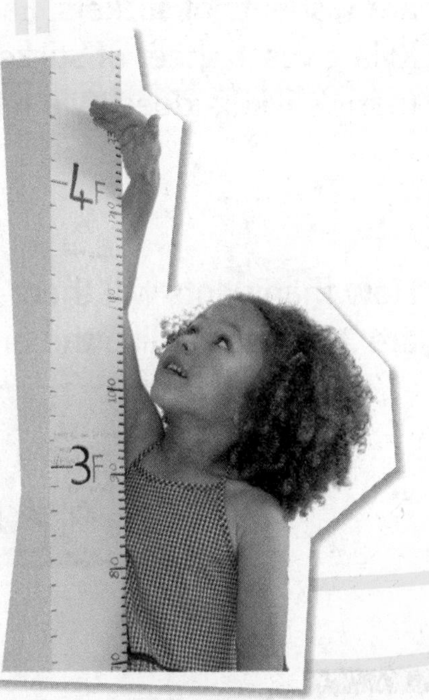

Review the Strategies

Use any strategy to solve each problem.

- Determine extra or missing information.
- Make a table.
- Look for a pattern.
- Use models.

My Work!

4. Ten of Eduardo's baseball cards are All Star cards. His friend has four times as many All Star cards. How many All Star baseball cards does his friend have?

5. The third grade class had four chicks hatch every day for 5 days. Nine of the chicks were yellow, and the rest were brown. How many chicks hatched in all?

6. Four friends bought a total of 24 computer games. Each friend bought the same number of games. How many games did each friend buy?

7. Mathematical PRACTICE 1 **Make Sense of Problems** There are 4 sheets of stickers. Each sheet has 7 stickers. Kyla gives 1 sheet of stickers to her friend. How many stickers does Kyla have left?

8. How many dots will there be altogether if there are 2 dominoes like the one below?

Name

MY Homework

Lesson 6

Problem Solving: Extra or Missing Information

Homework Helper

eHelp

Need help? connectED.mcgraw-hill.com

Arthur will cut a 12-foot board into 4 equal pieces. If it takes him 10 minutes to cut the board how long will each piece be?

1 Understand

What facts do you know?

Arthur has a 12-foot long board.

He will cut the board into 4 equal pieces.

It will take 10 minutes to cut the board.

What do you need to find?

the length of the 4 pieces

2 Plan

Decide which facts are important.

- the length of the board being cut
- the number of pieces being cut

3 Solve

Divide the board's length by the number of pieces being cut.

$12 \div 4 = 3$.

So, each piece will be 3 feet long.

4 Check

Does the answer make sense?

Use the inverse operation, multiplication, to check.

$3 \times 4 = 12$

So, the answer is reasonable.

Real World!

Problem Solving

Determine if there is extra or missing information to solve each problem. Then solve if possible.

My Work!

1. Brandy ate 9 corn chips for a snack. She ate twice as many raisins. She also drank a juice box. How many raisins did Brandy eat?

2. Miguel bought a movie ticket that cost $5. He also bought popcorn and a bottle of water. How much did Miguel spend altogether?

3. Annie has basketball practice from 2:30 P.M. to 4:00 P.M. She makes 10 free throws every practice. How many free throws does she make in 4 practices?

4. Naya has 12 flowers. She gives away 6 to Jane and 3 to Heather. Hannah does not have any flowers. How many flowers does Naya have left?

5. Mathematical PRACTICE 3 **Draw a Conclusion** Juan bought two new tires for his bike. He also had the bike tuned up. The total bill was $65. How much did Juan spend on the 2 tires?

Name ....

Multiply by 0 and 1

Lesson 7

ESSENTIAL QUESTION
What strategies can be used to learn multiplication and division facts?

Math in My World

Tools Watch Tutor

Example 1

There are 4 flower pots. Each has 1 daisy. How many daisies are there in all? Find the unknown.

$4 \times 1 = \blacksquare$ ← unknown

Circle 4 equal groups of 1.

1 + 1 + 1 + 1 = ______

So, 4 groups of ______ is ______. The unknown is ______.

There are ______ daisies in all.

The **Identity Property of Multiplication** says that when any number is multiplied by ______, the product is that number.

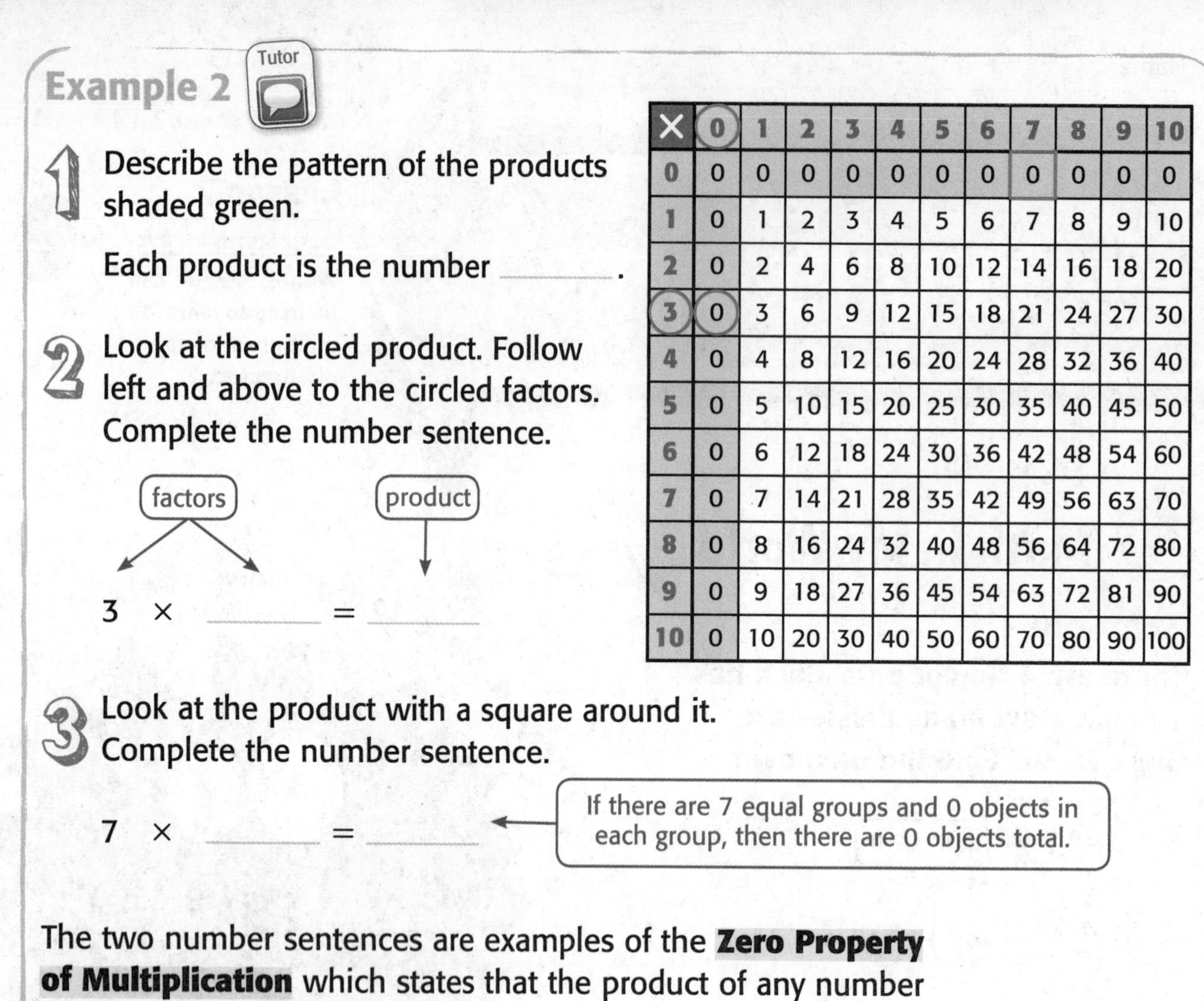

Example 2

Tutor

1. Describe the pattern of the products shaded green.

   Each product is the number ______.

2. Look at the circled product. Follow left and above to the circled factors. Complete the number sentence.

   factors   product

   3 × ______ = ______

×	0	1	2	3	4	5	6	7	8	9	10
0	0	0	0	0	0	0	0	0	0	0	0
1	0	1	2	3	4	5	6	7	8	9	10
2	0	2	4	6	8	10	12	14	16	18	20
3	0	3	6	9	12	15	18	21	24	27	30
4	0	4	8	12	16	20	24	28	32	36	40
5	0	5	10	15	20	25	30	35	40	45	50
6	0	6	12	18	24	30	36	42	48	54	60
7	0	7	14	21	28	35	42	49	56	63	70
8	0	8	16	24	32	40	48	56	64	72	80
9	0	9	18	27	36	45	54	63	72	81	90
10	0	10	20	30	40	50	60	70	80	90	100

3. Look at the product with a square around it. Complete the number sentence.

   7 × ______ = ______

   If there are 7 equal groups and 0 objects in each group, then there are 0 objects total.

The two number sentences are examples of the **Zero Property of Multiplication** which states that the product of any number and zero is zero.

Guided Practice

Look at the row and column of products shaded yellow. Describe the pattern.

1. When a number is multiplied by ______, the product is that number.

2. This is an example of the ______ Property of Multiplication.

×	0	1	2	3	4	5
0	0	0	0	0	0	0
1	0	1	2	3	4	5
2	0	2	4	6	8	10
3	0	3	6	9	12	15
4	0	4	8	12	16	20
5	0	5	10	15	20	25

Talk MATH

If 100 is multiplied by 0, what will be the product? Explain your reasoning.

Name ....

Independent Practice

Algebra **Circle equal groups to find the unknown. Write the unknown.**

3. $2 \times 1 = ■$ or $\begin{array}{r} 2 \\ \times\ 1 \\ \hline ■ \end{array}$

4. $5 \times 1 = ■$ or $\begin{array}{r} 5 \\ \times\ 1 \\ \hline ■ \end{array}$

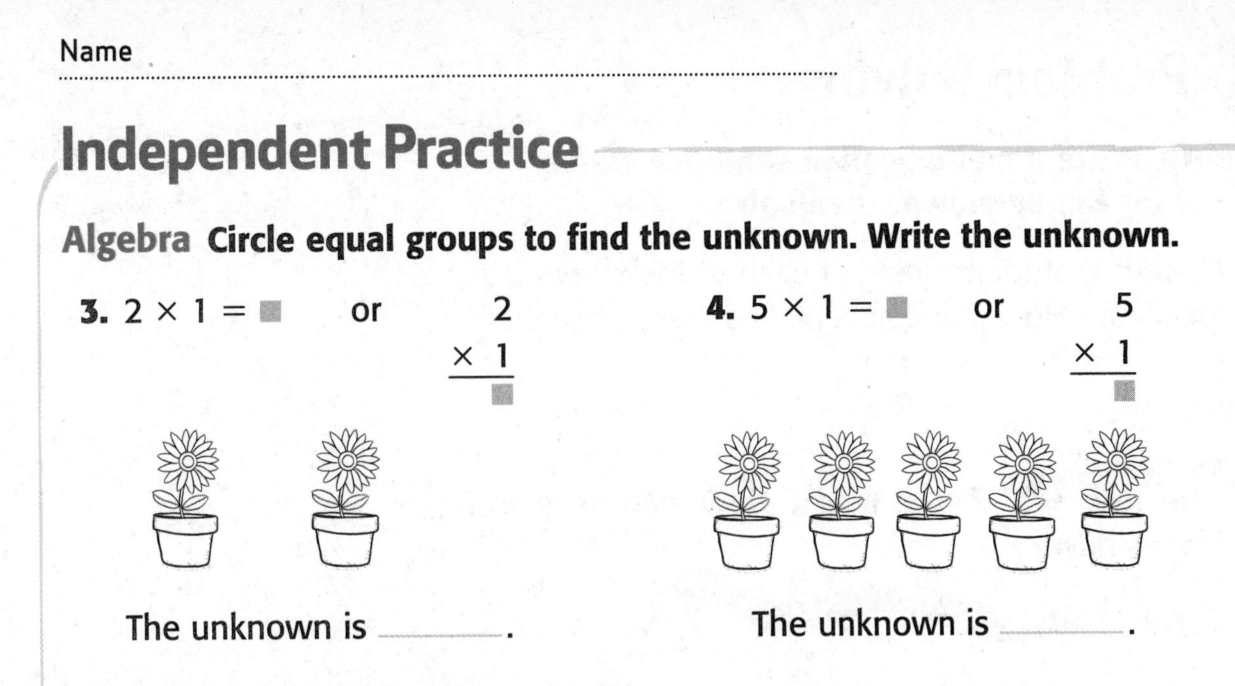

The unknown is ______.

The unknown is ______.

Write an addition sentence to help find each product. Check using the Identity Property of Multiplication.

5. $8 \times 1 =$ ____

6. $7 \times 1 =$ ____

Use the Zero Property of Multiplication to find each product.

7. $10 \times 0 =$ ______ **8.** $6 \times 0 =$ ______ **9.** $3 \times 0 =$ ______

10. **Algebra** **Draw a line to match each number sentence to its unknown.**

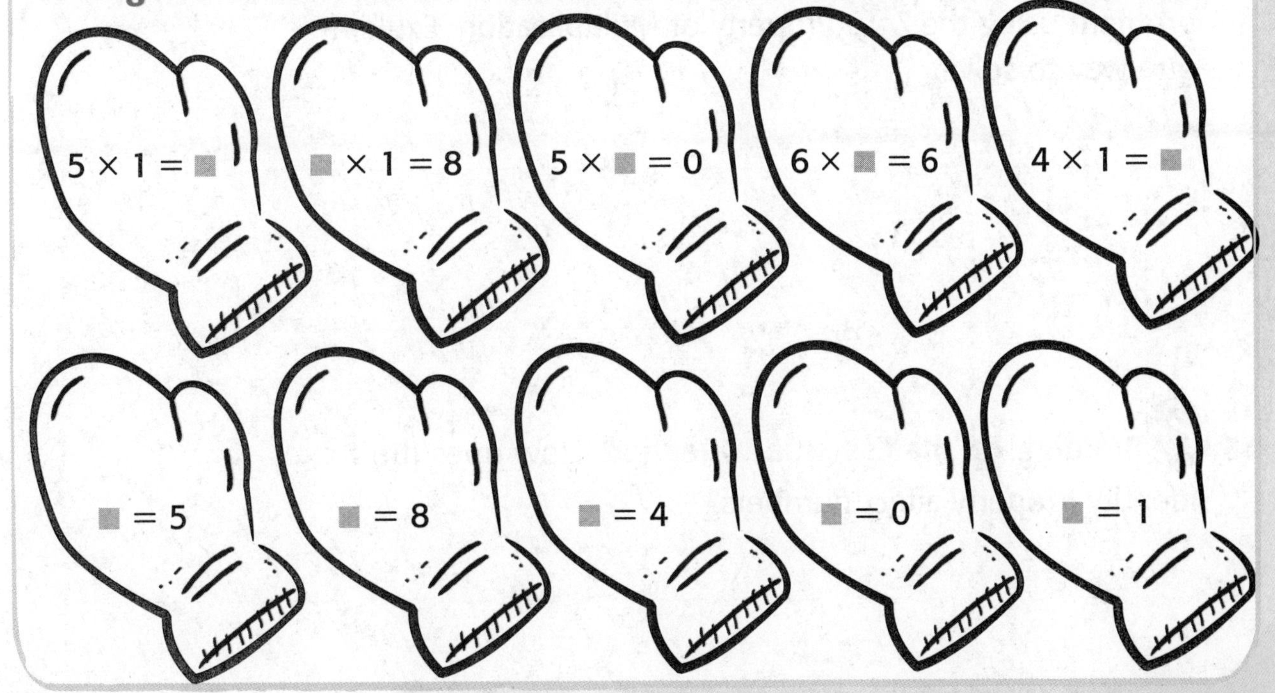

Problem Solving

Algebra Write a multiplication sentence with a symbol for the unknown. Then solve.

11. There is 1 student sitting at each of the 9 tables in the library. How many students are there altogether?

12. Mathematical PRACTICE 4 **Model Math** How many legs do 8 snakes have?

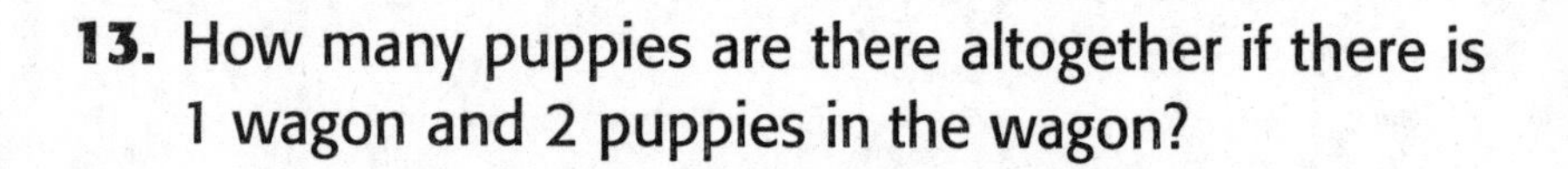

13. How many puppies are there altogether if there is 1 wagon and 2 puppies in the wagon?

My Work!

HOT Problems

14. Mathematical PRACTICE 7 **Identify Structure** Write a real-world problem using the Zero Property of Multiplication. Explain one way to solve.

15. **Building on the Essential Question** How does the Identity Property affect numbers?

Name ....

MY Homework

Lesson 7
Multiply by 0 and 1

Homework Helper

eHelp Need help? connectED.mcgraw-hill.com

Three customers each bought a cone with 1 scoop of frozen yogurt. How many scoops did the customers buy in all?

Find 3×1.

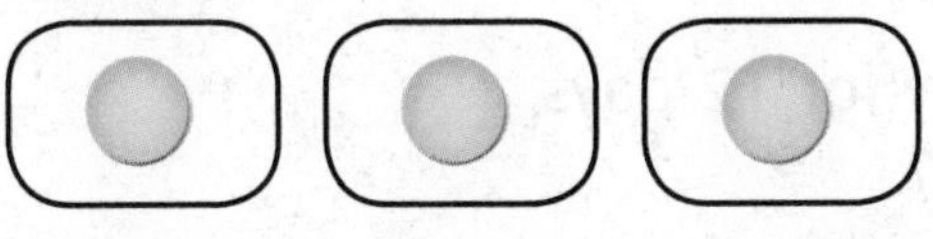

The Identity Property of Multiplication states that when any number is multiplied by 1, the product is that number.
So, $3 \times 1 = 3$.

The 3 customers have eaten their frozen yogurt cones. Now how many scoops do the customers have in all?

Find 3×0.

The Zero Property of Multiplication states that when any number is multiplied by 0, the product is 0.
So, $3 \times 0 = 0$.

Practice

Algebra Circle equal groups to find the unknown. Write the unknown.

1. $8 \times 1 = ■$

The unknown is ______.

2. $5 \times 1 = ■$

The unknown is ______.

Use the Identity Property of Multiplication or the Zero Property of Multiplication to find each product.

3. $4 \times 0 =$ ____ **4.** $7 \times 1 =$ ____ **5.** $7 \times 0 =$ ____

6. $6 \times 1 =$ ____ **7.** $1 \times 0 =$ ____ **8.** $9 \times 1 =$ ____

9. $2 \times 1 =$ ____ **10.** $8 \times 1 =$ ____ **11.** $5 \times 0 =$ ____

Problem Solving

Algebra Write a multiplication sentence with a symbol for the unknown for Exercises 12 and 13. Then solve.

12. Jordan collects stamps. If he gets 1 stamp a day for 12 days, how many stamps will he add to his collection?

____

13. Mathematical PRACTICE 6 **Be Precise** There are no pockets on any of Henry's shirts. How many pockets do 6 of Henry's shirts have?

____

Vocabulary Check

Vocab

Write the correct word to complete the sentence.

Identity Zero

14. The ________ Property of Multiplication states that any number multiplied by 0 equals 0.

15. The ________ Property of Multiplication states that any number multiplied by 1 is that number.

Test Practice

16. Which multiplication sentence shows how to find the number of wings two alligators have altogether?

Ⓐ $1 + 1 = 2$ Ⓒ $1 \times 1 = 1$

Ⓑ $2 \times 2 = 4$ Ⓓ $2 \times 0 = 0$

Name ....

Divide with 0 and 1

Lesson 8

ESSENTIAL QUESTION
What strategies can be used to learn multiplication and division facts?

There are rules you can use when you divide with 0 or 1.

Math in My World

Watch Tutor

Example 1

There are 3 friends and only 1 bean bag chair. How many friends will be on the bean bag chair?

Find 3 ÷ 1. Use Xs to draw the friends on the chair.

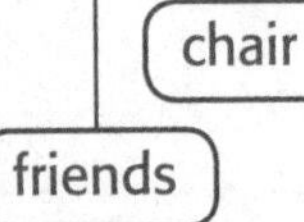

So, 3 ÷ 1 = ______.

There will be ______ friends on the bean bag chair.

Helpful Hint
When a number is divided by 1, the quotient is that number.

Example 2

There are 3 friends and 3 bean bag chairs. How many friends will be on each bean bag chair?

Find 3 ÷ 3.
Use Xs to equally partition the friends on the chairs.

So, 3 ÷ 3 = ______.

There will be ______ friend on each bean bag chair.

Helpful Hint
When a number is divided by itself, the quotient is 1.

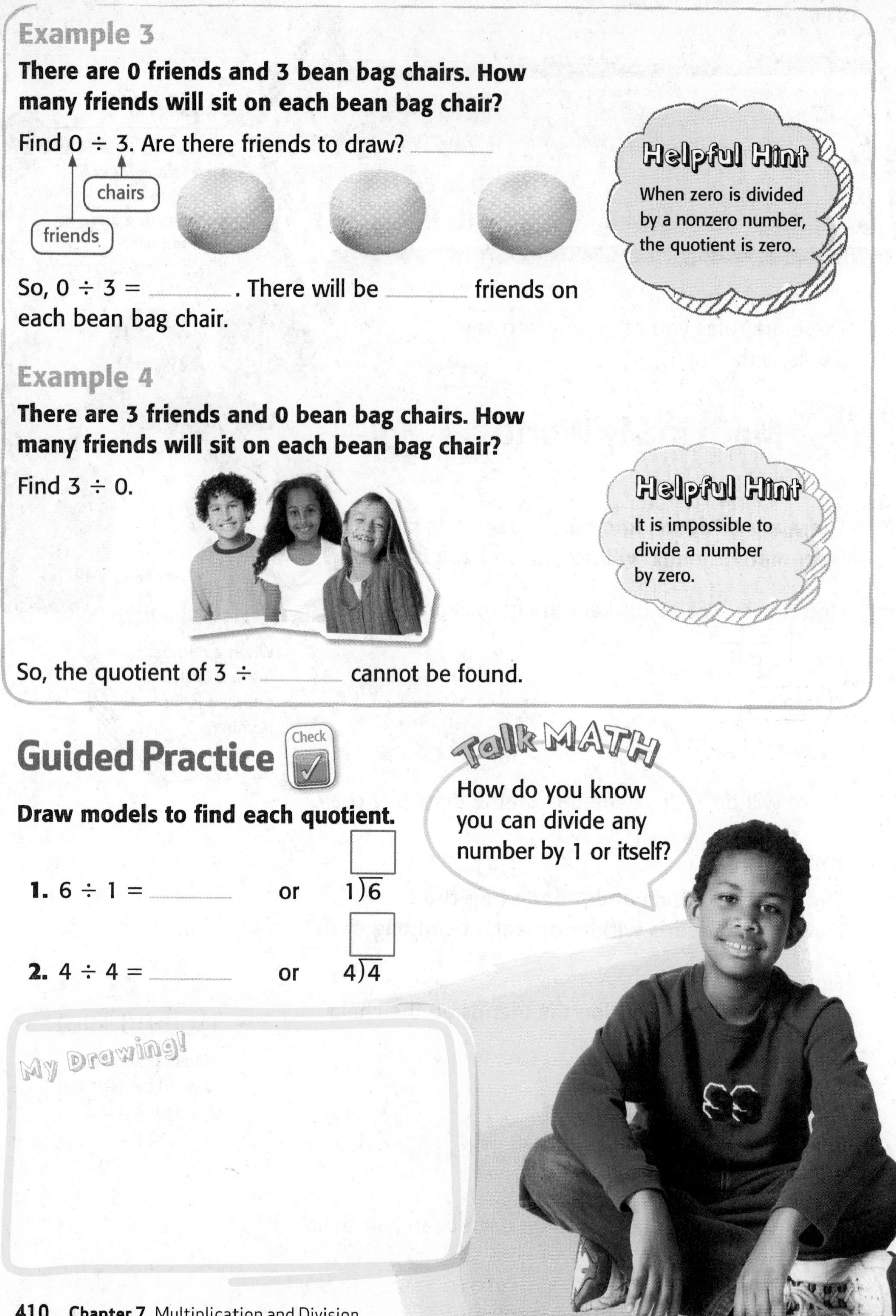

Example 3

There are 0 friends and 3 bean bag chairs. How many friends will sit on each bean bag chair?

Find 0 ÷ 3. Are there friends to draw? _______

So, 0 ÷ 3 = _______. There will be _______ friends on each bean bag chair.

Helpful Hint
When zero is divided by a nonzero number, the quotient is zero.

Example 4

There are 3 friends and 0 bean bag chairs. How many friends will sit on each bean bag chair?

Find 3 ÷ 0.

So, the quotient of 3 ÷ _______ cannot be found.

Helpful Hint
It is impossible to divide a number by zero.

Guided Practice

Check

Draw models to find each quotient.

1. 6 ÷ 1 = _______ or $1\overline{)6}$ ☐

2. 4 ÷ 4 = _______ or $4\overline{)4}$ ☐

My Drawing!

Talk MATH
How do you know you can divide any number by 1 or itself?

Name

Independent Practice

Find each quotient. Draw lines to match each division sentence to its model.

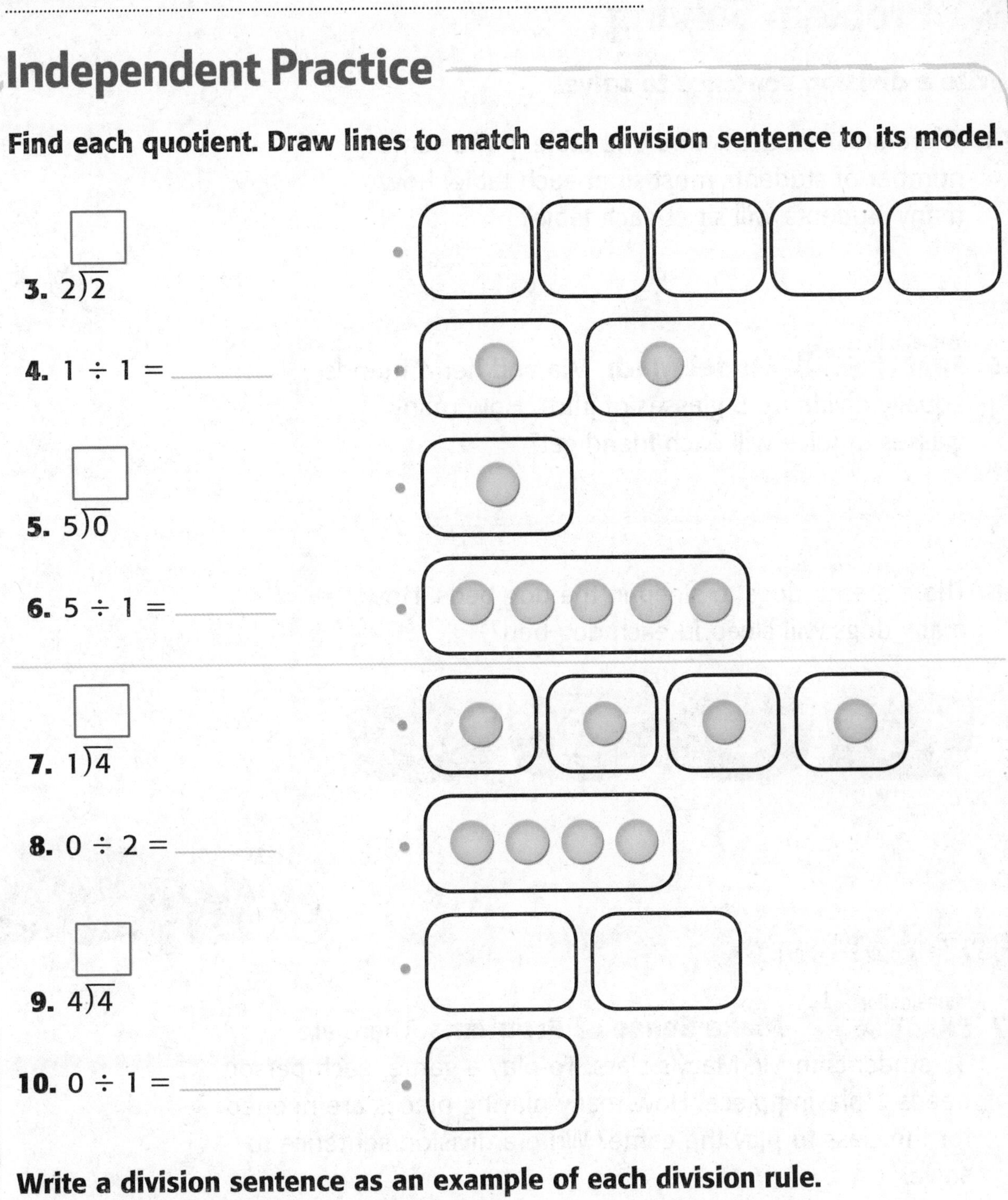

3. $2\overline{)2}$

4. $1 \div 1 =$ ______

5. $5\overline{)0}$

6. $5 \div 1 =$ ______

7. $1\overline{)4}$

8. $0 \div 2 =$ ______

9. $4\overline{)4}$

10. $0 \div 1 =$ ______

Write a division sentence as an example of each division rule.

11. When a number is divided by itself, the quotient is 1. ______

12. When 0 is divided by a nonzero number, the quotient is 0. ______

13. When a number is divided by 1, the quotient is that number. ______

Problem Solving

Write a division sentence to solve.

14. There are 7 students and one table. If the same number of students must sit at each table, how many students will sit at each table?

My Work!

15. Mathematical PRACTICE 4 **Model Math** Mia and her 4 friends equally divide up 5 glasses of juice. How many glasses of juice will each friend get?

16. There are no dogs to sleep in the dog beds. How many dogs will sleep in each dog bed?

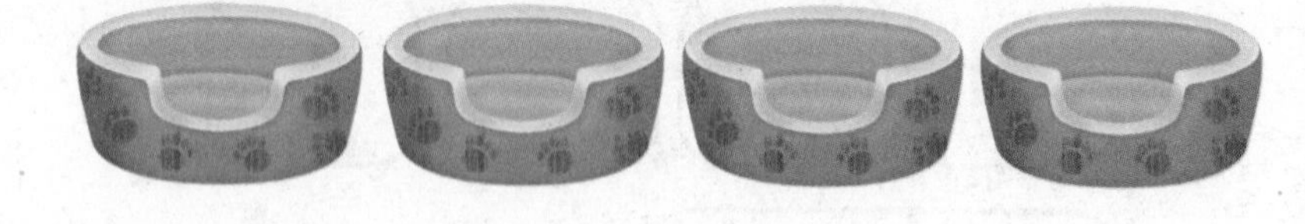

HOT Problems

17. Mathematical PRACTICE 1 **Make Sense of Problems** There are 35 students in Mr. Macy's class. To play a game, each person needs 1 playing piece. How many playing pieces are needed for the class to play the game? Write a division sentence to solve.

18. **Building on the Essential Question** How do division rules help me learn division facts more quickly?

Name

Lesson 8
Divide with 0 and 1

Homework Helper

Need help? connectED.mcgraw-hill.com

Larry has 3 keys. He divides them evenly among 1 keychain. How many keys are on each keychain?

Find 3 ÷ 1.

Use counters to model.

Divide 3 counters evenly into 1 group.

So, 3 ÷ 1 = 3.

There are 3 keys on the keychain.

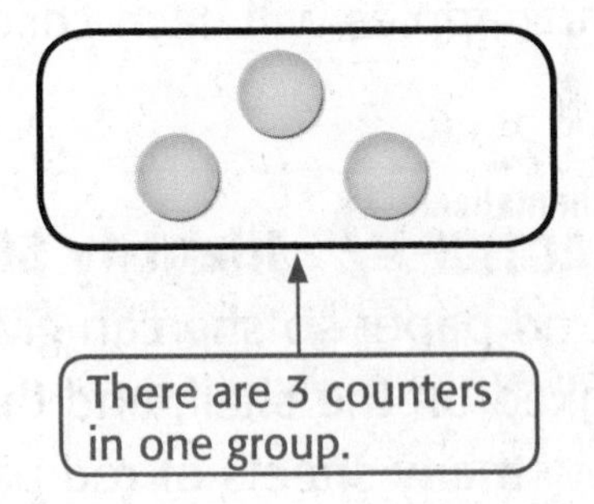

Practice

Complete the division sentence for each model.

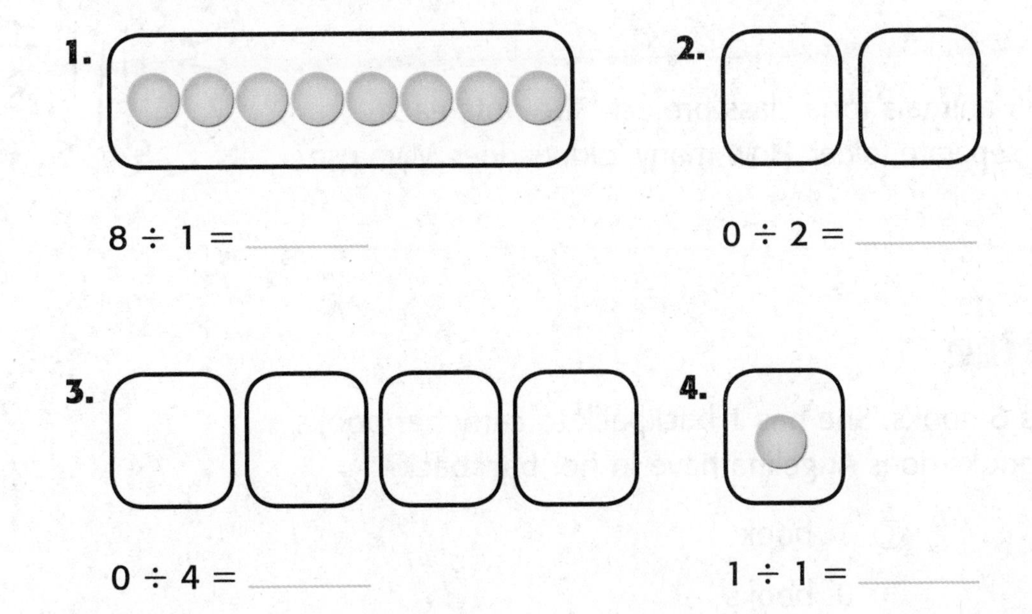

1. 8 ÷ 1 = ______

2. 0 ÷ 2 = ______

3. 0 ÷ 4 = ______

4. 1 ÷ 1 = ______

Algebra **Use a related multiplication fact to find the unknown.**

5. $9 \div 9 = ■$

$9 \times$ ______ $= 9$

The unknown is ______.

6. $0 \div 6 = ■$

$6 \times$ ______ $= 0$

The unknown is ______.

7. $0 \div 8 = ■$

$8 \times$ ______ $= 0$

The unknown is ______.

8. $2 \div 2 = ■$

$2 \times$ ______ $= 2$

The unknown is ______.

Real World

Problem Solving

Write a division sentence to solve.

9. There are 15 children who want to share 15 apples. How many apples will each child get?

______

10. Mathematical PRACTICE 7 **Identify Structure** Mrs. Perkins needs 24 sheets of red paper so she can give one to each student in her class. She looked on the shelf, and there are no sheets of red paper left. How many sheets of red paper can Mrs. Perkins hand out?

______

11. Lionel bought 3 model rockets. He shares them equally between himself and 2 friends. How many rockets does each boy have?

______

12. Myra draws 5 animals for a class project. She puts each drawing in a separate folder. How many folders does Myra use?

______

Test Practice

13. Angelina has 6 books. She has 1 backpack to carry her books. How many books does Angelina have in her backpack?

Ⓐ 7 books
Ⓑ 6 books
Ⓒ 1 book
Ⓓ 0 books

Name

Fluency Practice

Mathematical PRACTICE 6

Multiply.

1. $4 \times 9 =$ ______
2. $5 \times 3 =$ ______
3. $4 \times 6 =$ ______
4. $3 \times 6 =$ ______
5. $3 \times 2 =$ ______
6. $4 \times 4 =$ ______
7. $2 \times 2 =$ ______
8. $0 \times 7 =$ ______
9. $4 \times 5 =$ ______
10. $2 \times 5 =$ ______
11. $3 \times 7 =$ ______
12. $1 \times 2 =$ ______
13. $\begin{array}{r} 1 \\ \times\ 3 \\ \hline \end{array}$
14. $\begin{array}{r} 4 \\ \times\ 3 \\ \hline \end{array}$
15. $\begin{array}{r} 3 \\ \times\ 8 \\ \hline \end{array}$
16. $\begin{array}{r} 0 \\ \times\ 4 \\ \hline \end{array}$
17. $\begin{array}{r} 4 \\ \times\ 7 \\ \hline \end{array}$
18. $\begin{array}{r} 3 \\ \times\ 9 \\ \hline \end{array}$
19. $\begin{array}{r} 4 \\ \times\ 8 \\ \hline \end{array}$
20. $\begin{array}{r} 1 \\ \times\ 8 \\ \hline \end{array}$
21. $\begin{array}{r} 4 \\ \times\ 2 \\ \hline \end{array}$
22. $\begin{array}{r} 3 \\ \times\ 1 \\ \hline \end{array}$
23. $\begin{array}{r} 0 \\ \times\ 9 \\ \hline \end{array}$
24. $\begin{array}{r} 1 \\ \times\ 5 \\ \hline \end{array}$

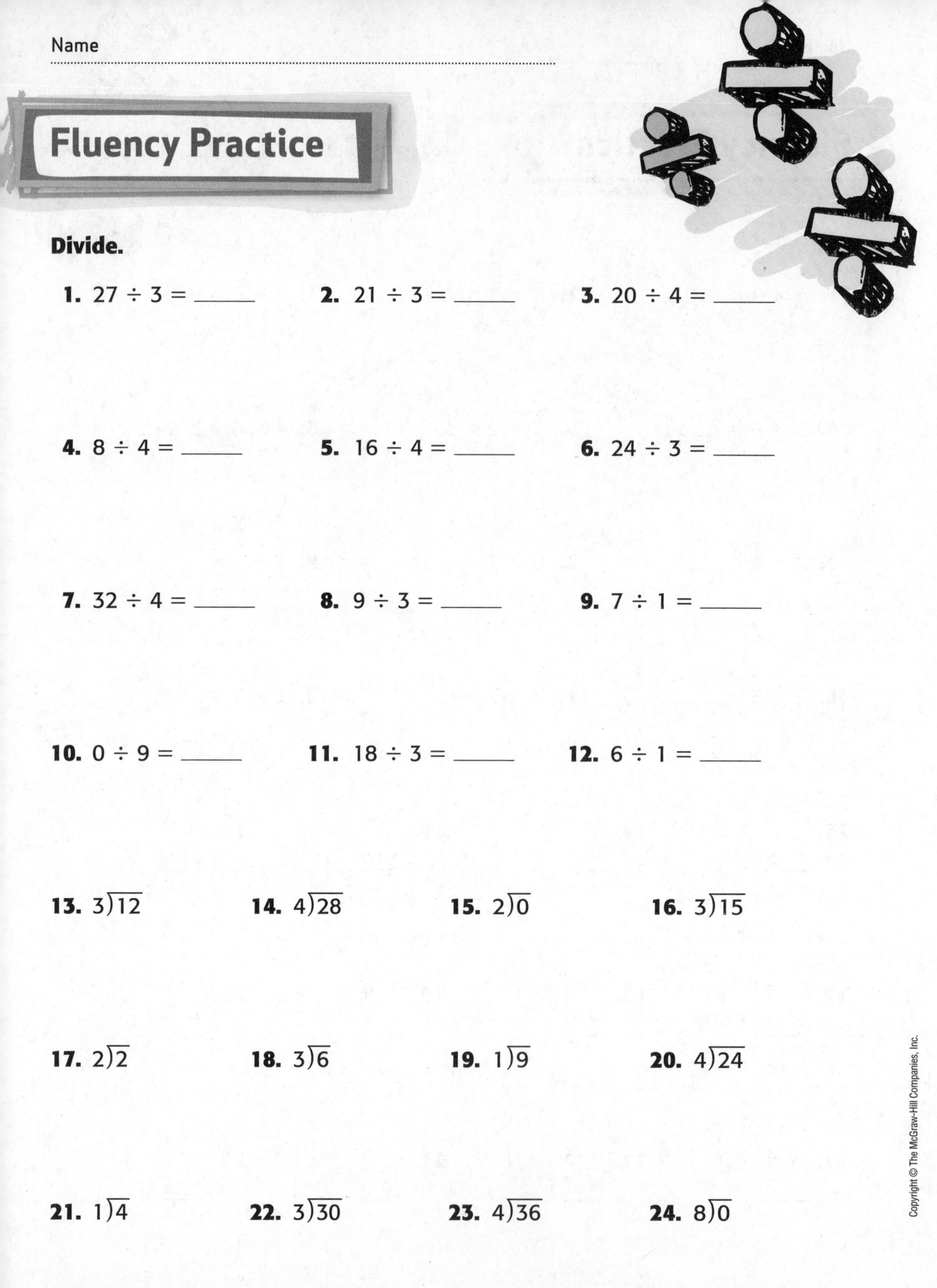

Name

Fluency Practice

Divide.

1. $27 \div 3 =$ ____ **2.** $21 \div 3 =$ ____ **3.** $20 \div 4 =$ ____

4. $8 \div 4 =$ ____ **5.** $16 \div 4 =$ ____ **6.** $24 \div 3 =$ ____

7. $32 \div 4 =$ ____ **8.** $9 \div 3 =$ ____ **9.** $7 \div 1 =$ ____

10. $0 \div 9 =$ ____ **11.** $18 \div 3 =$ ____ **12.** $6 \div 1 =$ ____

13. $3\overline{)12}$ **14.** $4\overline{)28}$ **15.** $2\overline{)0}$ **16.** $3\overline{)15}$

17. $2\overline{)2}$ **18.** $3\overline{)6}$ **19.** $1\overline{)9}$ **20.** $4\overline{)24}$

21. $1\overline{)4}$ **22.** $3\overline{)30}$ **23.** $4\overline{)36}$ **24.** $8\overline{)0}$

Review

Chapter 7

Multiplication and Division

Vocabulary Check

Use the word bank below to complete each clue.

decompose	**Identity Property of Multiplication**	
inverse operations	**known fact**	**Zero Property of Multiplication**

1. The ______________________________ says that when any number is multiplied by 1, the product is that number.

2. A ____________________ is a fact you have memorized.

3. Multiplication and division are ____________________ because they undo each other.

4. ____________________ means to take a number apart.

5. The ______________________________ says that when you multiply a number by 0, the product is zero.

Concept Check

Algebra Use a related multiplication fact to find the unknown.

6. $16 \div 4 = \blacksquare$

$4 \times \blacksquare = 16$

The unknown is ____.

7. $24 \div 3 = \blacksquare$

$3 \times \blacksquare = 24$

The unknown is ____.

8. $20 \div 4 = \blacksquare$

$4 \times \blacksquare = 20$

The unknown is ____.

Use counters to model a known fact that will help you find the first product. Draw the model two times.

9. $6 \times 6 =$ ____

Known fact:

____ $\times$ ____ $=$ ____

Double the product:

____ $+$ ____ $=$ ____

10. $7 \times 4 =$ ____

Known fact:

____ $\times$ ____ $=$ ____

Double the product:

____ $+$ ____ $=$ ____

Algebra Find each unknown. Double a known fact.

11. $7 \times 6 =$ ■

The unknown is ____.

12. $9 \times 4 =$ ■

The unknown is ____.

Write an addition sentence to help find each product.

13. $6 \times 1 =$ ____

$1 + 1 +$ ____ $+$ ____ $+$ ____ $+$ ____ $=$ ____

14. $7 \times 1 =$ ____

$1 + 1 + 1 +$ ____ $+$ ____ $+$ ____ $+$ ____ $=$ ____

Use the Zero Property of Multiplication to find each product.

15. $7 \times 0 =$ ____

16. $9 \times 0 =$ ____

17. $6 \times 0 =$ ____

Write a division sentence as an example of each division rule.

18. When a number is divided by 1, the quotient is that number. ____

19. When a number is divided by itself, the quotient is 1. ____

20. When 0 is divided by a nonzero number, the quotient is 0. ____

Name

Problem Solving

My Work!

21. There are a total of 12 slices of pizza. Each pizza is cut into 4 slices. How many pizzas are there? Write a number sentence to solve.

22. A van has 4 rows of seats, and each row holds 3 people. How many people can the van hold? Write a number sentence to solve.

23. Mika rides his bike 3 miles each way to his friend's house. He leaves his house at 4 P.M. How many miles does he ride his bike there and back? What information is extra?

24. Four campers divide 24 large marshmallows. Each s'more uses 2 marshmallows. How many s'mores can each camper have?

Test Practice

25. Mr. Thompson bought 3 of the same item. He paid a total of $21. Which item did he buy?

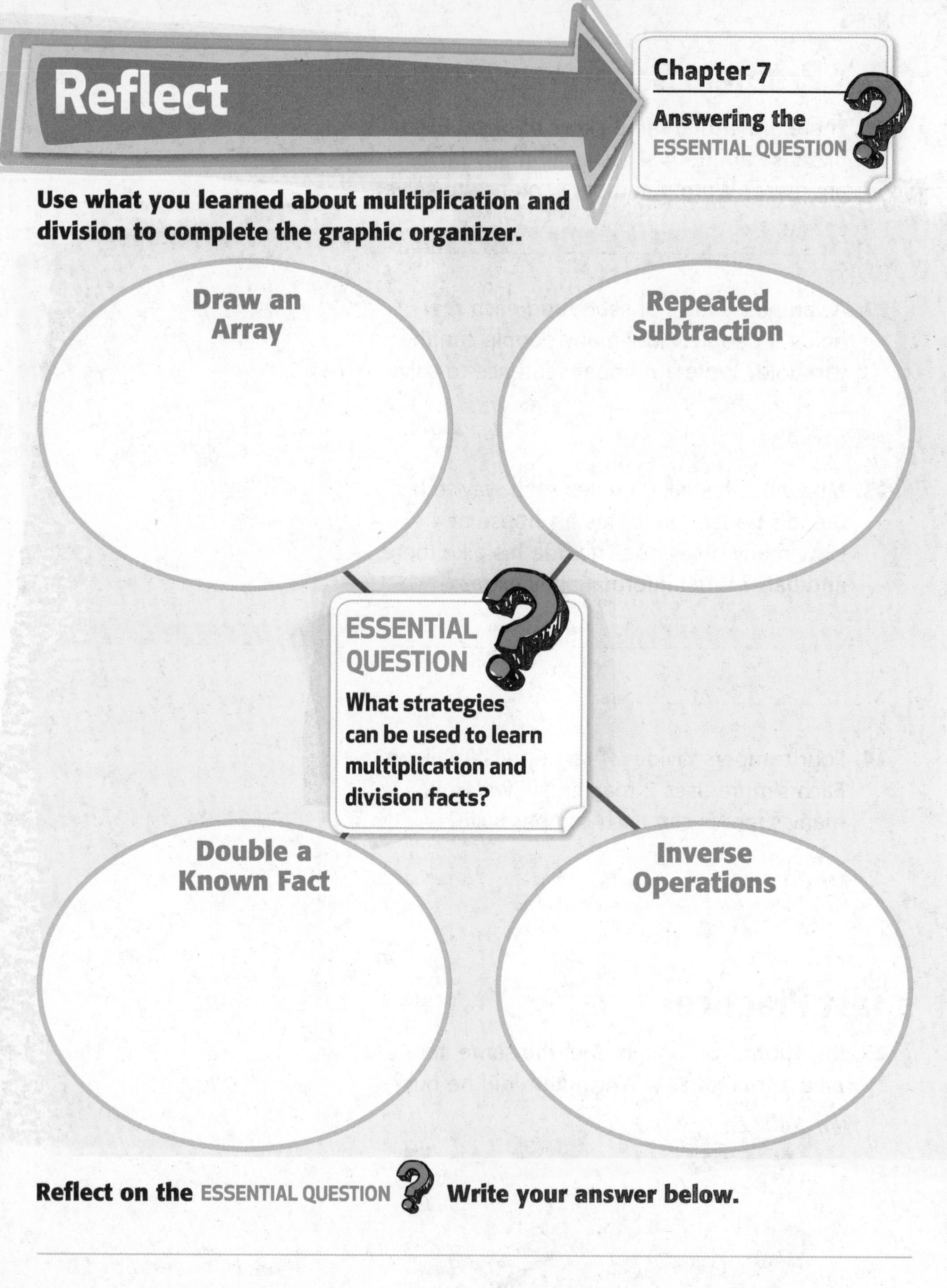

Reflect on the ESSENTIAL QUESTION **Write your answer below.**

Chapter

8 Apply Multiplication and Division

ESSENTIAL QUESTION

How can multiplication and division facts with smaller numbers be applied to larger numbers?

Watch a video!

Watch

MY Common Core State Standards

Operations and Algebraic Thinking

CCSS

3.OA.1 Interpret products of whole numbers, e.g., interpret 5×7 as the total number of objects in 5 groups of 7 objects each.

3.OA.2 Interpret whole-number quotients of whole numbers, e.g., interpret $56 \div 8$ as the number of objects in each share when 56 objects are partitioned equally into 8 shares, or as a number of shares when 56 objects are partitioned into equal shares of 8 objects each.

3.OA.3 Use multiplication and division within 100 to solve word problems in situations involving equal groups, arrays, and measurement quantities, e.g., by using drawings and equations with a symbol for the unknown number to represent the problem.

3.OA.4 Determine the unknown whole number in a multiplication or division equation relating three whole numbers.

3.OA.5 Apply properties of operations as strategies to multiply and divide.

3.OA.6 Understand division as an unknown-factor problem.

3.OA.7 Fluently multiply and divide within 100, using strategies such as the relationship between multiplication and division (e.g., knowing that $8 \times 5 = 40$, one knows $40 \div 5 = 8$) or properties of operations. By the end of Grade 3, know from memory all products of two one-digit numbers.

3.OA.9 Identify arithmetic patterns (including patterns in the addition table or multiplication table), and explain them using properties of operations.

Looks like I'll be doing a lot in this chapter!

Standards for Mathematical PRACTICE

1. Make sense of problems and persevere in solving them.
2. Reason abstractly and quantitatively.
3. Construct viable arguments and critique the reasoning of others.
4. Model with mathematics.
5. Use appropriate tools strategically.
6. Attend to precision.
7. Look for and make use of structure.
8. Look for and express regularity in repeated reasoning.

= focused on in this chapter

Name ....

Am I Ready?

Check — Go online to take the Readiness Quiz

Use counters to find the number of equal groups, or how many are in each group.

1. 9 counters
3 equal groups
_______ in each group
So, $9 \div 3 =$ _______.

2. 15 counters
5 equal groups
_______ in each group
So, $15 \div 5 =$ _______.

Algebra **Use the array to complete each pair of number sentences.**

3. $4 \times$ ___ $=$ ___
$24 \div$ ___ $=$ ___

4. ___ $\times$ ___ $=$ ___
___ $\div$ ___ $=$ ___

Multiply.

5. $5 \times 6 =$ _______

6. 10×2

7. 2×8

Divide.

8. $3\overline{)18}$

9. $24 \div 3 =$ _______

10. $10\overline{)50}$

11. There are a total of 15 prizes. Each child gets to pick 3 prizes. How many children will be picking prizes?

____

How Did I Do? Shade the boxes to show the problems you answered correctly.

1	2	3	4	5	6	7	8	9	10	11

Name ______________________________

Review Vocabulary

factors   known fact   pattern   product

Making Connections

Look at the partial multiplication table below. Label each section with the correct review vocabulary word.

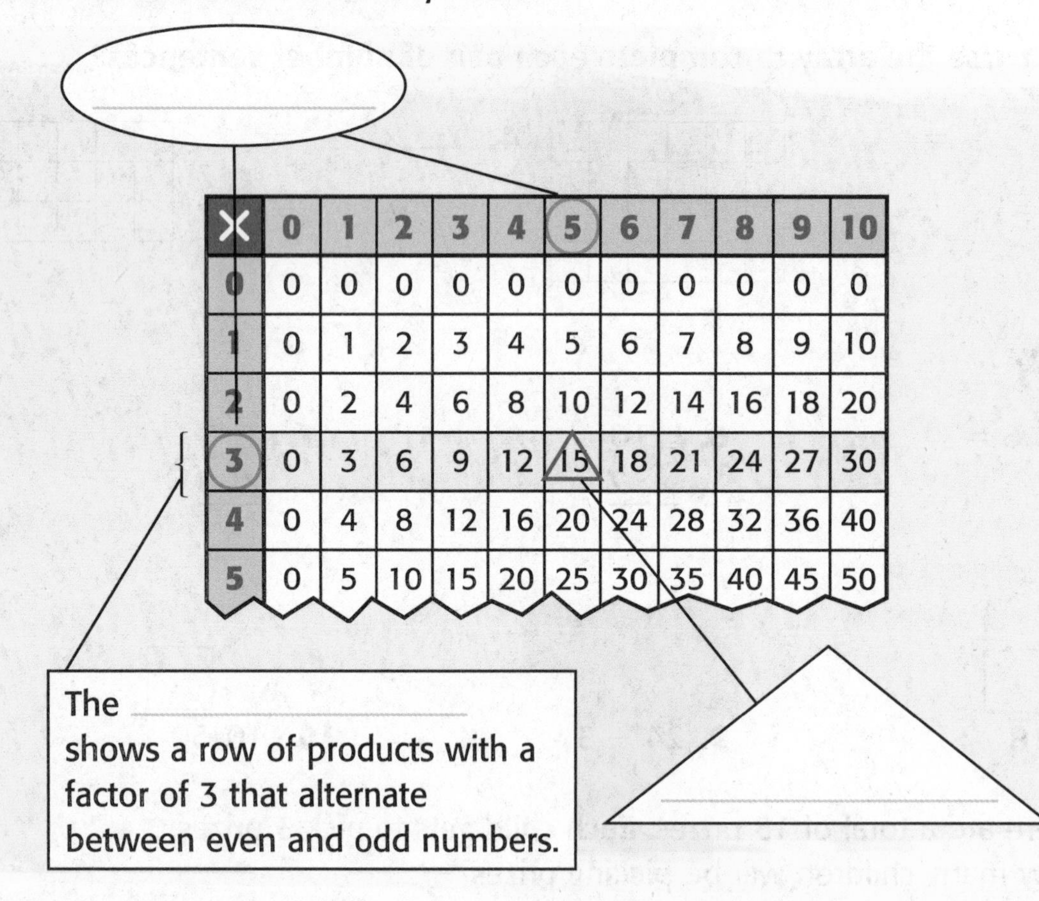

×	0	1	2	3	4	5	6	7	8	9	10
0	0	0	0	0	0	0	0	0	0	0	0
1	0	1	2	3	4	5	6	7	8	9	10
2	0	2	4	6	8	10	12	14	16	18	20
3	0	3	6	9	12	15	18	21	24	27	30
4	0	4	8	12	16	20	24	28	32	36	40
5	0	5	10	15	20	25	30	35	40	45	50

Complete the sentence with the vocabulary word not used in the activity.

A ______________________ is a fact that you have memorized.

MY Vocabulary Cards
Vocab
abc
Mathematical PRACTICE

Ideas for Use

- Use the blank cards to write words from previous chapters you would like to review. On the back of the card, draw or write a definition for each word.
- Use the blank cards to write multiplication facts horizontally. On the back of the cards, write the same facts vertically.
- Use the blank cards to write the division facts you have learned so far using the ÷ symbol. On the backs of the cards, write the same facts using the $\overline{)\quad}$ symbol.
- Use the blank cards to write basic multiplication and division facts. Write the answers on the backs of the cards.

MY Foldable

FOLDABLES® Follow the steps on the back to make your Foldable.

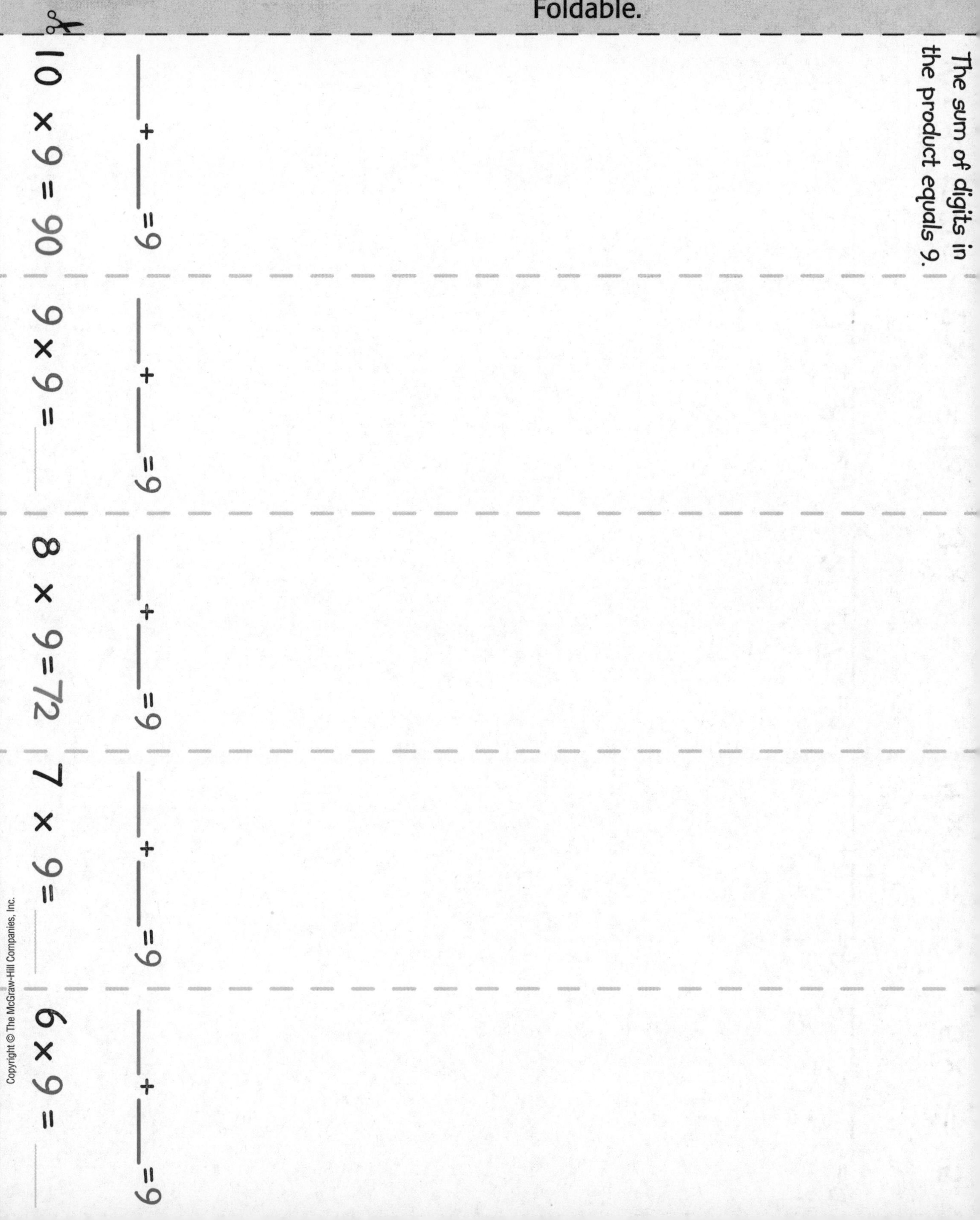

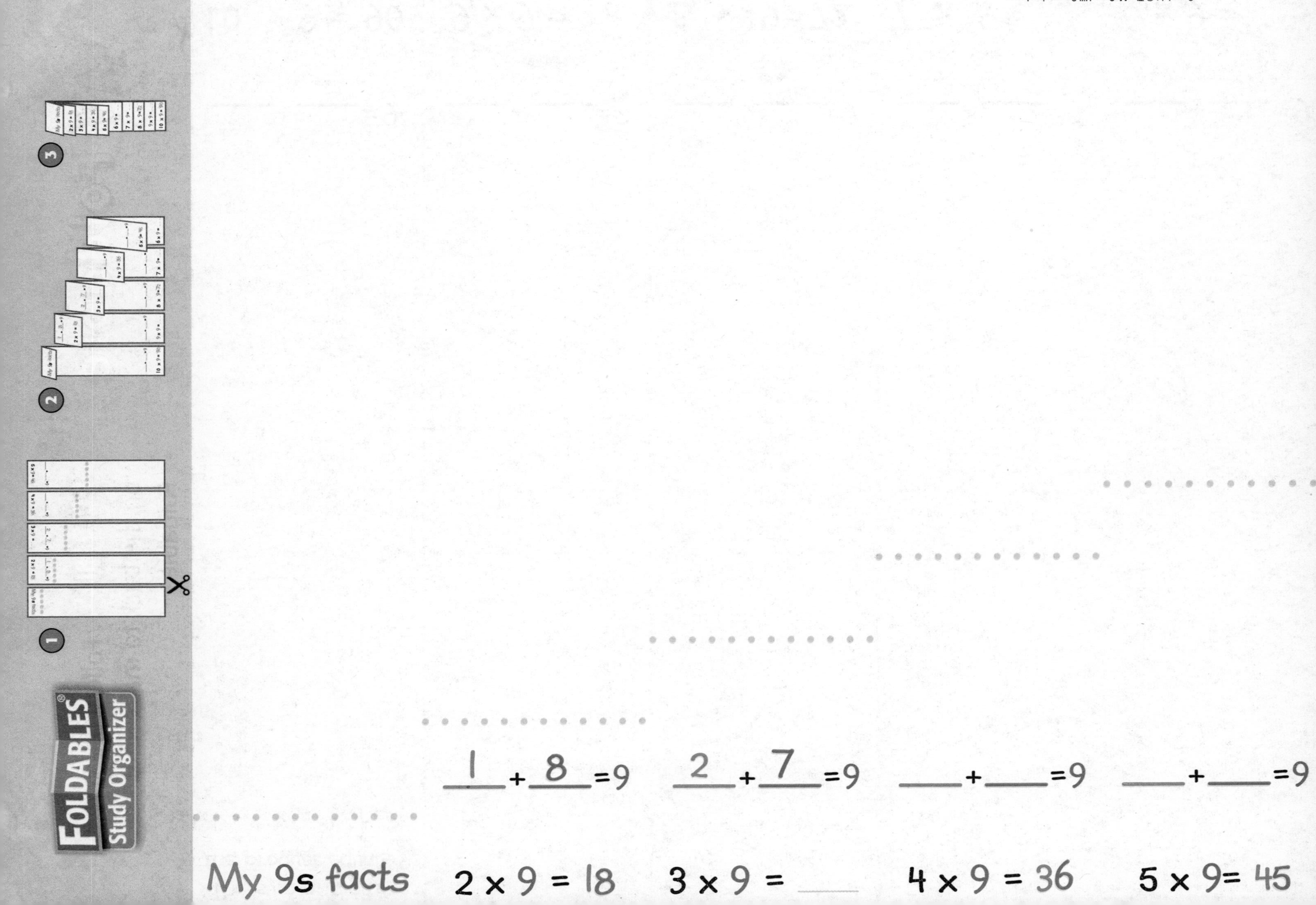
FOLDABLES®
Study Organizer
My 9s facts
2 × 9 = 18
1 + 8 = 9
3 × 9 =
2 + 7 = 9
4 × 9 = 36
+ = 9
5 × 9 = 45
+ = 9
Copyright © The McGraw-Hill Companies, Inc.

Name ....

Multiply by 6

Lesson 1

ESSENTIAL QUESTION
How can multiplication and division facts with smaller numbers be applied to larger numbers?

Math in My World

Watch Tutor

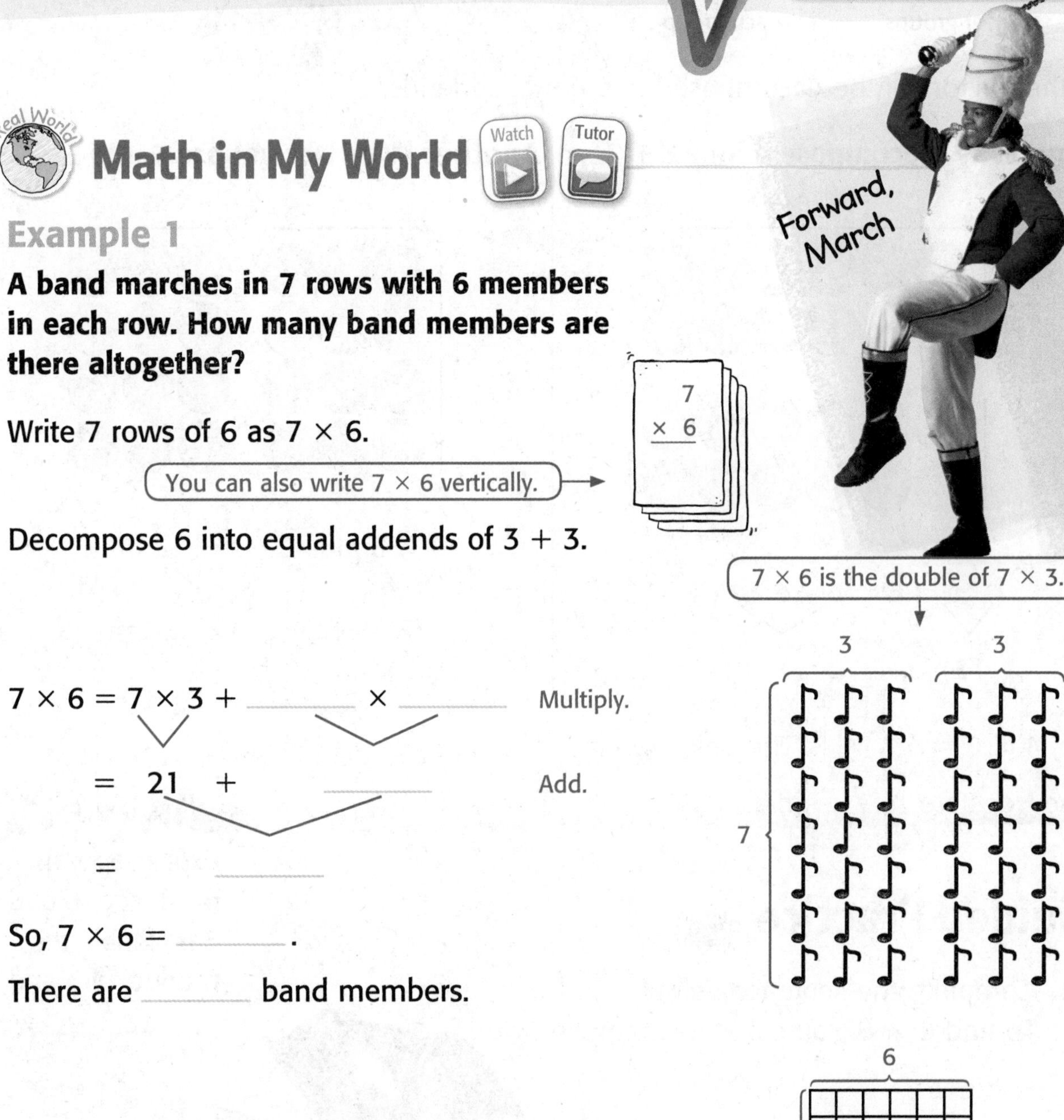

Example 1

A band marches in 7 rows with 6 members in each row. How many band members are there altogether?

Write 7 rows of 6 as 7×6.

You can also write 7×6 vertically.

$$\begin{array}{r} 7 \\ \times\ 6 \\ \hline \end{array}$$

Decompose 6 into equal addends of $3 + 3$.

7×6 is the double of 7×3.

$7 \times 6 = 7 \times 3 +$ _____ $\times$ _____ Multiply.

$= 21 +$ _____ Add.

$=$ _____

So, $7 \times 6 =$ _____.

There are _____ band members.

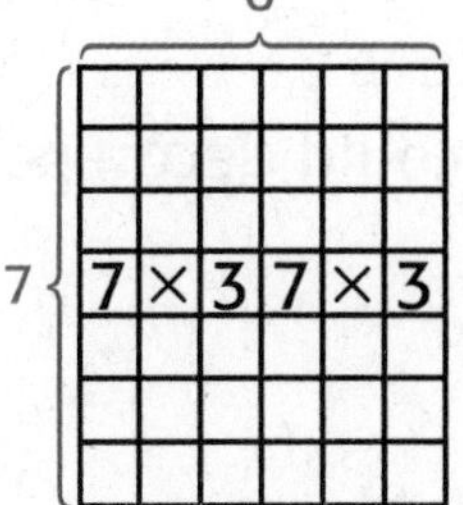

Check Shade part of the array yellow to show 7×3. Shade the remaining part green to show 7×3. The shaded array shows the known fact doubled.

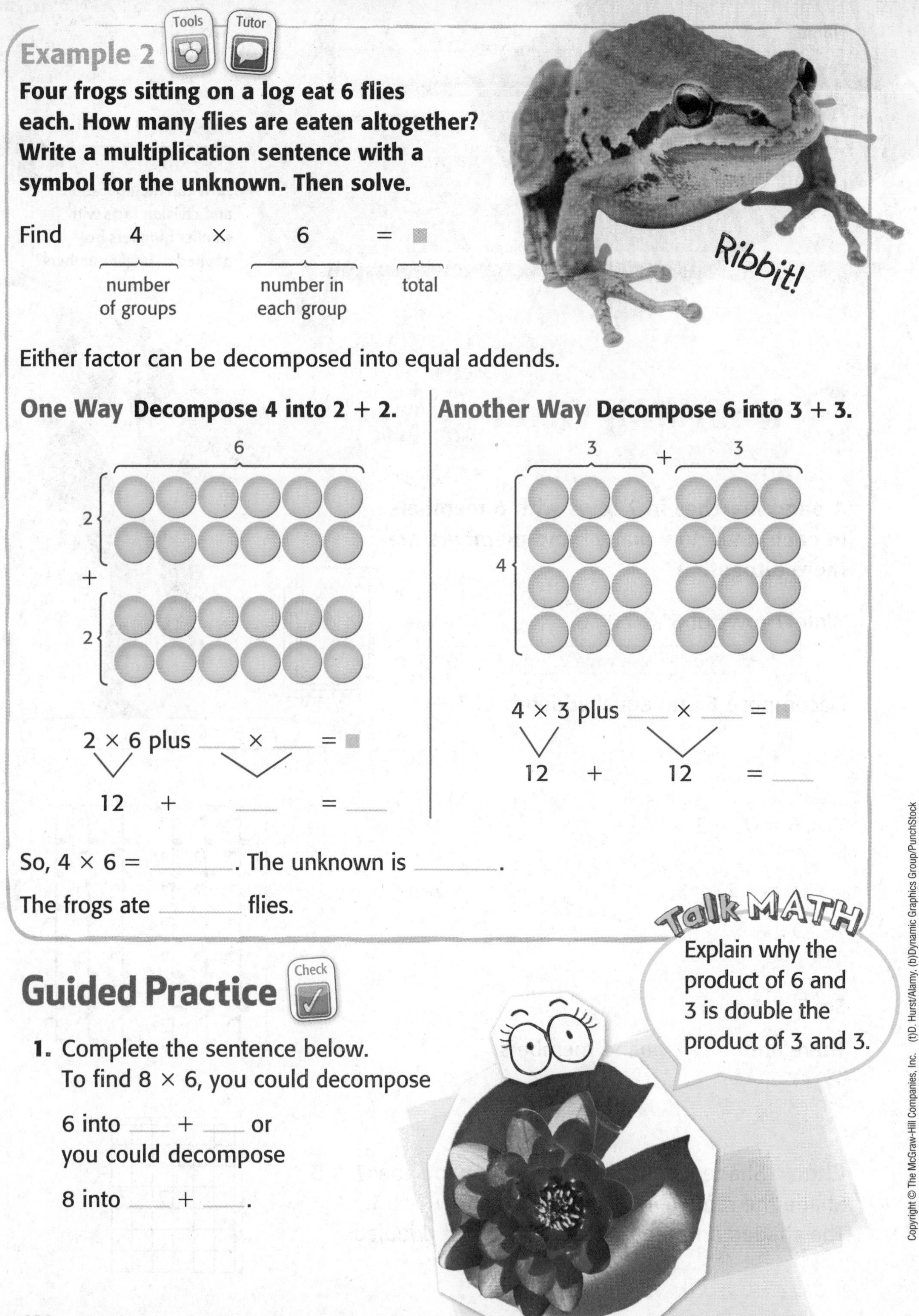

Example 2

Four frogs sitting on a log eat 6 flies each. How many flies are eaten altogether? Write a multiplication sentence with a symbol for the unknown. Then solve.

Find 4 × 6 = ■

4: number of groups; 6: number in each group; ■: total

Ribbit!

Either factor can be decomposed into equal addends.

One Way **Decompose 4 into 2 + 2.**

6

2 + 2

2 × 6 plus ____ × ____ = ■

12 + ____ = ____

Another Way **Decompose 6 into 3 + 3.**

3 + 3

4

4 × 3 plus ____ × ____ = ■

12 + 12 = ____

So, 4 × 6 = ______. The unknown is ______.

The frogs ate ______ flies.

Talk MATH

Explain why the product of 6 and 3 is double the product of 3 and 3.

Guided Practice

1. Complete the sentence below. To find 8 × 6, you could decompose

6 into ____ + ____ or you could decompose

8 into ____ + ____.

Name ________________________________

Independent Practice

Double a known fact to find each product. Draw an array.

2. $5 \times 6 =$ ______

$5 \times 3 = 15$

______ $\times 3 =$ ______

$15 +$ ______ $=$ ______

3. $9 \times 6 =$ ______

4. 3×6

5. 8×6

Algebra **Find each unknown. Double a known fact.**

6. $4 \times ■ = 24$

The unknown is ______.

7. $10 \times ■ = 60$

The unknown is ______.

8. $6 \times 6 = ■$

The unknown is ______.

9. $■ \times 6 = 42$

The unknown is ______.

Multiply. Use the multiplication table.

10. $1 \times 6 =$ ______

11. $7 \times 6 =$ ______

12. $6 \times 4 =$ ______

13. $6 \times 3 =$ ______

14. $2 \times 6 =$ ______

15. $6 \times 0 =$ ______

Problem Solving

Algebra Write a multiplication sentence with a symbol for the unknown. Solve.

My Work!

16. In the morning, 6 eggs hatched. By the evening, nine times as many had hatched. What is the total number of eggs that hatched?

____

17. Mathematical PRACTICE 6 **Be Precise** If Ida has 6 ten-dollar bills, does she have enough money for 8 bags of rabbit food that cost $6 each? Explain.

____

____

HOT Problems

18. Mathematical PRACTICE 2 **Reason** Anna forgot her 6s facts. She used the 5s fact $5 \times 6 = 30$ to find the product of 6×5. What property of multiplication allows her to do this?

____

19. Mathematical PRACTICE 8 **Look for a Pattern** Part of the multiplication table is shown. Study the pattern in the products of 6. Will the products of 6 always be even or always be odd? Explain.

×	0	1	2	3	4	5	6	7
0	0	0	0	0	0	0	0	0
1	0	1	2	3	4	5	6	7
2	0	2	4	6	8	10	12	14
3	0	3	6	9	12	15	18	21
4	0	4	8	12	16	20	24	28
5	0	5	10	15	20	25	30	35
6	0	6	12	18	24	30	36	42
7	0	7	14	21	28	35	42	49

____

____

20. **Building on the Essential Question** How can doubling a known fact be helpful when finding products mentally?

____

____

Name

Lesson 1
Multiply by 6

Homework Helper

eHelp Need help? connectED.mcgraw-hill.com

Tyrone spent 8 minutes playing each level of a video game. The video game had 6 levels. How many minutes did he spend altogether playing the video game?

Find 8×6.

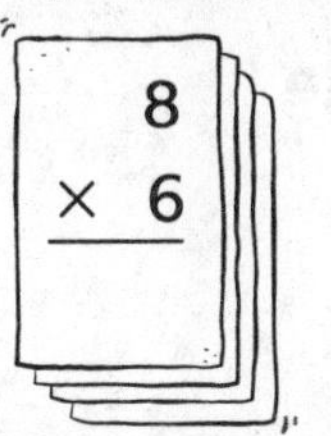

Decompose 6 into two equal addends of $3 + 3$.

6 is the double of 3. So, 8×6 is the double of 8×3.

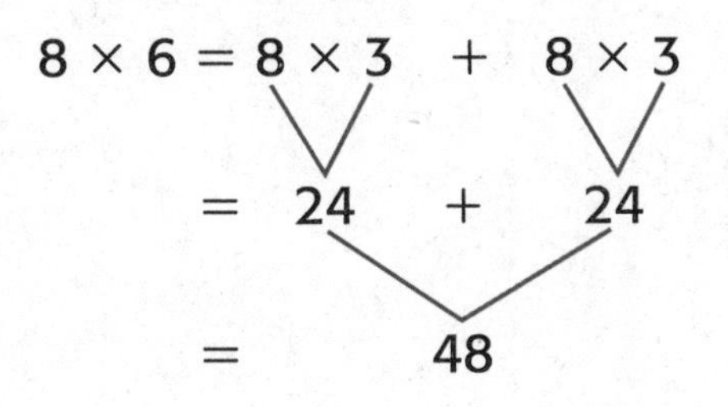

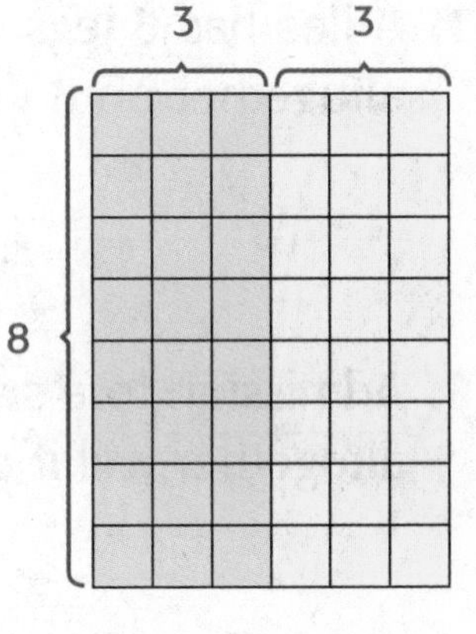

So, $8 \times 6 = 48$.

Tyrone spent 48 minutes playing the video game.

Practice

Double a known fact to find each product. Draw an array.

1. $2 \times 6 =$ ______

2. $\begin{array}{r} 9 \\ \times\ 6 \\ \hline \end{array}$

Algebra **Find each unknown. Double a known fact.**

3. 5 × ■ = 30

The unknown is ______.

4. ■ × 6 = 60

The unknown is ______.

5. 6 × ■ = 36

The unknown is ______.

6. ■ × 6 = 42

The unknown is ______.

Problem Solving

Mathematical PRACTICE 2 **Use Algebra For Exercises 7–8, write a multiplication sentence with a symbol for the unknown. Then solve.**

My Work!

7. A flea has 6 legs. How many legs are there altogether on 8 fleas?

____

8. Admission to a science museum is $9. How much altogether will it cost for 6 people?

____

9. Mathematical PRACTICE 1 **Make Sense of Problems** Gina's kitten weighs 5 ounces. If the kitten gains 3 ounces every week, how many ounces will the kitten weigh in 6 weeks?

____

Test Practice

10. Which number sentence represents the array shown at the right?

Ⓐ 4 × 6 = 24

Ⓑ 3 × 6 = 18

Ⓒ 4 + 6 = 10

Ⓓ 8 × 3 = 24

Name ....

Multiply by 7

Lesson 2

ESSENTIAL QUESTION
How can multiplication and division facts with smaller numbers be applied to larger numbers?

You can decompose larger facts into smaller facts.

Math in My World

Tutor Watch

It's SHOWTIME!

Example 1

A museum has a display of 9 kinds of beetles. There are 7 of each kind of beetle. How many beetles are on display? Write a multiplication sentence with a symbol for the unknown.

$9 \times 7 = \blacksquare$.

Decompose the factor 7 into addends of $5 + 2$.

Use the known facts of 9×5 and 9×2.

$9 \times 7 =$ 9×5 + 9×2 Multiply.

= ______ + ______ Add.

= ______

7
5 2
9

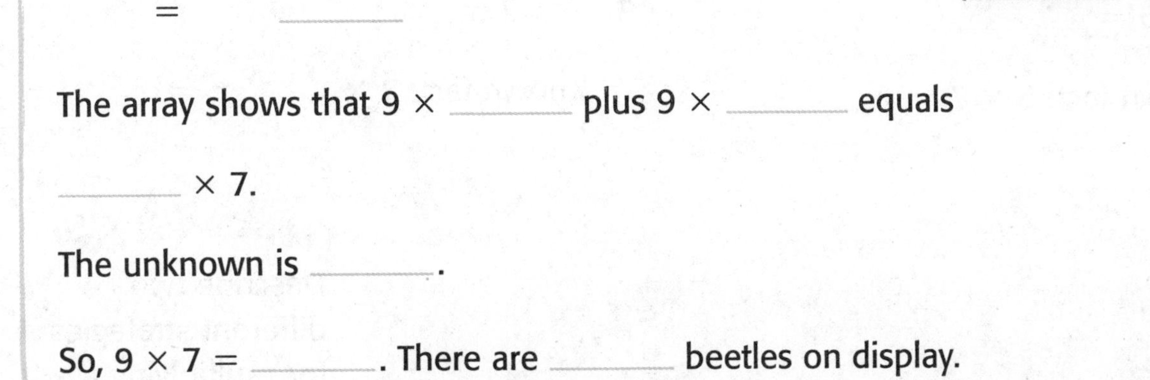

The array shows that 9 × ______ plus 9 × ______ equals ______ × 7.

The unknown is ______.

So, 9 × 7 = ______. There are ______ beetles on display.

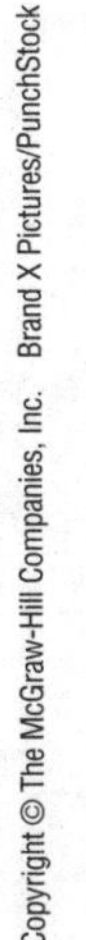

To multiply by 7, you can also use a related fact.

Example 2

Tutor

A pet store sold 3 gerbils. Each gerbil cost $7. How much money did the pet store make selling the gerbils?

Write 3 groups of $7 as 3 × $7 or you can write it vertically.

$$\begin{array}{r} 3 \\ \times\ \$7 \\ \hline \end{array}$$

Take Me Home

Use the Commutative Property of Multiplication.

You know 7 × 3 = ______. Turn the array. 3 × 7 = ______

3

7

Commutative Property

7

3

So, 3 × 7 = ______.

The pet store made $______ on the sale of the gerbils.

Guided Practice

Check

Use a known fact and the Commutative Property to find each product.

1. 7 × 5 = ______

Known fact: 5 × 7 = ______

2. 7 × 2 = ______

Known fact: 2 × ______ = ______

Talk MATH

Describe two different strategies for multiplying a number by 7.

Name

CCSS

Independent Practice

Algebra Find each unknown. Decompose the factor 7 into 5 + 2.

3. $7 \times 7 = \blacksquare$

Known facts: $7 \times 5 =$ ______

$7 \times 2 =$ ______

The unknown is ______.

4. $8 \times 7 = \blacksquare$

Known facts: $8 \times 5 =$ ______

$8 \times 2 =$ ______

The unknown is ______.

Use a known fact and the Commutative Property to find each product.

5. $7 \times 1 =$ ______

Known fact:

______

6. $7 \times 2 =$ ______

Known fact:

______

7. $7 \times 10 =$ ______

Known fact:

______

8. $7 \times 0 =$ ______

Known fact:

______

9. $7 \times 3 =$ ______

Known fact:

______

10. $7 \times 6 =$ ______

Known fact:

______

Algebra Find each unknown. Use the Commutative Property.

11. $5 \times \blacksquare = 35$

$\blacksquare \times 5 = 35$

The unknown is

______.

12. $3 \times 7 = \blacksquare$

$7 \times 3 = \blacksquare$

The unknown is

______.

13. $7 \times \blacksquare = 70$

$\blacksquare \times 7 = 70$

The unknown is

______.

Multiply.

14. $\begin{array}{r} 7 \\ \times\ 3 \\ \hline \end{array}$

15. $\begin{array}{r} 7 \\ \times\ 1 \\ \hline \end{array}$

16. $\begin{array}{r} 7 \\ \times\ 4 \\ \hline \end{array}$

17. $\begin{array}{r} 7 \\ \times\ 8 \\ \hline \end{array}$

Problem Solving

Algebra For Exercises 18 and 19, write a multiplication sentence with a symbol for the unknown. Solve.

18. Ryan and his 5 friends scored 7 points each while playing basketball. What is their total points?

______

19. Inez has 8 CDs. How many songs are there if each CD has 7 songs?

______

My Work!

HOT Problems

20. Mathematical PRACTICE 8 **Look for a Pattern** Look at the multiplication table. Color the row and column of products of 7. Describe a pattern.

×	0	1	2	3	4	5	6	7	8	9	10
0	0	0	0	0	0	0	0	0	0	0	0
1	0	1	2	3	4	5	6	7	8	9	10
2	0	2	4	6	8	10	12	14	16	18	20
3	0	3	6	9	12	15	18	21	24	27	30
4	0	4	8	12	16	20	24	28	32	36	40
5	0	5	10	15	20	25	30	35	40	45	50
6	0	6	12	18	24	30	36	42	48	54	60
7	0	7	14	21	28	35	42	49	56	63	70
8	0	8	16	24	32	40	48	56	64	72	80
9	0	9	18	27	36	45	54	63	72	81	90
10	0	10	20	30	40	50	60	70	80	90	100

______

______

21. Mathematical PRACTICE 3 **Find the Error** Circle the multiplication sentence which is incorrect. Explain.

$7 \times 7 = 48$ $\quad$ $7 \times 9 = 63$ $\quad$ $5 \times 7 = 35$

______

22. **Building on the Essential Question** Compare finding products using the Commutative Property of Multiplication and using related multiplication facts.

______

______

Name

Lesson 2
Multiply by 7

Homework Helper

eHelp Need help? connectED.mcgraw-hill.com

Jared will go on vacation for 8 weeks this summer. For how many days will he be on vacation?

Find 8 × 7.

Decompose the factor 7 into the addends 5 + 2.

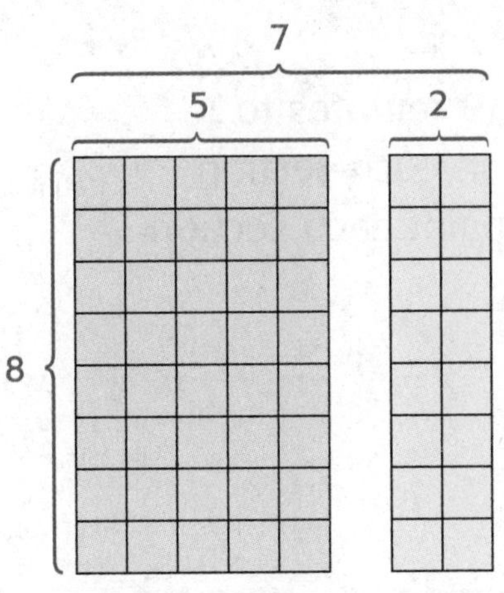

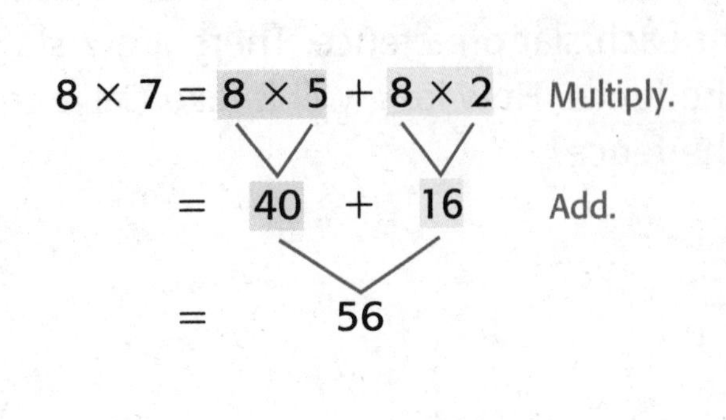

Practice

Algebra Find each unknown. Decompose the factor 7 into 5 + 2.

1. 7 × 10 = ■

Known facts: 5 × 10 = ______

2 × 10 = ______

The unknown is ______.

2. 5 × 7 = ■

Known facts: 5 × 5 = ______

5 × 2 = ______

The unknown is ______.

Algebra Find each unknown. Use the Commutative Property.

3. $7 \times 3 = \blacksquare$

$3 \times 7 = \blacksquare$

The unknown is _______.

4. $7 \times \blacksquare = 28$

$\blacksquare \times 7 = 28$

The unknown is _______.

5. $\blacksquare \times 7 = 49$

$7 \times \blacksquare = 49$

The unknown is _______.

6. $7 \times \blacksquare = 14$

$\blacksquare \times 7 = 14$

The unknown is _______.

Problem Solving

Algebra Write a multiplication sentence with a symbol for the unknown. Then solve.

7. Mathematical PRACTICE 4 **Model Math** It takes Callie 9 minutes to paint each slat on a fence. There are 7 slats in each section of the fence. How long will it take Callie to paint each section of the fence?

____

8. Each house on Mulberry Street has 7 front windows. There are 3 houses on each side of the street. How many front windows are there in all?

____

Test Practice

9. A bicycle shop is replacing both tires on 7 bikes. How many tires will be replaced altogether?

Ⓐ 2 tires

Ⓑ 7 tires

Ⓒ 9 tires

Ⓓ 14 tires

Name ....

Divide by 6 and 7

Lesson 3

ESSENTIAL QUESTION
How can multiplication and division facts with smaller numbers be applied to larger numbers?

Math in My World

DING! DING! Come and get it!

Example 1

Paco set each picnic table with 6 dinner plates. He used 24 plates to set the tables. How many tables did he set?

Find $24 \div 6$, or $6\overline{)24}$.

One Way **Draw an array.**

Draw an array. Think of a related multiplication fact.

Each column represents one table with ______ plates.

There are ______ columns.

You know that $6 \times$ ______ $= 24$. So, there will be ______ tables.

My Drawing!

Another Way **Use repeated subtraction.**

Skip count backwards. Draw the arrows to represent equal groups of 6.

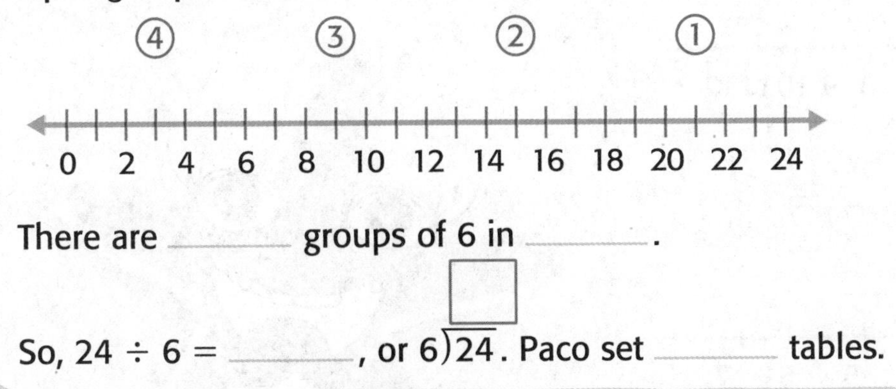

There are ______ groups of 6 in ______.

So, $24 \div 6 =$ ______, or $6\overline{)24}$ with answer box ☐. Paco set ______ tables.

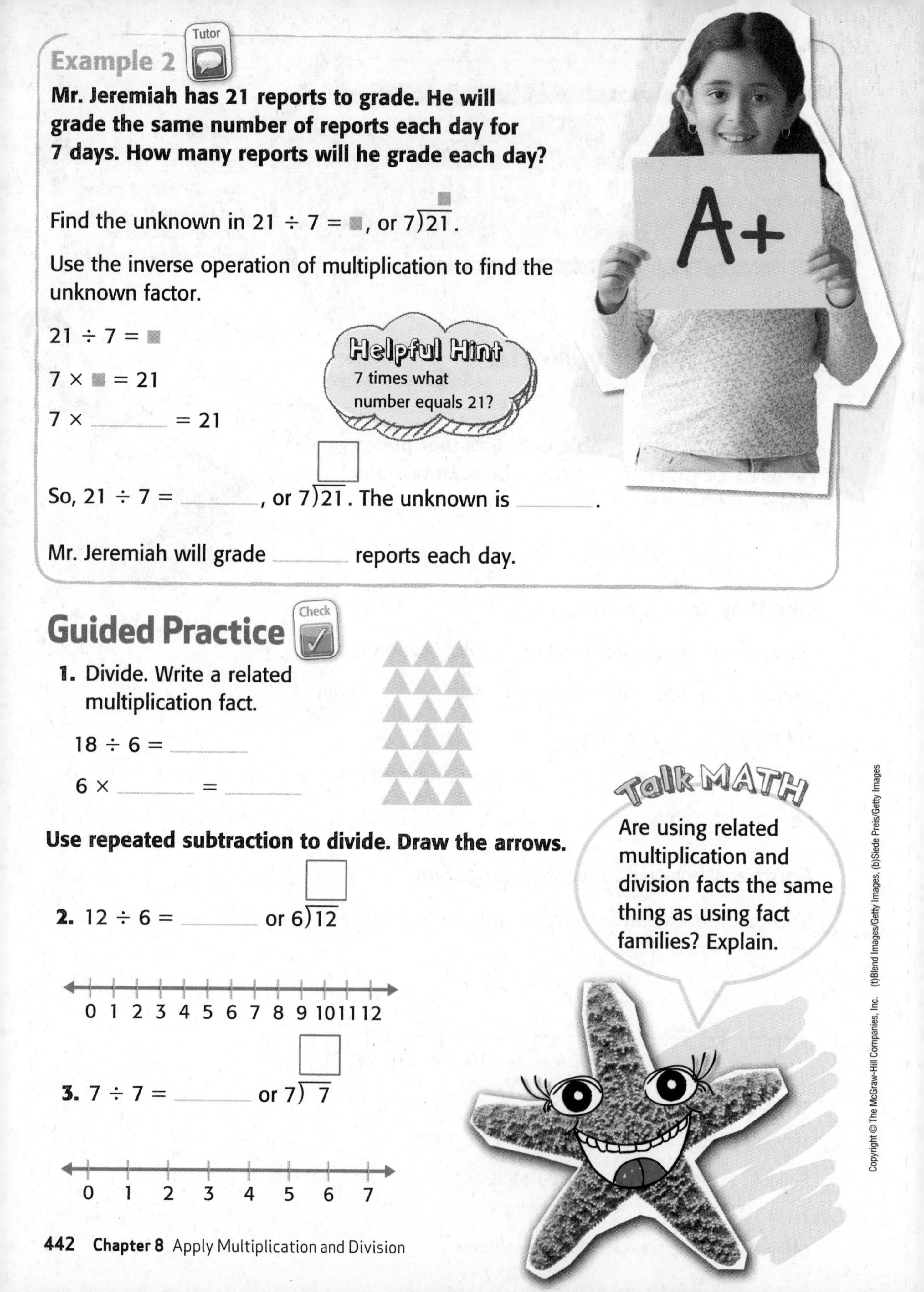

Example 2

Mr. Jeremiah has 21 reports to grade. He will grade the same number of reports each day for 7 days. How many reports will he grade each day?

Find the unknown in $21 \div 7 = \blacksquare$, or $7\overline{)21}$.

Use the inverse operation of multiplication to find the unknown factor.

$21 \div 7 = \blacksquare$

$7 \times \blacksquare = 21$

$7 \times$ ______ $= 21$

Helpful Hint
7 times what number equals 21?

So, $21 \div 7 =$ ______, or $7\overline{)21}$. The unknown is ______.

Mr. Jeremiah will grade ______ reports each day.

Guided Practice

1. Divide. Write a related multiplication fact.

   $18 \div 6 =$ ______

   $6 \times$ ______ $=$ ______

Use repeated subtraction to divide. Draw the arrows.

2. $12 \div 6 =$ ______ or $6\overline{)12}$

   0 1 2 3 4 5 6 7 8 9 10 11 12

3. $7 \div 7 =$ ______ or $7\overline{)7}$

   0 1 2 3 4 5 6 7

Talk MATH

Are using related multiplication and division facts the same thing as using fact families? Explain.

Name ....

Independent Practice

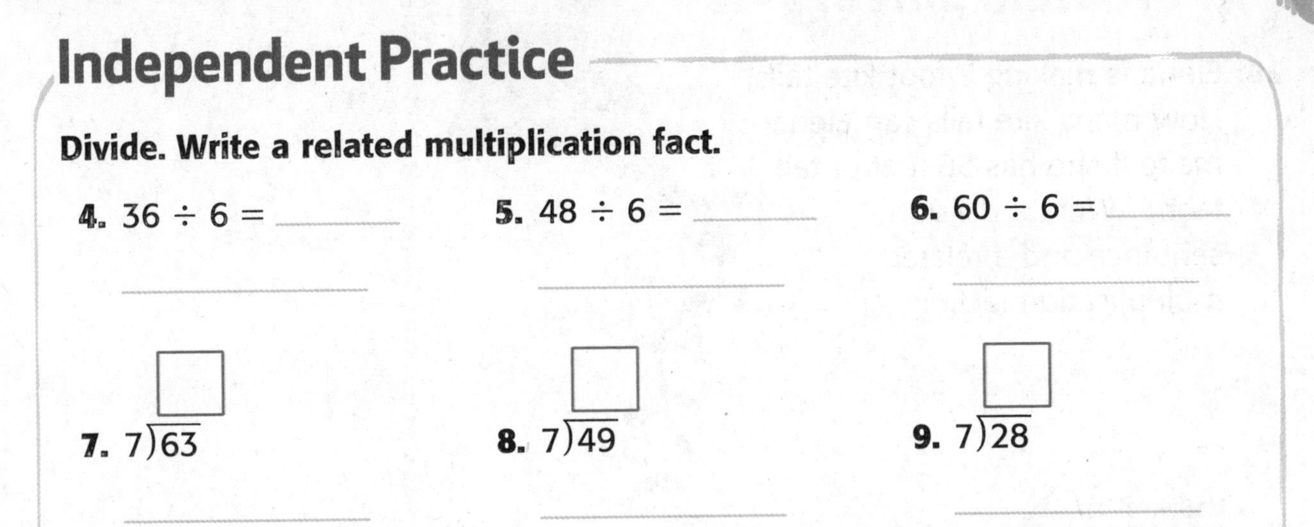

Divide. Write a related multiplication fact.

4. $36 \div 6 =$ ______

______

5. $48 \div 6 =$ ______

______

6. $60 \div 6 =$ ______

______

7. $7\overline{)63}$ ☐

______

8. $7\overline{)49}$ ☐

______

9. $7\overline{)28}$ ☐

______

Algebra **Draw an array and use the inverse operation to find each unknown.**

10. $42 \div ? = 7$
$6 \times ■ = 42$

$? =$ ______

$■ =$ ______

11. $30 \div ? = 6$
$5 \times ■ = 30$

$? =$ ______

$■ =$ ______

12. $54 \div ? = 9$
$6 \times ■ = 54$

$? =$ ______

$■ =$ ______

13. $35 \div 7 = ?$
$■ \times 5 = 35$

$? =$ ______

$■ =$ ______

Use a related multiplication fact to find each quotient. Draw a line to match.

14. $42 \div 6 =$ ______ • $7 \times 10 = 70$

15. $63 \div 7 =$ ______ • $6 \times 1 = 6$

16. $70 \div 7 =$ ______ • $8 \times 7 = 56$

17. $48 \div 6 =$ ______ • $7 \times 6 = 42$

18. $56 \div 7 =$ ______ • $8 \times 6 = 48$

19. $6 \div 6 =$ ______ • $9 \times 7 = 63$

Problem Solving

20. Elena is making 7-foot kite tails. How many kite tails can Elena make if she has 56 feet of tail fabric? Write a division sentence and a related multiplication fact.

My Work!

21. Mathematical PRACTICE 6 **Explain to a Friend** There are 35 students with 7 students at each table in Cafeteria 1. Cafeteria 2 has 35 students with 5 students at each table. Which cafeteria has more tables? Explain.

HOT Problems

22. Mathematical PRACTICE 3 **Which One Doesn't Belong?** Identify the division problem that does not belong with the others by circling it. Explain.

$56 \div 7$	$7\overline{)48}$	$49 \div 7$	$7\overline{)63}$

23. **Building on the Essential Question** How does learning multiplication and division facts at the same time help learn them quicker?

Name ....

Lesson 3
Divide by 6 and 7

Homework Helper

eHelp Need help? connectED.mcgraw-hill.com

Mariah sells jewelry. She has 18 pieces to deliver to 6 customers. Each customer bought the same number of pieces. How many pieces of jewelry did Mariah deliver to each customer?

You need to find the unknown in $18 \div 6 = \blacksquare$.

Use repeated subtraction.
Start at 18 on a number line and skip count backward by 6.

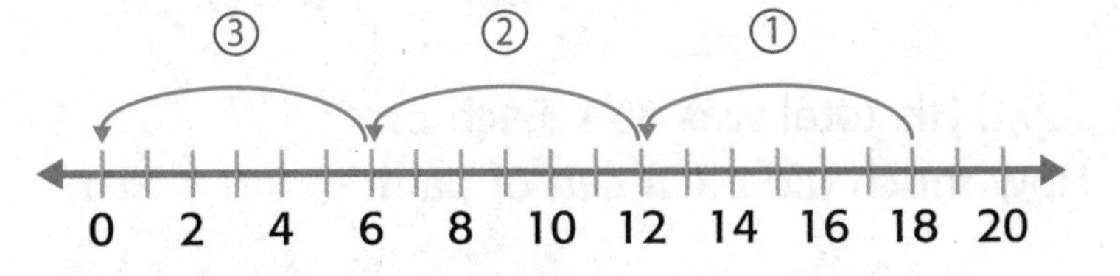

Helpful Hint
6 times what number equals 18?

There are 3 groups of 6 in 18.

So, Mariah delivered 3 pieces of jewelry to each customer.

Practice

Use repeated subtraction to divide. Draw the arrows.

1. $28 \div 7 =$ ______

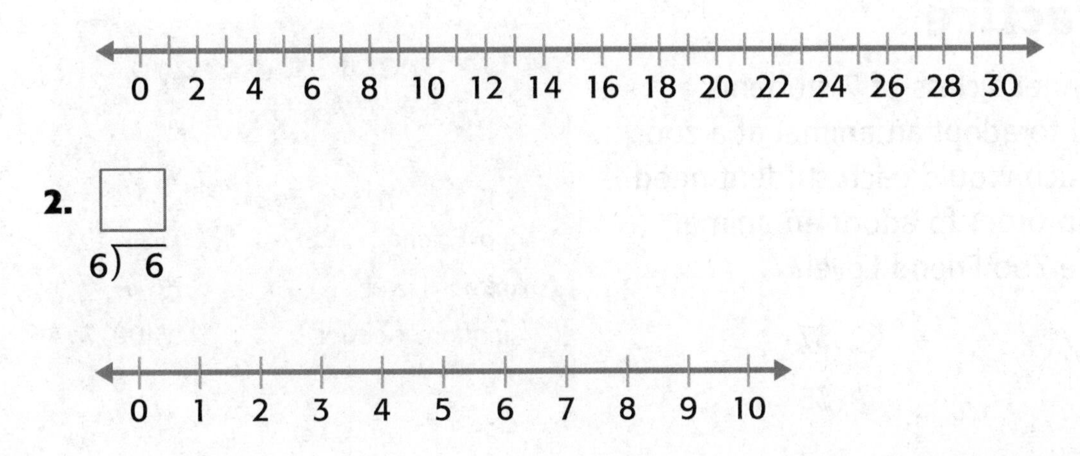

2. $6\overline{)\,6}$ ☐

0 1 2 3 4 5 6 7 8 9 10

Divide. Write a related multiplication fact.

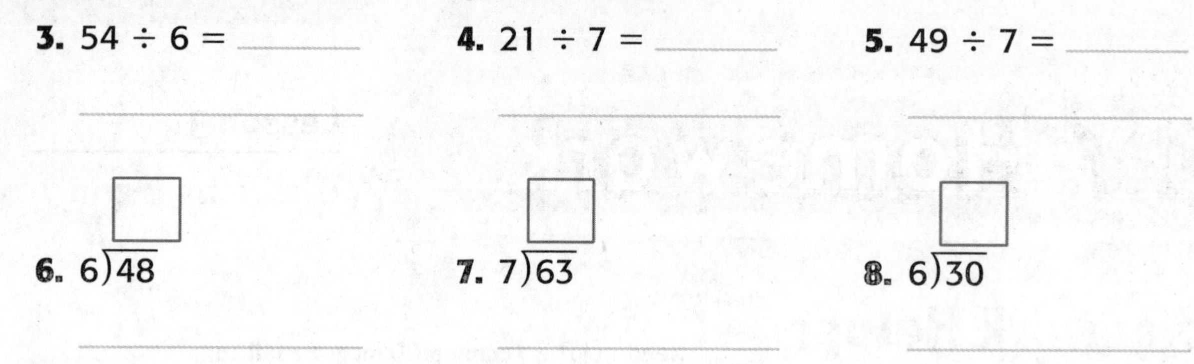

3. $54 \div 6 =$ ______

______

4. $21 \div 7 =$ ______

______

5. $49 \div 7 =$ ______

______

6. $6\overline{)48}$

______

7. $7\overline{)63}$

______

8. $6\overline{)30}$

______

Problem Solving

Mathematical PRACTICE 4 Model Math Write a division sentence to solve. Then write a related multiplication sentence.

9. There are 42 cards dealt to the players in a card game. Each player gets 7 cards. How many players are in the game?

______

10. Mr. Clancy bought 9 cans of paint. His total was $54. Each can of paint was the same price. How much did each can of paint cost?

______

11. Franklin's mother is making 6 snack bags for his campout. She will put 18 cherry fruit rolls and 18 grape fruit rolls into the bags. If she puts the same number in each bag, how many fruit rolls will be in each snack bag?

______

Test Practice

12. Mrs. Tanner's class of 7 students decided to adopt an animal at a zoo. How much would each student need to pay in order to adopt an animal from the Zoo Friend Level?

Ⓐ $35 Ⓒ $7

Ⓑ $8 Ⓓ $5

Central Florida Zoo Adopt An Animal

Adoption Level	Price
Zoo Friend	$35
Animal Lover	$56
Kingdom Keeper	$100

Check My Progress

Vocabulary Check

In Exercises 1–3 choose the correct word(s) to complete each sentence.

Commutative Property **known fact** **related facts**

1. A ______________________ is a fact you know by memory.

2. The ______________________ states that the order in which two numbers are multiplied does not change the product.

3. Facts that use the same three numbers are ______________________.

4. Write two multiplication sentences that are examples of the Commutative Property of Multiplication.

______________________ ______________________

5. Write a related multiplication fact for $48 \div 6 = 8$. ______________________

Concept Check

Double a known fact to find each product. Draw an array.

6. $4 \times 6 =$ _____

$2 \times$ _____ $=$ _____

_____ $\times 6 =$ _____

_____ $+$ _____ $=$ _____

7. $7 \times 6 =$ _____

Use a known fact and the Commutative Property to find each product.

8. $7 \times 4 =$ _____

Known fact: _____ $\times$ _____ $=$ _____

9. $7 \times 3 =$ _____

Known fact: _____ $\times$ _____ $=$ _____

Algebra Find each unknown. Decompose the factor 7 into 5 + 2.

10. $9 \times 7 = ■$

Known facts: $9 \times$ ____ = ____

$9 \times$ ____ = ____

The unknown is ____.

11. $7 \times 7 = ■$

Known facts: ____________

____

The unknown is ____.

Multiply.

12. $5 \times 6 =$ ____ **13.** $8 \times 7 =$ ____ **14.** $9 \times 6 =$ ____

Divide. Write a related multiplication fact.

15. $14 \div 7 =$ ____ **16.** $7\overline{)56}$ **17.** $70 \div 7 =$ ____ **18.** $48 \div 8 =$ ____

____________ ____________ ____________ ____________

Problem Solving

Algebra Write a division sentence with a symbol for the unknown. Then solve.

19. The zoo has 18 monkeys and 6 trees. Each tree has the same number of monkeys. How many monkeys are in each tree?

____

20. When a tree is cut down, 7 new trees are planted. If 56 new trees have been planted, how many trees have been cut down?

____

Test Practice

21. The picture shows the number of carrots Aisha's guinea pigs eat each day. She has 21 carrots. For how many days will the carrots last if they eat an an equal number each day?

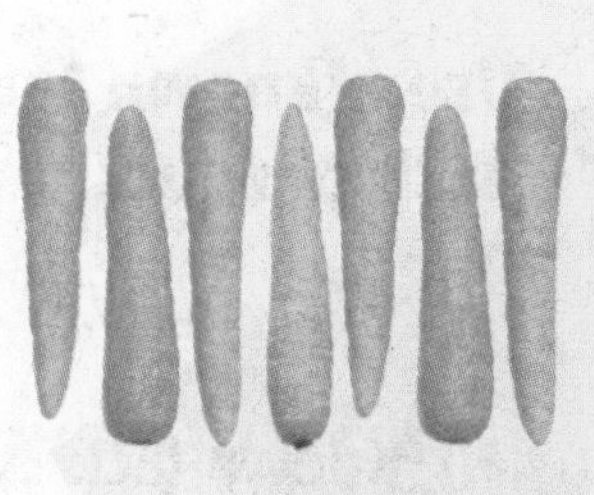

Ⓐ 2 days Ⓒ 4 days

Ⓑ 3 days Ⓓ 5 days

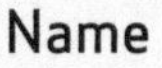

Name ....

Multiply by 8

Lesson 4

ESSENTIAL QUESTION
How can multiplication and division facts with smaller numbers be applied to larger numbers?

Math in My World

Watch Tutor

Example 1

There are 6 trees along a street. In each tree, there are 8 birds. How many birds are there altogether?

Chirp!

Find 6 × 8.

Each tree has a group of 8 birds.

One Way **Draw an array.**

Another Way **Draw a picture.**
Use an X for each bird.

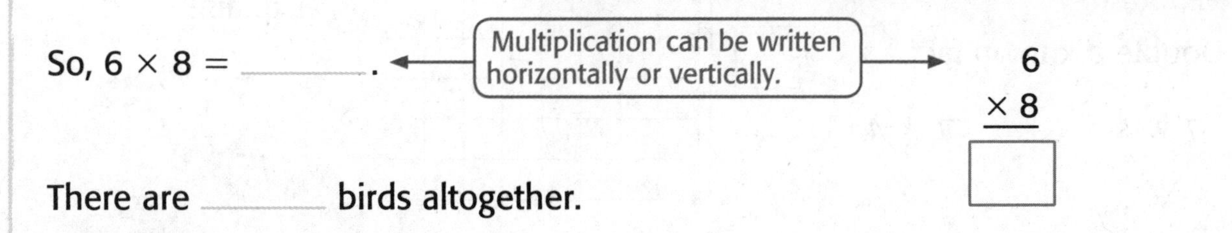

So, 6 × 8 = ______. ← Multiplication can be written horizontally or vertically. →

$$\begin{array}{r} 6 \\ \times\ 8 \\ \hline \square \end{array}$$

There are ______ birds altogether.

Check

The Commutative Property shows that 6 × 8 has the same product as 8 × 6. Since 8 × 6 = ______, then 6 × 8 = ______.

The 4s facts are helpful in remembering the 8s facts.

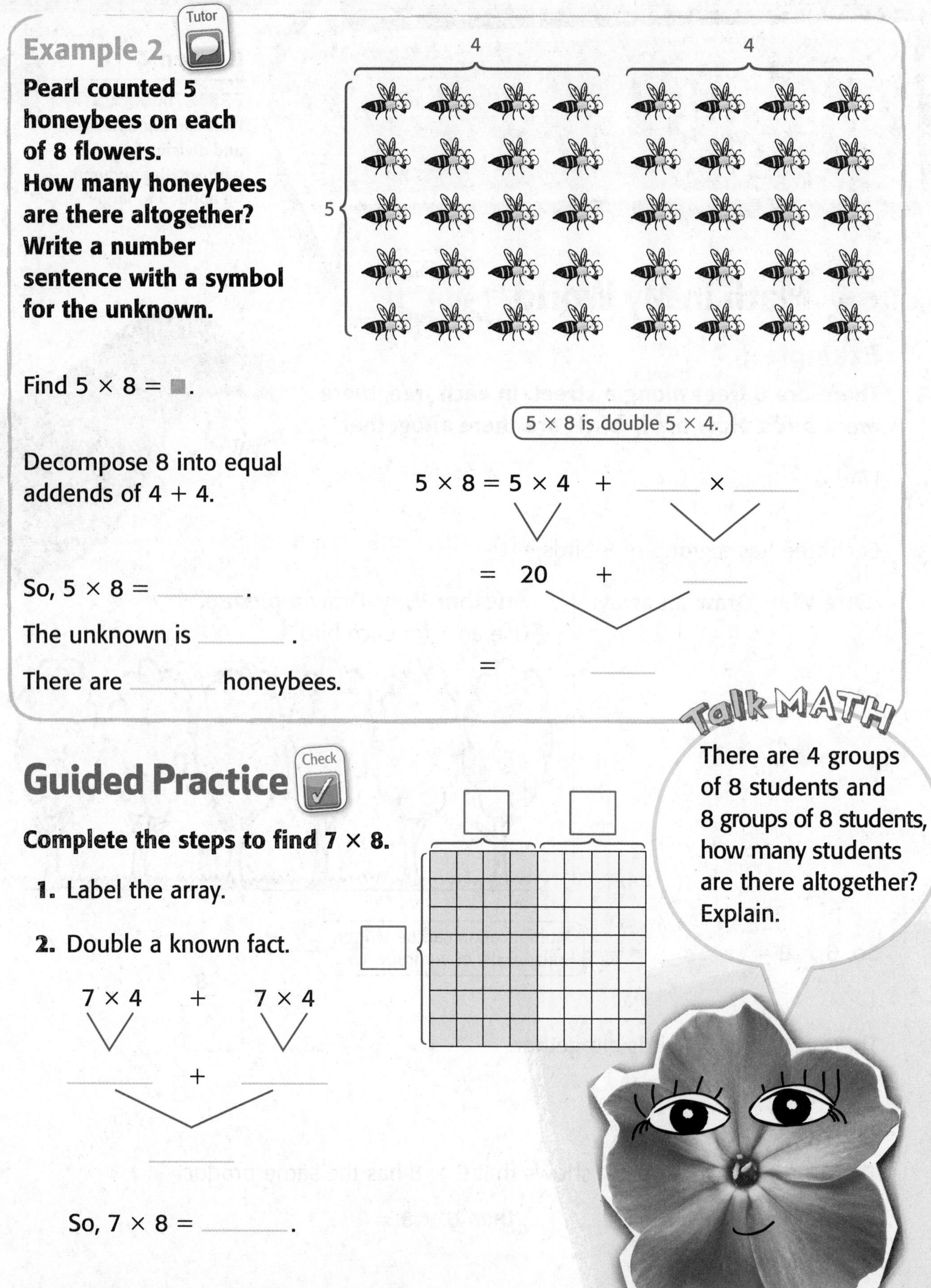

Example 2

Tutor

Pearl counted 5 honeybees on each of 8 flowers. How many honeybees are there altogether? Write a number sentence with a symbol for the unknown.

Find $5 \times 8 = \blacksquare$.

5 × 8 is double 5 × 4.

Decompose 8 into equal addends of 4 + 4.

$5 \times 8 = 5 \times 4 + \_\_\_ \times \_\_\_$

$= 20 + \_\_\_$

$= \_\_\_$

So, $5 \times 8 =$ ______.

The unknown is ______.

There are ______ honeybees.

Guided Practice

Check

Complete the steps to find 7×8.

1. Label the array.

2. Double a known fact.

$7 \times 4 + 7 \times 4$

______ + ______

______

So, $7 \times 8 =$ ______.

Name ________________________________

Independent Practice

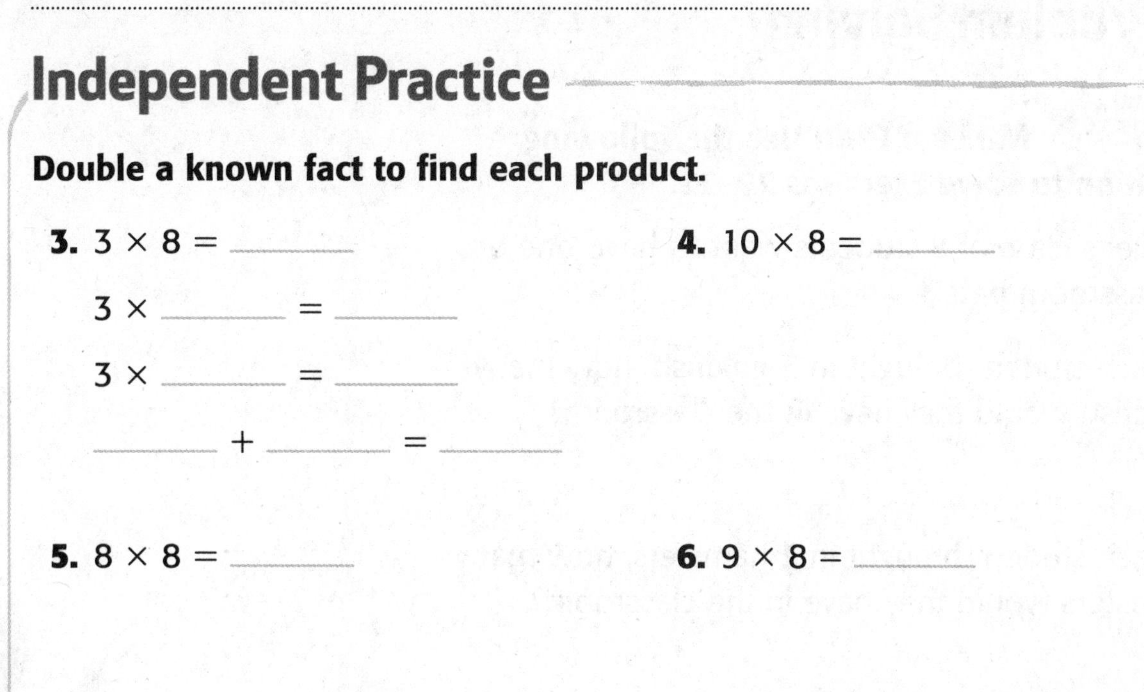

Double a known fact to find each product.

3. $3 \times 8 =$ ______

$3 \times$ ______ $=$ ______

$3 \times$ ______ $=$ ______

______ $+$ ______ $=$ ______

4. $10 \times 8 =$ ______

5. $8 \times 8 =$ ______

6. $9 \times 8 =$ ______

Use the Commutative Property to find each product. Write a related multiplication fact.

7. $1 \times 8 =$ ______

____

8. $0 \times 8 =$ ______

____

9. $6 \times 8 =$ ______

____

10. $7 \times 8 =$ ______

____

11. $2 \times 8 =$ ______

____

12. $4 \times 8 =$ ______

____

Algebra **Find each unknown factor. Use the Commutative Property.**

13. $8 \times ■ = 64$

$■ \times 8 = 64$

The unknown is ____.

14. $■ \times 1 = 8$

$1 \times ■ = 8$

The unknown is ____.

15. $8 \times ■ = 72$

$■ \times 8 = 72$

The unknown is ____.

Multiply.

16. $\begin{array}{r} 0 \\ \times 8 \\ \hline \end{array}$

17. $\begin{array}{r} 8 \\ \times 3 \\ \hline \end{array}$

18. $\begin{array}{r} 5 \\ \times 8 \\ \hline \end{array}$

19. $\begin{array}{r} 6 \\ \times 8 \\ \hline \end{array}$

Problem Solving

My Work!

Mathematical PRACTICE 1 **Make a Plan** **Use the following information to solve Exercises 20–22.**

Mrs. Miller's class of 8 students want to have one or more classroom pets.

20. If each student brought in 3 goldfish, how many goldfish would they have in the classroom?

______

21. If each student brought in 2 hamsters, how many hamsters would they have in the classroom?

______

22. A pet store charges $10 for a lizard. If each student pays $5, how many lizards would they be able to purchase?

______

HOT Problems

23. Mathematical PRACTICE 2 **Use Number Sense** The row that represents the products of 8 from the multiplication table is shown below. Describe one pattern in the products of 8. Will this pattern continue? Explain.

0	8	16	24	32	40	48	56	64	72	80

______

______

24. **Building on the Essential Question** When would I choose to decompose a multiplication fact rather than draw a picture?

______

______

MY Homework

Lesson 4
Multiply by 8

Homework Helper

eHelp Need help? connectED.mcgraw-hill.com

Each ladybug has 6 legs. Elaine counted 8 ladybugs. How many legs is that altogether?

Find 6×8.

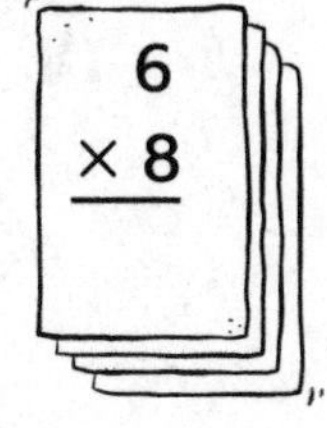

One Way **Draw an array.**

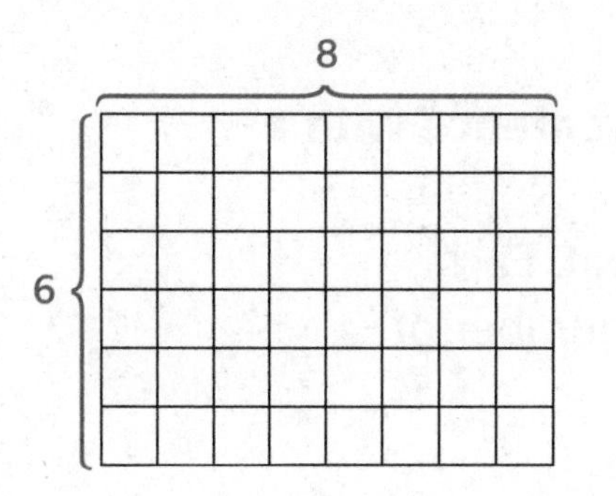

Another Way **Double a known fact.**

Decompose the number 8 into equal addends of $4 + 4$.

$$6 \times 8 = 6 \times 4 + 6 \times 4$$

$$24 + 24 = 48$$

$6 \times 8 = 48$. So, 8 ladybugs have 48 legs altogether.

Practice

Double a known fact to find each product.

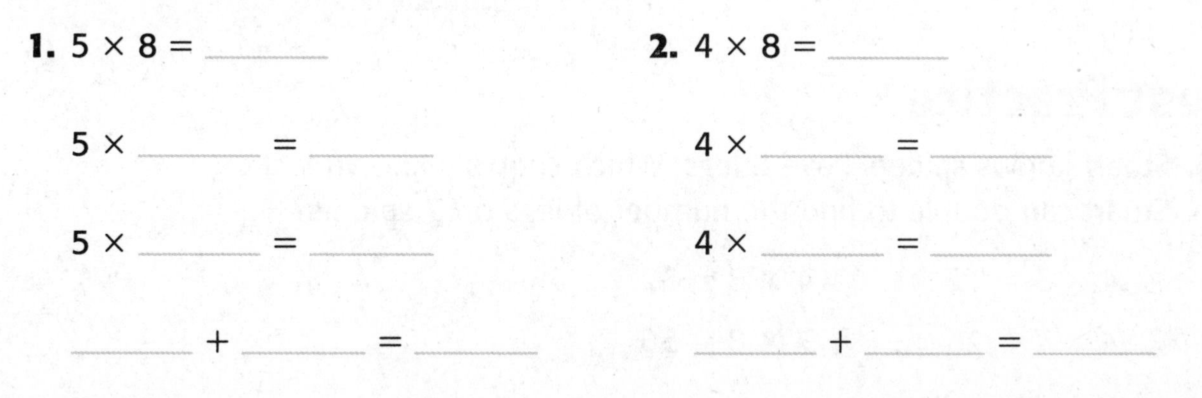

1. $5 \times 8 =$ ______

$5 \times$ ______ $=$ ______

$5 \times$ ______ $=$ ______

______ $+$ ______ $=$ ______

2. $4 \times 8 =$ ______

$4 \times$ ______ $=$ ______

$4 \times$ ______ $=$ ______

______ $+$ ______ $=$ ______

Algebra **Find each unknown. Use the Commutative Property.**

3. $8 \times \blacksquare = 40$
$\blacksquare \times 8 = 40$

The unknown is ______.

4. $\blacksquare \times 8 = 56$
$8 \times \blacksquare = 56$

The unknown is ______.

5. $2 \times 8 = \blacksquare$
$8 \times 2 = \blacksquare$

The unknown is ______.

6. $8 \times \blacksquare = 64$
$\blacksquare \times 8 = 64$

The unknown is ______.

Multiply.

7. $\begin{array}{r} 1 \\ \times 8 \\ \hline \end{array}$

8. $\begin{array}{r} 8 \\ \times 9 \\ \hline \end{array}$

9. $\begin{array}{r} 8 \\ \times 0 \\ \hline \end{array}$

10. $\begin{array}{r} 3 \\ \times 8 \\ \hline \end{array}$

Problem Solving

Mathematical PRACTICE 2 **Use Symbols Write a multiplication sentence with a symbol for the unknown. Then solve.**

11. There were 5 dolphins swimming around a tour boat. Each dolphin circled the boat 8 times. What is the total number of times all of the dolphins circled the boat?

____

12. Cameron worked 8 hours at the coffee shop. He earned the same amount in tips each hour. At the end of his shift, Cameron had \$32 in tips. How much money did he earn in tips each hour?

____

Test Practice

13. Stuart knows spiders have 8 legs. Which shows a known fact Stuart can double to find the number of legs on 7 spiders?

Ⓐ $4 \times 3 = 12$

Ⓑ $4 \times 7 = 28$

Ⓒ $4 \times 8 = 32$

Ⓓ $7 \times 8 = 56$

Name

Multiply by 9

Lesson 5

ESSENTIAL QUESTION
How can multiplication and division facts with smaller numbers be applied to larger numbers?

Use known facts to help you multiply by 9.

Math in My World

Watch Tutor

Example 1

Butterflies gathered on 5 branches of a tree. There are 9 butterflies on each branch. What is the total number of butterflies on the tree?

Find $5 \times 9 = ■$.

One Way Use the Commutative Property.

Think $9 \times 5 =$ _____. ← Use a known fact and Commutative Property.

The unknown is _____. There are _____ butterflies on the tree.

Another Way Subtract from a known 10s fact.

You know $5 \times 10 =$ _____.

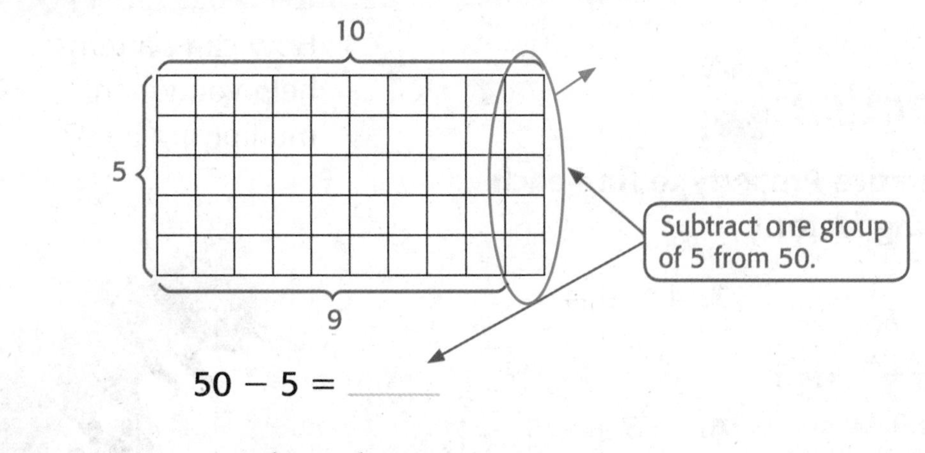

$50 - 5 =$ _____

So, $5 \times 9 =$ _____. The unknown is _____.

There are _____ butterflies on the tree.

Use patterns to help you remember the 9s facts.

Example 2

Tutor

Refer to the multiplication table. Describe the pattern among the 9s facts. Then use the pattern to find 8 × 9.

×	0	1	2	3	4	5	6	7	8	9	10
0	0	0	0	0	0	0	0	0	0	0	0
1	0	1	2	3	4	5	6	7	8	9	10
2	0	2	4	6	8	10	12	14	16	18	20
3	0	3	6	9	12	15	18	21	24	27	30
4	0	4	8	12	16	20	24	28	32	36	40
5	0	5	10	15	20	25	30	35	40	45	50
6	0	6	12	18	24	30	36	42	48	54	60
7	0	7	14	21	28	35	42	49	56	63	70
8	0	8	16	24	32	40	48	56	64	72	80
9	0	9	18	27	36	45	54	63	72	81	90
10	0	10	20	30	40	50	60	70	80	90	100

1 Shade the row green that gives the products with a factor of 9.

2 Beginning with 18, the tens digit in each product is 1 less than the factor that is not 9. The sum of the digits in each product is 9.

$2 - 1 = 1$ → $2 \times 9 = \square$ ← $1 + 8 = 9$

$3 - 1 = 2$ → $3 \times 9 = \square$ ← $2 + 7 = 9$

$4 - 1 = 3$ → $4 \times 9 = \square$ ← $3 + 6 = 9$

3 Use the pattern to find 8 × 9.

$8 - 1 = 7$ → $8 \times 9 = \square$ ← $7 + 2 = 9$

So, 8 × 9 = ____.

Talk MATH

How can patterns help you when multiplying by 9?

Guided Practice

Use the Commutative Property to find each product or missing factor.

1. 2 × 9 = ____

____ × ____ = ____

2. 4 × 9 = ____

____ × ____ = ____

3. $\begin{array}{r} 3 \\ \times\ 9 \\ \hline \end{array}$ $\begin{array}{r} 9 \\ \times\ \square \\ \hline \end{array}$

4. $\begin{array}{r} 5 \\ \times\ 9 \\ \hline \end{array}$ $\begin{array}{r} \square \\ \times\ 5 \\ \hline \end{array}$

Name ________

Independent Practice

Use the Commutative Property to find each product. Write a related multiplication fact.

5. $6 \times 9 =$ ____

____ $\times$ ____ $=$ ____

6. $10 \times 9 =$ ____

____

7. $7 \times 9 =$ ____

____

8. 8×9 ; $9 \times \square$

9. 1×9 ; $\square \times 1$

10. 3×9 ; $\square \times 3$

Draw an array for a known 10s fact. Then subtract 1 from each row to find each product.

11. $4 \times 9 =$ ____

Known fact: ____ $\times$ ____ $=$ ____

$40 -$ ____ $=$ ____

12. $5 \times 9 =$ ____

Known fact: ____ $\times$ ____ $=$ ____

$50 -$ ____ $=$ ____

Algebra **Find each unknown. Use the Commutative Property.**

13. $9 \times 10 = \blacksquare$

$? \times 9 = 90$

$\blacksquare =$ ____

$? =$ ____

14. $9 \times 2 = \blacksquare$

$? \times 9 = 18$

$\blacksquare =$ ____

$? =$ ____

15. $9 \times 8 = \blacksquare$

$? \times 9 = 72$

$\blacksquare =$ ____

$? =$ ____

Problem Solving

Mathematical PRACTICE 4 **Model Math Write a multiplication sentence with a symbol for the unknown. Then solve.**

My Work!

16. Lyle caught 3 buckets of crayfish. He put 9 crayfish in each bucket. How many crayfish did Lyle catch?

17. Cecilia needs to make 8 color copies of her babysitting flyer. The copy machine charges 9¢ a copy. How much will the 8 copies cost Cecilia?

18. There were 4 car races on Saturday and 3 on Sunday. If there were 9 cars racing in each race, how many cars raced over the two days?

HOT Problems

19. Mathematical PRACTICE 3 **Find the Error** Eva says she can find the product of 9×9 by finding $9 \times 8 = 72$, then adding 8 more. So, she says $9 \times 9 = 80$. Find and correct her mistake.

20. **Building on the Essential Question** How can the 10s facts help me solve the 9s facts? Explain.

Name ....

Lesson 5
Multiply by 9

Homework Helper

eHelp Need help? connectED.mcgraw-hill.com

Delia counted 9 petals on each flower she picked. If she picked 3 flowers, how many petals are there in all?

Find 3×9.

One Way **Subtract from a known 10s fact.**

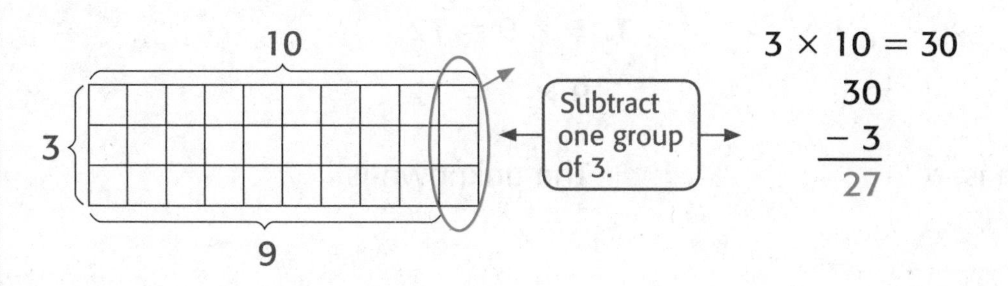

$3 \times 10 = 30$

$$30 - 3 = 27$$

Another Way **Use patterns.**

Starting with the product 18, the multiples of 9 follow a pattern. The tens digit in each product is 1 less than the factor that is not 9. The sum of the digits in the product is 9.

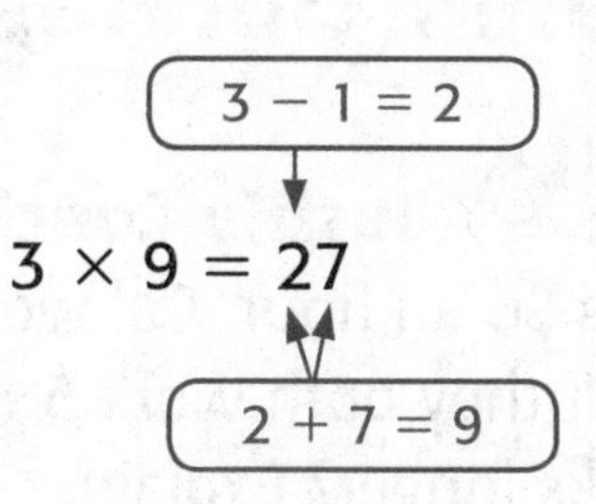

So, there are 27 petals in all.

Practice

Use the Commutative Property to find each product or missing factor.

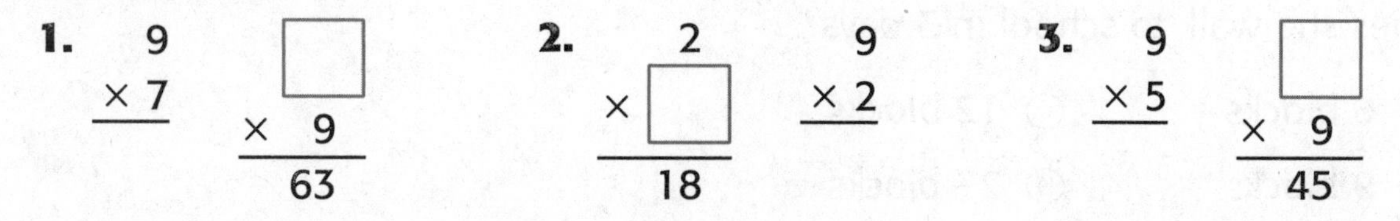

1. 9×7 ; $\square \times 9 = 63$

2. $2 \times \square = 18$; 9×2

3. 9×5 ; $\square \times 9 = 45$

Draw an array for a known 10s fact. Then subtract 1 from each row to find the product.

4. $6 \times 9 =$ ____

Known fact: ____ $\times$ ____ $=$ ____

$60 -$ ____ $=$ ____

5. $4 \times 9 =$ ____

Known fact: ____ $\times$ ____ $=$ ____

____ $-$ ____ $=$ ____

Algebra **Use the Commutative Property to find each unknown.**

6. $9 \times ■ = 36$

$■ \times 9 = 36$

The unknown is ____.

7. $■ \times 9 = 72$

$9 \times ■ = 72$

The unknown is ____.

Problem Solving

8. Mathematical PRACTICE 3 **Justify Conclusions** Ty works 9 hours a day and earns \$6 an hour. Cal works 6 hours a day and earns \$9 an hour. If they both work 5 days, who earns more money? Who works longer? Explain.

____

____

Test Practice

9. Anna lives 9 blocks from school. How many blocks does she walk to school in 3 days?

Ⓐ 6 blocks

Ⓑ 9 blocks

Ⓒ 12 blocks

Ⓓ 27 blocks

Name

Divide by 8 and 9

Lesson 6

ESSENTIAL QUESTION

How can multiplication and division facts with smaller numbers be applied to larger numbers?

Math in My World

Real World

Tools Watch Tutor

Example 1

Kyra and 8 of her friends made 63 paper airplanes. They will each take home an equal number. How many paper airplanes will each take home?

Find 63 ÷ 9.

My Drawing!

One Way **Use counters.**

Partition 63 counters into 9 equal groups. Draw the equal groups.

There are ______ counters in each group.

My drawing shows that 63 ÷ 9 = ______.

Another Way **Use repeated subtraction.**

Use repeated subtraction to find 63 ÷ 9, or $9\overline{)63}$.

①	②	③	④	⑤	⑥	⑦
63	54	45	36	27	18	9
− 9	− 9	− 9	− 9	− 9	− 9	− 9
☐	☐	☐	☐	☐	☐	☐

☐

9 is subtracted ______ times. So, 63 ÷ 9 = 7, or $9\overline{)63}$.

Each friend will take home ______ airplanes.

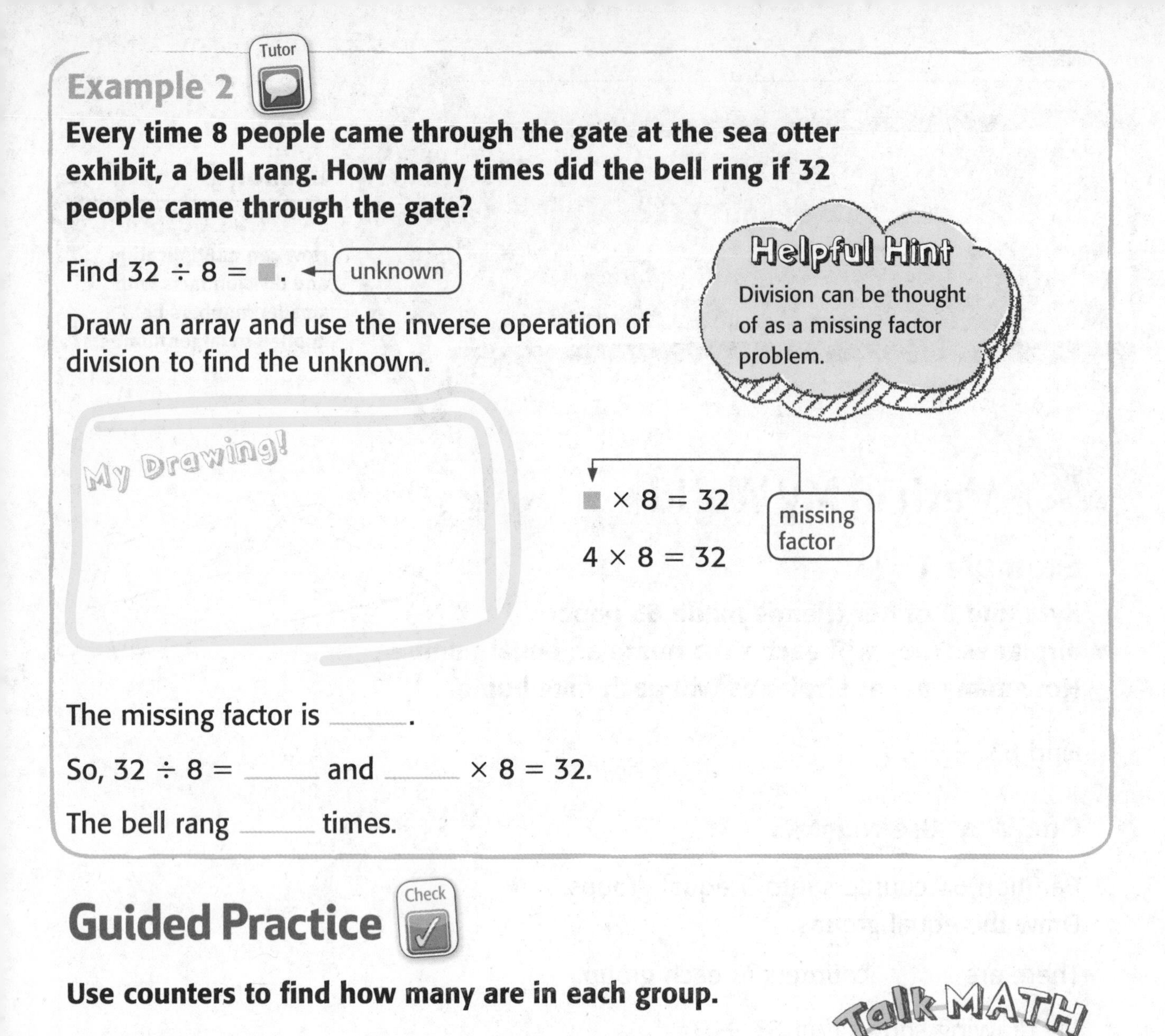

Example 2

Every time 8 people came through the gate at the sea otter exhibit, a bell rang. How many times did the bell ring if 32 people came through the gate?

Find $32 \div 8 = ■$. ← unknown

Draw an array and use the inverse operation of division to find the unknown.

Helpful Hint
Division can be thought of as a missing factor problem.

My Drawing!

$■ \times 8 = 32$ ← missing factor

$4 \times 8 = 32$

The missing factor is ______.

So, $32 \div 8 =$ ______ and ______ $\times 8 = 32$.

The bell rang ______ times.

Guided Practice

Use counters to find how many are in each group.

1. 40 counters
5 equal groups

______ in each group

So, $40 \div 5 =$ ______.

2. 54 counters
9 equal groups

______ in each group

So, $54 \div 9 =$ ______.

Talk MATH
How can multiplication facts help you to check if your division is correct?

3. Use repeated subtraction to find $48 \div 8$.

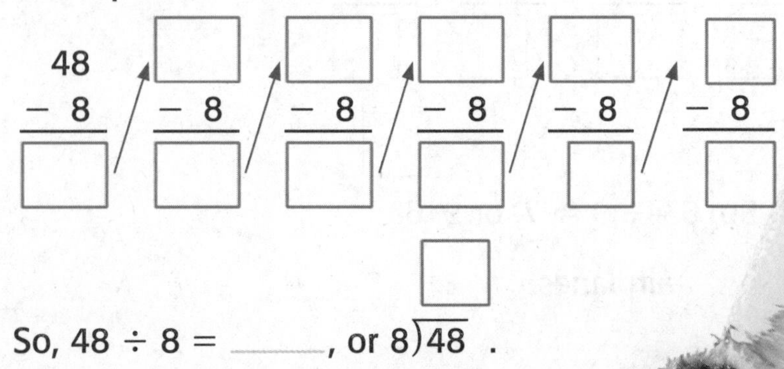

So, $48 \div 8 =$ ______, or $8\overline{)48}$.

Name ________________

Independent Practice

Use counters to find the number of equal groups or the number in each group.

4. 36 counters
9 equal groups
_____ in each group
So, $36 \div 9 =$ _____.

5. 45 counters
_____ equal groups
5 in each group
So, $45 \div$ _____ $= 5$.

6. 56 counters
8 equal groups
_____ in each group
So, $56 \div 8 =$ _____.

7. Use repeated subtraction to divide.

$64 \div 8 =$ _____ or $8\overline{)64}$ ☐

Algebra **Use the inverse operation to find each unknown. Draw an array.**

8. $40 \div 8 = ?$

$5 \times ■ = 40$

$? =$ _____

$■ =$ _____

9. $27 \div 9 = ?$

$3 \times ■ = 27$

$? =$ _____

$■ =$ _____

10. $48 \div 8 = ?$

$6 \times ■ = 48$

$? =$ _____

$■ =$ _____

Real World Problem Solving

Algebra For Exercises 11–13, write a division sentence with a symbol for the unknown. Then solve.

11. Each art project uses 9 tiles. There are 81 tiles. How many art projects can be made?

12. Forty-eight students visited the petting zoo. The students were divided evenly into eight groups. How many students were in each group?

13. Amy traveled 72 miles by bike along the coast in 9 days. She traveled the same number of miles each day. How many miles did Amy travel each day?

14. Mathematical PRACTICE 1 **Keep Trying** One baseball game has 9 innings. If 36 out of the 54 innings for the season have been played, how many games remain?

My Work!

HOT Problems

15. Mathematical PRACTICE 2 **Use Number Sense** Write 2 numbers that cannot be divided evenly by 8 or 9.

16. **Building on the Essential Question** Explain how finding a quotient can be thought of as an unknown, or missing, factor problem.

Name ....

Lesson 6

Divide by 8 and 9

Homework Helper

eHelp Need help? connectED.mcgraw-hill.com

Samantha bought a set of silverware with 48 pieces.
She divides the pieces evenly among 8 sections of a tray.
How many pieces of silverware are in each section of the tray?

One Way **Use counters to partition.**

Use 48 counters to model dividing evenly among 8 groups.

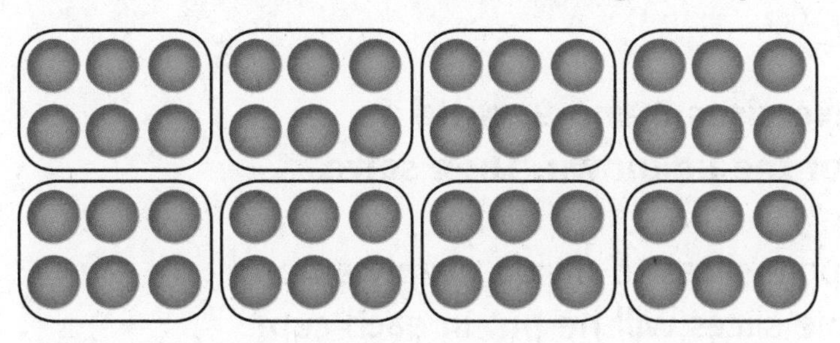

There are 6 counters in each group.

Another Way **Use repeated subtraction.**

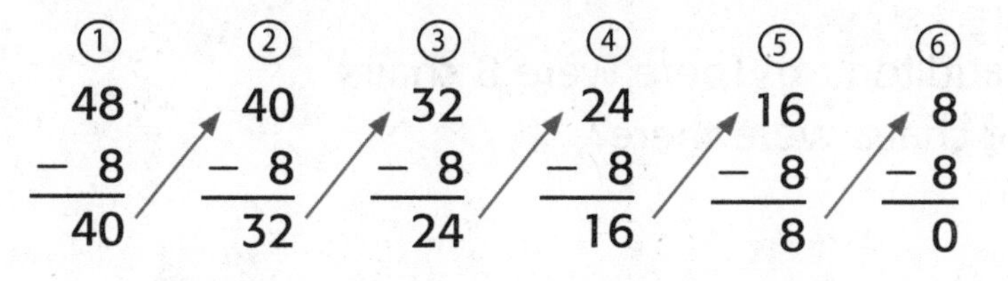

8 is subtracted 6 times.

$48 \div 8 = 6$. So, there were 6 pieces of silverware in each section.

Practice

Use counters to find the number of equal groups or the number in each group.

1. 27 counters
9 equal groups

_____ in each group

So, $27 \div 9 =$ _____.

2. 54 counters

_____ equal groups
6 in each group

So, $54 \div$ _____ $= 6$.

3. 32 counters
8 equal groups

_____ in each group

So, $32 \div 8 =$ _____.

4. Use repeated subtraction to divide.

$63 \div 9 =$ ______

Algebra Use the inverse operation to find each unknown.

5. $16 \div 8 = ■$

$■ \times 8 = 16$

$■ =$ ______

6. $■ \div 9 = 4$

$4 \times 9 = ■$

$■ =$ ______

7. $64 \div 8 = ■$

$■ \times 8 = 64$

$■ =$ ______

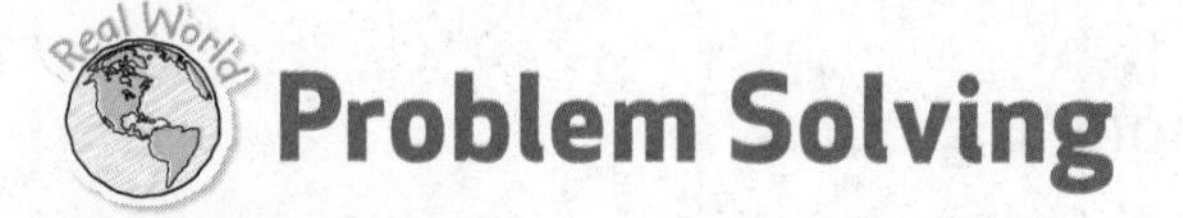

Problem Solving

Mathematical PRACTICE 2 **Use Algebra For Exercises 8 and 9, write a division sentence with a symbol for the unknown. Then solve.**

8. Michael, the chef, has 18 pineapple slices to divide evenly among 9 fruit cups. How many pineapple slices will he put in each cup?

______

9. Kayla counted 40 chairs in the auditorium. There were 8 chairs in each row. How many rows of chairs were there?

______

10. Simon sold 72 packages of popcorn for the fundraiser. There are 9 packages in each box. If he has delivered 27 packages, how many boxes does Simon have left to deliver?

______

Test Practice

11. Which number sentence uses the inverse operation to find the unknown in the division sentence $81 \div 9 = ■$?

Ⓐ $90 - 9 = 81$

Ⓑ $72 + 9 = 81$

Ⓒ $8 \times 9 = 72$

Ⓓ $9 \times 9 = 81$

Check My Progress

Vocabulary Check

1. Use the **pattern** created by the 9s facts to complete.

$1 \times 9 = 9$

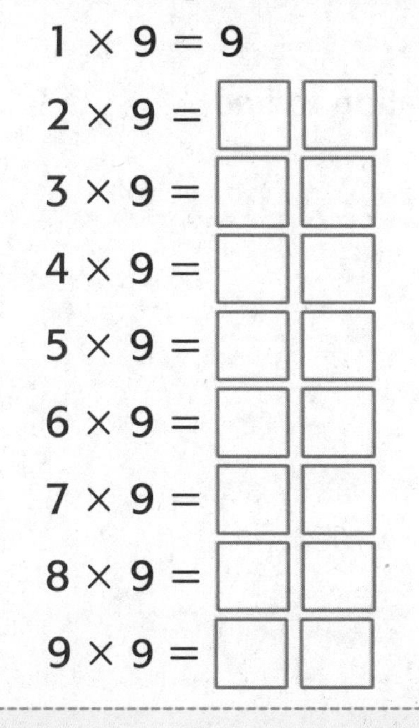

$2 \times 9 =$ ☐☐
$3 \times 9 =$ ☐☐
$4 \times 9 =$ ☐☐
$5 \times 9 =$ ☐☐
$6 \times 9 =$ ☐☐
$7 \times 9 =$ ☐☐
$8 \times 9 =$ ☐☐
$9 \times 9 =$ ☐☐

The tens digit of the product is always _____ less than the factor that is not 9.

The sum of the digits in the product is _____.

Concept Check

Double a known fact to find each product.

2. $4 \times 8 =$ _____

_____ $\times 8 =$ _____

_____ $\times 8 =$ _____

_____ + _____ = _____

3. $10 \times 8 =$ _____

Use the Commutative Property to find each product. Write a related multiplication fact.

4. $7 \times 9 =$ _____

____

5. $8 \times 5 =$ _____

____

6. $6 \times 8 =$ _____

____

Multiply.

7. $\begin{array}{r} 9 \\ \times\ 2 \\ \hline \end{array}$

8. $\begin{array}{r} 8 \\ \times\ 6 \\ \hline \end{array}$

9. $\begin{array}{r} 9 \\ \times\ 7 \\ \hline \end{array}$

10. $\begin{array}{r} 8 \\ \times\ 4 \\ \hline \end{array}$

Divide. Write a related multiplication fact.

11. $27 \div 9 =$ ____

____

12. $48 \div 8 =$ ____

____

13. $90 \div 9 =$ ____

____

14. $8\overline{)24}$ ☐

____

15. Algebra Draw an array and use the inverse operation to find each unknown.

$45 \div 9 = ?$

■ $\times 9 = 45$

$? =$ ____

■ $=$ ____

Problem Solving

16. Algebra Each side of a stadium is bordered with 8 flags. There are a total of 40 flags. How many sides are there to the stadium? Write a division sentence with a symbol for the unknown. Then solve.

____

Test Practice

17. Henry equally divided 54 pieces of paper among 9 people. To help you find the number of pieces of paper each person received, which related fact could you use?

Ⓐ $9 \times 9 = 81$

Ⓑ $9 \times 6 = 54$

Ⓒ $6 \times 3 = 18$

Ⓓ $6 + 9 = 15$

Name

Real World

Problem-Solving Investigation

STRATEGY: Make an Organized List

Lesson 7

ESSENTIAL QUESTION
How can multiplication and division facts with smaller numbers be applied to larger numbers?

Learn the Strategy

Tutor

Eva is giving away 8 stamps. Each friend will get an equal number of stamps. How many friends could get the stamps?

1 Understand

What facts do you know?

Eva is giving away _____ stamps to her friends.

She will give each friend an equal number of stamps.

What do you need to find?

the number of _______ to which she could give stamps

2 Plan

I will make an organized list to see the ways I can divide _____ evenly.

3 Solve

Divide the total number of stamps by the numbers 1 through 8.

So, Eva can give

_____, _____, _____, or _____ friends an equal number of stamps.

Number of Friends	Number of Stamps
1	$8 \div 1 = 8$
2	$8 \div 2 = 4$
3	$8 \div 3$ not possible
4	$8 \div 4 = 2$
5	$8 \div 5$ not possible
6	$8 \div 6$ not possible
7	$8 \div 7$ not possible
8	$8 \div 8 = 1$

4 Check

Does your answer make sense? Explain.

____

Practice the Strategy

Ian is numbering the pages of his journal 1 through 48. He wants to start a new section after every 8 pages. On what pages will each new section begin?

1 Understand

What facts do you know?

____

____

What do you need to find?

____

2 Plan

____

____

3 Solve

4 Check

Does your answer make sense? Explain.

____

Name

Apply the Strategy

Solve each problem by making an organized list.

1. Mathematical PRACTICE 5 **Use Math Tools** Gracie bought a goldfish at the pet store. She had only one nickel, one dime, and one quarter in her wallet. How much could her goldfish have cost?

____

2. Mathematical PRACTICE 8 **Look for a Pattern** Stuart wants to know how many times he gets an even number as a product of his 6s multiplication facts. When multiplying by 6, are the products odd or even?

____

____

Is the same true about quotients when dividing by 6? Explain.

____

____

____

My Work!

CCSS

Review the Strategies

Use any strategy to solve each problem.

- Determine extra or missing information.
- Make a table.
- Look for a pattern.
- Use models.

3. Paula put 6 books on one side of a balance scale that is 3 feet long. On the other side, she put 5 books and her baseball glove. The sides balanced. Each book weighs 3 pounds. How much does her glove weigh?

____

My Work!

4. Jonas has 6 fish tanks with 6 fish in each tank. After he sold some, he had 27 fish left. How much did each fish cost if he made $63?

____

5. Mathematical PRACTICE 4 **Model Math** Angelina's mother knits gloves and mittens that are red, blue, green, or brown. How many different colors of gloves and mittens can she knit? Explain.

____

____

6. Mathematical PRACTICE 1 **Make a Plan** A group of 16 people want to go to the zoo. It costs $30 for each group of 6 people. Otherwise, it costs $6 per person. How much does it cost for 16 people?

____

____

____

MY Homework

Lesson 7
Problem Solving: Make an Organized List

Homework Helper

eHelp

Need help? connectED.mcgraw-hill.com

Harold, Nina, Adam, and Rachel sit at the same table. Students must go to the drinking fountain in groups of 3. What possible combinations of these students can go to the drinking fountain together?

1 Understand

What facts do you know?

Harold, Nina, Adam, and Rachel sit together.
Students go to the drinking fountain in groups of 3.

What do you need to find?

the possible combinations of students that could go to the drinking fountain together

2 Plan

I will make an organized list of the possible combinations.

3 Solve

I will list the students in different groups of 3. So, there are four possible combinations of students who can go to the drinking fountain together.

Harold, Nina, Adam
Nina, Adam, Rachel
Harold, Adam, Rachel
Harold, Nina, Rachel

4 Check

Does the answer make sense?

Checking my list, I see that each student's name is listed the same number of times, and one is left out each time.
So, the answer is reasonable.

Real World!

Problem Solving

Solve each problem by making an organized list.

My Work!

1. Paul needs 34 cents. He has only dimes and pennies. How many ways can he make 34 cents using both kinds of coin? Explain.

2. Camille rides a bus to work. To get downtown, she can ride any bus number between 11 and 34, that can be divided evenly by 3, and is an even number. Which numbers are the buses that Camille could ride to work?

3. Bruce is grocery shopping. He can go to the deli, the bakery, and the dairy section in any order. How many possibilities are there for the order in which Bruce can do his shopping?

4. Flora has 5 boxes that increase in size. In the first box she packs 4 books. In each box after that, she packs 3 more books than the box before. How many books does Flora pack in the last box?

5. Mathematical PRACTICE 3 **Justify Conclusions** A mouse makes itself a new nest every 2 weeks. It uses 8 large leaves to line each nest. How many leaves will the mouse have used after 6 weeks? Explain.

Name ....

Multiply by 11 and 12

Lesson 8

ESSENTIAL QUESTION
How can multiplication and division facts with smaller numbers be applied to larger numbers?

Math in My World

Tools Watch Tutor

Example 1

There are 11 straws in a package. Helen bought 4 packages. How many straws does Helen have?

Find 4 × 11.

One Way Use patterns.

Study the pattern in the table.

Multiply by 11	
Factors	**Product**
1 × 11	11
2 × 11	22
3 × 11	33
4 × 11	44
5 × 11	55
6 × 11	66
7 × 11	77
8 × 11	88
9 × 11	99

The product of 11 and a single-digit factor has two digits. Each digit in the product is the same as the factor that is not 11.

So, 4 × 11 = ______.

Another Way Use models.

Model 4 rows of 11 counters.

Draw your result.

My Drawing!

Use repeated addition.

11 + 11 + 11 + 11 = ______

The model shows 4 × 11 = ______.

Helen has ______ straws.

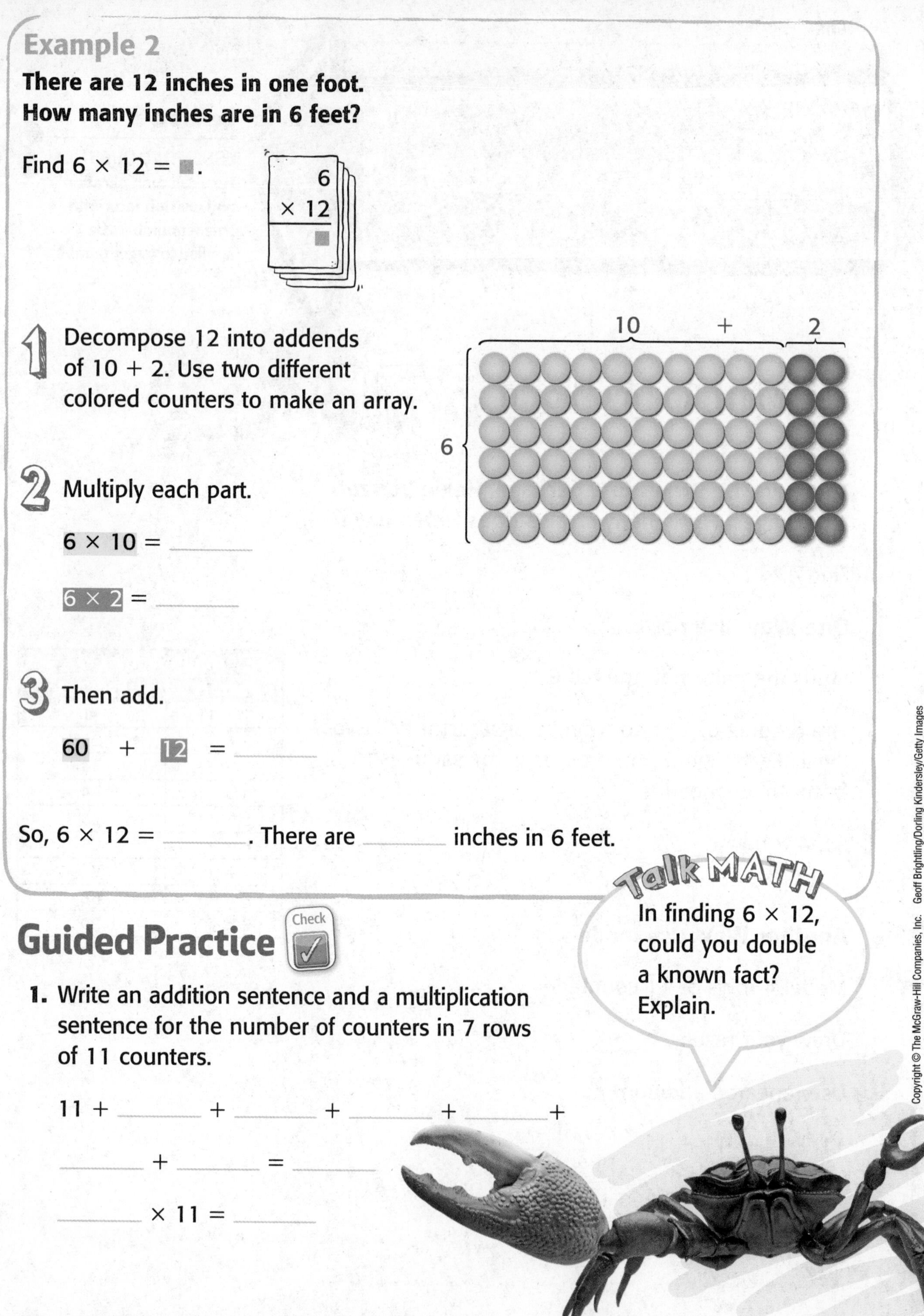

Example 2

There are 12 inches in one foot.
How many inches are in 6 feet?

Find $6 \times 12 = ■$.

$$\begin{array}{r} 6 \\ \times\ 12 \\ \hline ■ \end{array}$$

1 Decompose 12 into addends of 10 + 2. Use two different colored counters to make an array.

2 Multiply each part.

$6 \times 10 =$ ______

$6 \times 2 =$ ______

3 Then add.

60 + 12 = ______

So, $6 \times 12 =$ ______. There are ______ inches in 6 feet.

Talk MATH

In finding 6×12, could you double a known fact? Explain.

Guided Practice

Check

1. Write an addition sentence and a multiplication sentence for the number of counters in 7 rows of 11 counters.

11 + ______ + ______ + ______ + ______ +

______ + ______ = ______

______ $\times 11 =$ ______

Name ....

CCSS

Independent Practice

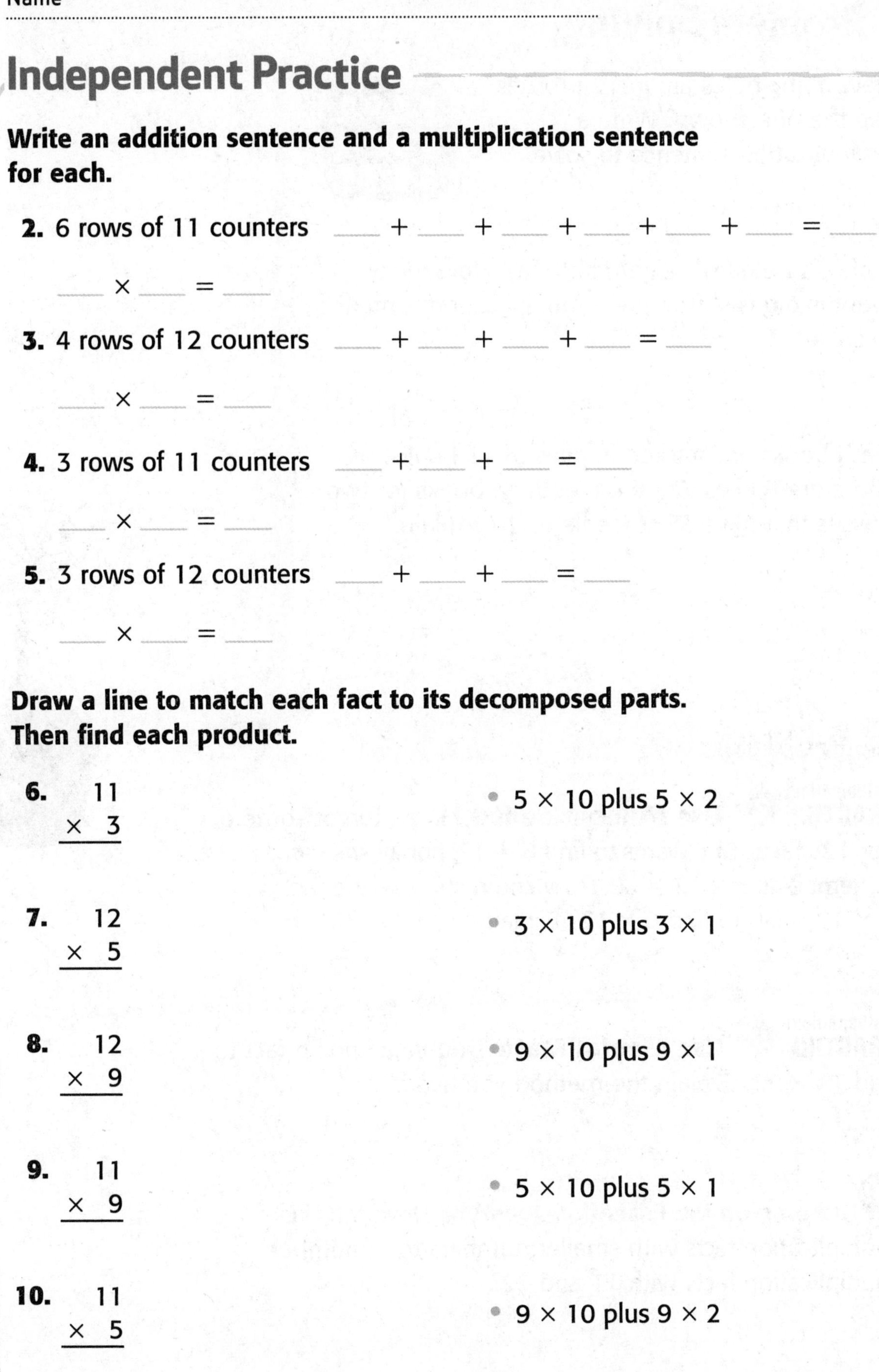

Write an addition sentence and a multiplication sentence for each.

2. 6 rows of 11 counters ___ + ___ + ___ + ___ + ___ + ___ = ___

___ × ___ = ___

3. 4 rows of 12 counters ___ + ___ + ___ + ___ = ___

___ × ___ = ___

4. 3 rows of 11 counters ___ + ___ + ___ = ___

___ × ___ = ___

5. 3 rows of 12 counters ___ + ___ + ___ = ___

___ × ___ = ___

Draw a line to match each fact to its decomposed parts. Then find each product.

6. 11×3 • 5×10 plus 5×2

7. 12×5 • 3×10 plus 3×1

8. 12×9 • 9×10 plus 9×1

9. 11×9 • 5×10 plus 5×1

10. 11×5 • 9×10 plus 9×2

Problem Solving

11. How many holes are in 12 pretzels like the one shown? Write a multiplication sentence to solve.

12. Today is Bethany's eighth birthday. How many months old is she? Write a multiplication sentence to solve.

13. Math books are stacked in piles of 11 books. There are 6 piles. Are there enough books for two classes that have 35 students each? Explain.

My Work!

HOT Problems

14. Mathematical PRACTICE 2 **Use Number Sense** Maren forgot some of her 12s facts. She wants to find 6×12, but all she can remember is $5 \times 12 = 60$. How could she use the fact $5 \times 12 = 60$ to find 6×12? Explain.

15. Mathematical PRACTICE 5 **Use Mental Math** Double a known fact to find 12×11. Explain the method you used.

16. **Building on the Essential Question** How can I use multiplication facts with smaller numbers to remember multiplication facts with 11 and 12?

MY Homework

Lesson 8
Multiply by 11 and 12

Homework Helper

eHelp Need help? connectED.mcgraw-hill.com

Felisa can put 6 photos on each page of her scrapbook. How many photos can she place altogether on 11 pages?

Find 6×11. Write multiplication vertically or horizontally.

One Way **Use repeated addition.**

$6 \times 11 =$

$11 + 11 + 11 + 11 + 11 + 11 = 66$

Another Way **Decompose 11 into 10 + 1.**

Decompose 11 into the addends 10 + 1.

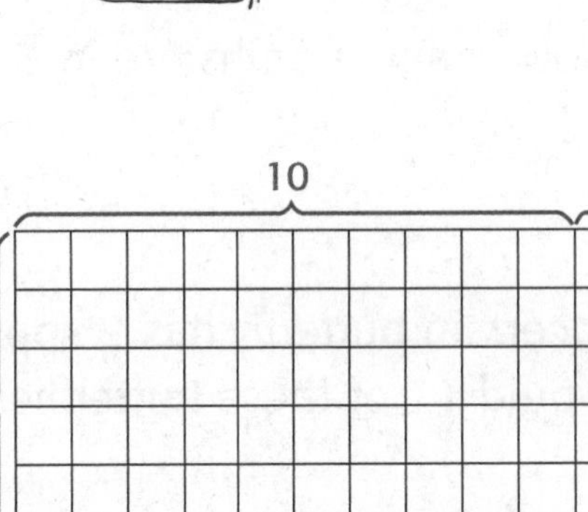

1. Multiply each part.
$6 \times 10 = 60$
$6 \times 1 = 6$

2. Add.
$60 + 6 = 66$

So, $6 \times 11 = 66$.
Felisa can place 66 photos on 11 pages in her scrapbook.

Practice

Write an addition sentence and a multiplication sentence for each.

1. 5 rows of 11 counters ____ + ____ + ____ + ____ + ____ = ____

____ × ____ = ____

2. 3 rows of 12 counters ____ + ____ + ____ = ____

____ × ____ = ____

Use repeated addition to find each product.

3. $3 \times 11 =$ ______

4. $8 \times 12 =$ ______

Decompose one factor to find each product.

5. $5 \times 12 =$ ______

6. $7 \times 11 =$ ______

Real World

Problem Solving

7. How many eggs are there altogether in 7 dozen eggs? (*Hint*: 1 dozen = 12)

______

8. How many months are in 6 years?

______

9. A certain butterfly has 9 spots. How many spots would 11 of these butterflies have?

______

10. Mathematical PRACTICE 1 **Keep Trying** Luke can run a mile in 7 minutes. Colleen can run a mile in 5 minutes. At this rate, how much longer would it take Luke to run 11 miles than it would take Colleen to run 11 miles?

______

My Work!

Test Practice

11. Which number sentence does *not* belong with the other three?

Ⓐ $4 \times 12 = 48$

Ⓑ $12 \times 4 = 48$

Ⓒ $4 + 12 = 16$

Ⓓ $12 + 12 + 12 + 12 = 48$

Name ....

Divide by 11 and 12

Lesson 9

ESSENTIAL QUESTION
How can multiplication and division facts with smaller numbers be applied to larger numbers?

Math in My World

Tools Tutor

Example 1

For a field trip, 33 students went to the science museum. There were 11 microscopes. Each was used by an equal number of students in a group. How many students were in each group?

Find $33 \div 11$, or $11\overline{)33}$.

Partition 33 counters into 11 equal groups. Draw the equal groups.

Helpful Hint
Division can be thought of as partitioning into equal groups.

My Drawing!

There are ______ counters in each group. ☐

My drawing shows that $33 \div 11 =$ ☐, or $11\overline{)33}$.

There were ______ students in each group.

When you divide by 11 and 12, it is often quicker to use the inverse operation of multiplication.

Example 2

Maurice bought 48 eggs. They were packaged in cartons of 12. How many cartons did Maurice buy?

Find the unknown in 48 ÷ 12 = ■.

Think of division as a missing factor problem.

12 × ■ = 48

The missing factor is 4.

12 × 4 = 48

So, 48 ÷ 12 = ______. The unknown is ______.

Maurice bought ______ cartons of eggs.

Guided Practice

Use counters to find the number in each group.

1. 44 counters

11 equal groups

______ in each group

So, 44 ÷ 11 = ______.

2. 36 counters

12 equal groups

______ in each group

So, 36 ÷ 12 = ______.

Talk MATH

Describe the pattern seen in the quotients when numbers such as 66, 55, and 44 are divided by 11.

3. Use repeated subtraction to divide.

60 ÷ 12 = ______

60 − 12 = ☐	☐ − 12 = ☐	☐ − 12 = ☐	☐ − 12 = ☐	☐ − 12 = ☐

Name ____________

Independent Practice

Use counters to find the number of equal groups or the number in each group.

4. 22 counters

11 equal groups

____ in each group

So, $22 \div 11 =$ ____.

5. 72 counters

12 equal groups

____ in each group

So, $72 \div$ ____ $=$ ____.

6. 84 counters

____ equal groups

7 in each group

So, ____ $\div$ ____ $= 7$.

Use repeated subtraction to divide.

7. $55 \div 11 =$ ____

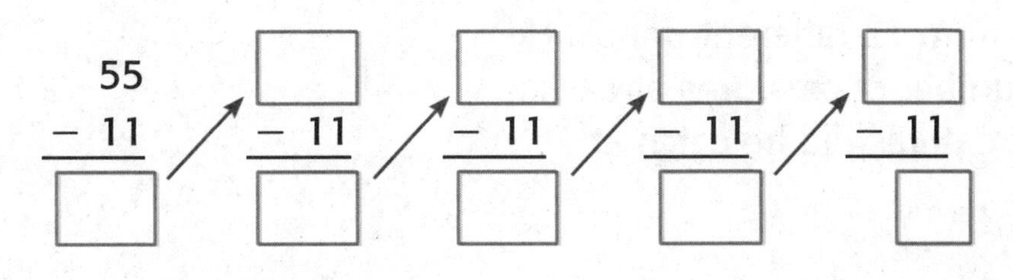

Algebra **Use the inverse operation to find each unknown.**

8. $77 \div 11 = \blacksquare$

$11 \times \blacksquare = 77$

The unknown is ____.

9. $99 \div 11 = \blacksquare$

$11 \times \blacksquare = 99$

The unknown is ____.

10. $44 \div 11 = \blacksquare$

$11 \times \blacksquare = 44$

The unknown is ____.

11. $12\overline{)48}$ (quotient: $\blacksquare$)

$12 \times \blacksquare = 48$

The unknown is ____.

12. $12\overline{)96}$ (quotient: $\blacksquare$)

$12 \times \blacksquare = 96$

The unknown is ____.

13. $11\overline{)88}$ (quotient: $\blacksquare$)

$11 \times \blacksquare = 88$

The unknown is ____.

14. $33 \div 3 = \blacksquare$

The unknown is ____.

15. $66 \div 11 = \blacksquare$

The unknown is ____.

16. $36 \div 12 = \blacksquare$

The unknown is ____.

Problem Solving

17. Mathematical PRACTICE 4 **Model Math** Sharon traveled 96 miles in her truck on 12 gallons of gasoline. How many miles did she drive on each gallon? Write a division sentence to solve.

My Work!

18. Vickram took 33 photographs of his pet ferret. He sent an equal number of photographs to each of his 11 friends. How many photographs will each friend receive? Write a division sentence to solve.

19. A river otter caught 4 frogs, 19 crayfish, and 13 other small fish from 12 different ponds. He caught the same number of creatures at each pond. How many creatures did he catch at each pond?

HOT Problems

20. Mathematical PRACTICE 1 **Make Sense of Problems** How could the multiplication fact $4 \times 12 = 48$ be used to find $96 \div 12$?

21. **Building on the Essential Question** How can I think of dividing by 11 or 12 as an unknown factor problem?

Name ....

Lesson 9
Divide by 11 and 12

Homework Helper

eHelp Need help? connectED.mcgraw-hill.com

Jolene's little sister, Camille, is 36 months old. How old is Camille in years?

Find $36 \div 12$.

Think of division as a missing factor problem.

$12 \times ? = 36$

The missing factor is 3.

$12 \times 3 = 36$

So, $36 \div 12 = 3$. Camille is 3 years old.

Check using models. Partitioning 36 counters into 12 groups will result in 3 counters in each group.

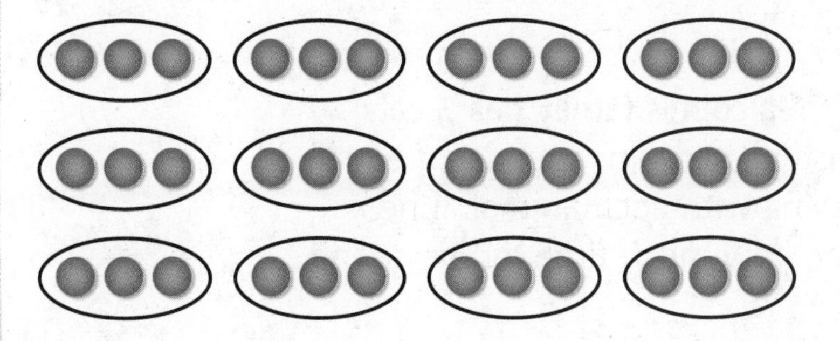

Practice

Find the number of equal groups.

1. 77 counters
11 in each group

There will be ______ groups.

2. 60 counters
12 in each group

There will be ______ groups.

Use repeated subtraction to divide.

3. 48 ÷ 12 = ______ **4.** 33 ÷ 11 = ______

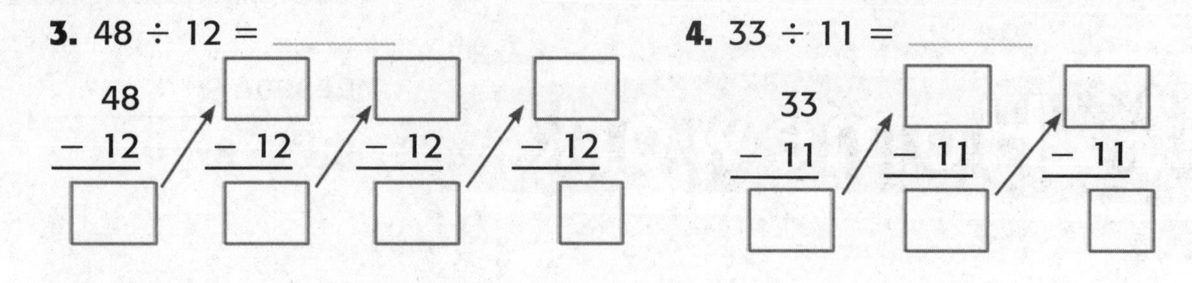

Algebra Use the inverse operation to find each unknown.

5. 88 ÷ 11 = ■

11 × ______ = 88

The unknown is ______.

6. $12\overline{)72}$ ■

12 × ______ = 72

The unknown is ______.

Problem Solving

7. Tim is saving to buy a new cell phone that costs $84. If he saves $12 each month, in how many months will he have $84?

______

8. A grocery store has 60 boxes of cereal. There are 12 different kinds of cereal. If there are an equal number of boxes of each kind, how many boxes of each kind are there?

______

9. Mathematical PRACTICE 1 **Keep Trying** Malcolm's family has 3 cats, 2 dogs, 2 rabbits, and 4 hamsters. Malcolm spends an equal amount of time each day playing with each animal. If he spends 55 minutes altogether, how much time did he spend with each animal?

______

Test Practice

10. Which number sentence can you use to check your answer when finding 44 ÷ 11?

Ⓐ 4 + 11 = 15

Ⓑ 44 − 11 = 33

Ⓒ 4 × 11 = 44

Ⓓ 44 + 11 = 55

Review

Chapter 8

Apply Multiplication and Division

Vocabulary Check

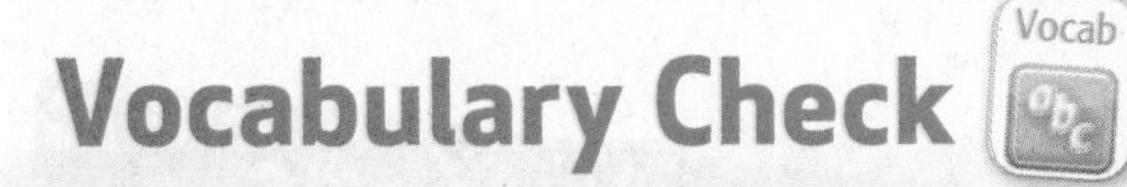

Use the clues and the word bank below to solve the puzzle.

Commutative **known facts** **pattern** **related facts**

Across

1. The 11s facts show a ______. When a single digit number is multiplied by 11, the product is the digit repeated.
2. Basic facts using the same three numbers. Sometimes called a fact family.

Down

3. Facts which you have memorized.
4. The property of multiplication which states that the order in which two numbers are multiplied does not change the product.

Concept Check

Check

Algebra Use the inverse operation to find each unknown.

5. $30 \div 6 = ■$

$6 \times ■ = 30$

The unknown is ____.

6. $28 \div 7 = ■$

$7 \times ■ = 28$

The unknown is ____.

7. $48 \div 6 = ■$

$6 \times ■ = 48$

The unknown is ____.

Double a known fact to find each product. Draw an array.

8. $8 \times 7 =$ ____

$4 \times 7 =$ ____

____ × ____ = ____

28 + ____ = ____

9. $6 \times 9 =$ ____

Write an addition sentence and a multiplication sentence for each.

10. 5 rows of 11 counters ____ + ____ + ____ + ____ + ____ = ____

____ × ____ = ____

11. 6 rows of 12 counters ____ + ____ + ____ + ____ + ____ + ____ = ____

____ × ____ = ____

12. 3 rows of 8 counters ____ + ____ + ____ = ____

____ × ____ = ____

13. 7 rows of 11 counters

____ + ____ + ____ + ____ + ____ + ____ + ____ = ____

____ × ____ = ____

Name

Problem Solving

14. Bev notices this heart-shaped button has 4 holes. She needs 11 of these buttons for a project. How many button holes will there be altogether? Write a multiplication sentence to solve.

15. The array is a model for $5 \times 9 = 45$. Write the division sentence that is modeled by the array.

16. Chandler works 4 hours each week. How many weeks will it take him to work 36 hours? Write a division sentence to solve.

My Work!

17. Mrs. King took 12 crates packed with lunch boxes on a field trip. Each crate had 6 lunch boxes inside. How many lunch boxes were there altogether? Write a multiplication sentence to solve.

Test Practice

18. How much will these 4 paperback books cost altogether?

Ⓐ \$7 Ⓑ \$14 Ⓒ \$21 Ⓓ \$28

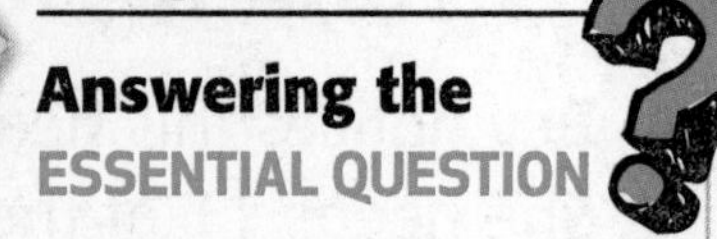

Use what you learned about multiplication and division to complete the graphic organizer.

Double a Known Fact

Repeated Subtraction

ESSENTIAL QUESTION

How can multiplication and division facts with smaller numbers be applied to larger numbers?

Models

Properties

Reflect on the ESSENTIAL QUESTION **Write your answer below.**

Chapter

9 Properties and Equations

ESSENTIAL QUESTION

How are properties and equations used to group numbers?

Watch a video!

Watch

MY Common Core State Standards

CCSS

Operations and Algebraic Thinking

3.OA.5 Apply properties of operations as strategies to multiply and divide.

3.OA.7 Fluently multiply and divide within 100, using strategies such as the relationship between multiplication and division (e.g., knowing that $8 \times 5 = 40$, one knows $40 \div 5 = 8$) or properties of operations. By the end of Grade 3, know from memory all products of two one-digit numbers.

3.OA.8 Solve two-step word problems using the four operations. Represent these problems using equations with a letter standing for the unknown quantity. Assess the reasonableness of answers using mental computation and estimation strategies including rounding.

Hmm, I don't think this will be too bad!

Standards for Mathematical PRACTICE

1. Make sense of problems and persevere in solving them.
2. Reason abstractly and quantitatively.
3. Construct viable arguments and critique the reasoning of others.
4. Model with mathematics.
5. Use appropriate tools strategically.
6. Attend to precision.
7. Look for and make use of structure.
8. Look for and express regularity in repeated reasoning.

= focused on in this chapter

Name ....

Check — Go online to take the Readiness Quiz

Algebra **Find each unknown.**

1. 8 + ■ = 11

   The unknown is ______.

2. ■ × 5 = 20

   The unknown is ______.

3. 36 ÷ 6 = ■

   The unknown is ______.

4. $\begin{array}{r} 15 \\ -\ ■ \\ \hline 6 \end{array}$

   The unknown is ______.

5. $\begin{array}{r} 9 \\ \times\ ■ \\ \hline 27 \end{array}$

   The unknown is ______.

6. $7\overline{)42}$ with ■ as the quotient

   The unknown is ______.

7. Use the number sentence 12 + 15 + ■ = 36 to find how many books Tony read in August.

Summer Reading Club	
Month	Number of Books Read
June	12
July	15
August	■

   The unknown is ______ books.

8. Circle the property which is represented by 6 + 5 = 5 + 6.

   Associative Property of Addition

   Commutative Property of Addition

   Identity Property of Addition

9. Abia sold 1 more candle than Mel. Together, they sold 15 candles. Draw a picture that shows how many candles they each sold.

10. Daniela spent $20 at the grocery store and $15 at the gas station. How much did she spend in all? Write a number sentence with a symbol for the unknown. Solve.

    ____

Shade the boxes to show the problems you answered correctly.

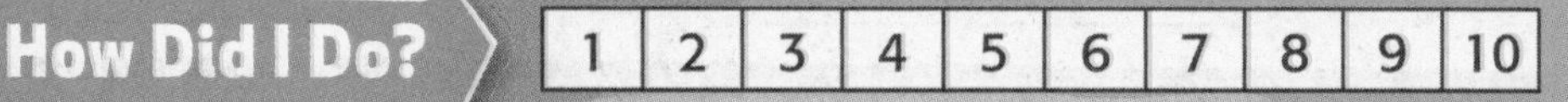

Name

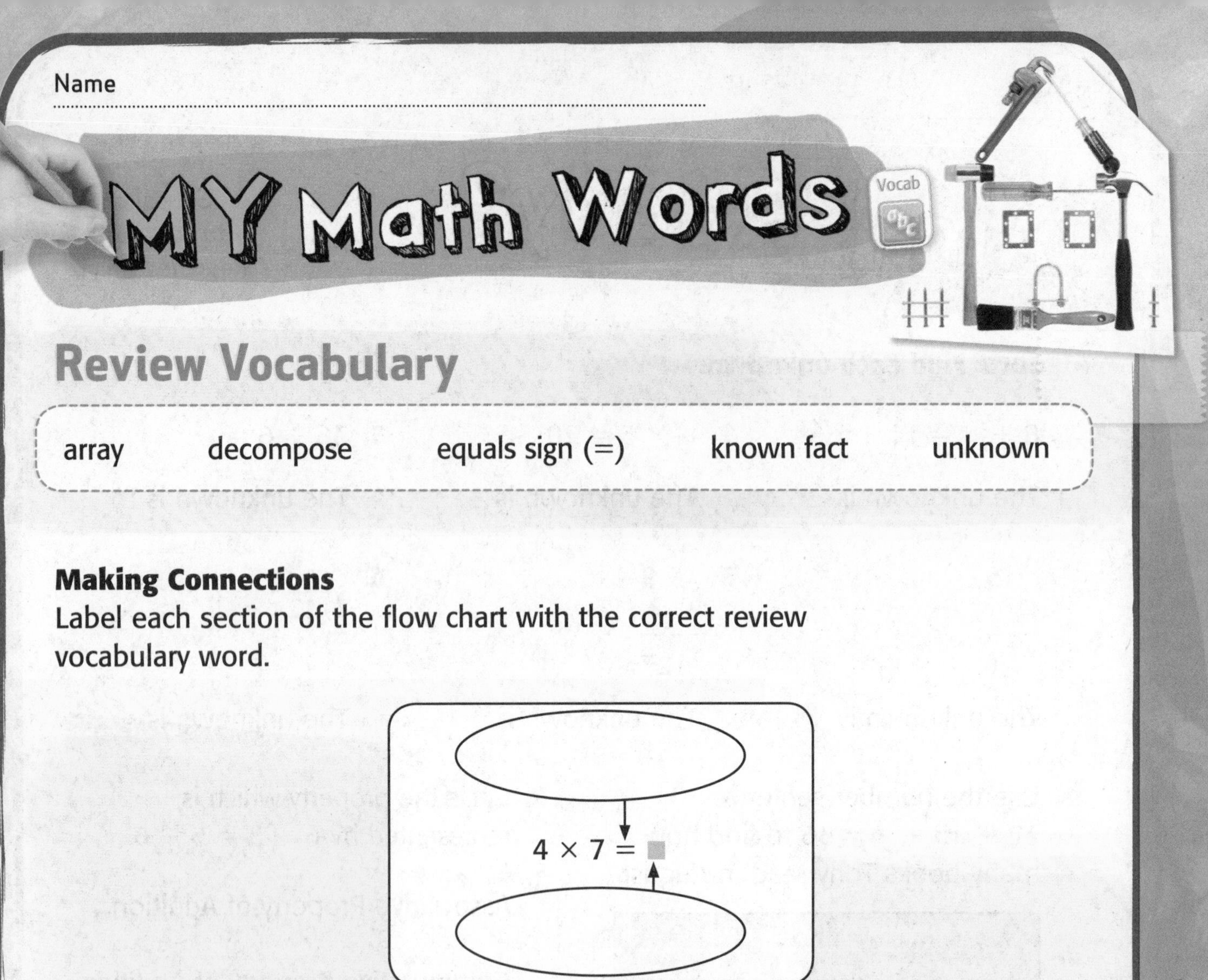

Review Vocabulary

array decompose equals sign (=) known fact unknown

Making Connections
Label each section of the flow chart with the correct review vocabulary word.

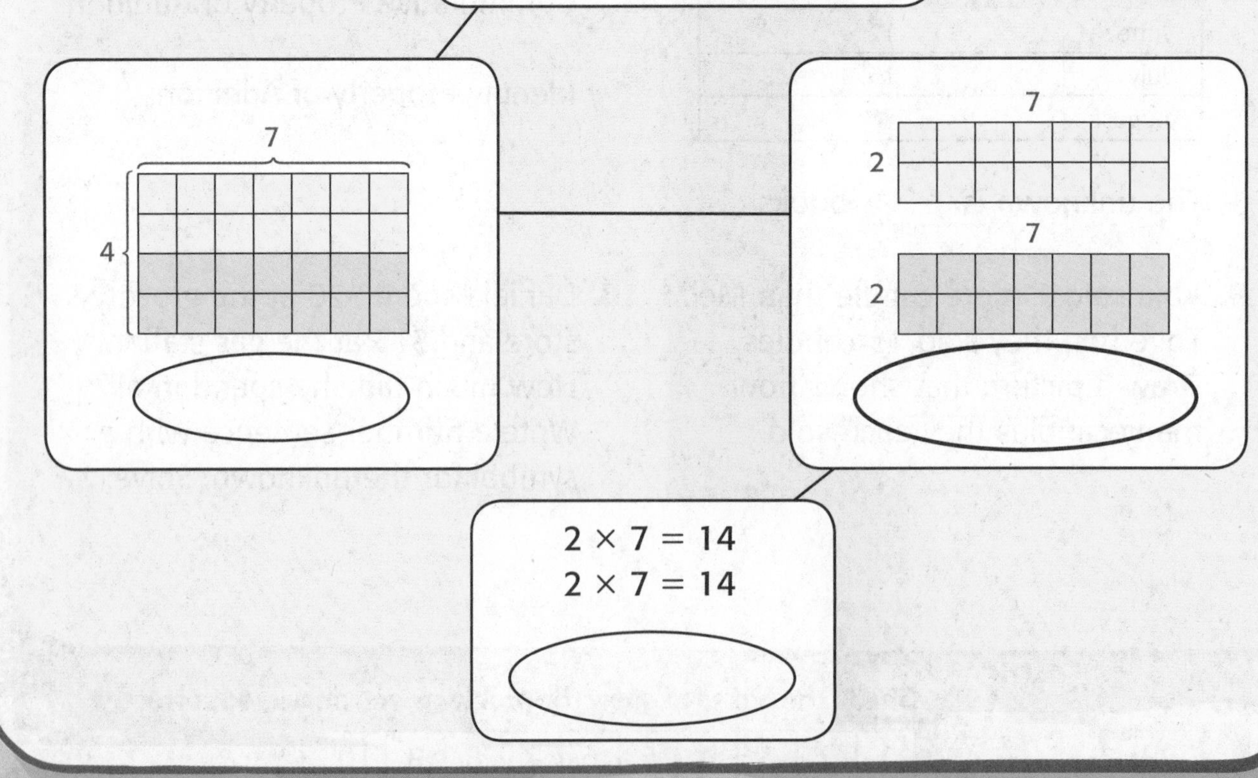

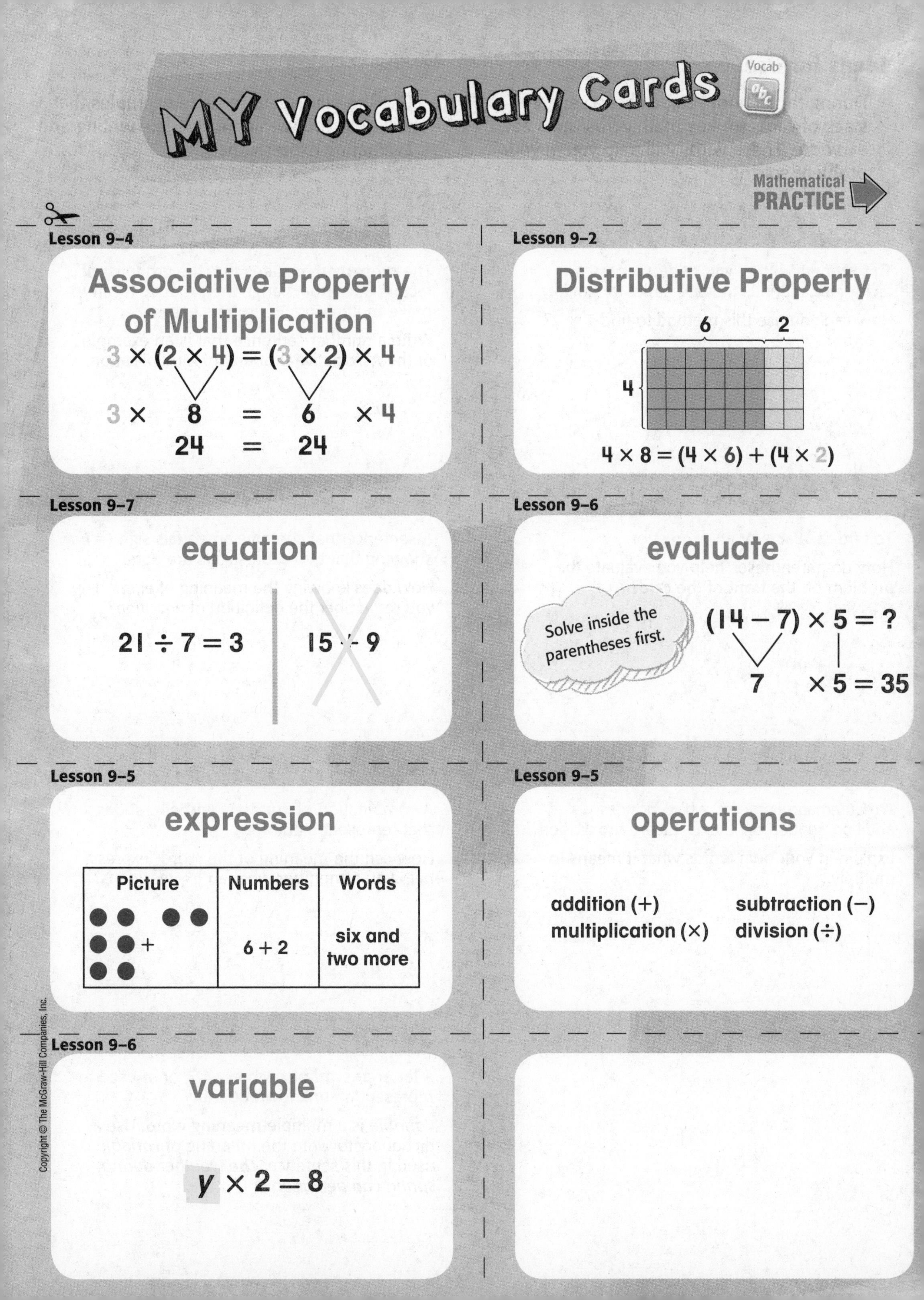

MY Vocabulary Cards
Vocab
abc
Mathematical PRACTICE
Lesson 9-4
Associative Property of Multiplication
3 × (2 × 4) = (3 × 2) × 4
3 × 8 = 6 × 4
24 = 24
Lesson 9-2
Distributive Property
6
2
4
4 × 8 = (4 × 6) + (4 × 2)
Lesson 9-7
equation
21 ÷ 7 = 3
15 ÷ 9
Lesson 9-6
evaluate
Solve inside the parentheses first.
(14 − 7) × 5 = ?
7 × 5 = 35
Lesson 9-5
expression
Picture
Numbers
Words
6 + 2
six and two more
Lesson 9-5
operations
addition (+)
subtraction (−)
multiplication (×)
division (÷)
Lesson 9-6
variable
y × 2 = 8
Copyright © The McGraw-Hill Companies, Inc.

Ideas for Use

- During this school year, create a separate stack of cards for key math verbs, such as *evaluate.* These verbs will help you in your problem solving.
- Use the blank card to write examples that will help you with concepts like writing and evaluating expressions.

This property allows you to decompose one factor into addends that are easier to multiply.

How can you use this method to find 7×2?

The property that states that the grouping of factors does not change the product. It can make multiplying 3 numbers easier.

Write a number sentence that is an example of the Associative Property of Multiplication.

To find the value of an expression.

How do parentheses help you evaluate the problem on the front of the card?

A sentence that contains an equals sign (=), showing that two expressions are equal.

How does knowing the meaning of *equal* help you remember the definition of *equation?*

A mathematical process, which includes addition, subtraction, multiplication, and division.

Explain in your own words what it means to multiply.

A combination of numbers and operations that represent a quantity.

How can the meaning of the word "express" help you remember what an expression is?

A letter or symbol, such as ■, ?, or *m* used to represent an unknown quantity.

Variable is a multiple-meaning word. Use a dictionary to write the meaning of *variable* used in this sentence. *The weather during spring can be variable.*

FOLDABLES® Follow the steps on the back to make your Foldable.

Commutative Property of Multiplication

Distributive Property

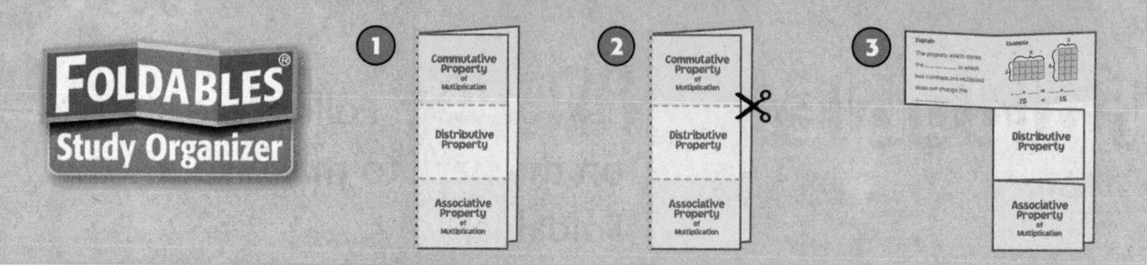

Explain

The property which states the ________ in which two numbers are multiplied does not change the ________.

Example

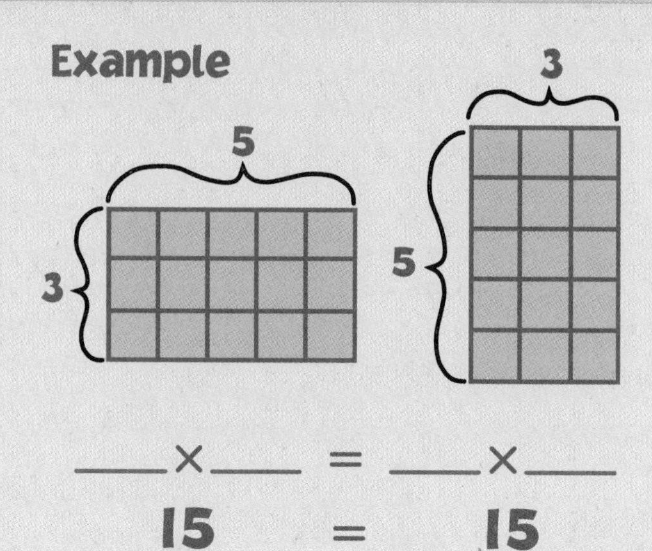

Explain

The property which allows me to ____________ factors into addends that are easier to work with.

Example

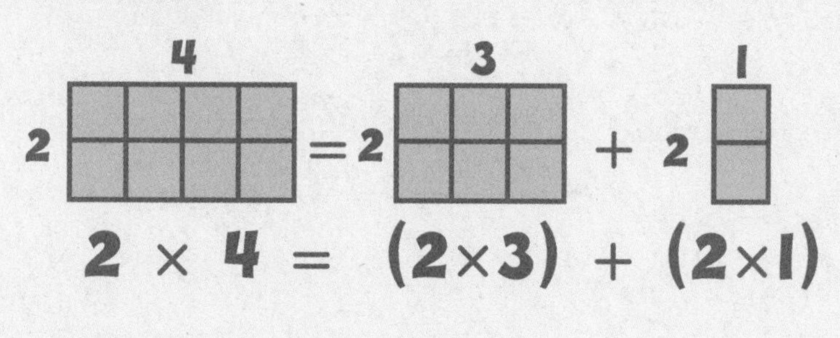

$= \_\_\_ + \_\_\_$

$= \_\_\_$

Explain

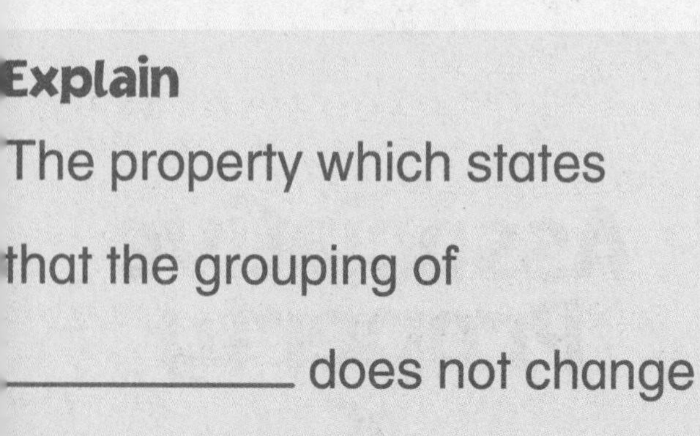

The property which states that the grouping of ________ does not change the ________.

Example

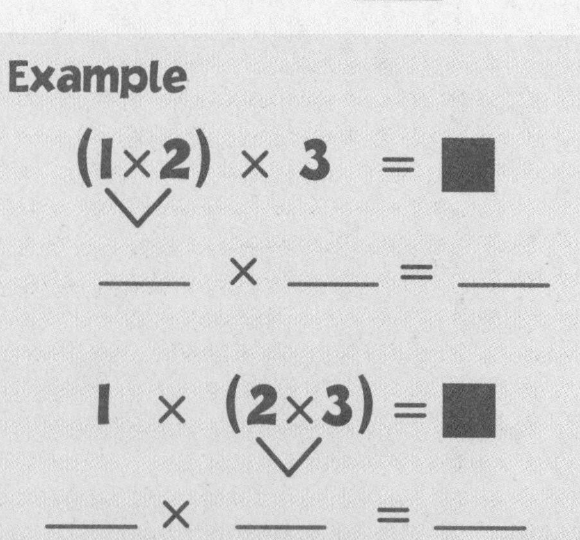

Name ....

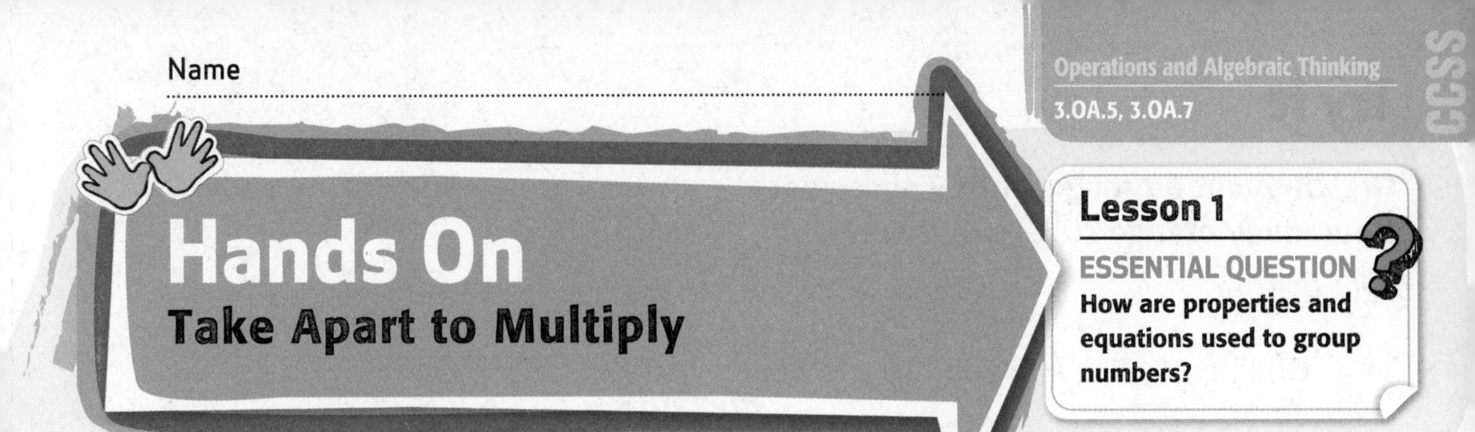

Hands On
Take Apart to Multiply

Lesson 1

ESSENTIAL QUESTION
How are properties and equations used to group numbers?

When you take apart, or decompose, a factor, you have smaller numbers that are easier to multiply.

Build It

Find 4×7.

1 Model 4×7.

Use color tiles to make a 4×7 array. Draw the array.

My Drawing!

2 Decompose one factor.

- Take apart the 7.
- Separate 7 columns into 5 columns + 2 columns.

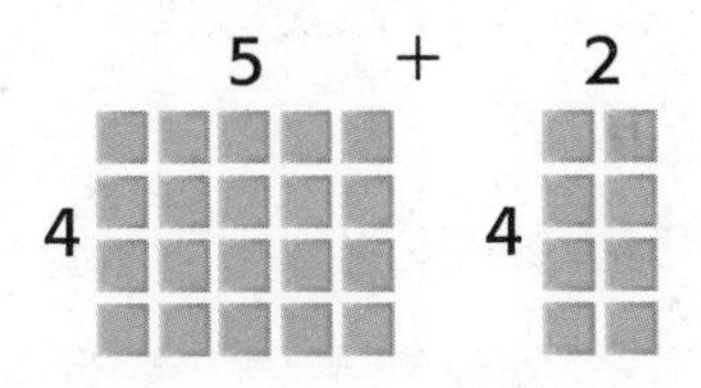

3 Find the products of each part. Then add.

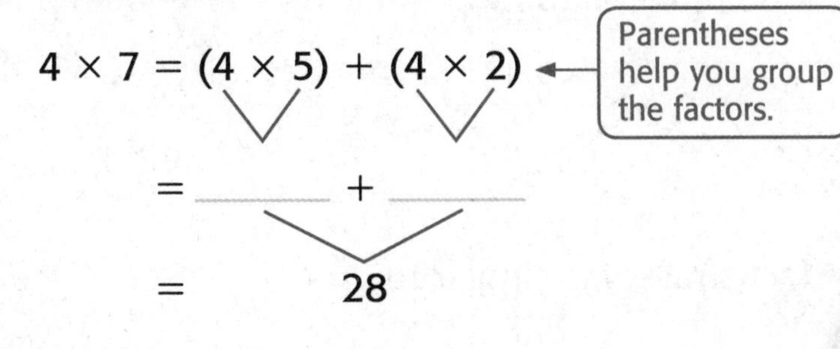

$4 \times 7 = (4 \times 5) + (4 \times 2)$ ← Parentheses help you group the factors.

= ______ + ______

= 28

So, $4 \times 7 =$ ______.

Try It

Gretchen cut 6 oranges into 9 slices each. How many orange slices are there?

Find 6×9.

Outline a 6×9 array on the grid paper.

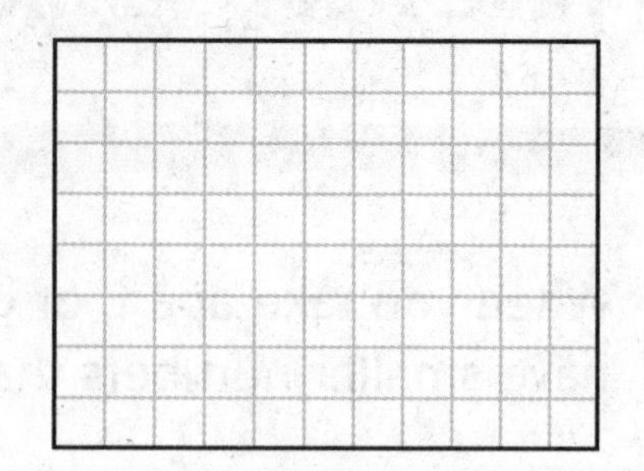

Decompose one factor.

Draw a vertical line through the array to decompose the factor 9 into $5 + 4$. Write the addends above.

Find the product of each part.

Multiply. Then add the products.

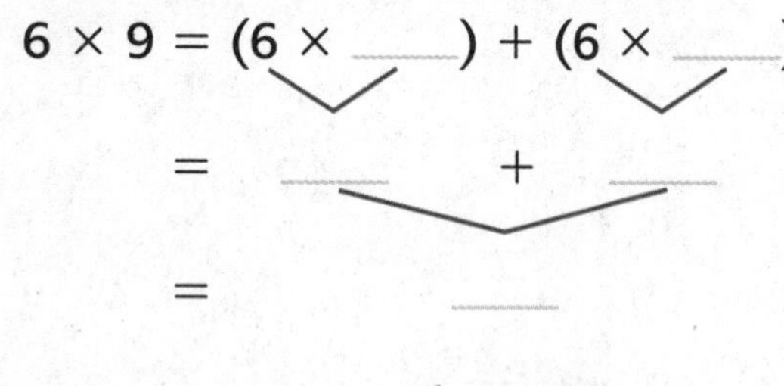

$6 \times 9 = (6 \times \_\_\_) + (6 \times \_\_\_)$

$= \_\_\_ + \_\_\_$

$= \_\_\_$

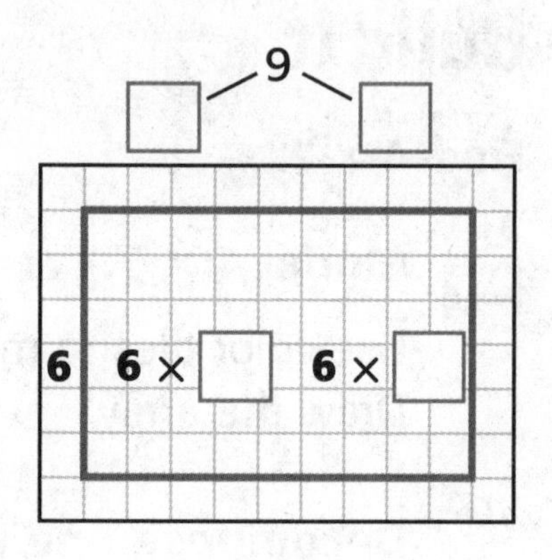

So, $6 \times 9 =$ ______. There are ______ orange slices.

Talk About It

1. Mathematical PRACTICE 3 **Justify Conclusions** In the example above, could the 6 have been decomposed instead of the 9? Explain.

____

2. How is decomposing a factor helpful in finding products?

____

____

3. Explain how using a known fact strategy is similar to decomposing a factor.

____

Name ....

CCSS

Practice It

Use color tiles to model the array. Decompose one factor. Then find the product for each part and add.

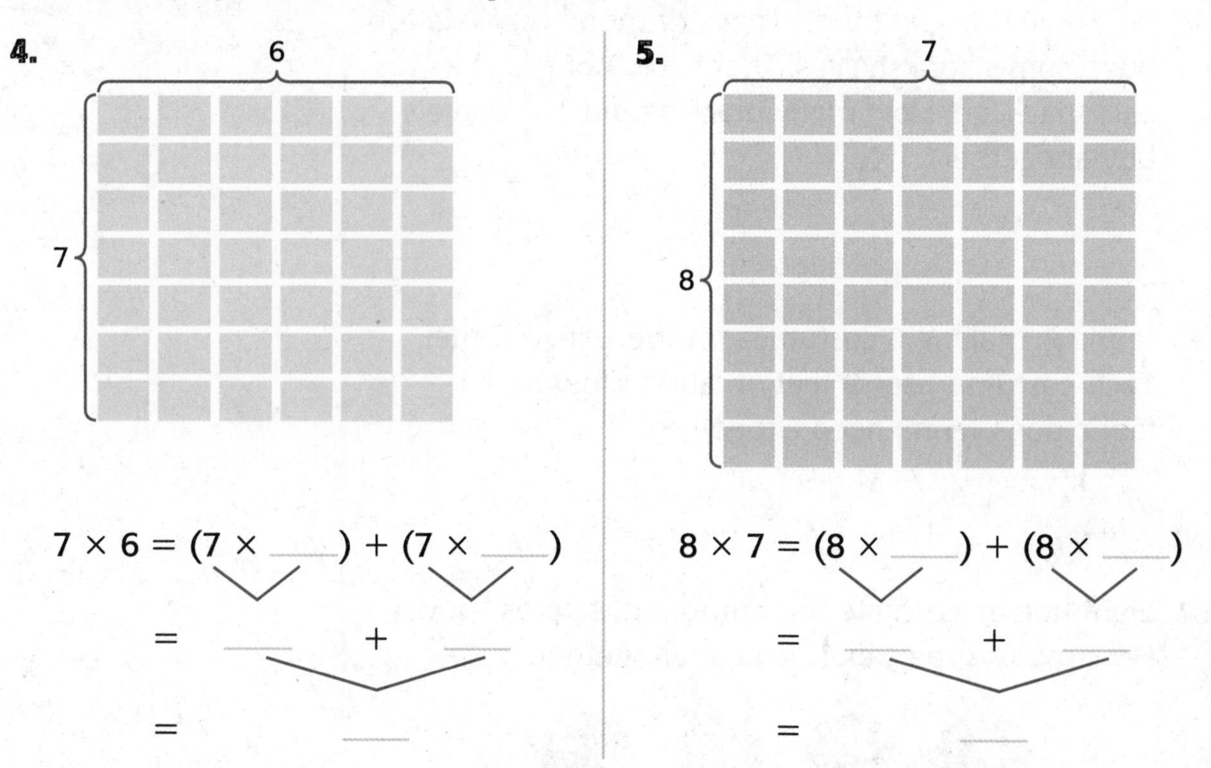

4. $7 \times 6 = (7 \times \_\_\_) + (7 \times \_\_\_)$

$= \_\_\_ + \_\_\_$

$= \_\_\_$

5. $8 \times 7 = (8 \times \_\_\_) + (8 \times \_\_\_)$

$= \_\_\_ + \_\_\_$

$= \_\_\_$

6. Decompose one factor. Color the array two colors to represent your numbers. Then find the product for each part and add.

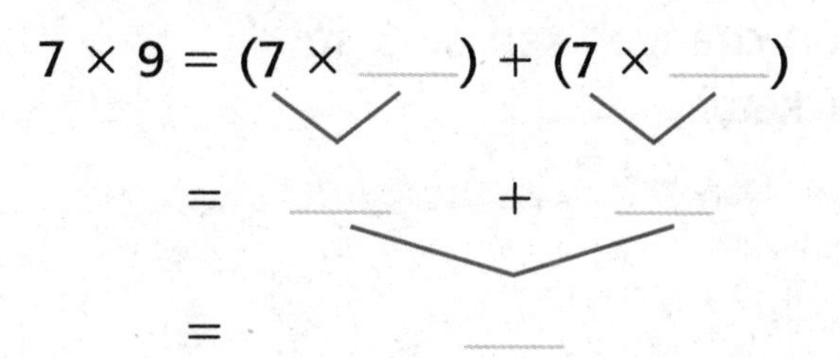

$7 \times 9 = (7 \times \_\_\_) + (7 \times \_\_\_)$

$= \_\_\_ + \_\_\_$

$= \_\_\_$

7. Decompose the fact a different way.

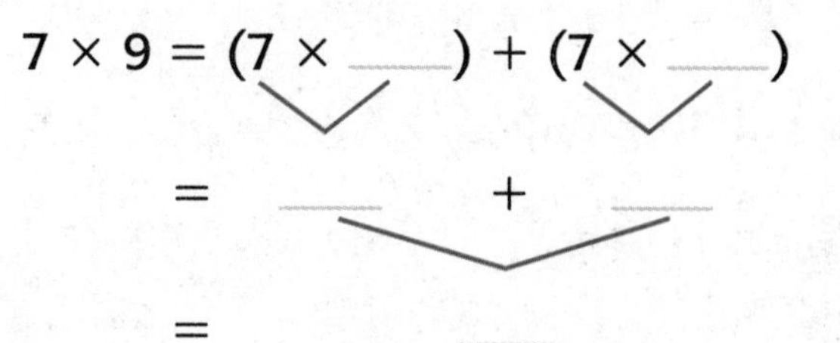

$7 \times 9 = (7 \times \_\_\_) + (7 \times \_\_\_)$

$= \_\_\_ + \_\_\_$

$= \_\_\_$

Apply It

Mathematical PRACTICE 7 Identify Structure Decompose one factor. Find each product. Then add.

8. Mr. Daniels bought 9 packages of metal brackets to build some bookshelves. There are 8 brackets in each package. How many brackets did Mr. Daniels buy altogether?

My Work!

9. Jenna is making 6 costumes for the dance recital. Each costume uses 9 feet of fabric. How much fabric does Jenna need altogether?

10. Eight horses each ate the number of apples shown. How many apples did they eat altogether?

11. Mathematical PRACTICE 2 **Reason** How could you change Exercise 8 so that Mr. Daniels buys a total of 81 brackets?

Write About It

12. How does decomposing a factor allow you to group numbers differently?

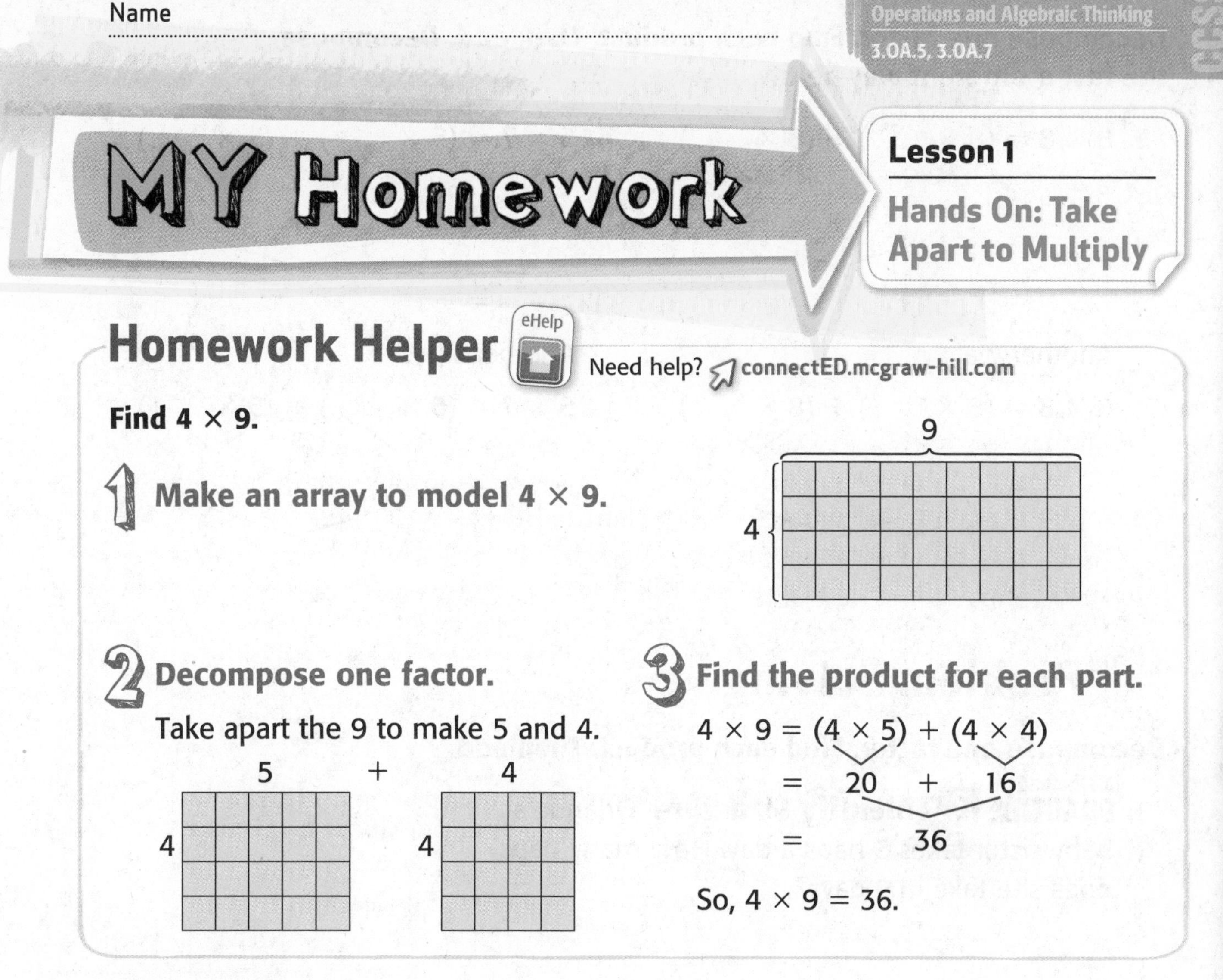

Practice

Decompose one factor. Color the array two colors to represent your numbers. Then find the product for each part and add.

1.

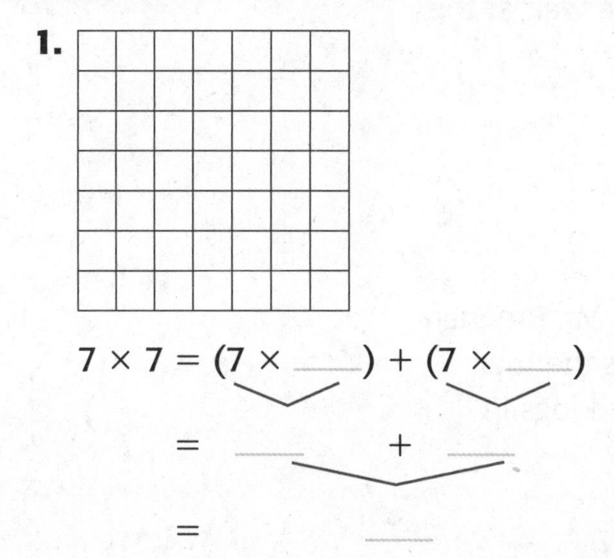

$7 \times 7 = (7 \times \_\_\_) + (7 \times \_\_\_)$

$= \_\_\_ + \_\_\_$

$= \_\_\_$

2.

$6 \times 8 = (6 \times \_\_\_) + (6 \times \_\_\_)$

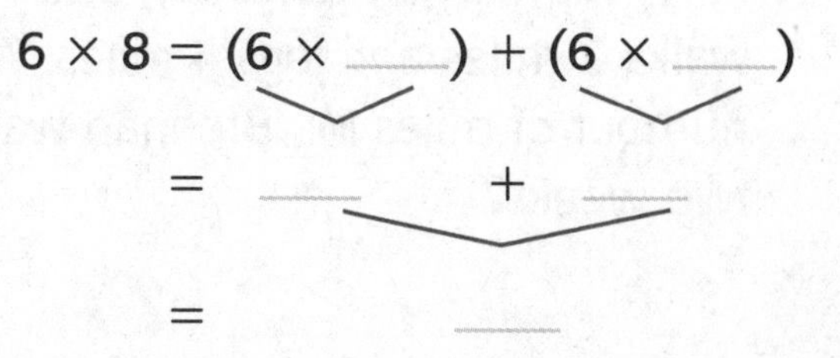

$= \_\_\_ + \_\_\_$

$= \_\_\_$

Decompose one factor. Find each product. Then add. Decompose the fact a different way below.

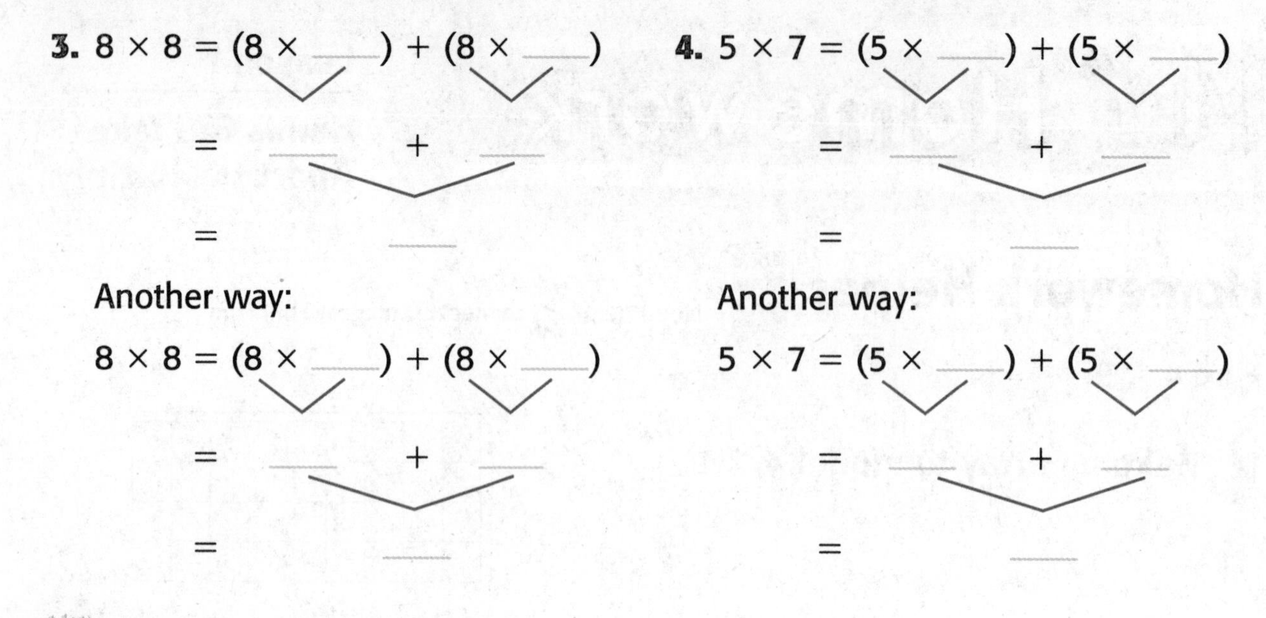

3. $8 \times 8 = (8 \times \_\_\_) + (8 \times \_\_\_)$

$= \_\_\_ + \_\_\_$

$= \_\_\_$

Another way:

$8 \times 8 = (8 \times \_\_\_) + (8 \times \_\_\_)$

$= \_\_\_ + \_\_\_$

$= \_\_\_$

4. $5 \times 7 = (5 \times \_\_\_) + (5 \times \_\_\_)$

$= \_\_\_ + \_\_\_$

$= \_\_\_$

Another way:

$5 \times 7 = (5 \times \_\_\_) + (5 \times \_\_\_)$

$= \_\_\_ + \_\_\_$

$= \_\_\_$

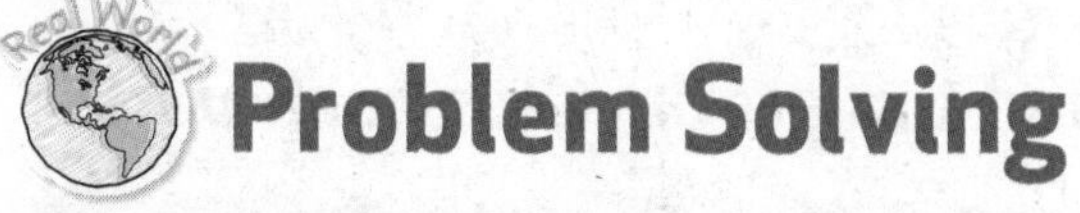

Problem Solving

Decompose one factor. Find each product. Then add.

5. Mathematical PRACTICE 7 **Identify Structure** Orlando's baby sister takes 3 naps a day. How many naps does she take in 9 days?

____

6. Carli gets to the bus stop 5 minutes early each morning. How many minutes does she wait at the bus stop in 5 days?

____

7. Every Monday, Wednesday, and Friday, Mr. Brennan walks 2 miles and jogs 4 miles. What is the total number of miles Mr. Brennan walks and jogs in two weeks?

____

My Work!

The Distributive Property

Lesson 2

ESSENTIAL QUESTION
How are properties and equations used to group numbers?

The **Distributive Property** allows you to decompose one factor. Then you can use smaller known facts to find products.

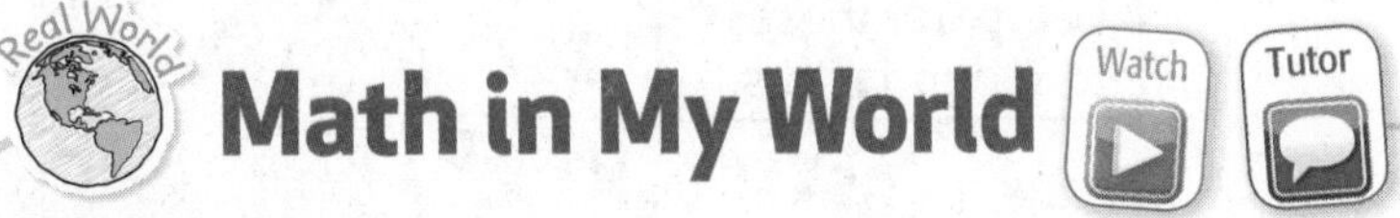

Example 1

Henry's Hardware Store sells wrench sets. Each set holds 6 wrenches. How many wrenches are there in 8 sets?

Find 8 × 6.

Decompose one factor.
One way is to decompose 6 into 5 + 1.

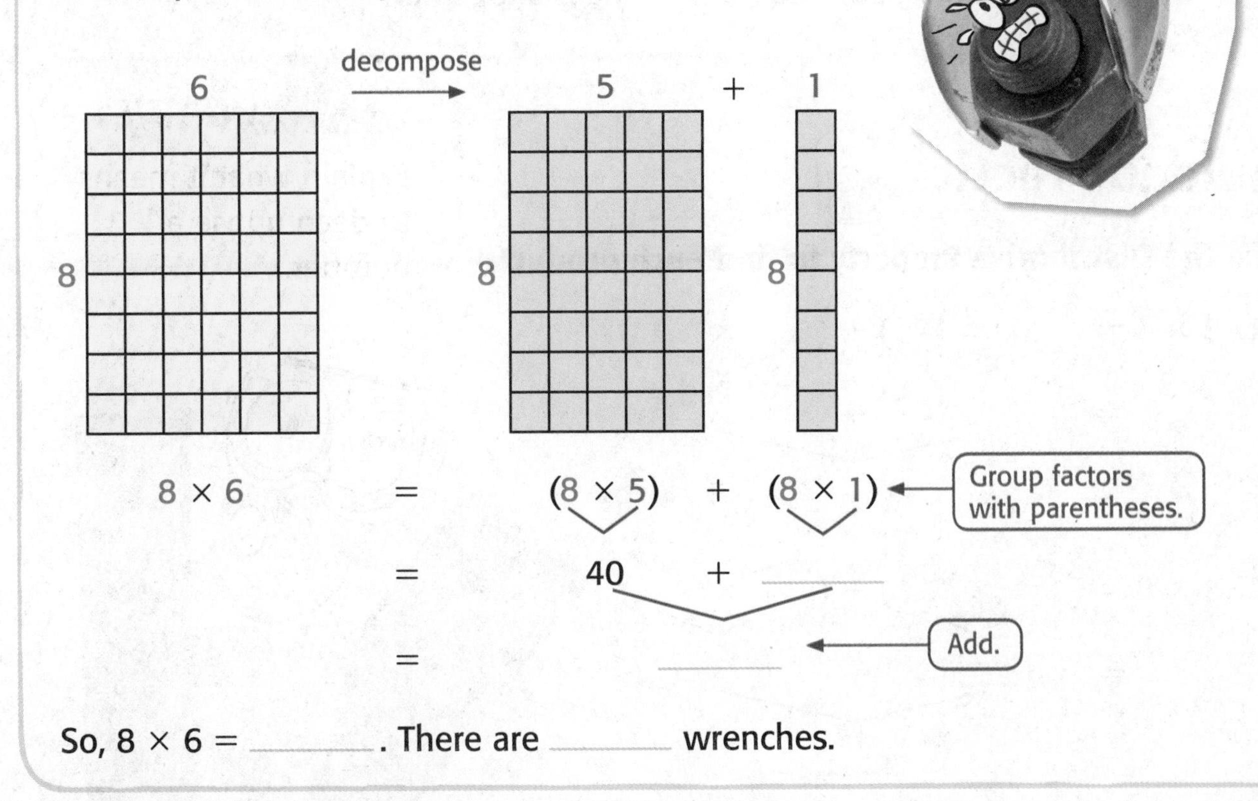

Example 2

Karina's dad used 7 boards to build a picnic table. How many nails were used if each board needed 7 nails?

Find 7×7.

Decompose one factor.
One way is to decompose 7 into $5 + 2$.

Use parentheses to group factors.

$7 \times 7 = (\_\_ \times \_\_) + (\_\_ \times \_\_)$ Multiply.

$= \_\_ + \_\_$ Add.

$= \_\_$

So, Karina and her dad used ______ nails altogether.

Guided Practice

Check

Talk MATH

Explain what it means to decompose a number.

Use the Distributive Property to find each product.

1. $8 \times 3 = (\_\_ \times \_\_) + (\_\_ \times \_\_)$

$= \_\_ + \_\_$

$= \_\_$

2. $8 \times 8 = (\_\_ \times \_\_) + (\_\_ \times \_\_)$

$= \_\_ + \_\_$

$= \_\_$

Name ....

Independent Practice

Use the Distributive Property to find each product.

3. $4 \times 6 =$ ______

4. $6 \times 6 =$ ______

5. $8 \times 9 =$ ______

6. $10 \times 4 =$ ______

7. $12 \times 4 =$ ______

8. $11 \times 8 =$ ______

9. $10 \times 10 =$ ______

10. $12 \times 6 =$ ______

Problem Solving

11. **Mathematical PRACTICE 7** **Identify Structure** The Fix It Right Hardware Store is open 12 hours every day. How many hours are they open altogether from Monday through Friday?

____

My Work!

12. A restaurant orders 9 dozen eggs. The picture shows the number of eggs from each dozen that broke during shipping. How many eggs are unbroken? (Hint: 1 dozen = 12)

____

13. Each aquarium has 10 clown fish and 6 puffer fish. There are 7 aquariums. How many fish are there in all?

____

HOT Problems

14. **Mathematical PRACTICE 1** **Make Sense of Problems** There are 12 inches in one foot and 3 feet in one yard. How many inches are in 2 yards?

____

15. **Building on the Essential Question** How are parentheses used when grouping factors?

____

____

Name ....

Lesson 2
The Distributive Property

Homework Helper

eHelp Need help? connectED.mcgraw-hill.com

Melanie ran 6 laps around a track each day for 7 days. How many laps did Melanie run that week?

Find 6×7.

One Way **Decompose 7 into 5 + 2.**

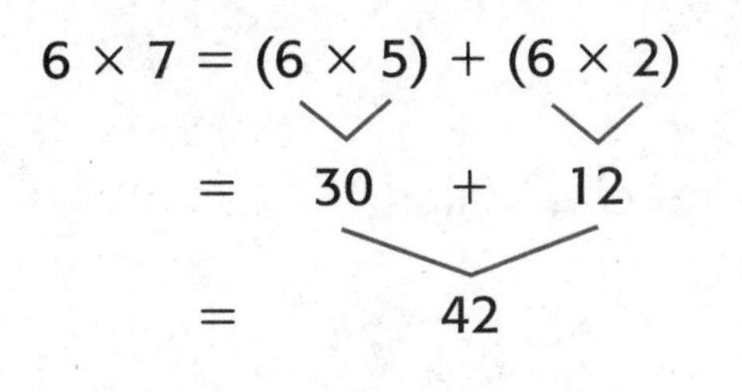

$6 \times 7 = (6 \times 5) + (6 \times 2)$
$= 30 + 12$
$= 42$

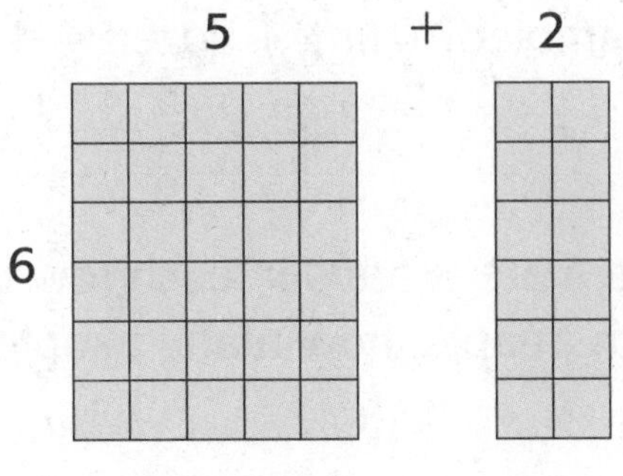

Another Way **Decompose 7 into 3 + 4.**

$6 \times 7 = (6 \times 3) + (6 \times 4)$
$= 18 + 24$
$= 42$

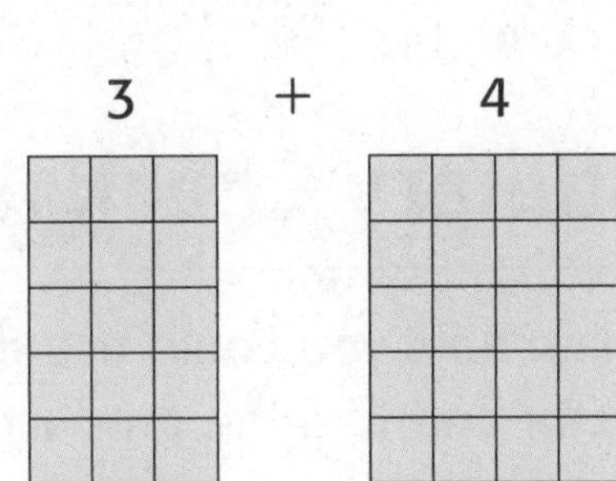

$6 \times 7 = 42$

So, Melanie ran 42 laps in one week.

Practice

Use the Distributive Property to find each product.

1. $4 \times 9 =$ ______

2. $5 \times 6 =$ ______

Use the Distributive Property to find each product.

3. $5 \times 11 =$ ______

4. $12 \times 7 =$ ______

Problem Solving

5. Milly bought 4 bags of apples at the grocery store. Each bag contains 6 apples. How many apples does Milly have in all?

______

6. Mathematical PRACTICE 7 **Identify Structure** Byron scrambled 8 dozen eggs for the campers. What is the total number of eggs Byron scrambled? (*Hint:* 1 dozen = 12)

______

7. There are 6 seats in each row in the theater. If 8 rows are filled with people, how many people are in the theater?

______

Vocabulary Check

Vocab

8. Explain how you could use the Distributive Property to decompose a factor and find the product of 5×9.

______

______

______

Test Practice

9. Which shows the correct use of the Distributive Property to find 4×12?

Ⓐ $(2 \times 6) + (2 \times 6)$

Ⓑ $(4 \times 10) + (4 \times 2)$

Ⓒ $(4 \times 6) + (2 \times 6)$

Ⓓ $(4 \times 8) + (4 \times 3)$

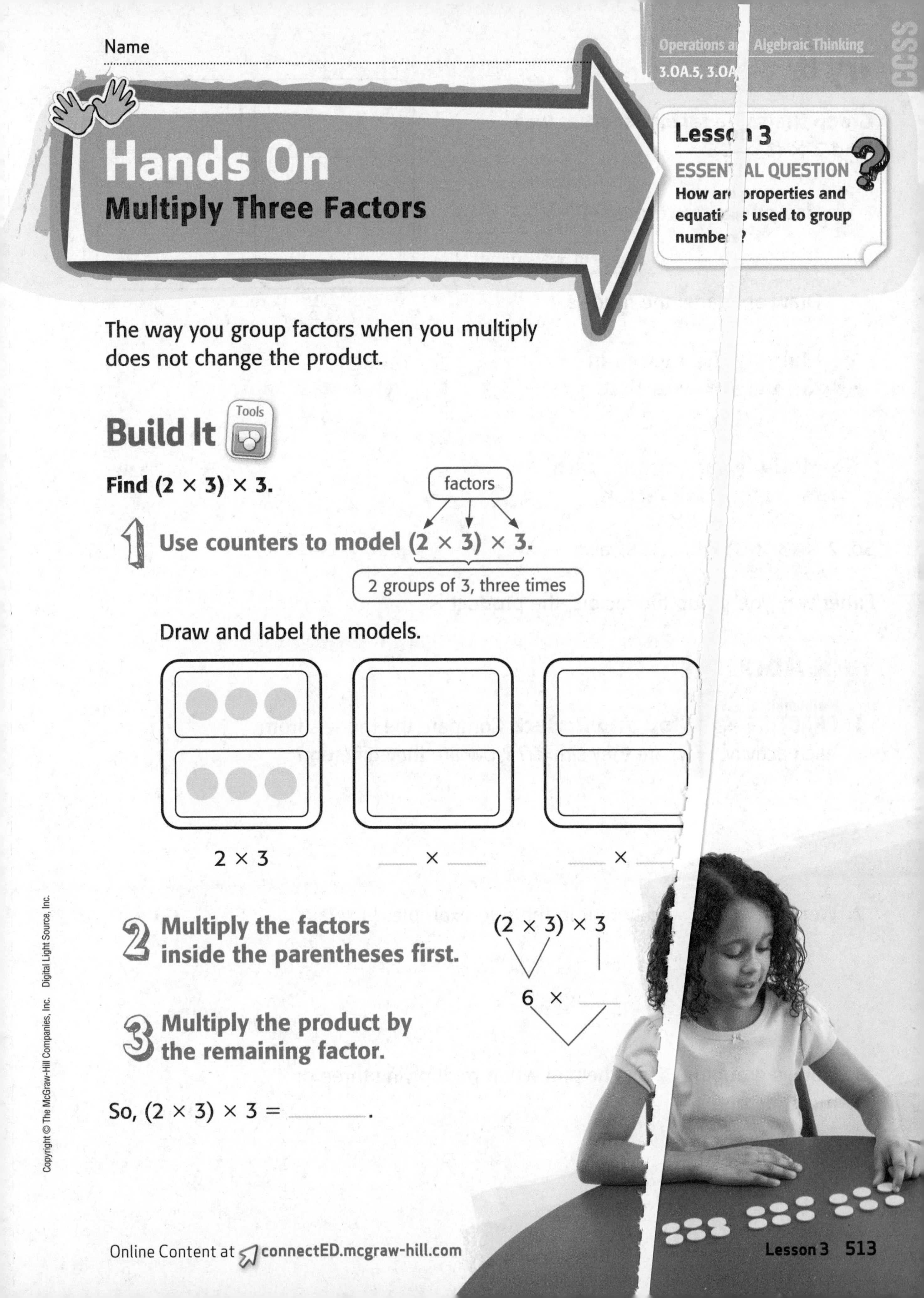

Name

Hands On
Multiply Three Factors

Lesso[n] 3
ESSEN[TI]AL QUESTION
How ar[e] properties and equati[on]s used to group numbe[rs]?

The way you group factors when you multiply does not change the product.

Build It

Tools

Find (2 × 3) × 3.

1 Use counters to model (2 × 3) × 3.

factors

2 groups of 3, three times

Draw and label the models.

2 × 3 ____ × ____ ____ × ____

2 Multiply the factors inside the parentheses first.

3 Multiply the product by the remaining factor.

(2 × 3) × 3

6 × ____

____

So, (2 × 3) × 3 = ______.

Try It

Group the same factors another way.
Find 2 × (3 × 3).

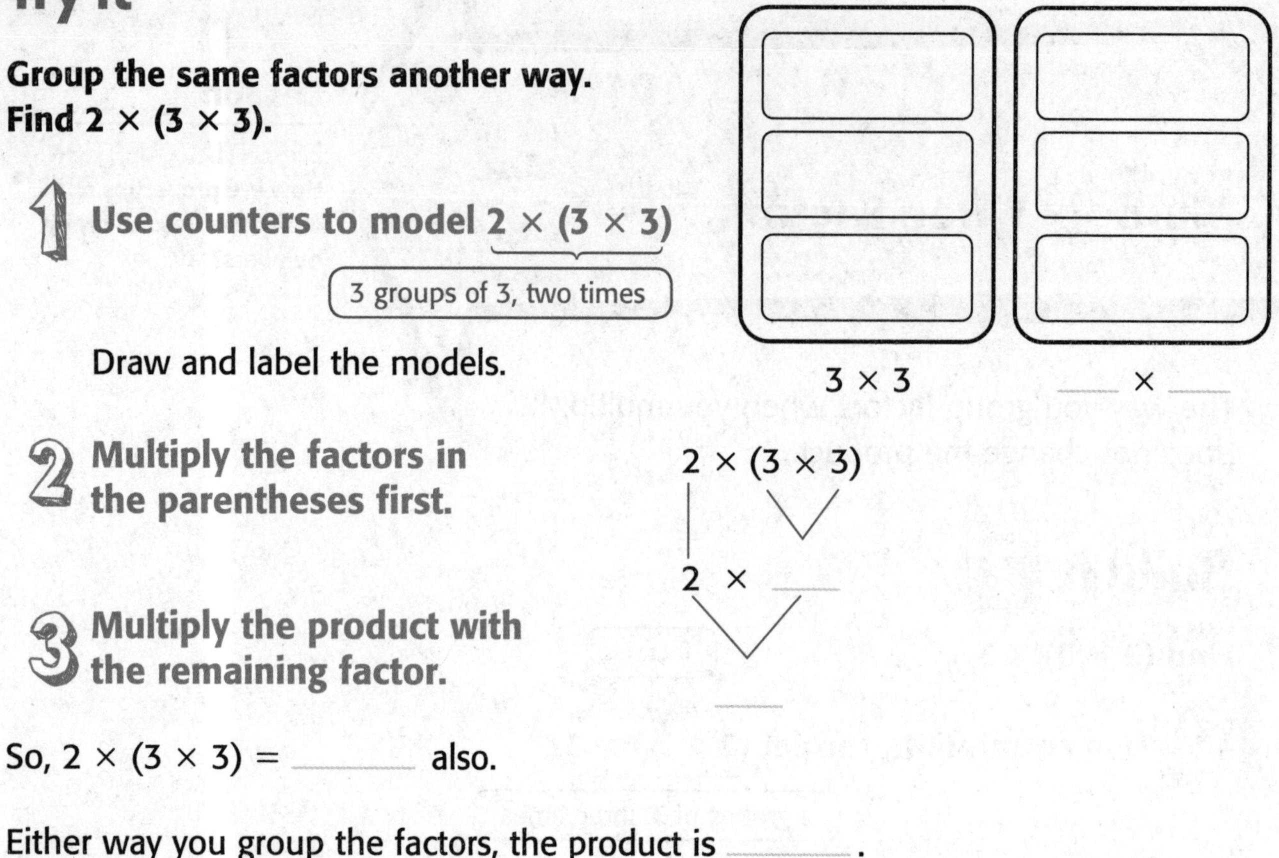

1 **Use counters to model 2 × (3 × 3)**

Draw and label the models.

2 **Multiply the factors in the parentheses first.**

3 **Multiply the product with the remaining factor.**

So, 2 × (3 × 3) = ______ also.

Either way you group the factors, the product is ______.

Talk About It

1. Mathematical PRACTICE 2 **Stop and Reflect** Compare the models from each activity. How are they similar? How are they different?

____

____

2. Were the products different in the two examples? Explain.

____

____

3. How is grouping factors helpful when multiplying three or more factors?

____

____

Name

The Associative Property

Lesson 4

ESSENTIAL QUESTION
How are properties and equations used to group numbers?

The **Associative Property of Multiplication** states that the grouping of factors does not change the product.

Example 1

Chris and Katie each received 4 smile stickers a week for 3 weeks. How many smile stickers did they earn altogether?

Find the unknown in $2 \times 3 \times 4 = \blacksquare$.

When there are no parentheses, multiply in order from left to right. Or, use parentheses to group factors.

One Way **Multiply 2 and 3 first.**

$(2 \times 3) \times 4 = \blacksquare$

$6 \times 4 = \blacksquare$

The unknown is ______.

Another Way **Multiply 3 and 4 first.**

$2 \times (3 \times 4) = \blacksquare$

$2 \times 12 = \blacksquare$

The unknown is ______.

Either way $2 \times 3 \times 4 =$ ______.

The ______________ Property shows that grouping does not change the product.

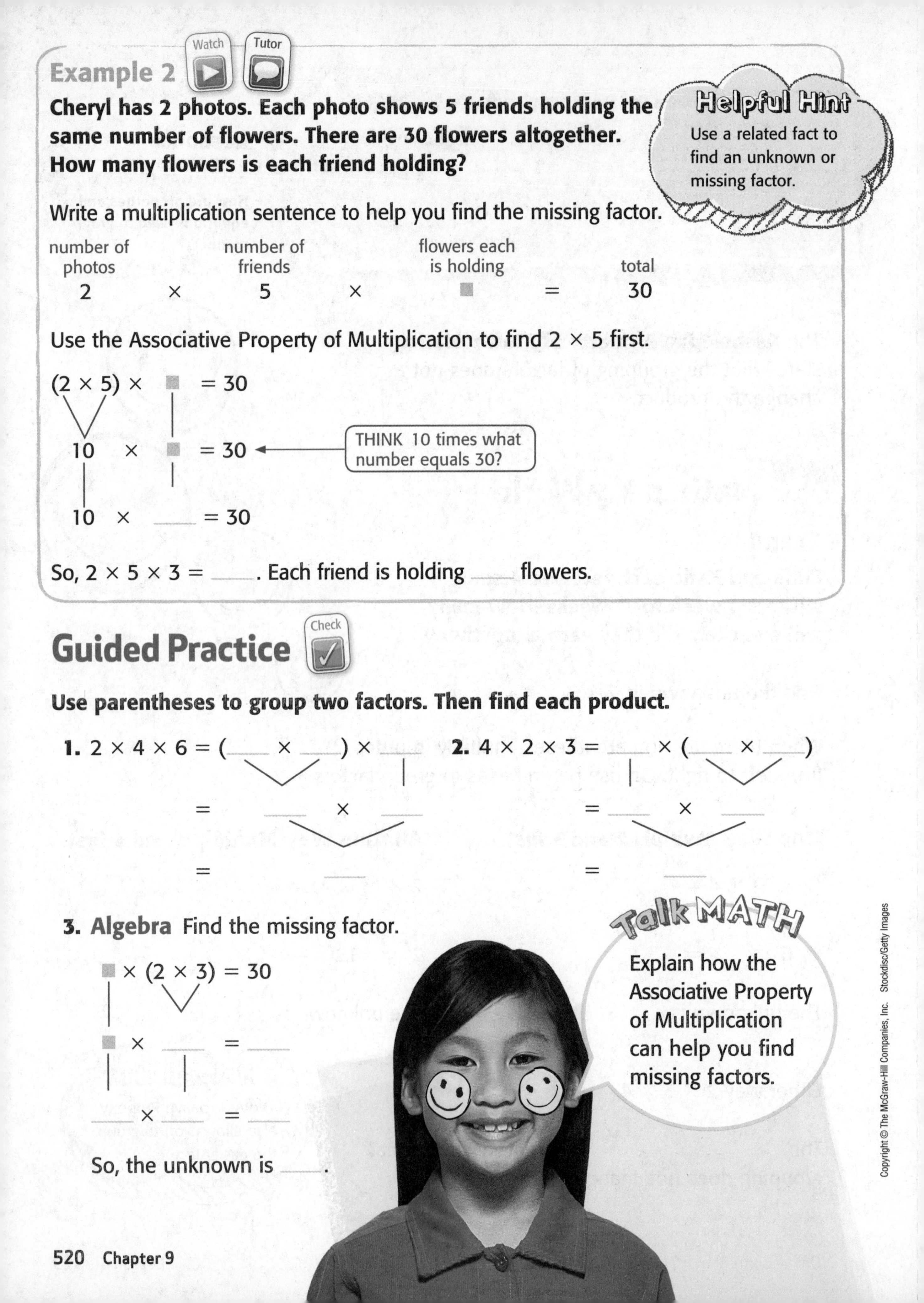

Example 2

Watch Tutor

Cheryl has 2 photos. Each photo shows 5 friends holding the same number of flowers. There are 30 flowers altogether. How many flowers is each friend holding?

Helpful Hint
Use a related fact to find an unknown or missing factor.

Write a multiplication sentence to help you find the missing factor.

number of photos		number of friends		flowers each is holding		total
2	×	5	×	■	=	30

Use the Associative Property of Multiplication to find 2 × 5 first.

(2 × 5) × ■ = 30

10 × ■ = 30 ← THINK 10 times what number equals 30?

10 × ____ = 30

So, 2 × 5 × 3 = ____. Each friend is holding ____ flowers.

Guided Practice

Check

Use parentheses to group two factors. Then find each product.

1. 2 × 4 × 6 = (____ × ____) × ____
= ____ × ____
= ____

2. 4 × 2 × 3 = ____ × (____ × ____)
= ____ × ____
= ____

3. Algebra Find the missing factor.

■ × (2 × 3) = 30

■ × ____ = ____

____ × ____ = ____

So, the unknown is ____.

Talk MATH

Explain how the Associative Property of Multiplication can help you find missing factors.

Name ....

Independent Practice

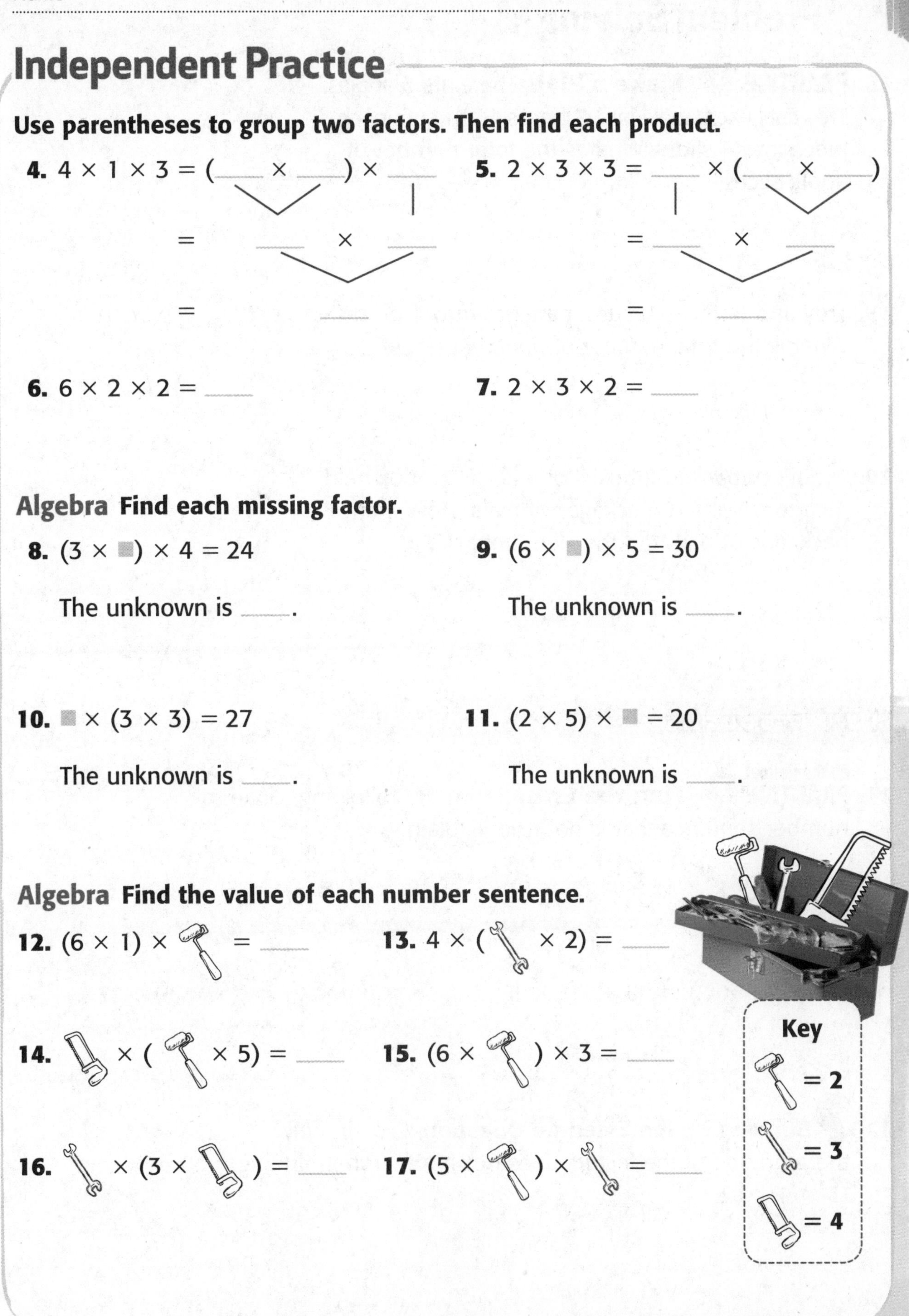

Use parentheses to group two factors. Then find each product.

4. $4 \times 1 \times 3 = (\_\_\_ \times \_\_\_) \times \_\_\_$

$= \_\_\_ \times \_\_\_$

$= \_\_\_$

5. $2 \times 3 \times 3 = \_\_\_ \times (\_\_\_ \times \_\_\_)$

$= \_\_\_ \times \_\_\_$

$= \_\_\_$

6. $6 \times 2 \times 2 = \_\_\_$

7. $2 \times 3 \times 2 = \_\_\_$

Algebra Find each missing factor.

8. $(3 \times \blacksquare) \times 4 = 24$

The unknown is ___.

9. $(6 \times \blacksquare) \times 5 = 30$

The unknown is ___.

10. $\blacksquare \times (3 \times 3) = 27$

The unknown is ___.

11. $(2 \times 5) \times \blacksquare = 20$

The unknown is ___.

Algebra Find the value of each number sentence.

12. (6 × 1) × = ___

13. 4 × (× 2) = ___

14. × (× 5) = ___

15. (6 ×) × 3 = ___

16. × (3 ×) = ___

17. (5 ×) × = ___

Key

= 2

= 3

= 4

Real World!

Problem Solving

18. Mathematical PRACTICE 1 **Make a Plan** There are 5 apples. Troy cuts each apple into 2 pieces. Beth cuts each piece into 4 slices. What is the total number of apple slices?

My Work!

19. Troy and Beth each cut 2 bananas into 4 pieces. What is the total number of banana pieces?

20. A clerk unpacked 2 boxes of nails. Each box held 4 cartons with 10 packages of nails. How many packages of nails did the clerk unpack?

HOT Problems

21. Mathematical PRACTICE 3 **Find the Error** From the following, circle the number sentence that is not true. Explain.

$(2 \times 3) \times 3 = 2 \times (3 \times 3)$

$3 \times (1 \times 5) = (3 \times 1) \times 5$

$4 \times (4 \times 2) = (3 \times 4) \times 4$

$6 \times (4 \times 2) = (6 \times 4) \times 2$

22. **Building on the Essential Question** Explain why the grouping of the factors does not matter when finding $(3 \times 4) \times 2$.

Check My Progress

Vocabulary Check

Choose the correct word(s) to complete each sentence.

Associative Property of Multiplication **Distributive Property**

decompose **parentheses**

1. 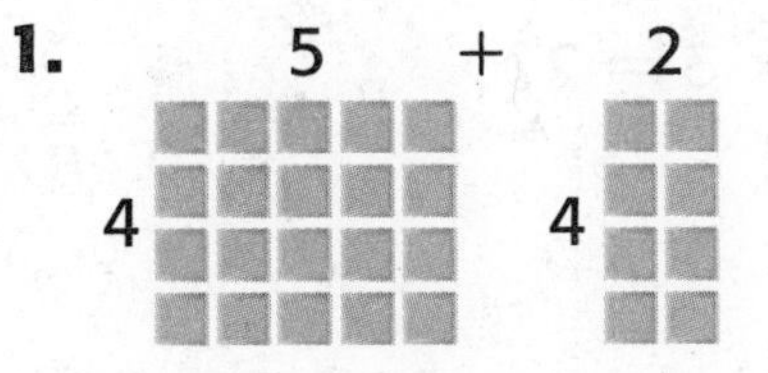

One way to find 4 × 7 with models is to ____________ the factor 7 into addends of 5 + 2.

2. 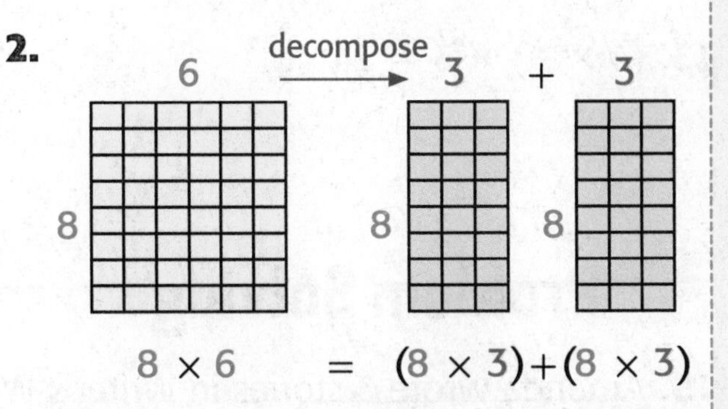

The ____________ allows you to decompose one factor into addends that are easier to multiply. Then you can use smaller known facts to find products.

3. (2 × 3) × 4 = 24
2 × (3 × 4) = 24

The ____________ ____________ states that the grouping of the factors does not change the product.

4. 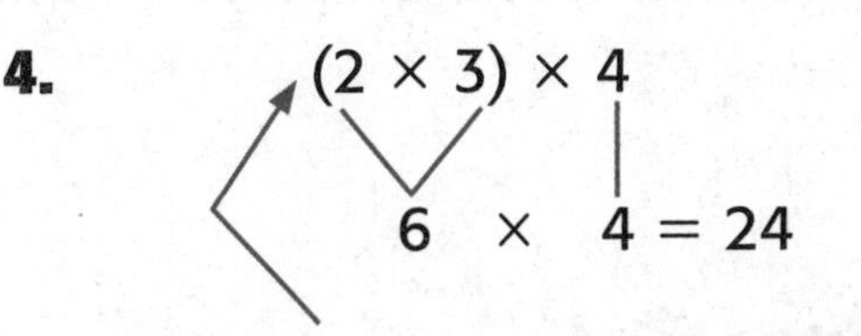

The ____________ show the grouping of the factors to multiply first.

Concept Check

Use the Distributive Property to find each product.

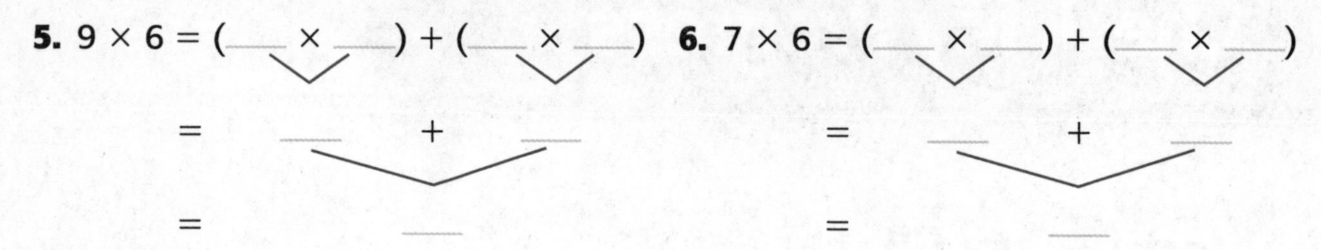

5. $9 \times 6 = (\_\_ \times \_\_) + (\_\_ \times \_\_)$
$= \_\_ + \_\_$
$= \_\_$

6. $7 \times 6 = (\_\_ \times \_\_) + (\_\_ \times \_\_)$
$= \_\_ + \_\_$
$= \_\_$

Find each product.

7. $3 \times (4 \times 2) = \_\_$

8. $2 \times (3 \times 2) = \_\_$

9. $(5 \times 2) \times 1 = \_\_$

10. $(2 \times 3) \times 3 = \_\_$

11. $4 \times (2 \times 3) = \_\_$

12. $(3 \times 3) \times 2 = \_\_$

Algebra Find each missing factor.

13. $(4 \times ■) \times 3 = 24$

■ = __

14. $(3 \times ■) \times 3 = 27$

■ = __

Problem Solving

15. Amanda wrote 3 stories in Writer's Workshop. Each story was 6 pages long. She drew 2 illustrations on each page. How many illustrations did Amanda draw altogether?

____

16. Mrs. Andrew's classroom has 4 rows of desks with 3 desks in each row. She placed 2 pencils on each desk. How many pencils did Mrs. Andrew place on the desks altogether?

____

Test Practice

17. Becka made 2 cards. She drew 3 balloons on each card. Each balloon had 3 stars. How many stars did Becka use on her cards altogether?

Ⓐ 15 stars
Ⓑ 16 stars
Ⓒ 17 stars
Ⓓ 18 stars

Name ....

Write Expressions

Lesson 5

ESSENTIAL QUESTION
How are properties and equations used to group numbers?

The four **operations** are addition, subtraction, multiplication, and division. An **expression** is a number or a combination of numbers and operations. An expression does not have an equals sign.

Example 1

Alice invited three friends to play in her backyard. Write an expression to represent the total number of friends.

Use pictures. ○ + ● ● ● ← three friends
Alice

Use numbers. 1 + ______

Use words. ______ plus three, or ______ more than one

Example 2

Five nails were hammered in the wood. One nail bent. Write an expression to represent the number of good nails left.

Use pictures. ○ ○ ○ ○ ⊗ ← One nail bent.
Five nails were hammered.

Use numbers. 5 − ______

Use words. five minus ______, or ______ less than five

Example 3

Tracy bought 3 magnets. Scott has two times as many. Then Tracy buys one more. Write an expression to represent the total number of magnets.

Use pictures.

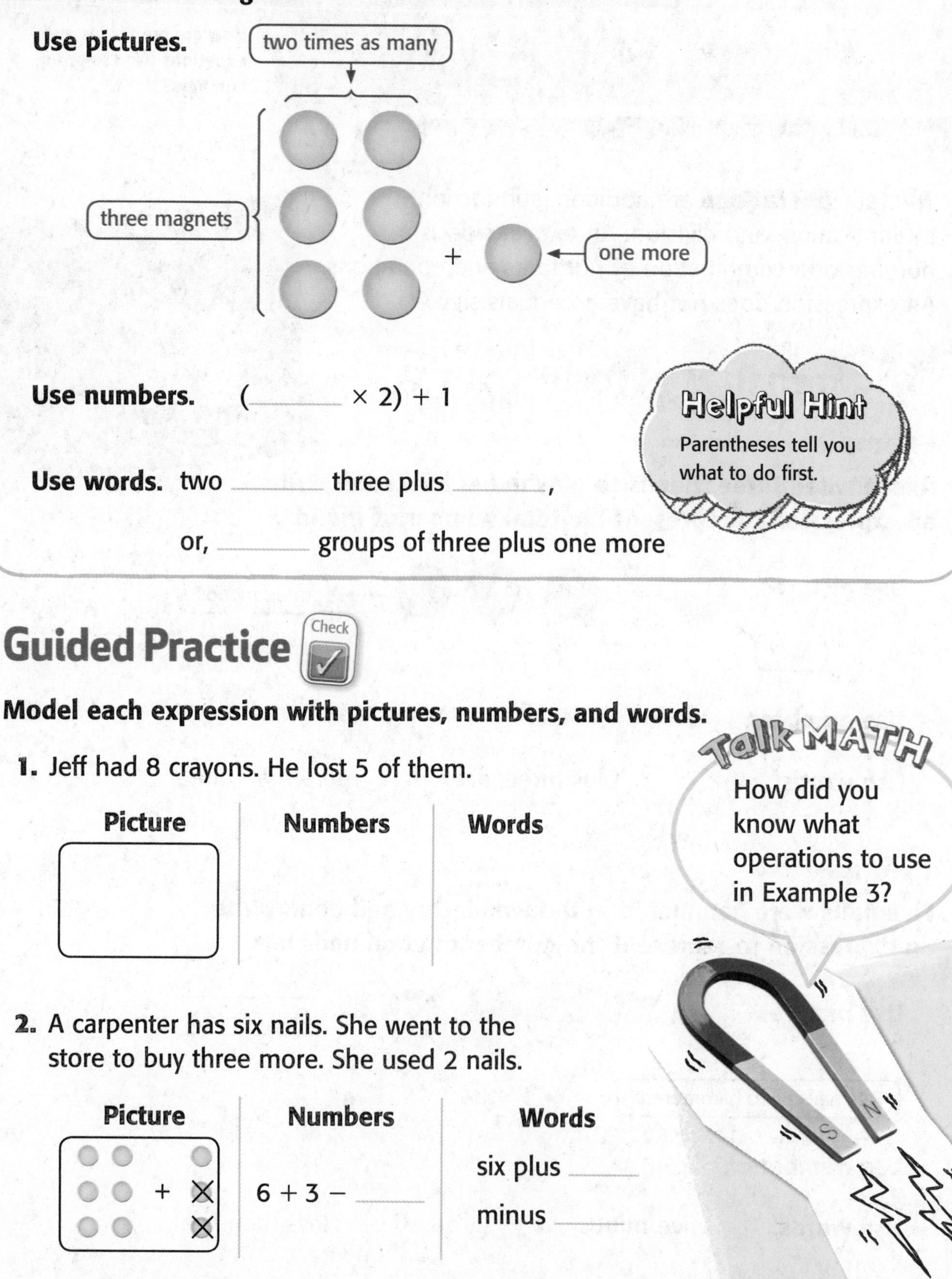

Use numbers. (______ × 2) + 1

Use words. two ______ three plus ______,

or, ______ groups of three plus one more

Helpful Hint
Parentheses tell you what to do first.

Guided Practice

Check

Model each expression with pictures, numbers, and words.

Talk MATH
How did you know what operations to use in Example 3?

1. Jeff had 8 crayons. He lost 5 of them.

Picture	Numbers	Words

2. A carpenter has six nails. She went to the store to buy three more. She used 2 nails.

Picture	Numbers	Words
	6 + 3 − ______	six plus ______ minus ______

Name

Independent Practice

Use numbers and operations to write each phrase as an expression.

3. 4 more than 7

4. the total of 5 rows of 6 chairs

5. half of 18

6. 3 people equally divide $21

7. difference between 89 and 80

8. 6 groups with 6 people in each

There are 6 nails in the toolbox. Write an expression to tell how many there will be when there are:

9. 2 fewer nails

10. 4 times as many nails

11. half as many nails

12. 10 more nails

13. 3 equal groups of nails

Write an expression for each.

14. the cost of 5 bottles of glue

Hardware Sale	
Glue	10¢
Tape Measure	95¢
Spool of Wire	89¢
Nails	10¢

15. the number of nails for 90¢

16. the total cost of a spool of wire, tape measure, and a bottle of glue

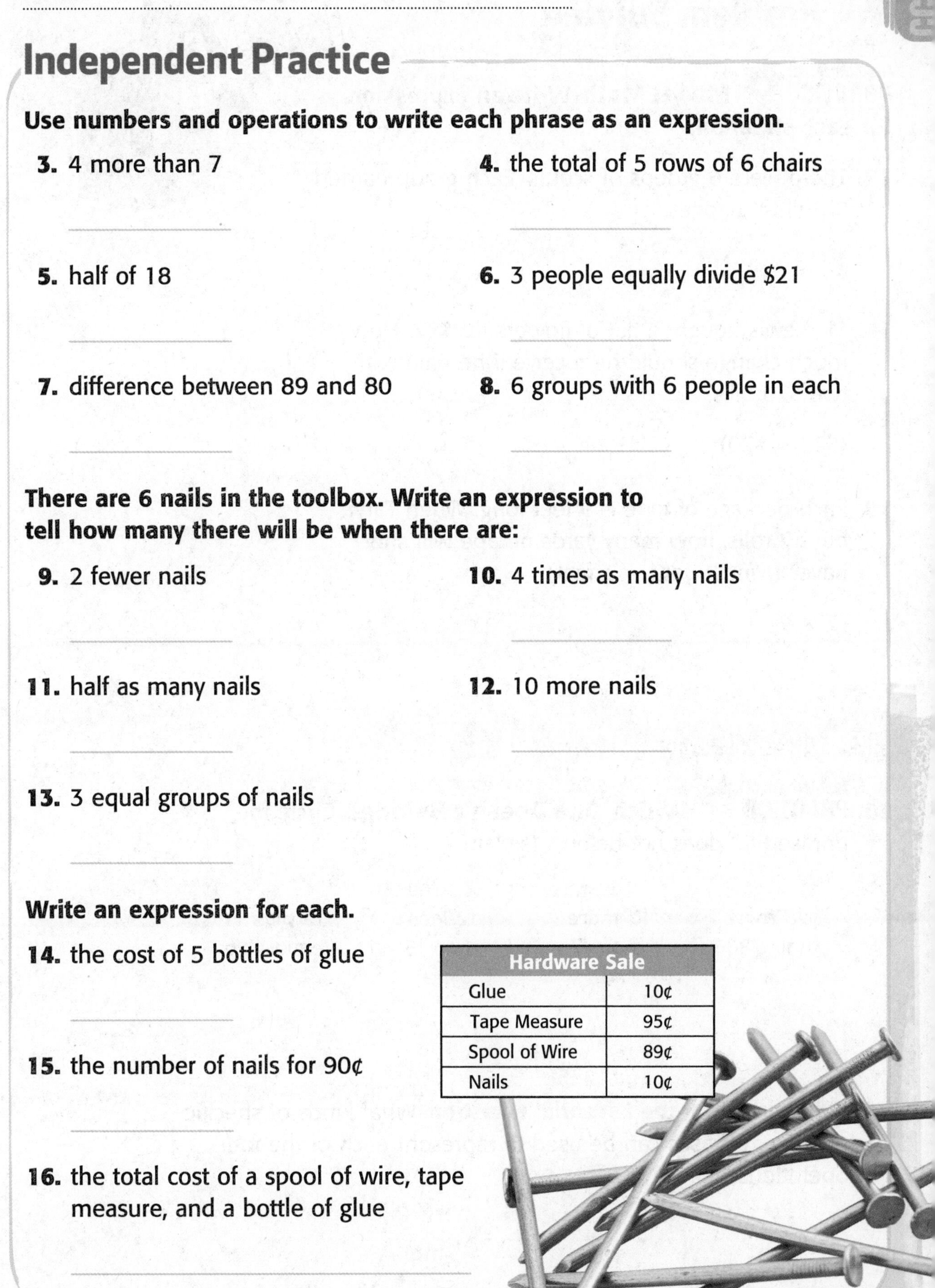

Problem Solving

Mathematical PRACTICE 4 **Model Math** **Write an expression for each situation.**

17. There were 6 groups of scouts. Each group earned 9 Build-It badges.

____

18. Mr. Lewis bought a flat of flowers for $22. How much change should he receive if he paid with two $20 bills?

($20 + $20) ______

19. Each package of tape is 9 feet long. When Taryn buys 2 rolls, how many yards of tape will she have? (*Hint:* 1 yard = 3 feet)

(2 × 9) ______

My Work!

HOT Problems

20. Mathematical PRACTICE 3 **Which One Doesn't Belong?** Circle the phrase that does not belong. Explain.

$25 more than $30	16 more than 17	12 less than 15	12 plus 14 equals 26

____

21. **Building on the Essential Question** What kinds of specific words or phrases can be used to represent each of the four operations?

____

____

Name ....

Lesson 5
Write Expressions

Homework Helper

eHelp Need help? connectED.mcgraw-hill.com

Charles filled 4 balloons for the party. Model each of the following situations with pictures, numbers, and words.

Charles filled 2 more balloons.

Numbers: 4 + 2
Words: four plus two

Charles filled twice as many balloons.

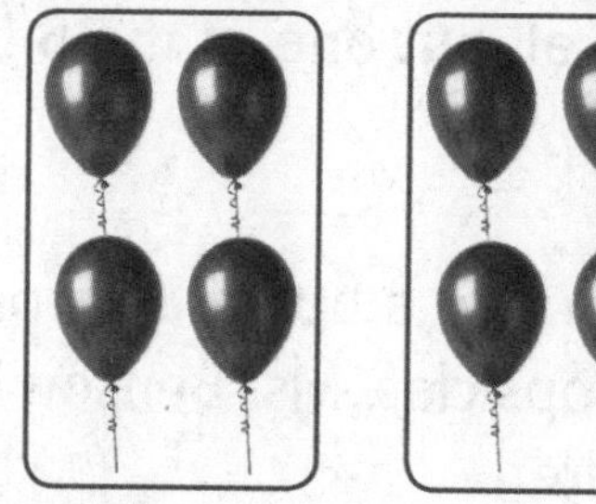

Numbers: 4 × 2
Words: four times two

One balloon floats away.

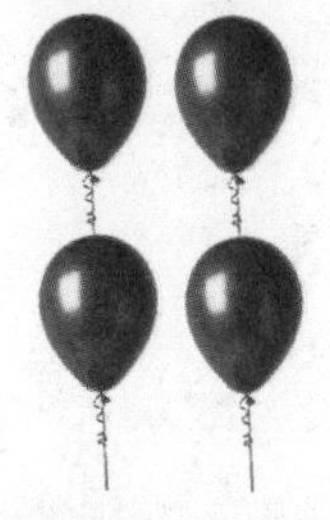

Numbers: 4 − 1
Words: one less than four

Charles gives half of the balloons to Lia.

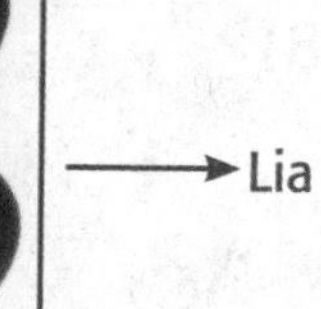

Numbers: 4 ÷ 2
Words: half of four

Practice

1. Annette has 6 pencils. She divides them evenly among 3 friends. Model the expression with a picture, numbers, and words.

Picture	Numbers	Words
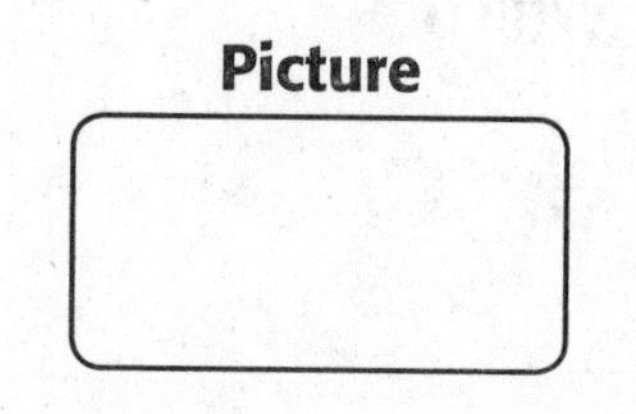		

Use numbers and operations to write each phrase as an expression.

2. 4 boxes with 2 shoes in each

____

3. the difference between 58 and 47

____

4. 5 more than 12

____

5. 30 books shared equally by 10 people

____

Problem Solving

Mathematical PRACTICE 4 **Model Math Write an expression for each situation.**

6. Stella read all but one of the 5 books she took on vacation.

____

7. Ms. Benson had a box of 8 popsicles. She bought another box of 4 popsicles. Ms. Benson divided the popsicles among her 2 children.

(______________) ÷ 2

8. Frieda bought 3 packs of 8 candles. Then she found 1 candle at home.

(3 × 8) ______

Vocabulary Check

Match each vocabulary word with its example.

9. expression • 7 × 4

10. operations • +, −, ×, and ÷

Test Practice

11. Zoe had 9 bracelets. She lost 1 and gave 3 to Blaire. Which expression matches the situation?

Ⓐ 9 − 3

Ⓑ (9 − 1) + (9 − 3)

Ⓒ 9 − 1 − 3

Ⓓ (9 − 1) + 3

Name ....

Evaluate Expressions

Lesson 6

ESSENTIAL QUESTION
How are properties and equations used to group numbers?

When a symbol, such as ? and ■, or a letter, such as x or y, is used to stand for an unknown, it is called a **variable.**

Example 1

Santiago unpacked 5 more boxes of light bulbs than boxes of flashlights.

Write an expression using the variable x for the unknown.

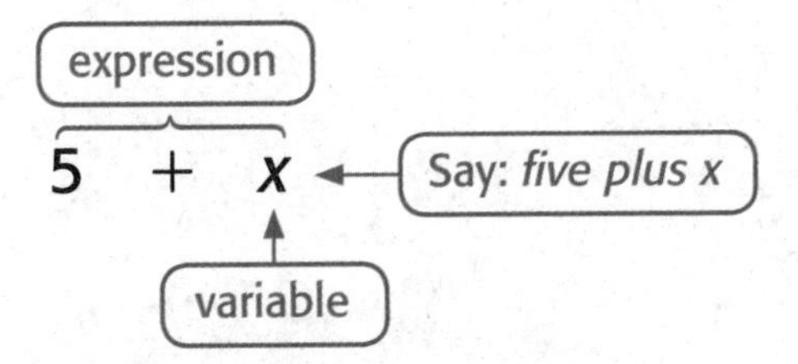

When you find the value of an expression by replacing the variable with a number, you **evaluate** the expression.

Example 2

The hardware store has 6 fewer step ladders than extension ladders. There are y extension ladders. Write an expression using the variable y. Then evaluate the expression if $y = 10$.

$y - 6$ Write the expression.

$10 - 6$ Replace y with ______.

______ Subtract.

There are ______ step ladders.

Sometimes an expression has more than one operation. When there are no parentheses, multiply and/or divide in order from left to right. Then add and/or subtract in order from left to right.

Example 3

Owen looked at a set of 4 pliers. His dad looked at a set that had *s* times as many pliers plus 3 more. If $s = 2$, how many pliers were in the set Owen's dad looked at?

Write an expression, then evaluate it.

1. Write the expression. $3 + s \times 4$

2. Replace *s* with 2. $3 + 2 \times 4$

3. When there are no parentheses, first multiply or divide, in order from left to right. $3 + 8$

4. Then, add or subtract, in order from left to right. 11

So, if $s = 2$, then $3 + s \times 4 =$ ______. There were ______ pliers in the set.

Guided Practice

Evaluate each expression if $a = 2$ and $b = 5$.

1. $3 + a$

$3 +$ ______ $=$ ______

2. $11 - b$

$11 -$ ______ $=$ ______

3. $b \times 4$

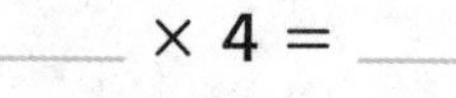

______ $\times 4 =$ ______

4. $12 \div a + 4$

$12 \div$ ______ $+ 4$

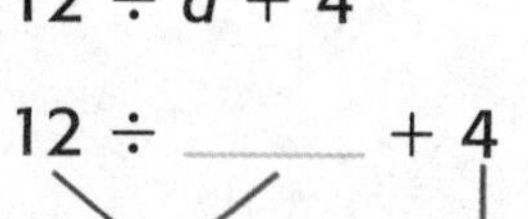

______ $+$ ______ $=$ ______

Talk MATH

Look back at Example 3. How would your answer be different if you evaluated the expression left to right? Explain.

Name ....

Independent Practice

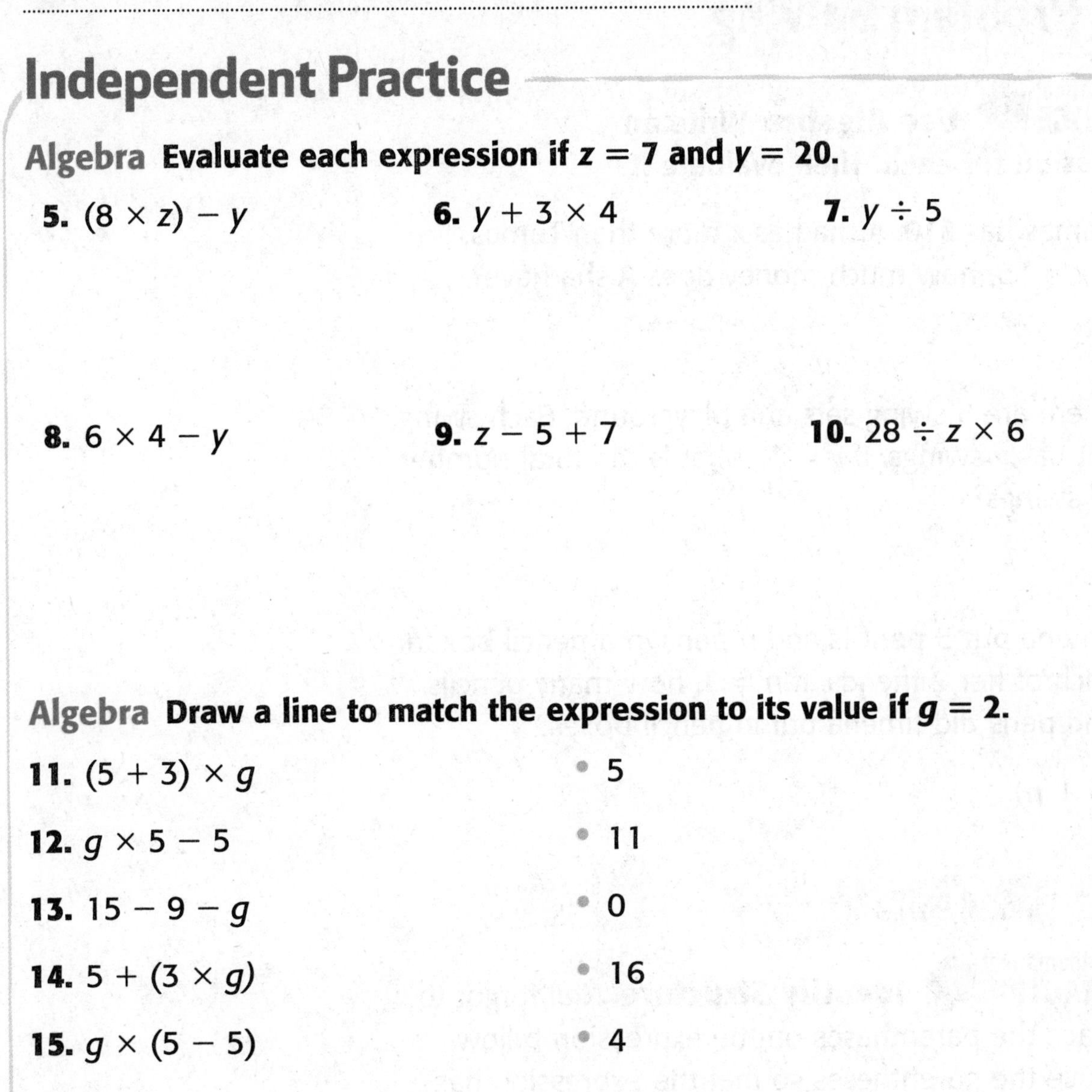

Algebra **Evaluate each expression if $z = 7$ and $y = 20$.**

5. $(8 \times z) - y$

6. $y + 3 \times 4$

7. $y \div 5$

8. $6 \times 4 - y$

9. $z - 5 + 7$

10. $28 \div z \times 6$

Algebra **Draw a line to match the expression to its value if $g = 2$.**

11. $(5 + 3) \times g$	• 5
12. $g \times 5 - 5$	• 11
13. $15 - 9 - g$	• 0
14. $5 + (3 \times g)$	• 16
15. $g \times (5 - 5)$	• 4

Algebra **Circle yes or no to tell whether the expression is evaluated correctly if $n = 12$.**

16. $n \div 4 \times 6$

$12 \div 4 \times 6$

$3 \times 6 = 18$

Yes No

17. $12 + n \div 4$

$12 + 12 \div 4$

$24 \div 4 = 6$

Yes No

18. Did you circle no for either Exercise 16 or 17? Explain.

____

____

Problem Solving

Mathematical PRACTICE 2 **Use Algebra Write an expression for each. Then evaluate it.**

19. Tomas has \$10. Aisha has x more than Tomas. If $x = \$5$, how much money does Aisha have?

____

20. There are 5 swing sets at a playground. Each swing set has v swings. If $v = 3$, what is the total number of swings?

____

21. Jimena put 5 pencils and n pens in a pencil box for each of her 2 friends. If $n = 3$, how many pencils and pens did Jimena put in pencil boxes?

$(5 + n)$ ______ ; ______________

HOT Problems

22. Mathematical PRACTICE 7 **Identify Structure** Neil forgot to place the parentheses on the expression below. Place the parentheses so that the expression has a value of 2.

$$12 - 4 + 6$$

Why are the parentheses important in this expression?

____

____

23. **Building on the Essential Question** When evaluating an expression with more than one operation and no parentheses, how should you proceed?

____

____

MY Homework

Lesson 6
Evaluate Expressions

Homework Helper

eHelp

Need help? connectED.mcgraw-hill.com

Kevin used half of the tools from his toolbox. An hour later he put 3 tools back. How may tools is Kevin still using if he had z tools in his toolbox? Write an expression. Then evaluate the expression if $z = 8$.

Write the expression. $z \div 2 - 3$

Replace z with 8. $8 \div 2 - 3$

When there are no parentheses, first multiply or divide, in order, from left to right. $4 - 3$

1

Kevin is still using 1 tool.

Practice

Algebra Evaluate each expression if $c = 4$ and $d = 7$.

1. $15 - d$

$15 - \_\_\_\_ = \_\_\_\_$

2. $16 + c$

$16 + \_\_\_\_ = \_\_\_\_$

3. $35 \div d$

$35 \div \_\_\_\_ = \_\_\_\_$

Algebra Evaluate each expression if $x = 14$ and $y = 6$.

4. $(x + y) \div 4$

5. $x - 2 \times 2$

6. $y + 24 \div 2$

Problem Solving

Mathematical PRACTICE 4 **Model Math** **Write an expression for each situation. Then evaluate it.**

7. Monica has 7 hats. Andrea has b fewer hats than Monica. If $b = 5$, how many hats does Andrea have?

____

8. There are 4 shelves with canned dog food. Each shelf has t cans. Then Tracy adds 2 cans to only 1 of the shelves. If $t = 8$, how many cans are on the shelves altogether?

$4 \times t$ ________ ; $4 \times$ ______________________

9. Valerie is making identical quilts for herself and her sister. For each quilt she buys 5 yards of solid fabric and w yards of printed fabric. If $w = 4$, how much fabric did Valerie buy to make both quilts?

(______________) $\times$ 2; ______________________

Vocabulary Check

Vocab

10. Explain what a variable is.

____

11. What does it mean to evaluate an expression?

____

____

Test Practice

12. Evaluate the expression $h + 8 \div 4$ if $h = 16$.

Ⓐ 20 Ⓒ 8

Ⓑ 18 Ⓓ 6

Name

Write Equations

Lesson 7

ESSENTIAL QUESTION
How are properties and equations used to group numbers?

An **equation,** or number sentence, shows that two expressions are equal. An equation contains an equals sign (=).

Math in My World

APPLES
Red 5
Yellow . . . 3
Green . . . 4

Example 1

Use the information shown to find the total number of red and green apples. Write an equation to represent the counters.

red apples + green apples = total

_____ + _____ = _____

Equation: _____ + _____ = _____

The equation _____ + 4 = _____ tells us there are _____ red and green apples.

In order to write an equation, you need to decide what operation to use. There are words and phrases that can suggest whether to add, subtract, multiply, or divide. Here are some examples.

Addition	Subtraction	Multiplication	Division
sum	difference	product	quotient
more	less than	times as many	divide
in all	left	twice	half
total	fewer than	in each	into equal groups

Example 2

Hayden used his tape measure to find the total length of a board needed to finish his tree fort. When he cuts the board, one piece will be 48 inches and the other will be 32 inches. What is the total length of the board?

Write an equation to represent the problem.
Use the letter b for the unknown.

The word *total* suggests adding.

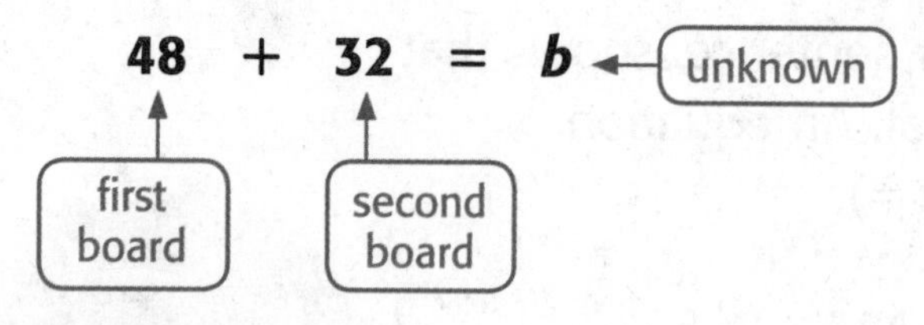

Example 3

A hardware store ordered 2 sets of monkey wrenches. There are 3 wrenches in each set. After the wrenches are shipped, the store will have a total of 7 wrenches. How many wrenches did they already have?

Write an equation to represent the problem. Use the letter w for the unknown.

The phrases *sets* and *in each set* suggest multiplying. The word *total* suggests adding.

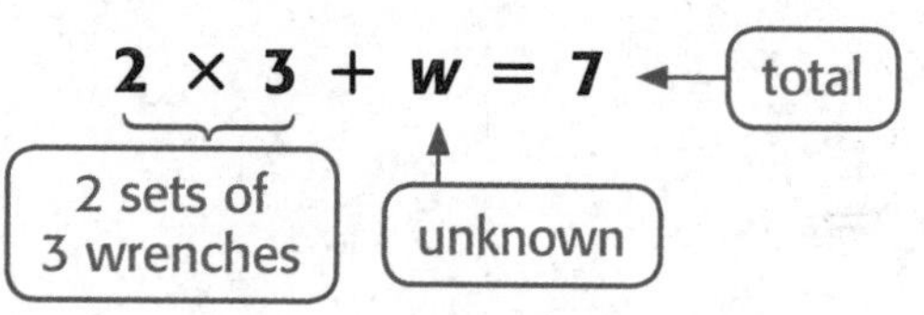

Talk MATH

What is the difference between an expression and an equation?

Guided Practice

Write an equation to represent each sentence.

1. The total of 3 letters plus 2 letters is x letters.

______ + ______ = ______

2. A group of 6 has x taken away and 2 are left.

______ − ______ = ______

Name ....

CCSS

Independent Practice

Underline the part of the phrase that suggests which operation to use. Circle the operation.

3. the difference between a pack of flashcards and a pack of pens

addition subtraction multiplication division

4. the total cost of glue, markers, and pencils

addition subtraction multiplication division

5. the number of crayons equally separated into each box

addition subtraction multiplication division

Algebra Write an equation to represent each sentence.

6. 9 inches less than 14 inches is y inches.

____

7. 24 hammers are divided into y equal sets of 3.

____

8. 12 fish minus y fish plus 4 more equals 9 fish.

____

9. 5 games plus two times as many is y games.

____

Algebra Use the numbers in the table for Exercises 10–12 to write an equation for each sentence.

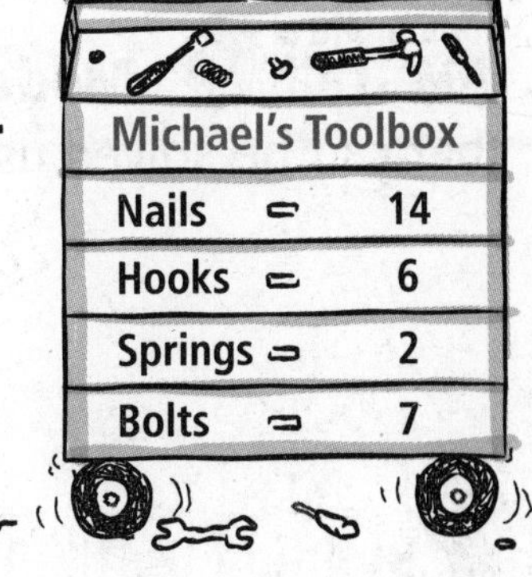

Michael's Toolbox	
Nails	14
Hooks	6
Springs	2
Bolts	7

10. The difference between the number of nails and hooks is m hooks.

____

11. The number of hooks, springs, and bolts altogether is t tools.

____

12. Half the number of hooks plus the number of nails is n tools.

____

Problem Solving

Mathematical PRACTICE 2 **Use Algebra Write an equation using any letter for the unknown.**

My Work!

13. Steph used some nails from the toolbox. Her dad used 9 nails. How many nails did Steph use if they used 17 nails altogether?

14. Twenty customers ordered sandwiches. Three ordered a ham sandwich. Thirteen ordered a chicken sandwich. The rest ordered a turkey sandwich. How many customers ordered a turkey sandwich?

15. Al gave his iguana 12 beans. The iguana ate half of them by noon. How many beans were left at the end of the day if the iguana ate 4 more?

HOT Problems

16. Mathematical PRACTICE 4 **Model Math** Write a real-world problem that can be solved using the equation $16 \div 2 - 3 = n$.

17. **Building on the Essential Question** How are letters and symbols used in equations?

MY Homework

Lesson 7
Write Equations

Homework Helper

eHelp Need help? connectED.mcgraw-hill.com

Use the numbers in the table to write an equation for each situation. Use x for the unknown.

Sammy's Pets	
Fish	12
Hamsters	4
Dogs	2
Birds	3

The difference between the number of fish and the number of birds is x.

$12 - 3 = x$

The total number of pets is x.

$12 + 4 + 2 + 3 = x$

Two times the number of hamsters minus x equals the number of dogs.

$2 \times 4 - x = 2$

The number of fish grouped equally into three aquariums is x.

$12 \div 3 = x$

Practice

Algebra **Write an equation to represent each sentence.**

1. Five more than 7 shells is s.

____

2. Four times as many as 4 pencils is p.

____

3. Half as many as 18 squirrels is x.

____

4. Eleven spoons minus s equals 9 spoons.

____

Algebra Write an equation to represent each sentence.

5. 3 more than 14 eggs divided into 2 equal groups is *e*.

____

6. 5 boxes of muffins with *m* number in each box equals 30.

____

7. The total of 13 cherries, 8 more cherries, and 2 more cherries is *c*.

____

8. 32 tennis balls shared equally by 4 players plus 3 more is *b*.

____

Problem Solving

Mathematical PRACTICE 2 Use Algebra Write an equation using any letter for the unknown.

9. Irving paid for his lunch with a \$10 bill and got \$6 back in change. How much did his lunch cost?

____

10. Erin's beagle weighs 35 pounds. Her Great Dane weighs twice as much as the beagle plus 2 more pounds. How much does the Great Dane weigh?

____

Vocabulary Check

11. Explain the difference between an expression and an equation.

____

____

Test Practice

12. Venus bought 3 loaves of bread that have 20 slices each. Then she used 2 slices to make a sandwich. There are *b* slices left. Which equation represents the situation?

Ⓐ $3 \times 20 - 2 = b$

Ⓑ $3 + 20 - 2 = b$

Ⓒ $(3 \times 20) \div 2 = b$

Ⓓ $3 + 20 - b = 2$

Name ______________________

Solve Two-Step Word Problems

Lesson 8

ESSENTIAL QUESTION
How are properties and equations used to group numbers?

Sometimes in order to solve a problem you need to do more than one step or use more than one operation.

Ahhhh!

Math in My World

Example 1

Gustavo bought some tools at the hardware store. He bought five tools for $6 each and one tool for $7. How much did he spend altogether on tools?

Write an equation with a letter for the unknown. Then solve.

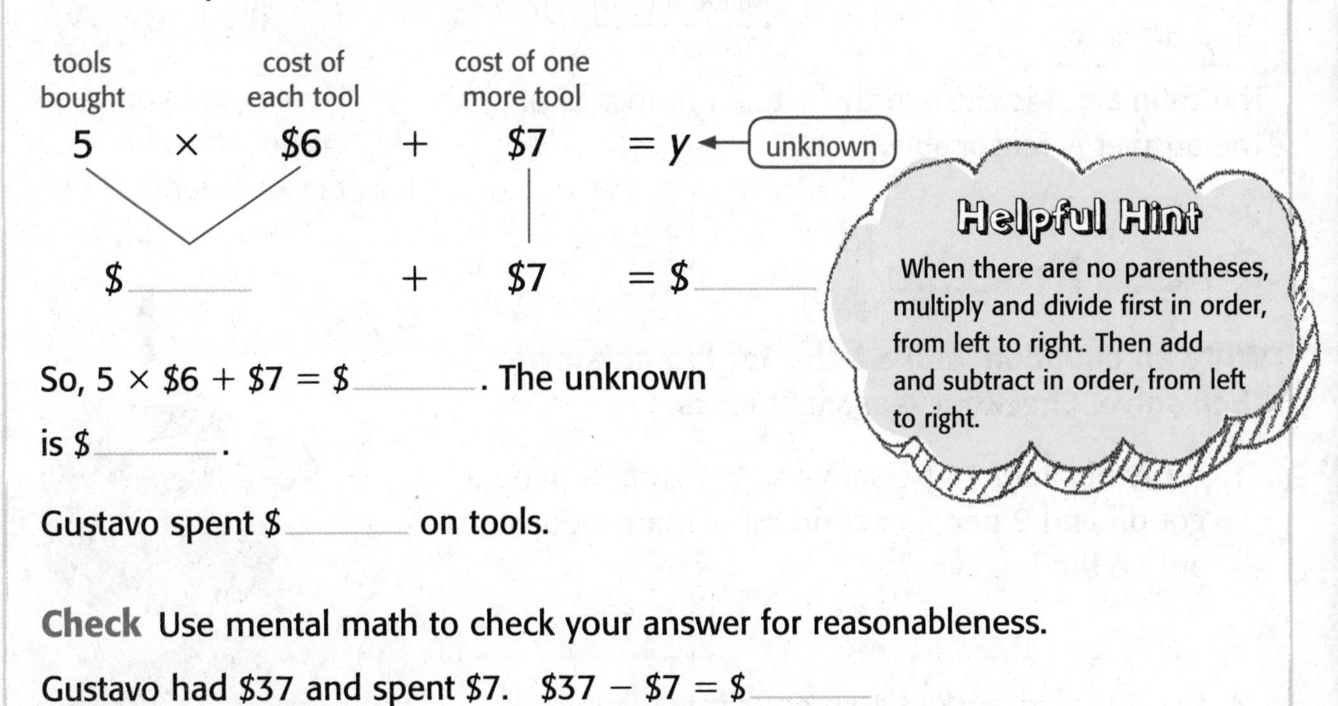

So, $5 \times \$6 + \$7 = \$$ ______. The unknown is $ ______.

Gustavo spent $ ______ on tools.

Check Use mental math to check your answer for reasonableness.

Gustavo had $37 and spent $7. $\$37 - \$7 = \$$ ______

Since $\$30 \div 5$ tools = $ ______ each, the answer is reasonable.

Example 2

Orlon has 48 comic books. He keeps 8 for himself and divides the rest equally among his friends. If each friend gets 8 comic books, to how many friends did he give comic books?

Write an equation with a letter for the unknown. Then solve.

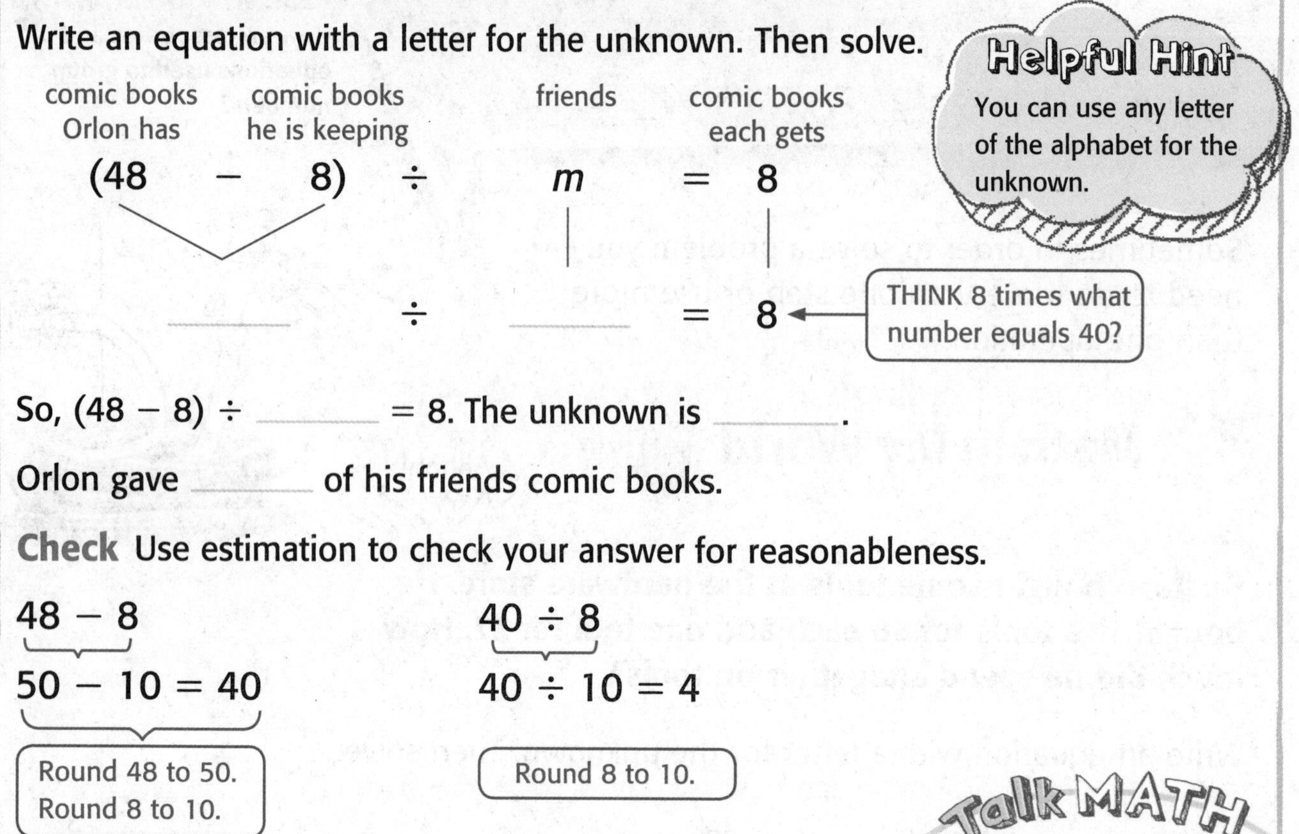

So, $(48 - 8) \div$ _____ $= 8$. The unknown is _____.

Orlon gave _____ of his friends comic books.

Check Use estimation to check your answer for reasonableness.

$48 - 8$ → $50 - 10 = 40$ (Round 48 to 50. Round 8 to 10.)

$40 \div 8$ → $40 \div 10 = 4$ (Round 8 to 10.)

The estimate 4 is close to the actual number of 5. The answer is reasonable.

Talk MATH

How could you check an equation for reasonableness?

Guided Practice

Write an equation with a letter for the unknown. Then solve. Check for reasonableness.

1. A city bus had 14 passengers. At a stop, 5 people got off and 8 people got on. How many people are on the bus now?

_____

2. Grandmother picked 4 times as many apples as pears. What is the difference in the number of apples and pears picked if she picked 8 pears?

_____

Name

Independent Practice

Algebra Write an equation with a letter for the unknown. Then solve. Check for reasonableness.

3. Whitney went to the hobby store. She bought 3 model airplanes for $4 each. She received $8 in change. How much money did she start with?

4. Mr. Robbins gave 9 students one pencil each. That afternoon, he gave 5 more students one pencil each. Now he has 15 pencils. How many pencils did he start with?

5. Look at the table. How many more pens does Carmen have than Pamela and Cesar have together?

Name	Pens
Pamela	7
Cesar	9
Carmen	20

Algebra Circle the correct equation. Then solve the problem.

6. Molly earns $10 each week for babysitting. She spends $3 of that each week and saves the rest. How much money does she save after 8 weeks?

$(\$10 - \$3) \times 8 = m$ $\qquad$ $\$10 - \$3 \times 8 = m$

7. The first 5 pages of Angel's photo album are filled with 8 photos each. The next page has only 7 photos. How many photos are on the 6 pages?

$5 \times 8 + 7 = p$ $\qquad$ $5 \times 8 + p = 40$

Algebra Find each unknown.

8. $k - 9 = 9$

$k =$ ______

9. $45 \div v = 5$

$v =$ ______

10. $9 + 2 = 12 - q$

$q =$ ______

Problem Solving

Mathematical PRACTICE 1 **Check for Reasonableness Write an equation with a letter for the unknown. Then solve. Check for reasonableness.**

My Work!

11. It rained 6 inches each month for the last 6 months. How much will it need to rain this month for the total rainfall to be 43 inches?

12. There were 48 oranges in 6 equal layers in a box. Mother took some oranges from the top layer for snacks. How many oranges did mother take if there are 5 oranges left on the top layer?

HOT Problems

13. Mathematical PRACTICE 1 **Make Sense of Problems** Amelia made 10 fruit kabobs. She divided 20 cherries equally among half of the kabobs. How many cherries did each kabob get?

14. Mathematical PRACTICE 3 **Find the Error** Look back at Exercise 6. Explain why the other choice is incorrect.

15. **Building on the Essential Question** Why is it important to perform the operations in an equation in a certain order?

Name ....

Lesson 8

Solve Two-Step Word Problems

Homework Helper

eHelp Need help? connectED.mcgraw-hill.com

Talia picked 8 quarts of strawberries. She picked half as many quarts of blueberries, then bought 1 more quart of blueberries. How many quarts of blueberries does Talia have?

Write an equation with a letter for the unknown. Then solve.

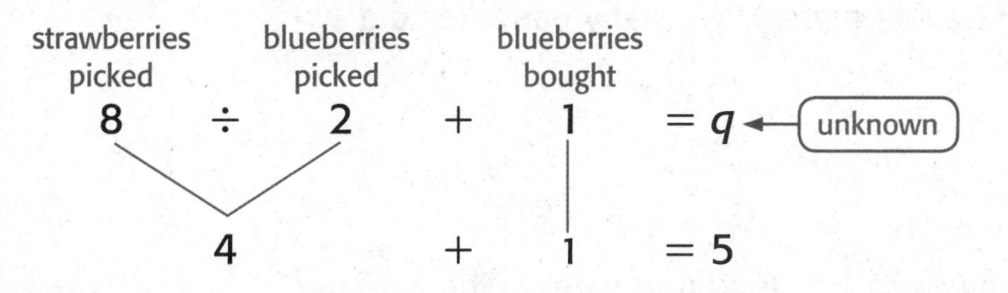

So, $8 \div 2 + 1 = 5$. The unknown is 5. Talia has 5 quarts of blueberries.

Check Use mental math to check your answer for reasonableness.

Subtract the one quart Talia bought from the total.
$5 - 1 = 4$ and 4 is half of 8.

The numbers make sense for the problem. The answer is reasonable.

Practice

Algebra Find each unknown.

1. $48 \div 6 + m = 11$

$m =$ ______

2. $37 - 9 = h \times 4$

$h =$ ______

3. $20 + 20 = 4 \times w$

$w =$ ______

4. $(4 + 2) \times r = 54$

$r =$ ______

Problem Solving

Mathematical PRACTICE 1 Check for Reasonableness Write an equation with a letter for the unknown. Then solve. Check for reasonableness.

5. The football team had its photo taken. There are 3 rows of 8 players each. The fourth row has 6 players. How many players are in the team photo?

6. Mrs. Dove made 15 pancakes. She divided them evenly among Kurt, Joan, and David. Kurt and Joan ate all of their pancakes, but David did not eat some of his. There were 2 pancakes left on David's plate. How many pancakes did he eat?

7. Keira has 83 spelling words to study in 8 weeks. She already knows 3 words. She will study the same number of words each week. How many spelling words will Keira study each week?

8. Grant bought 6 packs of stickers for \$2 each. How much change will Grant receive if he pays with three \$5 bills?

Test Practice

9. Isaac has taken five quizzes. He scored 8 points on each of the first 4 quizzes. He scored y points on the fifth quiz. He has scored a total of 41 points. Which equation represents the situation?

Ⓐ $41 \div 5 = y$

Ⓑ $8 \times 4 \div 5 = y$

Ⓒ $4 \times 8 + y = 41$

Ⓓ $41 \div 4 + y = 8$

Name ....

Real World

Problem-Solving Investigation

STRATEGY: Use Logical Reasoning

Lesson 9

ESSENTIAL QUESTION

How are properties and equations used to group numbers?

Learn the Strategy

Watch Tutor

Sara, Barb, and Erin each wrote a different expression. The expressions were $3 + 5 \times 2$, $(3 + 5) \times 2$, and $3 \times 5 + 2$. The value of Barb's expression is 13. The value of Erin's expression is an even number. Which expression did each person write?

1 Understand

What facts do you know?

The value of Barb's expression is ______.

The value of Erin's expression is an ______ number.

What do you need to find?

which expression each person wrote

2 Plan

I will use logical reasoning to solve the problem.

3 Solve

Find the value of each expression.

$3 + 5 \times 2 =$ ______ ← Barb

$(3 + 5) \times 2 =$ ______ ← ______

$3 \times 5 + 2 =$ ______ ← ______

	Sara	Barb	Erin
$3 + 5 \times 2$	X	yes	X
$(3 + 5) \times 2$	X	X	yes
$3 \times 5 + 2$	yes	X	X

4 Check

Does your answer make sense? Explain.

____

Practice the Strategy

Caleb, Thi, Joyce, and Lawanda each have one of four pets. Thi does not have a dog or a fish. Joyce does not have a bird or a fish. Caleb has a cat. What pet does each person have?

1 Understand

What facts do you know?

____

____

What do you need to find?

____

2 Plan

____

____

3 Solve

4 Check

Does your answer make sense? Explain.

____

____

Name

Apply the Strategy

Solve each problem by using logical reasoning.

My Work!

1. Marilee places her language book next to her science book. Her math book is next to her reading book, which is next to her language book. What is one possible order?

2. Mathematical PRACTICE 2 **Reason** Three friends will put their money together to buy a game that costs $5. Dexter has 5 quarters and 6 dimes. Belle has 6 quarters and 8 dimes. Emmett has 5 coins. If they have 10¢ left over, what coins does Emmett have?

3. Rod is less than 17 years old. The sum of the two digits in his age is even and greater than 4, but both digits are odd. How old is Rod?

4. Jan is 3 inches taller than Dan. Ellie is 2 inches taller than Jan. If Ellie is 54 inches tall, how tall are Jan and Dan?

Review the Strategies

Use any strategy to solve each problem.

- Determine extra or missing information.
- Make a table.
- Look for a pattern.
- Use models.

My Work!

5. Mathematical PRACTICE 1 **Plan a Solution** Hailey planted 30 tomato seeds. Three out of every 5 seeds grew into tomato plants. How many tomato plants does Hailey have?

6. There are 11 scouts in a troop. Their van has 4 rows of seats, and each row holds 3 scouts. How many scouts can the van hold?

7. Mathematical PRACTICE 2 **Use Number Sense** The amusement park sold ride tickets in packs of 5, 10, 15, and 20 tickets. What would a packet of 5 tickets cost if 20 tickets cost $4?

8. Morgan buys 8 packages of 5 bookmarks. Each package costs $2. How much did she spend on the bookmarks?

9. Mathematical PRACTICE 1 **Make a Plan** Madison can make two apple pies with the apples shown. If she has 9 times as many apples, how many pies can she make?

MY Homework

Lesson 9
Problem Solving: Use Logical Reasoning

Homework Helper

eHelp Need help? connectED.mcgraw-hill.com

Sarah, Parker, Kelly, and Nate each have favorite outfits. Nate wears shorts or pants. Parker always wears something green. Sarah wears shorts, but does not like the color blue. Kelly never wears shorts. Which clothing item could belong to each person?

1 Understand

What facts do you know?

I know the clothes and colors each person would wear.

What do you need to find?

I need to find which clothing item belongs to each person.

2 Plan

I will use logical reasoning to solve the problem.

3 Solve

	Red Shorts	Blue Shorts	Green Pants	Brown Pants
Sarah	yes	X	X	X
Parker	X	X	yes	X
Kelly	X	X	X	yes
Nate	X	yes	X	X

The red shorts could belong to Sarah, the blue shorts to Nate, the green pants to Parker, and the brown pants to Kelly.

4 Check

Does your answer make sense? Yes. The clues match the answer.

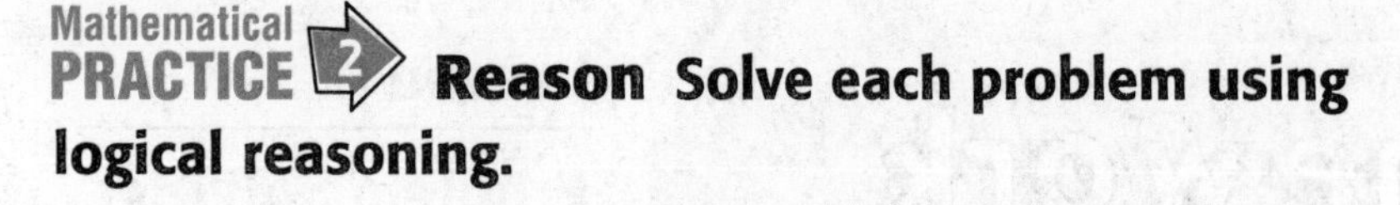

Mathematical PRACTICE 2 **Reason** **Solve each problem using logical reasoning.**

1. Granola bars cost 45¢, gum costs 35¢, and crackers cost 50¢. Lauren buys two different items. She pays with a $1 bill and receives 3 of the same type of coin as change. What did Lauren buy, and what did she receive as change?

2. There are four cars parked next to each other. The blue car is not in the fourth space. The silver car is in the third space. The black car is two spaces in front of the red car. In what order are the cars parked?

3. There are 21 wheels at the bike shop. The wheels will be used to build tricycles and bicycles. There will be half as many tricycles as bicycles. How many of each type of bike will be built?

4. Mitchell has $18 to spend. What is the greatest number of any one item he can buy?

cap	$9
baseball	$10
stopwatch	$9
yo-yo	$6
water bottle	$3

My Work!

Review

Chapter 9

Properties and Equations

Vocabulary Check

Use the word bank below to complete each clue.

Associative	**Distributive**	**equation**	**evaluate**
expression	**operations**	**variable**	

ACROSS

1. A number sentence that uses the equals sign.
3. The property which allows you to decompose a factor into smaller numbers.
5. A symbol or letter that stands for the unknown.
6. Addition, subtraction, multiplication, and division.

DOWN

1. A number or a combination of numbers and operations.
2. The property which states that the grouping of factors does not change the product.
4. To find the value of an expression.

Concept Check

Use the Distributive Property to find each product.

7. $9 \times 7 =$ ______

8. $7 \times 6 =$ ______

Use parentheses to group two factors. Then find each product.

9. $1 \times 3 \times 4 =$ ______

10. $2 \times 5 \times 3 =$ ______

Use numbers and operations to write each phrase as an expression.

11. 5 people equally divide $45

______

12. 6 tables with 4 legs each

______

Algebra Evaluate each expression if $a = 4$ and $b = 5$.

13. $3 + a$

$3 +$ ______ $=$ ______

14. $20 \div b + 5$

______ $+$ ______ $=$ ______

Algebra Write an equation to represent each sentence.

15. If there are 7 cars on a roller coaster with 3 seats each and 2 seats are empty, then m seats are full.

______

16. There are 2 vases with 3 flowers each and each flower has m petals, so there are a total of 30 petals.

______

17. Bree counted 51 birds at the park. Twenty-seven were geese and the rest were ducks. The ducks flew away in groups of 8. How many groups of ducks were there? Write an equation with a letter for the unknown. Then solve.

______

Name

Problem Solving

Algebra Write an equation with a letter for the unknown for Exercises 18–19. Then solve.

My Work!

18. The building manager put new door knobs on 4 doors in each apartment. There are 3 apartments on each floor and 3 floors in the apartment building. How many new door knobs did he install?

19. A soccer team scored 1 point. They scored four more points. The other team had twice as many points. How many points did the other team have?

20. Morgan needed to write an equation. Explain whether she actually wrote an equation.

Test Practice

21. Kayla is x years old. Kevin is 3 years younger than Kayla. If $x = 12$, how old is Kevin?

Ⓐ 7 years old

Ⓑ 8 years old

Ⓒ 9 years old

Ⓓ 10 years old

Reflect

Chapter 9

Answering the ESSENTIAL QUESTION

Use what you learned about properties and equations to complete the graphic organizer.

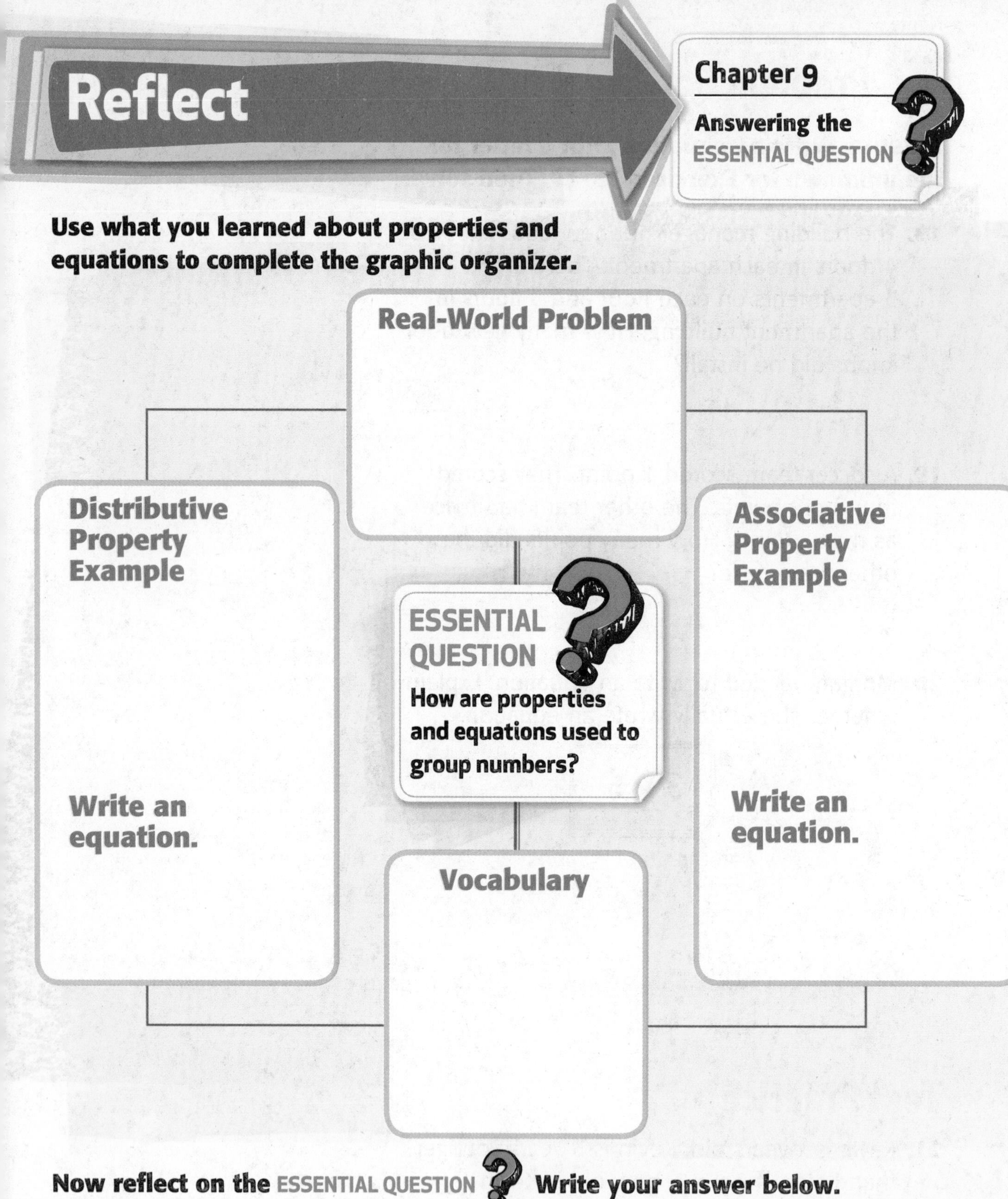

Now reflect on the ESSENTIAL QUESTION Write your answer below.

Glossary/Glosario

Vocab abc ← Go online for the eGlossary.

Go to the eGlossary to find out more about these words in the following 13 languages:

Arabic • Bengali • Brazilian Portuguese • Cantonese • English • Haitian Creole Hmong • Korean • Russian • Spanish • Tagalog • Urdu • Vietnamese

English | **Spanish/Español**

analog clock A clock that has an *hour* hand and a *minute* hand.

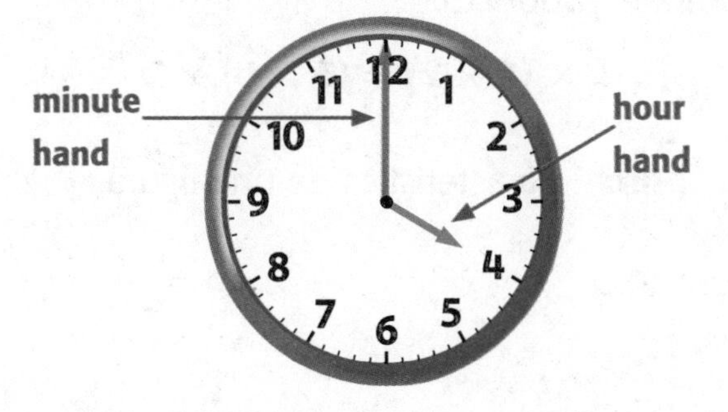

reloj analógico Reloj que tiene una manecilla *horaria* y un *minutero*.

analyze To break information into parts and study it.

analizar Separar la información en partes y estudiarla.

angle A figure that is formed by two *rays* with the same *endpoint*.

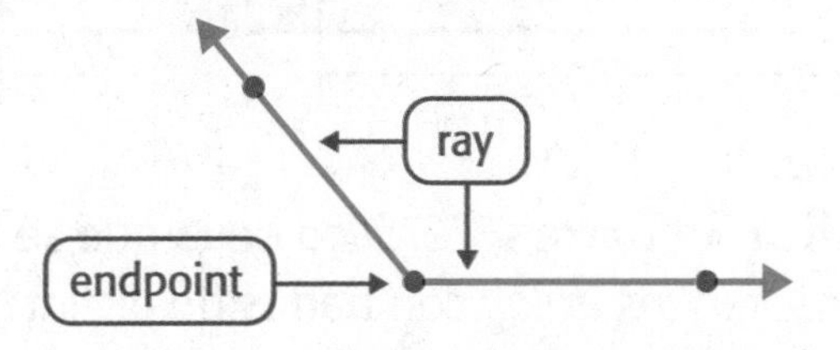

ángulo Figura formada por dos *semirrectas* con el mismo *extremo*.

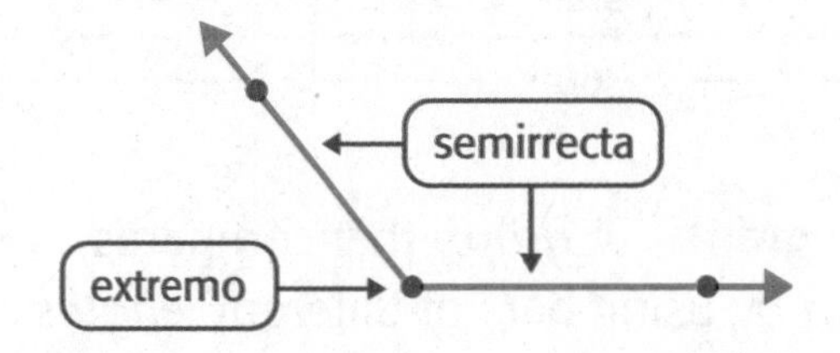

area The number of *square units* needed to cover the inside of a region or *plane figure*.

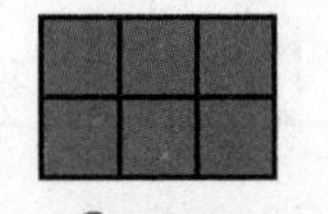

area = 6 square units

área Cantidad de *unidades cuadradas* necesarias para cubrir el interior de una región o *figura plana*.

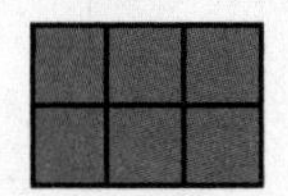

área = 6 unidades cuadradas

Aa

array Objects or symbols displayed in rows of the same *length* and columns of the same *length*.

column
row

Associative Property of Addition The property that states that the grouping of the addends does not change the sum.

$$(4 + 5) + 2 = 4 + (5 + 2)$$

Associative Property of Multiplication The property that states that the grouping of the *factors* does not change the *product*.

$$3 \times (6 \times 2) = (3 \times 6) \times 2$$

attribute A characteristic of a shape.

arreglo Objetos o símbolos organizados en filas y columnas de la misma *longitud*.

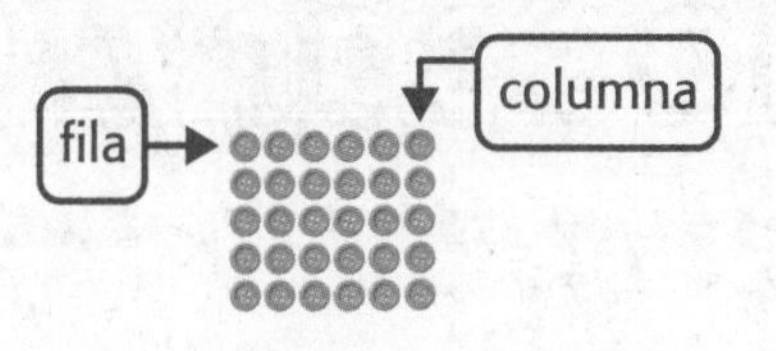

propiedad asociativa de la suma Propiedad que establece que la forma de agrupar los sumandos no altera la suma.

$$(4 + 5) + 2 = 4 + (5 + 2)$$

propiedad asociativa de la multiplicación Propiedad que establece que la forma de agrupar los *factores* no altera el *producto*.

$$3 \times (6 \times 2) = (3 \times 6) \times 2$$

atributo Característica de una figura.

Bb

bar diagram A problem-solving strategy in which bar models are used to visually organize the facts in a problem.

96 children

girls	boys
60	?

bar graph A *graph* that compares *data* by using bars of different *lengths* or heights to show the values.

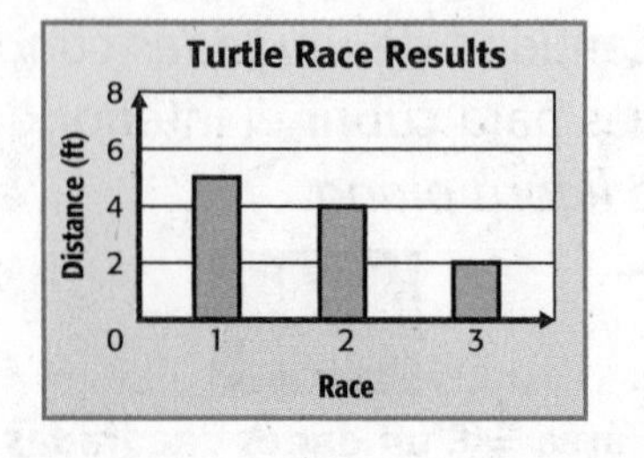

diagrama de barras Estrategia para la resolución de problemas en la cual se usan barras para modelos de organizar visualmente los datos de un problema.

96 estudiantes

niñas	niños
60	?

gráfica de barras *Gráfica* en la que se comparan *los datos* con barras de distintas *longitudes* o alturas para ilustrar los valores.

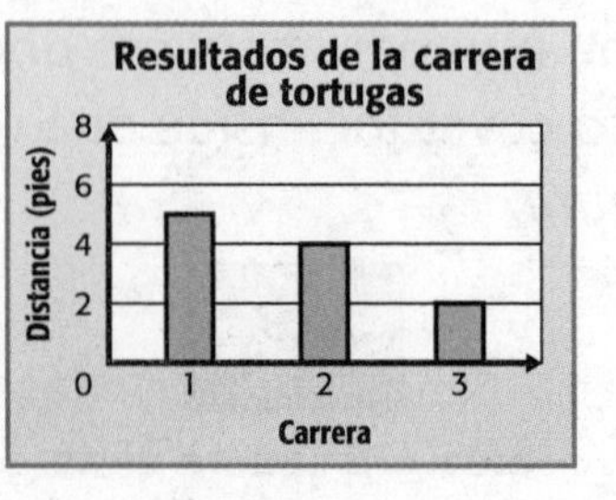

Cc

capacity The amount a container can hold, measured in *units* of dry or liquid measure.

capacidad Cantidad que puede contener un recipiente, medida en *unidades* líquidas o secas.

combination A new set made by combining parts from other sets.

combinación Conjunto nuevo que se forma al combinar partes de otros conjuntos.

Commutative Property of Addition The property that states that the order in which two numbers are added does not change the *sum.*

$$12 + 15 = 15 + 12$$

propiedad conmutativa de la suma Propiedad que establece que el orden en el cual se suman dos o más números no altera la *suma.*

$$12 + 15 = 15 + 12$$

Commutative Property of Multiplication The property that states that the order in which two numbers are multiplied does not change the *product.*

$$7 \times 2 = 2 \times 7$$

propiedad conmutativa de la multiplicación Propiedad que establece que el orden en el cual se multiplican dos o más números no altera el *producto.*

$$7 \times 2 = 2 \times 7$$

compose To form by putting together.

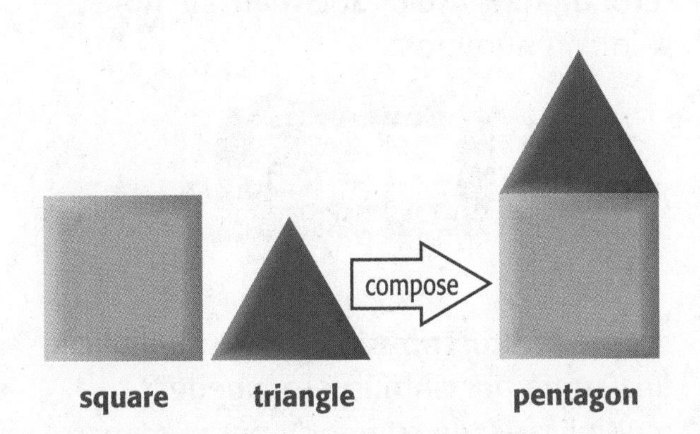

componer Juntar para formar.

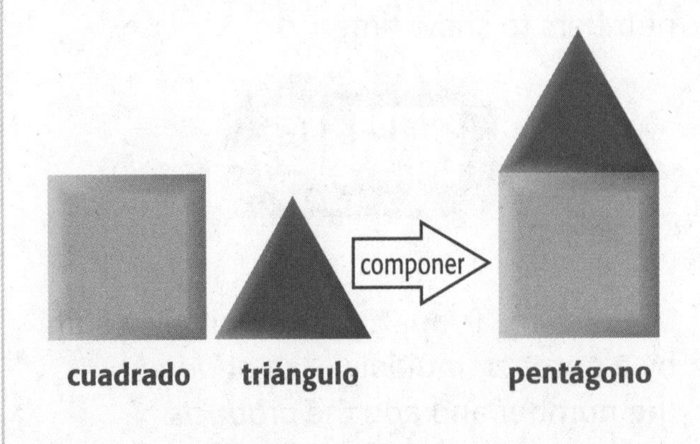

composite figure A figure made up of two or more shapes.

figura compuesta Figura conformada por dos o más figuras.

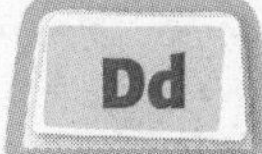

data Numbers or symbols sometimes collected from a *survey* or experiment to show information. *Datum* is singular; *data* is plural.

datos Números o símbolos que se recopilan mediante una *encuesta* o un experimento para mostrar información.

decagon A *polygon* with 10 sides and 10 *angles.*

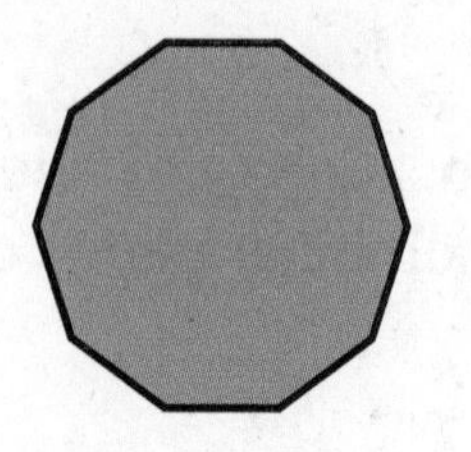

decágono *Polígono* con 10 lados y 10 *ángulos.*

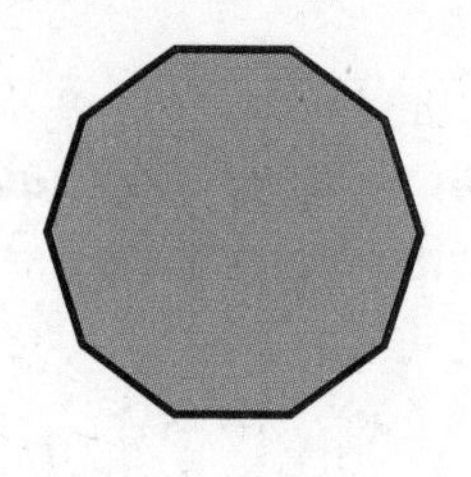

decompose To break a number into different parts.

descomponer Separar un número en differentes partes.

denominator The bottom number in a *fraction.*

In $\frac{5}{6}$, 6 is the denominator.

denominador El número de abajo en una *fracción.*

En $\frac{5}{6}$, 6 es el denominador.

digit A symbol used to write a number. The ten digits are 0, 1, 2, 3, 4, 5, 6, 7, 8, and 9.

dígito Símbolo que se usa para escribir un número. Los diez dígitos son 0, 1, 2, 3, 4, 5, 6, 7, 8 y 9.

digital clock A clock that uses only numbers to show time.

reloj digital Reloj que marca la hora solo con números.

Distributive Property To multiply a sum by a number, multiply each *addend* by the number and add the *products.*

$$4 \times (1 + 3) = (4 \times 1) + (4 \times 3)$$

propiedad distributiva Para multiplicar una suma por un número, puedes multiplicar cada *sumando* por el número y luego sumar los *productos.*

$$4 \times (1 + 3) = (4 \times 1) + (4 \times 3)$$

divide (division) To separate into equal groups, to find the number of groups, or the number in each group.

dividir (división) Separar en grupos iguales para hallar el número de grupos que hay, o el número de elementos que hay en cada grupo.

dividend A number that is being *divided*.

$3\overline{)9}$ **9 is the dividend.**

dividendo Número que se *divide*.

$3\overline{)9}$ **9 es el dividendo.**

division sentence A *number sentence* that uses the *operation* of *division*.

división Enunciado *numérico* que usa la *operación* de *dividir.*

divisor The number by which the *dividend* is being *divided*.

$3\overline{)9}$ **3 is the divisor.**

divisor Número entre el cual se *divide* el *dividendo.*

$3\overline{)9}$ **3 es el divisor.**

double Twice the number or amount.

doble Dos veces el número o la cantidad.

Ee

elapsed time The amount of time that has passed from the beginning to the end of an activity.

tiempo transcurrido Cantidad de tiempo que ha pasado entre el principio y el fin de una actividad.

endpoint The point at the beginning of a *ray*.

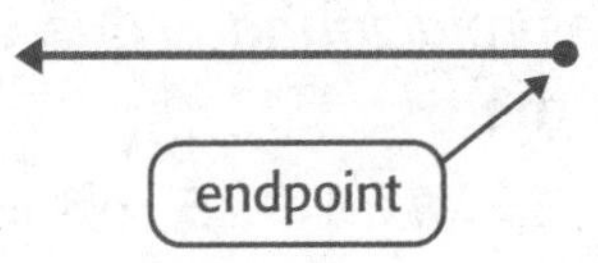

extremo Punto al principio de una *semirrecta.*

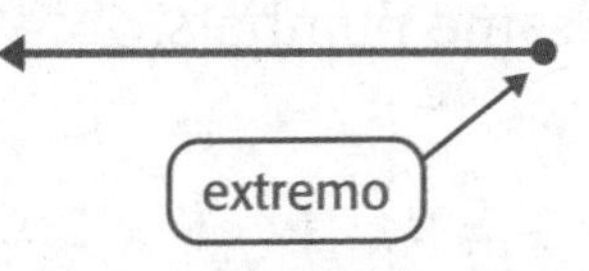

equal groups Groups that have the same number of objects.

grupos iguales Grupos que tienen el mismo número de objetos.

equation A *number sentence* that contains an equals sign, =, indicating that the left side of the equals sign has the same value as the right side.

ecuación *Enunciado* numérico que tiene un signo igual, =, e indica que el lado izquierdo del signo igual tiene el mismo valor que el lado derecho.

equivalent fractions *Fractions* that have the same value.

$\frac{2}{4} = \frac{1}{2}$

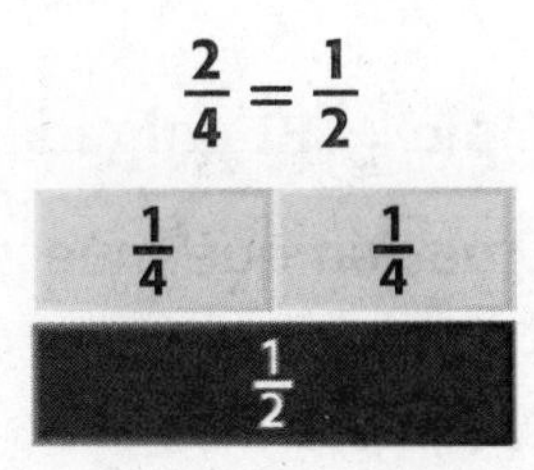

fracciones equivalentes *Fracciones* que tienen el mismo valor.

$\frac{2}{4} = \frac{1}{2}$

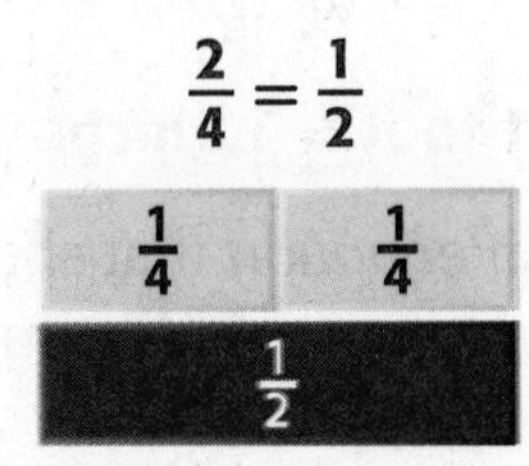

estimate A number close to an exact value. An estimate indicates *about* how much.

47 + 22 is about 70.

estimación Número cercano a un valor exacto. Una estimación indica una cantidad *aproximada.*

47 + 22 es aproximadamente 70.

evaluate To find the value of an *expression* by replacing *variables* with numbers.

evaluar Calcular el valor de una *expresión* reemplazando las *variables* por números.

expanded form/expanded notation The representation of a number as a sum that shows the value of each *digit.*

536 is written as 500 + 30 + 6.

forma desarrollada/notación desarrollada Representación de un número como la suma que muestra el valor de cada *dígito.*

536 se escribe como 500 + 30 + 6.

experiment To test an idea.

experimentar Probar una idea.

expression A combination of numbers and *operations.*

5 + 7

expresión Combinación de números y *operaciones.*

5 + 7

Ff

fact family A group of *related facts* using the same numbers.

5 + 3 = 8	**5 × 3 = 15**
3 + 5 = 8	**3 × 5 = 15**
8 − 3 = 5	**15 ÷ 5 = 3**
8 − 5 = 3	**15 ÷ 3 = 5**

familia de operaciones Grupo de *operaciones relacionadas* que tienen los mismos números.

5 + 3 = 8	**5 × 3 = 15**
3 + 5 = 8	**3 × 5 = 15**
8 − 3 = 5	**15 ÷ 5 = 3**
8 − 5 = 3	**15 ÷ 3 = 5**

factor A number that is *multiplied* by another number.

factor Número que se *multiplica* por otro número.

foot (ft) A customary unit for measuring *length.* Plural is *feet.*

1 foot = 12 inches

pie Unidad usual para medir la *longitud.*

1 pie = 12 pulgadas

formula An *equation* that shows the relationship between two or more quantities.

fórmula *Ecuación* que muestra la relación entre dos o más cantidades.

fraction A number that represents part of a whole or part of a set.

$\frac{1}{2}, \frac{1}{3}, \frac{1}{4}, \frac{3}{4}$

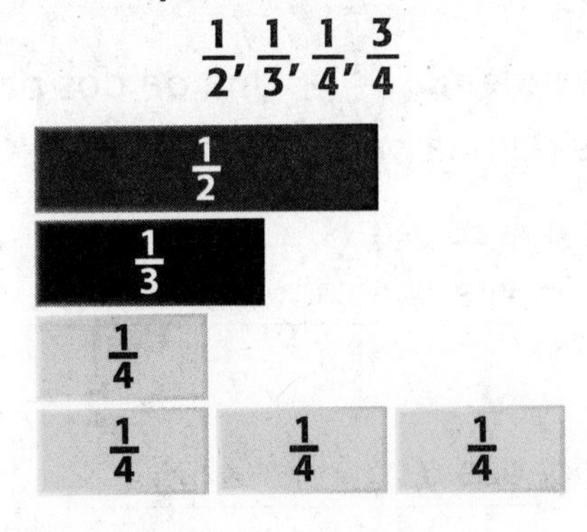

fracción Número que representa una parte de un todo o una parte de un conjunto.

$\frac{1}{2}, \frac{1}{3}, \frac{1}{4}, \frac{3}{4}$

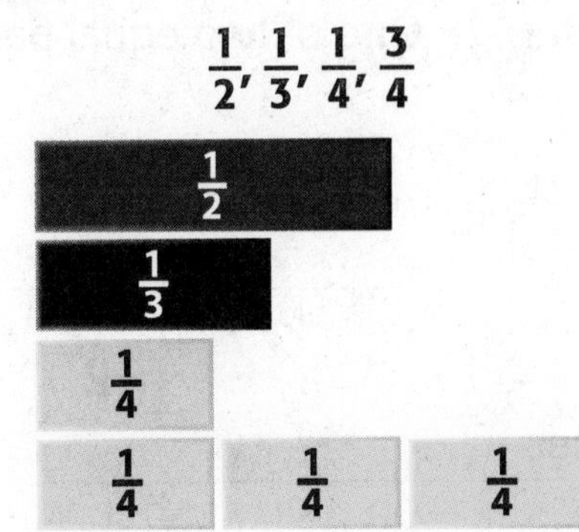

frequency table A *table* for organizing a set of *data* that shows the number of times each result has occurred.

Bought Lunch Last Month	
Name	**Frequency**
Julia	6
Martin	4
Lin	5
Tanya	4

tabla de frecuencias *Tabla* para organizar un conjunto de *datos* que muestra el número de veces que ha ocurrido cada resultado.

Compraron almuerzo el mes pasado	
Nombre	**Frecuencia**
Julia	6
Martín	4
Lin	5
Tanya	4

Gg

gram (g) A *metric unit* for measuring lesser *mass.*

gramo (g) *Unidad métrica* para medir la *masa.*

graph An organized drawing that shows sets of *data* and how they are related to each other. Also a type of chart.

bar graph

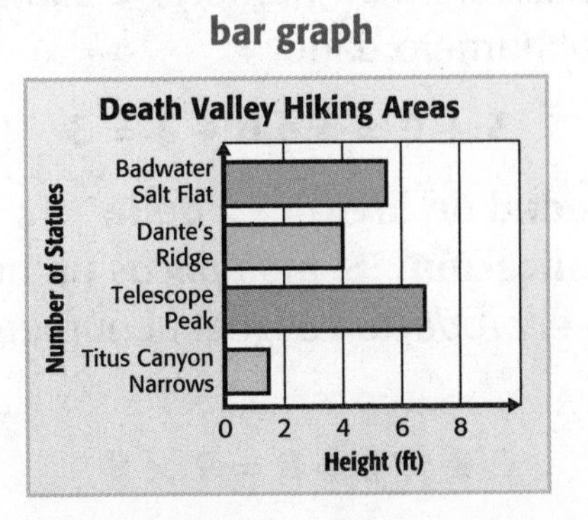

gráfica Dibujo organizado que muestra conjuntos de *datos* y cómo se relacionan. También, es un tipo de diagrama.

gráfica de barras

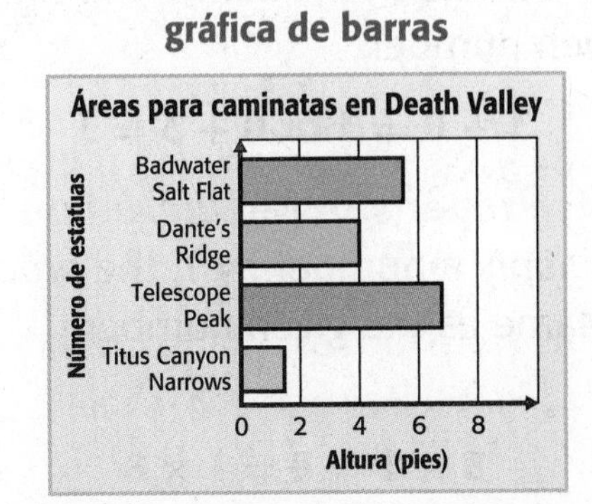

Hh

half inch $\left(\frac{1}{2}\right)$ One of two equal parts of an *inch.*

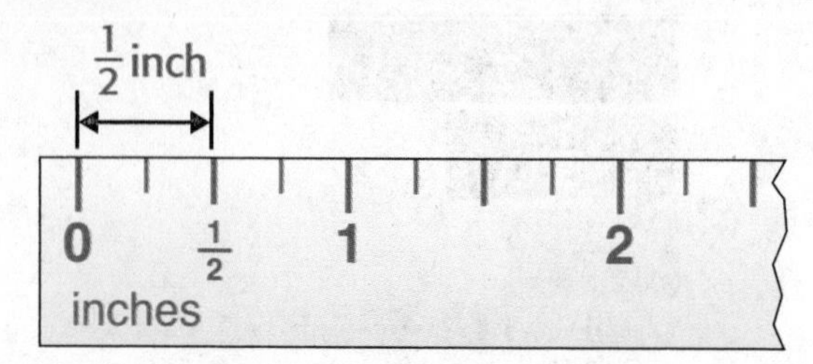

media pulgada $\left(\frac{1}{2}\right)$ Una de dos partes iguales de una *pulgada.*

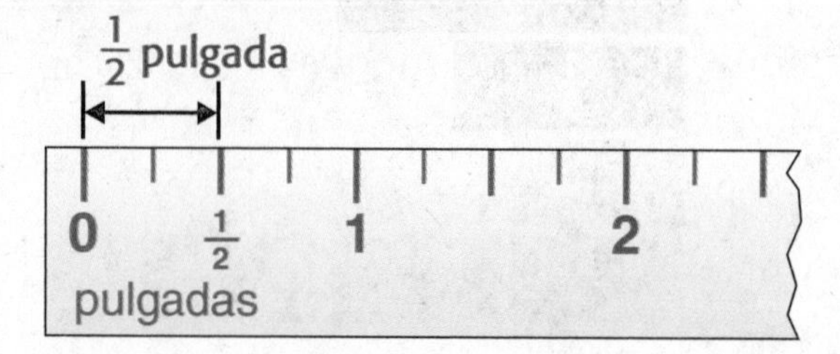

hexagon A *polygon* with six *sides* and six *angles.*

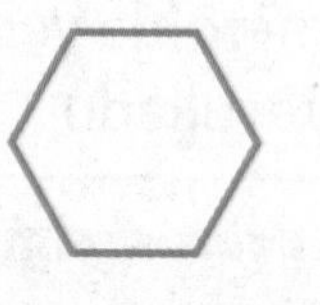

hexágono *Polígono* con seis *lados* y seis *ángulos.*

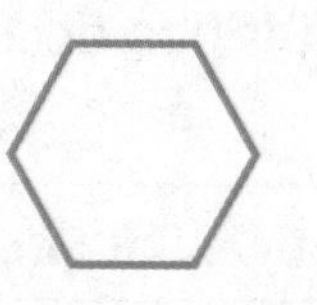

hour (h) A *unit* of time equal to 60 *minutes.*

1 hour = 60 minutes

hora (h) *Unidad* de tiempo igual a 60 *minutos.*

1 hora = 60 minutos

hundreds A position of *place value* that represents the numbers 100–999.

centenas *Valor posicional* que representa los números del 100 al 999.

Ii

Identity Property of Addition If you add zero to a number, the sum is the same as the given number.

3 + 0 = 3 or 0 + 3 = 3

propiedad de identidad de la suma Si sumas cero a un número, la suma es igual al número dado.

3 + 0 = 3 o 0 + 3 = 3

Identity Property of Multiplication If you *multiply* a number by 1, the *product* is the same as the given number.

8 × 1 = 8 = 1 × 8

propiedad de identidad de la multiplicación Si *multiplicas* un número por 1, el *producto* es igual al número dado.

8 × 1 = 8 = 1 × 8

interpret To take meaning from information.

inverse operations *Operations* that undo each other.

Addition and subtraction are inverse, or opposite, operations.

Multiplication and *division* are also inverse operations.

is equal to (=) Having the same value.

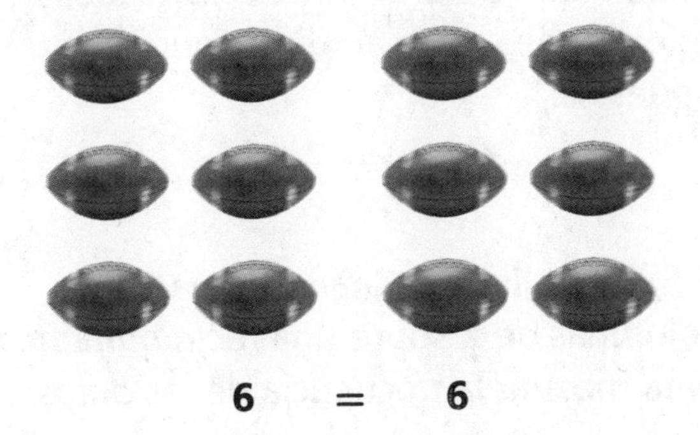

6 is equal to, or the same, as 6.

is greater than > An inequality relationship showing that the value on the left of the symbol is greater than the value on the right.

$5 > 3$ **5 is greater than 3.**

is less than < An inequality relationship showing that the value on the left side of the symbol is smaller than the value on the right side.

$4 < 7$ **4 is less than 7.**

interpretar Extraer significado de la información.

operaciones inversas *Operaciones* que se anulan entre sí.

La suma y la resta son operaciones inversas u opuestas.

La *multiplicación* y la *división* también son operaciones inversas.

es igual a (=) Que tienen el mismo valor.

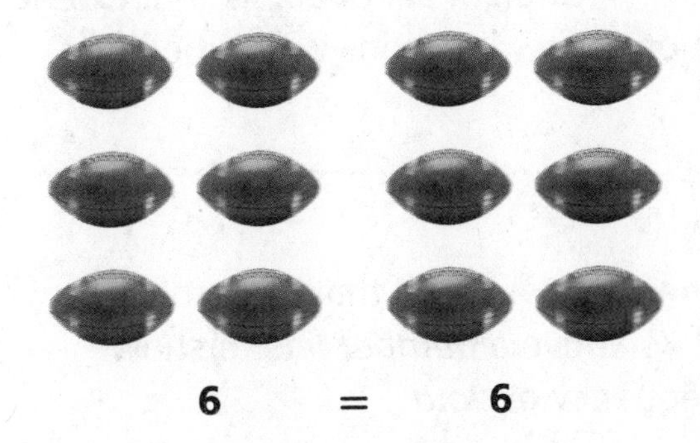

6 es igual o lo mismo que 6.

es mayor que > Relación de desigualdad que muestra que el valor a la izquierda del signo es más grande que el valor a la derecha.

$5 > 3$ **5 es mayor que 3.**

es menor que < Relación de desigualdad que muestra que el valor a la izquierda del signo es más pequeño que el valor a la derecha.

$4 < 7$ **4 es menor que 7.**

Kk

key Tells what or how many each symbol in a *graph* stands for.

kilogram (kg) A *metric unit* for measuring greater *mass.*

clave Indica qué significa o cuánto representa cada símbolo en una *gráfica*.

kilogramo (kg) *Unidad métrica* para medir la *masa.*

Kk

known fact A fact that you already know.

hecho conocido Hecho que ya sabes.

Ll

length Measurement of the distance between two *points*.

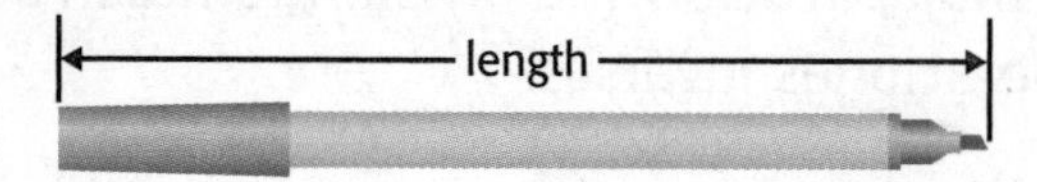

longitud Medida de la distancia entre dos *puntos*.

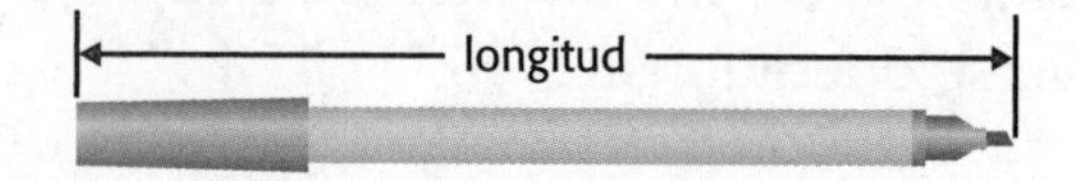

line A straight set of *points* that extend in opposite directions without ending.

recta Conjunto de *puntos* alineados que se extiende sin fin en direcciones opuestas.

line plot A graph that uses columns of Xs above a *number line* to show frequency of *data*.

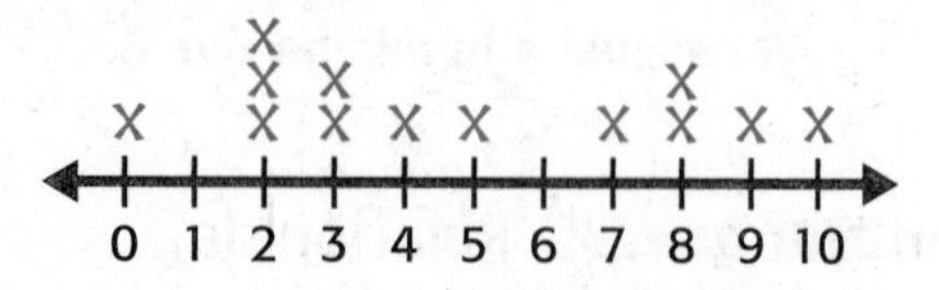

diagrama lineal Gráfica que usa columnas de X sobre una *recta numérica* para mostrar la frecuencia de los *datos*.

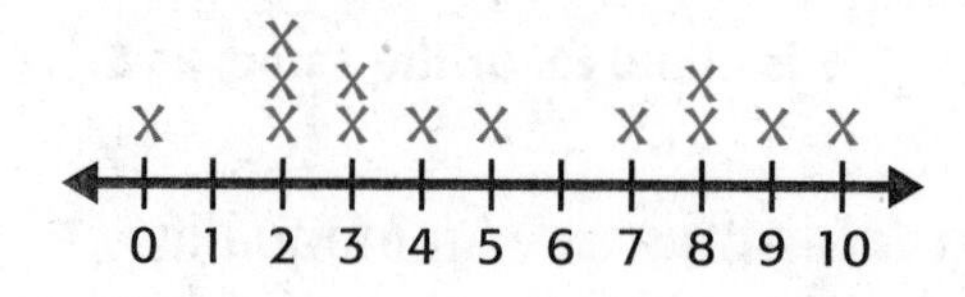

liquid volume The amount of liquid a container can hold. Also known as *capacity*.

volumen líquido Cantidad de líquido que puede contener un recipiente. También se conoce como *capacidad*.

liter (L) A *metric unit* for measuring greater *volume* or *capacity*.

1 liter = 1,000 milliliters

litro (L) *Unidad métrica* para medir el *volumen* o la *capacidad*.

1 litro = 1,000 mililitros

Mm

mass The amount of matter in an object. Two examples of *units* of mass are *gram* and *kilogram*.

masa Cantidad de materia en un cuerpo. Dos ejemplos de *unidades* de masa son el *gramo* y el *kilogramo*.

mental math Ordering or grouping numbers so that they are easier to compute in your head.

cálculo mental Ordenar o agrupar números de modo que sean más fáciles de operar mentalmente.

metric system (SI) The measurement system based on powers of 10 that includes *units* such as *meter, gram,* and *liter.*

sistema métrico (SI) Sistema decimal de medidas que se basa en potencias de 10 y que incluye *unidades* como el *metro,* el *gramo* y el *litro.*

metric unit A *unit* of measure in the *metric system.*

unidad métrica *Unidad* de medida del *sistema métrico.*

milliliter (mL) A *metric unit* used for measuring lesser *capacity.*

1,000 milliliters = 1 liter

mililitro (mL) *Unidad métrica* para medir las *capacidades* pequeñas.

1,000 mililitros = 1 litro

minute (min) A *unit* used to measure short periods of time.

1 minute = 60 seconds

minuto (min) *Unidad* que se usa para medir el tiempo.

1 minuto = 60 segundos

multiple A multiple of a number is the *product* of that number and any *whole number.*

15 is a multiple of 5 because 3 × 5 = 15.

múltiplo Un múltiplo de un número es el *producto* de ese número y cualquier otro *número natural.*

15 es múltiplo de 5 porque 3 × 5 = 15.

multiplication An *operation* on two numbers to find their *product.* It can be thought of as repeated addition.

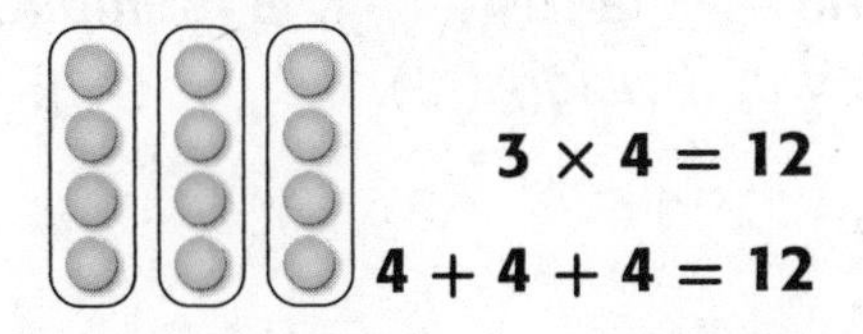

3 × 4 = 12

4 + 4 + 4 = 12

multiplicación *Operación* entre dos números para hallar su *producto.* Puede considerar como una suma repetida.

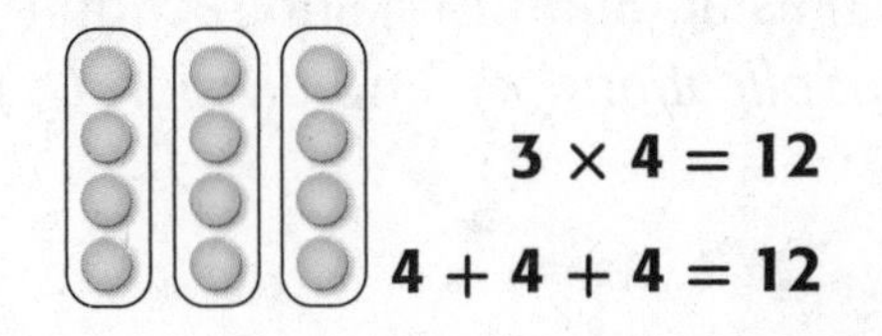

3 × 4 = 12

4 + 4 + 4 = 12

multiplication sentence A *number sentence* that uses the *operation* of *multiplication.*

multiplicación *Enunciado numérico* que usa la *operación* de *multiplicar.*

multiply To find the *product* of 2 or more numbers.

multiplicar Hallar el *producto* de 2 o más números.

number line A line with numbers marked in order and at regular intervals.

0 10 20 30 40 50

recta numérica Recta con números marca dos en orden y a intervalos regulares.

0 10 20 30 40 50

number sentence An *expression* using numbers and the =, <, or > sign.

$5 + 4 = 9; 8 > 5$

enunciado numérico *Expresión* que usa números y el signo =, <, o >.

$5 + 4 = 9; 8 > 5$

numerator The number above the bar in a *fraction;* the part of the *fraction* that tells how many of the equal parts are being used.

In the fraction $\frac{3}{4}$, 3 is the numerator.

numerador Número que está encima de la barra de *fracción;* la parte de la *fracción* que indica cuántas partes iguales se están usando.

En la fracción $\frac{3}{4}$, 3 es numerador.

observe A method of collecting *data* by watching.

observar Método que utiliza la observación para recopilar *datos.*

octagon A *polygon* with eight sides and eight *angles.*

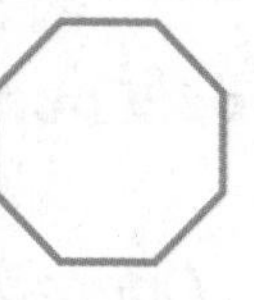

octágono *Polígono* de ocho lados y ocho ángulos.

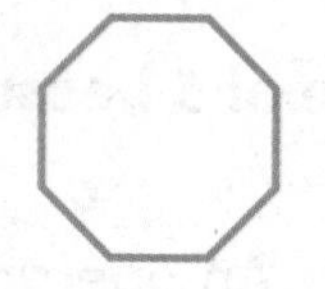

operation(s) A mathematical process such as addition (+), subtraction (−), *multiplication* (×), and *division* (÷).

operación Proceso matemático como la suma (+), la resta (−), la *multiplicación* (×) y la *división* (÷).

parallel (lines) *Lines* that are the same distance apart. Parallel lines do not meet.

rectas paralelas *Rectas* separadas por la misma distancia en cualquier punto. Las rectas paralelas no se intersecan.

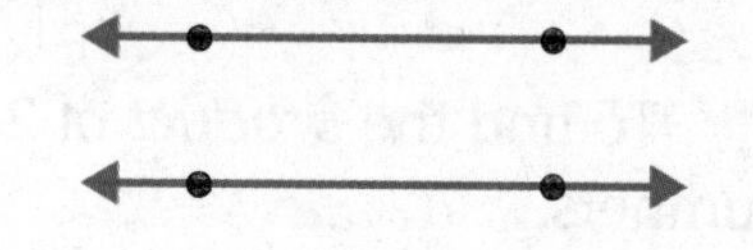

parallelogram A *quadrilateral* with four sides in which each pair of opposite sides is *parallel* and equal in *length.*

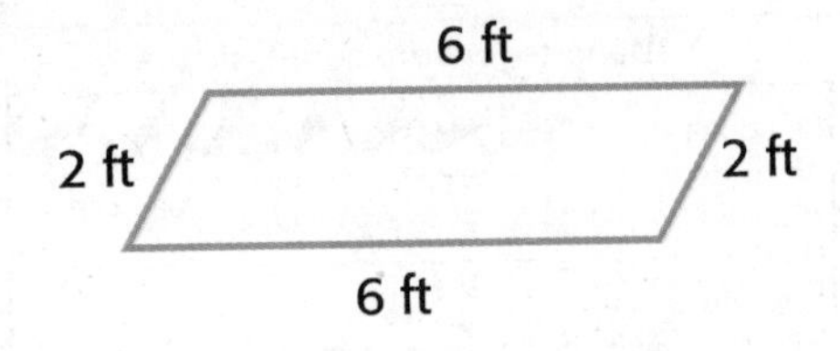

paralelogramo *Cuadrilátero* en el que cada par de lados opuestos son *paralelos* y tienen la misma *longitud.*

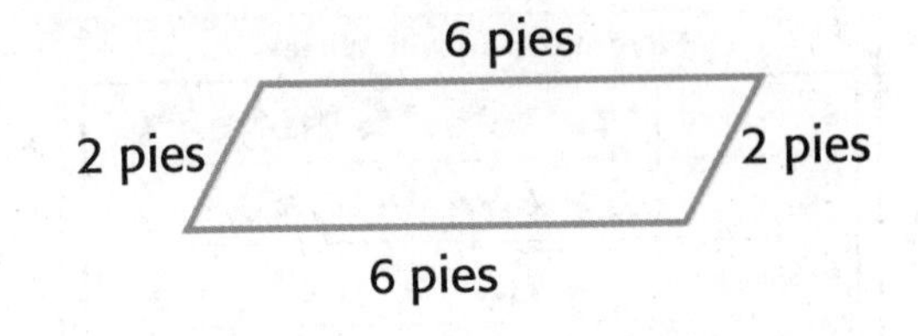

parentheses Symbols that are used to group numbers. They show which *operations* to complete first in a *number sentence*.

paréntesis Signos que se usan para agrupar números. Muestran cuáles *operaciones* se completan primero en un *enunciado numérico*.

partition To *divide* or "break up."

separar *Dividir* o desunir.

pattern A sequence of numbers, figures, or symbols that follow a rule or design.

2, 4, 6, 8, 10

patrón Sucesión de números, figuras o símbolos que sigue una regla o un diseño.

2, 4, 6, 8, 10

pentagon A *polygon* with five sides and five *angles.*

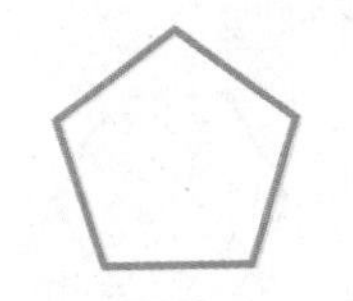

pentágono *Polígono* de cinco lados y cinco ángulos.

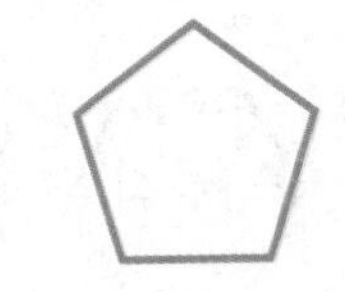

perimeter The distance around a shape or region.

perímetro Distancia alrededor de una figura o región.

period The name given to each group of three *digits* on a *place-value* chart.

período Nombre dado a cada grupo de tres *dígitos* en una tabla de valor *posicional*..

pictograph A *graph* that compares *data* by using pictures or symbols.

Books Read During Read-A-Thon	
Anita	
David	
Emma	
Jonah	
Mary	
Sam	

pictografía *Gráfica* en la que se comparan *datos* usando figuras o símbolos.

Libros leídos durante el maratón de lectura	
Anita	
David	
Emma	
Jonah	
Mary	
Sam	

Pp

picture graph A *graph* that has different pictures to show information collected.

Favorite Sport with Wheels

Skateboard	
Bicycle	
Scooter	
Inline Skate	

place value The value given to a *digit* by its place in a number.

plane figure A *two-dimensional figure* that lies entirely within one plane, such as a *triangle* or *square*.

point An exact location in space.

polygon A closed *plane figure* formed by *line* segments that meet only at their *endpoints*.

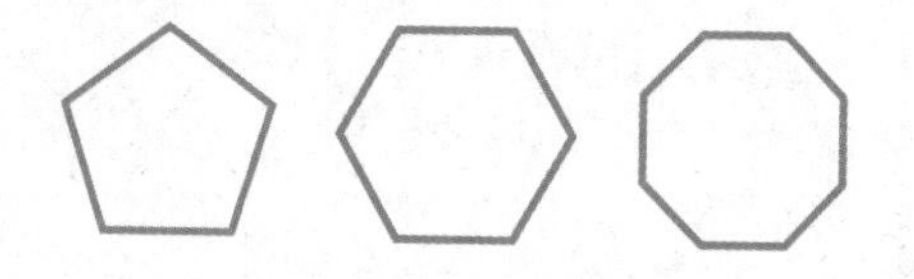

prediction Something you think will happen, such as a specific outcome of an *experiment*.

product The answer to a *multiplication* problem.

gráfica con imágenes *Gráfica* que tiene diferentes imágenes para ilustrar la información recopilada.

Deporte favorito sobre ruedas

Monopatín	
Bicicleta	
Patineta	
Patines en línea	

valor posicional Valor dado a un *dígito* según su lugar en el número.

figura plana *Figura bidimensional* que yace completamente en un plano, como un *triángulo* o un *cuadrado*.

punto Ubicación exacta en el espacio.

polígono *Figura plana* cerrada formada por segmentos de *recta* que solo se unen en sus *extremos*.

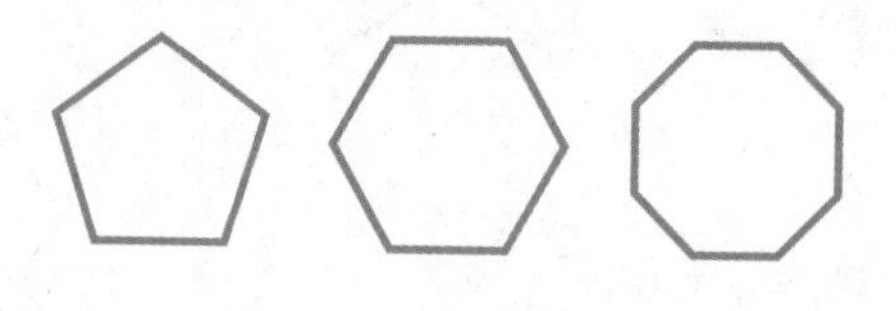

predicción Algo que crees que sucederá, como un resultado específico de un *experimento*.

producto Respuesta a un problema de *multiplicación*.

Qq

quadrilateral A shape that has 4 sides and 4 *angles*.

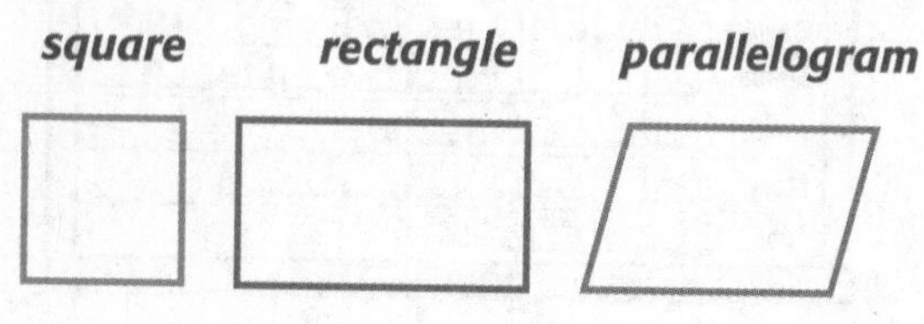

cuadrilátero Figura que tiene 4 lados y 4 *ángulos*.

cuadrado ***rectángulo*** ***paralelogramo***

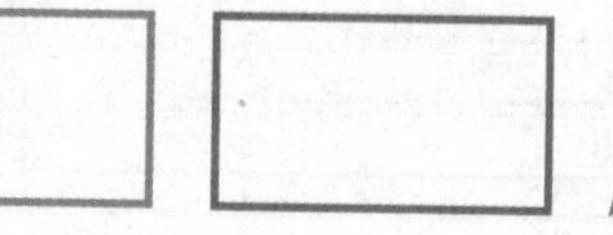

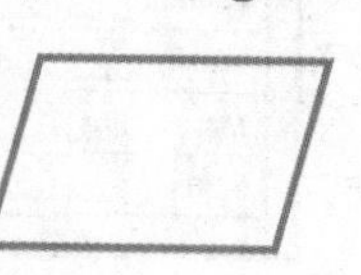

quarter hour One-fourth of an *hour*, or 15 *minutes*.

quarter inch $\left(\frac{1}{4}\right)$ One of four equal parts of an *inch*.

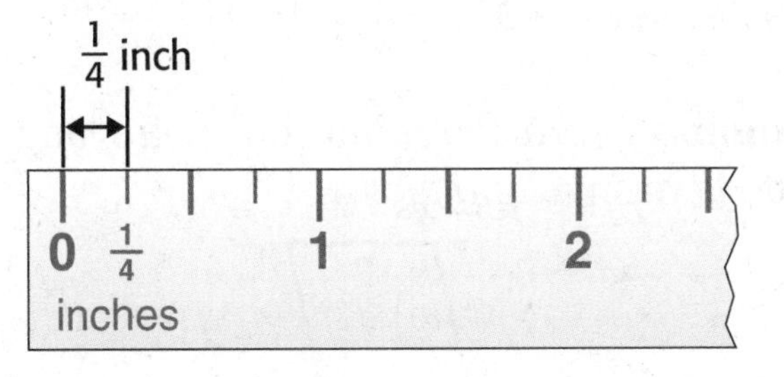

quotient The answer to a *division* problem.

15 ÷ 3 = 5 ← 5 is the quotient.

cuarto de hora La cuarta parte de una *hora* o 15 *minutos*.

cuarto de pulgada $\left(\frac{1}{4}\right)$ Una de cuatro partes iguales de una *pulgada*.

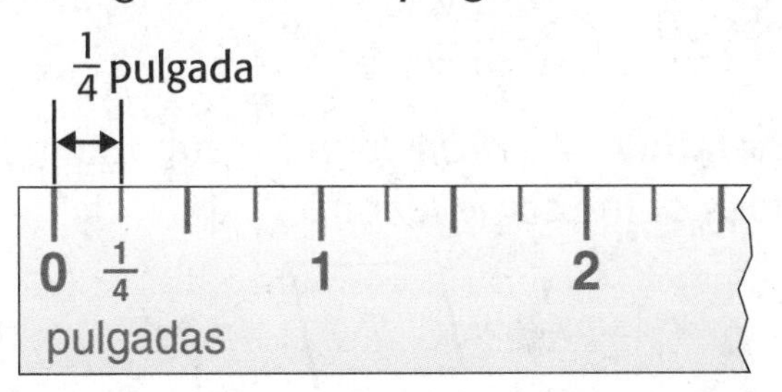

cociente Respuesta a un problema *de división*.

15 ÷ 3 = 5 ← 5 es el cociente.

ray A part of a *line* that has one *endpoint* and extends in one direction without ending.

reasonable Within the bounds of making sense.

rectangle A *quadrilateral* with four *right angles*; opposite sides are equal in *length* and are *parallel*.

regroup To use *place value* to exchange equal amounts when renaming a number.

semirrecta Parte de una *recta* que tiene un *extremo* y que se extiende sin fin en una dirección.

razonable Dentro de los límites de lo que tiene sentido.

rectangle *Cuadrilátero* con cuatro *ángulos rectos*; los lados opuestos son de igual *longitud* y *paralelos*.

reagrupar Usar el *valor posicional* para intercambiar cantidades iguales cuando se convierte un número.

related fact(s) Basic facts using the same numbers. Sometimes called a *fact family.*

$4 + 1 = 5$	$5 \times 6 = 30$
$1 + 4 = 5$	$6 \times 5 = 30$
$5 - 4 = 1$	$30 \div 5 = 6$
$5 - 1 = 4$	$30 \div 6 = 5$

relacionadas Operaciones básicas que tienen los mismos números. También se llaman *familia de operaciones.*

$4 + 1 = 5$	$5 \times 6 = 30$
$1 + 4 = 5$	$6 \times 5 = 30$
$5 - 4 = 1$	$30 \div 5 = 6$
$5 - 1 = 4$	$30 \div 6 = 5$

repeated subtraction To subtract the same number over and over until you reach 0.

resta repetida Procedimiento por el que se resta un número una y otra vez hasta llegar a 0.

rhombus A *parallelogram* with four sides of the same *length.*

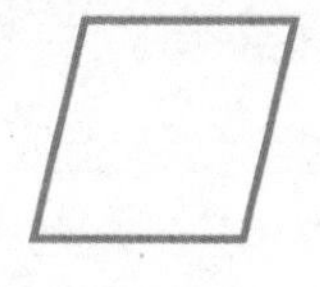

rombo *Paralelogramo* con cuatro lados de la misma *longitud.*

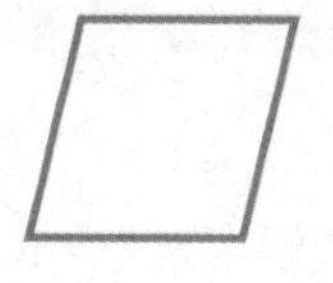

right angle An *angle* that forms a *square* corner.

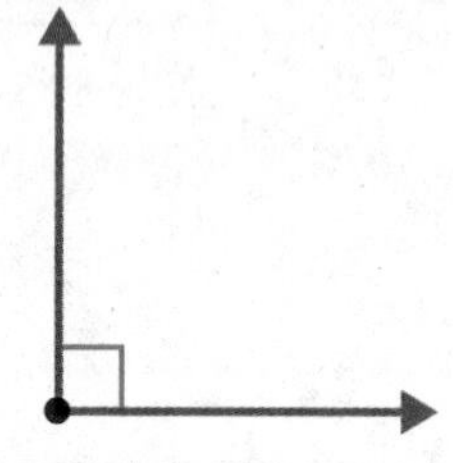

ángulo recto *Ángulo* que forma una esquina *cuadrada*.

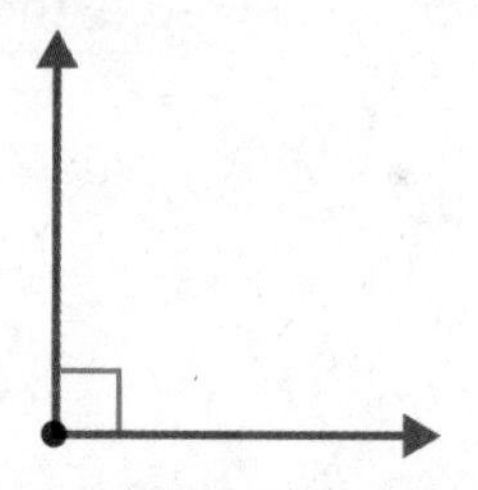

right triangle A *triangle* with one *right angle.*

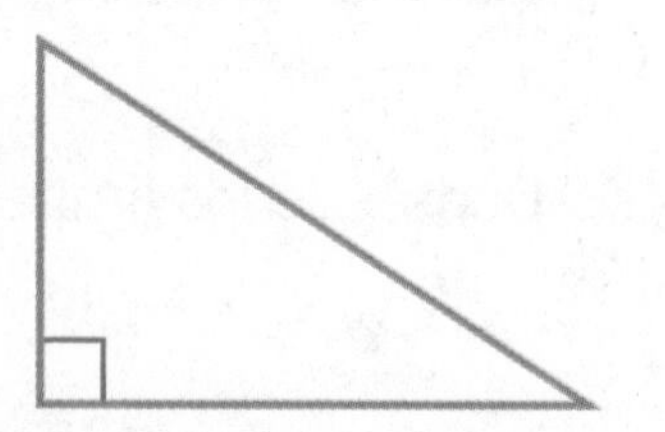

triángulo rectángulo *Triángulo* con un *ángulo recto*.

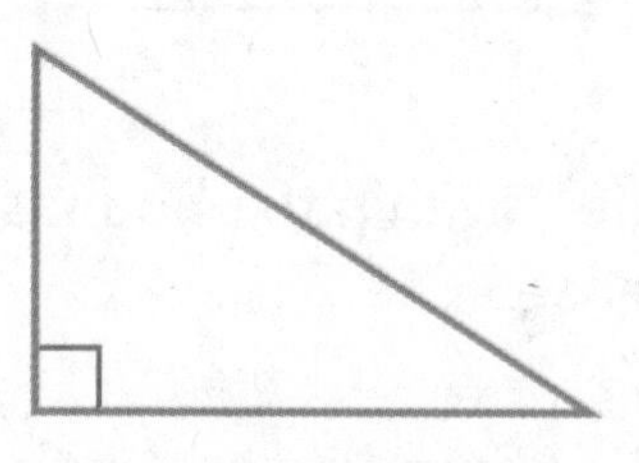

round To change the value of a number to one that is easier to work with. To find the nearest value of a number based on a given *place value.* 27 rounded to the nearest ten is 30.

redondear Cambiar el valor de un número a uno con el que es más fácil trabajar. Hallar el valor más cercano a un número con base en un *valor posicional* dado. 27 redondeado a la décima más cercana es 30.

Ss

scale A set of numbers that represents the *data* in a *graph.*

escala Conjunto de números que representa los *datos* en una *gráfica.*

square A *plane* shape that has four equal sides. Also a *rectangle*.

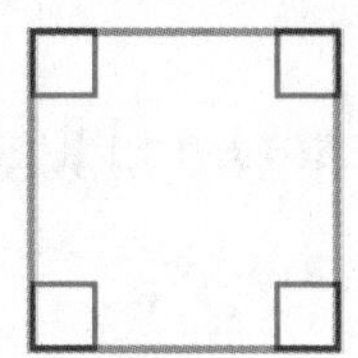

cuadrado *Figura plana* que tiene cuatro lados iguales. También es un *rectángulo*.

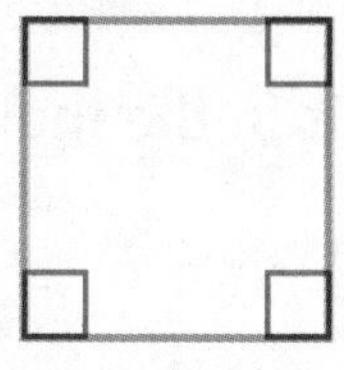

square unit A *unit* for measuring *area*.

unidad cuadrada *Unidad* para medir el *área*.

standard form/standard notation The usual way of writing a number that shows only its *digits,* no words.

537 89 1642

forma estándar/notación estándar Manera habitual de escribir un número usando solor sus *dígitos,* sin usar palabras.

537 89 1642

survey A method of collecting *data* by asking a group of people a question.

encuesta Método para recopilar *datos* haciendo una pregunta a un grupo de personas.

Tt

table A way to organize and display *data* in rows and columns.

tabla Manera de organizar y representar *datos* en filas y columnas.

tally chart A way to keep track of *data* using *tally marks* to record the results.

What is Your Favorite Color?	
Color	**Tally**
Blue	𝍸 III
Green	IIII

tabla de conteo Manera de llevar la cuenta de los *datos* usando *marcas de conteo* para anotar los resultados.

¿Cuál es tu color favorito?	
Color	**Conteo**
Azul	𝍸 III
Verde	IIII

tally mark(s) A mark made to record and display *data* from a *survey.*

marca de conteo Marca que se hace para anotar y presentar los *datos* de una.

thousands A position of *place value* that represents the numbers 1,000–9,999.

In 1,253, the **1** is in the thousands place.

millares *Valor posicional* que representa los números del 1,000 al 9,999.

En 1,253, el **1** está en el lugar de los millares.

time interval The time that passes from the start of an activity to the end of an activity.

intervalo de tiempo Tiempo que transcurre entre de el comienzo y el final de una actividad.

time line A *number line* that shows when and in what order events took place.

Jason's Time Line

Jason born 1999 — **First day of school** 2004 — **Sister born** 2007

1999 2001 2003 2005 2007 2009

línea cronológica *Recta numérica* que muestra cuándo y en qué orden ocurrieron los eventos.

Linea Cronológica de Jason

Nació Jason 1999 — **Primer día de escuela** 2004 — **Nació la hermanita** 2007

1999 2001 2003 2005 2007 2009

trapezoid A *quadrilateral* with exactly one pair of *parallel* sides.

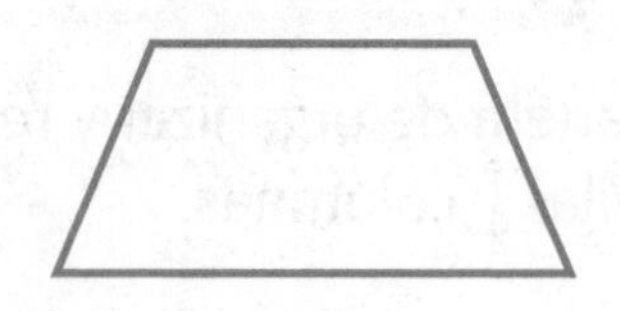

trapecio *Cuadrilátero* con exactamente un par de lados *paralelos.*

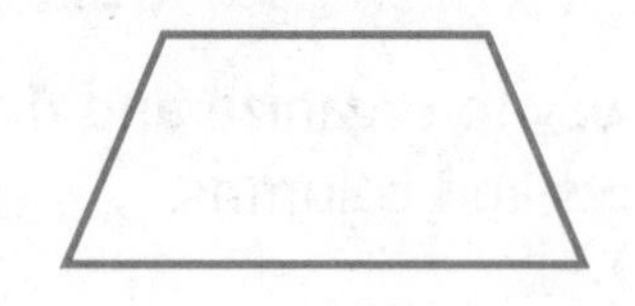

tree diagram A branching diagram that shows all the possible *combinations* when combining sets.

diagrama de árbol Diagrama con ramas que muestra todas las posibles *combinaciones* al reunir conjuntos.

triangle A *polygon* with three sides and three *angles.*

triángulo *Polígono* con tres lados y tres *ángulos.*

two-dimensional figure The outline of a shape—such as a *triangle*, *square*, or *rectangle*—that has only *length*, width, and *area*. Also called a *plane figure*.

figura bidimensional Contorno de una figura, como un *triángulo*, un *cuadrado* o un *rectángulo*, que solo tiene *largo, ancho* y *área*. También conocida como *figura plana*.

Uu

unit The quantity of 1, usually used in reference to measurement.

unidad Cantidad unitaria, que se usa para referirse a medidas.

unit fraction Any *fraction* with a *numerator* of 1.

$$\frac{1}{2}, \frac{1}{3}, \frac{1}{4}$$

fracción unitaria Cualquier *fracción* cuyo *numerador* es 1.

$$\frac{1}{2}, \frac{1}{3}, \frac{1}{4}$$

unit square A *square* with a side *length* of one *unit*.

cuadrado unitario *Cuadrado* cuya *longitud* de los lados es igual a una *unidad*.

unknown A missing number, or the number to be solved for.

incógnita Número que falta, o el número por el que hay que resolver algo.

Vv

variable A letter or symbol used to represent an *unknown* quantity.

variable Letra o símbolo que se usa para representar una cantidad *desconocida*.

vertex The *point* where two *rays* meet in an *angle*.

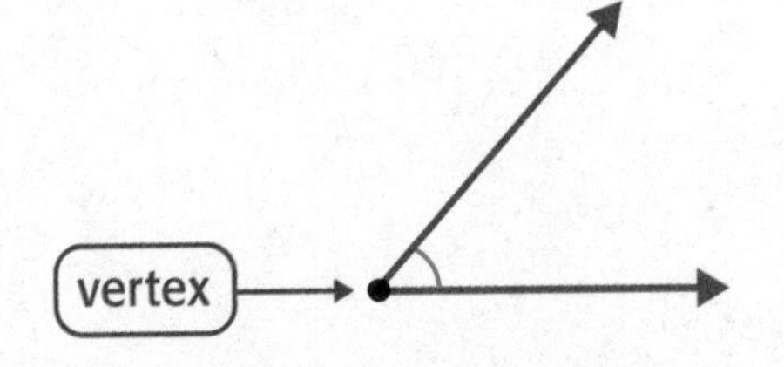

vértice *Punto* donde se unen dos semirrectas y forman un *ángulo*.

vértice

whole number The numbers 0, 1, 2, 3, 4 . . .

word form/word notation The form of a number that uses written words.

6,472

six thousand, four hundred seventy-two

número natural Los números 0, 1, 2, 3, 4 . . .

forma verbal/notación verbal Forma de un número que se escribe en palabras.

6,472

seis mil
cuatrocientos setenta y dos

yard (yd) A customary unit for measuring *length.*

1 yard = 3 feet or 36 inches

yarda (yd) Unidad usual para medir la *longitud.*

1 yarda = 3 pies o 36 pulgadas

Zero Property of Multiplication The property that states that any number multiplied by zero is zero.

$0 \times 5 = 0$ $5 \times 0 = 0$

propiedad del cero de la multiplicación Propiedad que establece que cualquier número multiplicado por cero es igual a cero.

$0 \times 5 = 0$ $5 \times 0 = 0$

Name ....

Work Mat 1: Thousands Place-Value Chart

thousands	hundreds	tens	ones

Work Mat 2: Number Lines

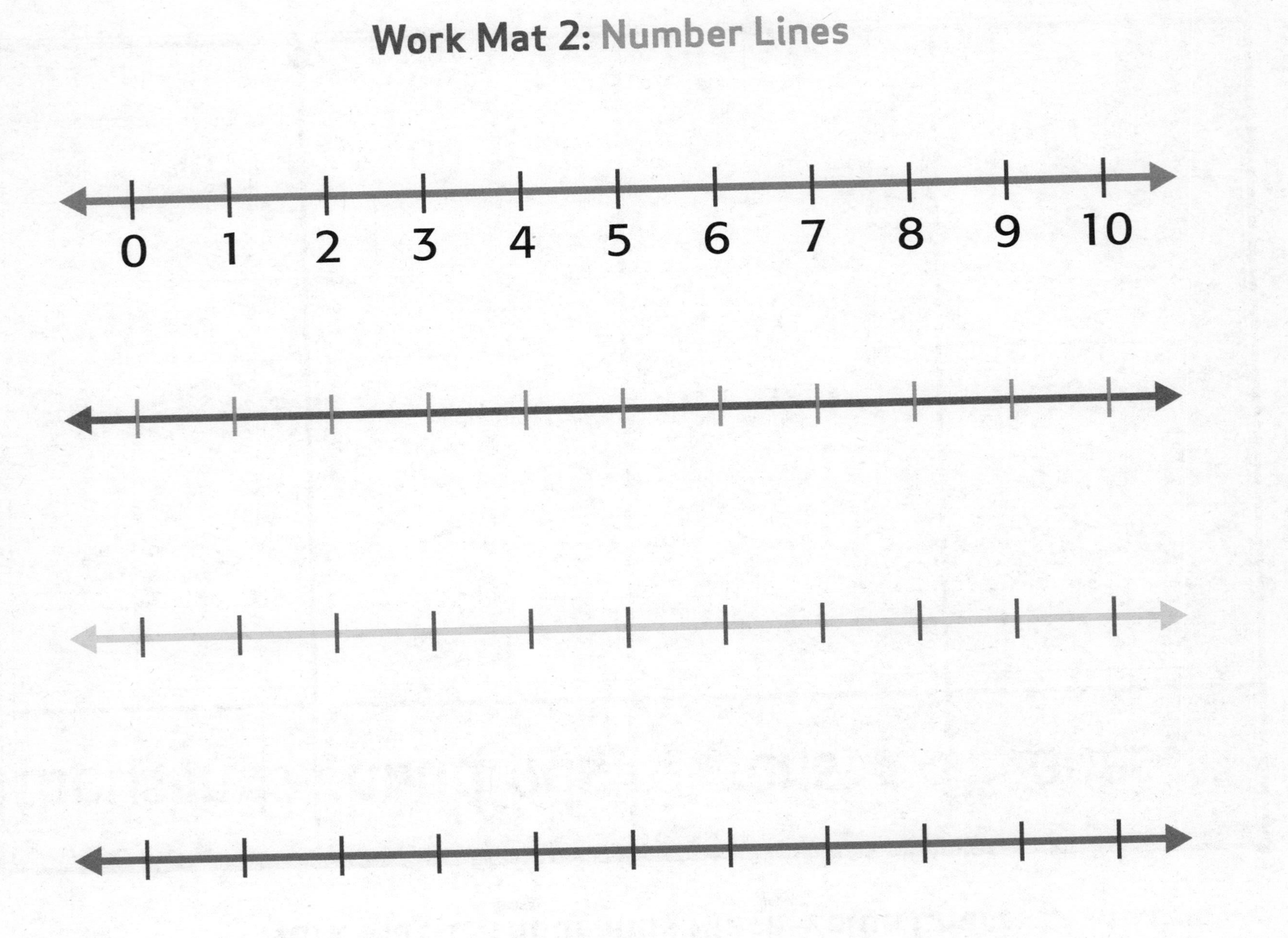

Name

Work Mat 3: Hundred Chart

1	2	3	4	5	6	7	8	9	10
11	12	13	14	15	16	17	18	19	20
21	22	23	24	25	26	27	28	29	30
31	32	33	34	35	36	37	38	39	40
41	42	43	44	45	46	47	48	49	50
51	52	53	54	55	56	57	58	59	60
61	62	63	64	65	66	67	68	69	70
71	72	73	74	75	76	77	78	79	80
81	82	83	84	85	86	87	88	89	90
91	92	93	94	95	96	97	98	99	100

Work Mat 4: Place-Value Chart

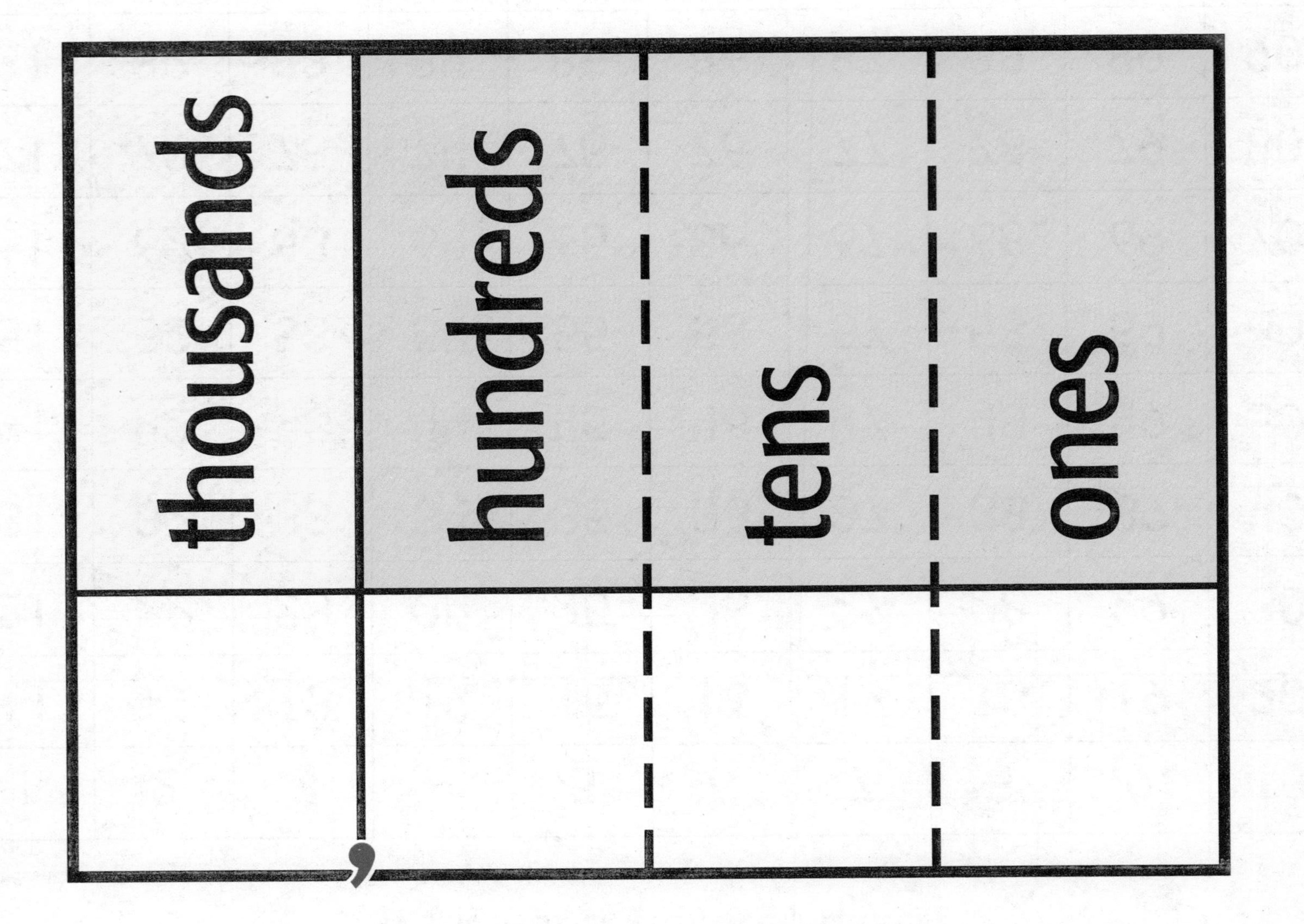

Name ....

Work Mat 5: Centimeter Grid

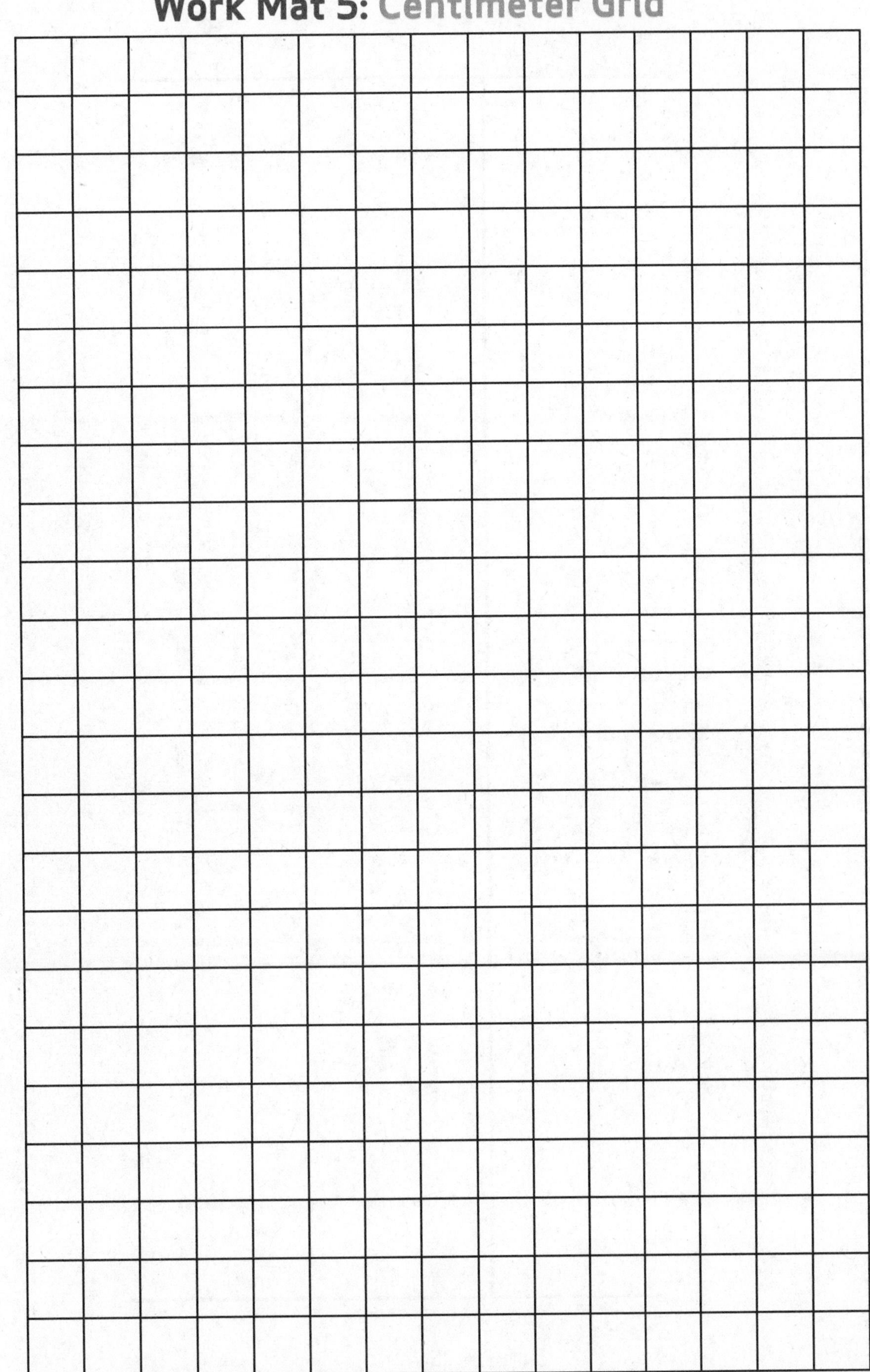

Work Mat 6: Bar Diagram

Name ....

Work Mat 7: Multiplication Fact Table, to 12

×	0	1	2	3	4	5	6	7	8	9	10	11	12
0	0	0	0	0	0	0	0	0	0	0	0	0	0
1	0	1	2	3	4	5	6	7	8	9	10	11	12
2	0	2	4	6	8	10	12	14	16	18	20	22	24
3	0	3	6	9	12	15	18	21	24	27	30	33	36
4	0	4	8	12	16	20	24	28	32	36	40	44	48
5	0	5	10	15	20	25	30	35	40	45	50	55	60
6	0	6	12	18	24	30	36	42	48	54	60	66	72
7	0	7	14	21	28	35	42	49	56	63	70	77	84
8	0	8	16	24	32	40	48	56	64	72	80	88	96
9	0	9	18	27	36	45	54	63	72	81	90	99	108
10	0	10	20	30	40	50	60	70	80	90	100	110	120
11	0	11	22	33	44	55	66	77	88	99	110	121	132
12	0	12	24	36	48	60	72	84	96	108	120	132	144

Work Mat 8: Algebra Mat

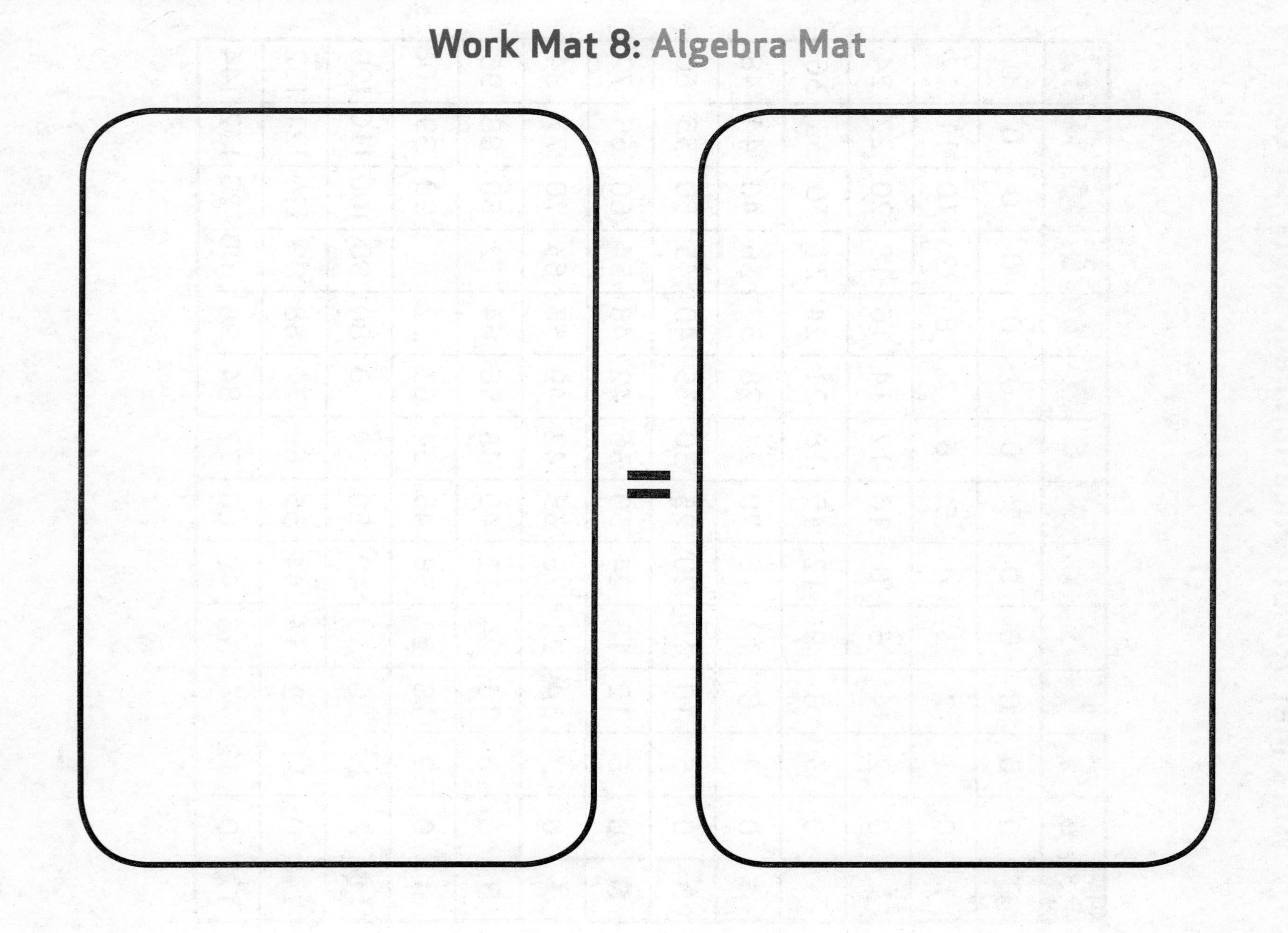